An Overview of BASIC THEORETICAL PHYSICS

K D ABHYANKAR
Formerly, Chairman
Department of Astronomy, Osmania University
Hyderabad

A W JOSHI
Formerly, Professor
Department of Physics, University of Pune
Pune

Universities Press

Universities Press (India) Private Limited

Registered Office
3-6-747/1/A & 3-6-754/1, Himayatnagar, Hyderabad 500 029 (A.P.), India
E-mail: *info@universitiespress.com*

Distributed by
Orient Blackswan Private Limited

Registered Office
3-6-752, Himayatnagar, Hyderabad 500 029 (A.P.), India

Other Offices
Bangalore / Bhopal / Bhubaneshwar / Chennai / Ernakulam / Guwahati
Hyderabad / Jaipur / Kolkata / Lucknow / Mumbai / New Delhi / Patna

© Universities Press (India) Private Limited 2009
First published 2009

Cover and book design
© Universities Press (India) Private Limited 2009

ISBN 978 81 7371 655 3

Set in Original Garamond 11/13 *by*
OSDATA, Hyderabad 500 049

Printed at
Graphica Printers & Binders, Hyderabad 500 013

Published by
Universities Press (India) Private Limited
3-6-747/1/A & 3-6-754/1, Himayatnagar, Hyderabad 500 029 (A.P.), India

Contents

Preface

This book is neither a monograph nor a treatise. It is an attempt to bring together various branches of theoretical physics on one platform to give a panoramic view of the subject. Its need is felt because modern physics courses gloss over the basic physics subjects in preference to specialised topics like solid state physics, electronics, plasma physics, nanotechnology, cosmology, astrophysics and computer science. The idea of such a book arose from a course taken by one of the authors, K D Abhyankar (KDA), from Professor L G Heneyey of Berkeley Astronomy Department. KDA taught such a course to the students of MSc Astronomy at Osmania University for more than two decades. During this period he had made several additions, deletions, and expansions of the material and topics. The second author, A W Joshi (AWJ), has given courses on classical mechanics, electromagnetism, quantum mechanics and statistical mechanics in four universities. It is hoped that the book will be useful to BSc (Hons) and MSc students of physics as well as those studying physics as part of an engineering course, or as a foundation for the allied subjects of astrophysics, geophysics, meteorology, laser and plasma physics. It can also serve as a stepping stone for teachers who want to go deeper into any of these topics according to their inclinations and needs. It is assumed that the reader is familiar with calculus of real and complex variables, vector calculus, and matrices.

The world of physics has seen three major paradigm shifts in the past few centuries. These have changed the physicist's way of thinking and describing physical phenomena. The first was the Galilean–Newtonian paradigm of classical mechanics which began with the seventeenth century, the second was the paradigm of quantum mechanics which began with Planck in 1900, and the third was the paradigm of theories of relativity enunciated by Einstein during 1905–16. These have been discussed in this book at appropriate places.

The book is essentially divided into two parts—prequantum physics and postquantum physics. Prequantum physics begins in Chapter 1 with Hamiltonian mechanics, which is the generalisation of Newtonian formulation of the motion of a point mass to a system of n particles. It forms the basis for later developments in quantum and statistical mechanics.

Then follows, in Chapter 2, a discussion of the revolutionary concepts of special and general theories of relativity, which changed the earlier view of absolute space, time and mass and introduced new ideas of spacetime continuum, mass–energy equivalence, and gravitation as curvature of space. Relativistic effects become important at speeds approaching the speed of light, and in intense gravitational fields. Relativistic physics plays an important role in atomic, molecular and particle physics, as well as in understanding the structure of stars and the universe at large.

Besides gravitation, electric and magnetic forces form other basic aspects of physical phenomena. Their unification by Maxwell and the subsequent development of electromagnetic theory of radiation is the subject of Chapter 3. There we discuss the properties of electromagnetic waves including their radiation, transmission, absorption and scattering.

Chapter 4 on thermodynamics deals with thermal effects on assemblies of a large number of molecules in equilibrium. In thermodynamics we are concerned with the effect of heat on bulk properties like volume, density, pressure, entropy, specific heats, etc. We wish to study how they are governed by the laws of thermodynamics. The chapter ends with a discussion of thermodynamics of radiation.

Postquantum physics starts with the study of atomic and molecular spectra which form the subjects of Chapters 5 and 6, respectively. After discussing the preliminary theories of Bohr and Sommerfeld, we introduce the ideas of quantum mechanics of Heisenberg and Schrödinger, and consider their applications to hydrogen and other atoms. The concept of spin and its treatment by Pauli and Dirac is elaborated. Effects of magnetic and electric fields on atomic spectra are explained. In the case of diatomic molecules, the structure of rotational, vibrational and electronic bands are discussed in detail and some typical examples are given. The concept of partition function and its use in determining the rotational and vibrational temperature is discussed.

While the classical electromagnetic theory of radiation deals with electromagnetic waves, one has to think of photons in quantum theory of radiation. So one has to quantise the electromagnetic field through the postulate of oscillators. After accomplishing this, one can calculate the transition probabilities for emission and absorption of quanta. This forms the topic of Chapter 7.

Chapter 8 on statistical mechanics considers the detailed distribution of atoms and molecules in various energy states as opposed to the bulk properties which were discussed in Chapter 4. Here we find that we have to distinguish between two kinds of particles, viz fermions and bosons.

The last chapter, Chapter 9, is devoted to the elements of nuclear and particle physics. After describing the properties of atomic nuclei, we discuss spontaneous, induced and thermonuclear reactions and their applications in astrophysics. The chapter ends with a historical review of particle physics.

The authors are indebted to Professor N Mukunda for reading the first draft of this book and making several valuable suggestions for improvement. AWJ is thankful to the National Centre for Radio Astrophysics (TIFR), Pune, for providing him all the facilities of the centre for three years, and its director Prof Rajaram Nityananda for several sessions of discussion which clarified many concepts. He expresses his thanks to Jayant Roy for lessons in the use of LaTeX, which did not bear much fruit, and to Anandita Mitra, Shilpa Bhave and Achala Sabane for typing the manuscript.

K D Abhyankar and A W Joshi

1

Hamiltonian Mechanics

IN this chapter we start with Newton's laws of motion and universal law of gravitation. We then deal with a system of n particles in Cartesian and spherical coordinate systems. We introduce generalised coordinates and forces and discuss Lagrangian and Hamiltonian equations of motion. Then we discuss the principle of least action and introduce Poisson brackets and contact transformations, and go on to the Hamilton–Jacobi equation. The chapter ends with a discussion on virial theorem.

1.1 Introduction

Foundations of mechanics were laid by Galileo through his experiments with bodies moving on inclined planes and falling objects. But it was given a mathematical foundation by Newton through his laws of motion and the universal law of gravitation. So we shall first take a look at these.

1.1.1 Laws of motion

The laws of motion enunciate the concepts of an inertial frame of reference, inertial mass, force, etc. Galileo began with the idea that we do have an intuitive concept of force as something which influences the motion of a body. The First Law connects the concept of an inertial frame of reference, uniform motion and force. It picks up inertial frames as special among all frames of reference—special because the description of motion becomes the simplest among descriptions in all the frames. It states that, if there is no force acting on a body at rest in an inertial frame of reference, then it remains at rest, and a body in uniform motion continues to move with that uniform motion. This goes right against Aristotle's theory which requires a force to keep a body moving with a uniform velocity. Newton's first law is based on Galileo's experiments and the insight which he gained from these.

It is worth noting that phenomena, observers and frames of reference are independent of each other. Physics deals with measurements of phenomena, and their analyses, theories and models to explain them. A frame of reference does not need an observer to be present—measurements could be carried out by instruments. Also, one must not think that a phenomenon occurs in a frame of reference. A phenomenon occurs in the universe, in spacetime, and it can be observed in any frame of reference.

Then comes another law of great quantitative significance, the Second Law, which states that the net force **F** applied on a body is proportional to the rate of change of its momentum **p**. If we take the constant of proportionality to be unity, then we have **F**=d**p**/d**t**. In the case when **p** = m**v**, where m is the mass and **v** the velocity, and if m is a constant, it reduces to **F**=m**a**, where **a** = d**v**$/d$**t** is the acceleration of the body. This law gives us a scale of forces independent of masses, or a scale of masses independent of forces.

As a special case, it turns out that if **F** = **0** and $m \neq 0$, then **a** = **0** so that **v** is a constant (including zero). Thus it is consistent with the first law. But it must not be thought that the first law is merely a special case of the second law, because both these laws, as well as the third law, are valid only in inertial frames of reference, and introducing this concept is the main purpose of the first law.

The Third Law states that every action has an equal and opposite reaction. It is important to note that, in Newton's formulation, this reaction is instantaneous. For example, if we press our finger on a table with a certain force, the table (after tolerating certain depression) also pushes the finger by an equal and opposite force (which also causes a depression in the finger). When the earth pulls a ball toward it, the ball also pulls the earth with an equal and opposite force. One should remember that the action-reaction pairs of forces *act on two different bodies*. This law is the genesis of conservation of momentum. If conservation of momentum fails, the third law fails, though the second law may still hold.

One may notice that there is an inherent circularity involved in the above formulation. What is an inertial frame? It is one in which Newton's laws of motion are valid. And when are Newton's laws of motion valid? They are valid in an inertial frame of reference. But then, at the most fundamental level, definitions, laws and measurements merge into one circular but mutually consistent scheme. They are based on a large number of observations, experiments, and logic.

It is also important to remember that there is no truly inertial frame of reference in the universe. Every frame of reference may be said to be approximately inertial for some phenomena but not for others. For example, our frame of reference on the earth is approximately inertial for local phenomena confined to small regions around us, not so good an inertial frame for global phenomena, and patently non-inertial for describing the motion of planets and stars. The sun's frame of reference is a better inertial frame to describe planetary motion, because the description is the simplest in this frame, though it is not so good a frame for stellar and galactic phenomena.

Finally, it is worth noting that Newton's laws do not contain velocity anywhere. This is the reason why observers in an inertial frame of reference cannot determine their own velocity by performing any experiment. But they contain acceleration, which allows one to determine the acceleration of the frame with respect to an inertial frame of reference from such experiments. This is an important aspect and will be commented upon in the next chapter in the context of Maxwell's equations and the theories of relativity.

1.1.2 Law of gravitation

Newton used his laws of motion to explain the motions of the planets around the sun and of satellites around their parent planets. In order to be consistent with Kepler's empirical laws of planetary motion and his own calculations of the motion of the moon around the earth, he had to postulate that every body in this universe attracts every other body with a force which is proportional to the product of their masses and inversely proportional to the square of the distance between them. Thus, if m_1 and m_2 are the masses of the two particles situated at a distance r from each other, the attractive 'gravitational' force between them would be $F = Gm_1m_2/r^2$, where G is a universal constant of gravitation. Further, according to the third law of motion, as m_1 attracts m_2 towards it with a force $\mathbf{F}$, so does m_2 attract m_1 towards it with the same force. If $\mathbf{r}$ is the position of m_2 with respect to m_1, then the force acting on m_2 due to m_1 will be $\mathbf{F_2} = -(Gm_1m_2/r^2)(\mathbf{r}/r)$, while that acting on m_1 due to m_2 will be $\mathbf{F_1} = (Gm_1m_2/r^2)\mathbf{r}/r$. Thus, $\mathbf{F_1} = -\mathbf{F_2}$; they are action–reaction pairs.

Now the solar system consists of the sun, eight (or more or less!) planets, thousands of asteroids and over a hundred satellites. And every one of them attracts every other body gravitationally. This makes the problem of studying their motion quite complicated. However, the problem can be simplified in the first approximation because the sun is a massive body, while the satellites are small compared to their mother planets. So we can neglect the attractions of all bodies other than the sun in the case of individual planets, and in the case of satellites consider only the attraction from the mother planet. In reality, though, for greater accuracy, we have to deal with n bodies interacting with one another under the forces of their mutual gravitation. This compels us to consider a system of n bodies, which is the subject of Hamiltonian mechanics.

It may be noted that every body in the solar system and in the universe is made of innumerable particles which attract each other by various interactions. But most of the bodies like the sun, planets and satellites have approximately spherical shape. A simple calculation shows[1] that a uniform spherical mass shell (i) produces zero gravitational field at points inside the shell, and (ii) for points outside the shell, behaves as if the whole mass of the shell is concentrated at the centre. Therefore a spherical body of uniform mass density behaves, for points outside the body, as if its total mass were concentrated at the centre. So they can be considered as point masses.

---

[1]See Abhyankar, K.D. (1999).

1.2 System of n Particles

Consider a system of n particles in Cartesian coordinates whose positions at time t are given by $\mathbf{r}_i(t) \equiv (x_i(t), y_i(t), z_i(t)), 1 \le i \le n$. Let $\mathbf{F}_i(t)$ be the net force acting on the particle i, with Cartesian components $(X_i(t), Y_i(t), Z_i(t))$. We shall now work out their equations of motion, kinetic energy and total energy of the system, and related quantities.

1.2.1 Equations of motion

The motion of the particle i is governed by the equation of motion

$$m_i \frac{d^2 x_i}{dt^2} = X_i, \, m_i \frac{d^2 y_i}{dt^2} = Y_i, \, m_i \frac{d^2 z_i}{dt^2} = Z_i. \tag{1.1}$$

We have $3n$ equation of the second order and their solution will contain $6n$ constants of integration. They could be $3n$ position coordinates, x_{i0}, y_{i0}, z_{i0}, and $3n$ velocity components, $\dot{x}_{i0}, \dot{y}_{i0}, \dot{z}_{i0}$ at time t_0. Here dots represent differentiation with respect to time. Since all $3n$ equations are similar, we can create a labelling system to write all the above equations as a single equation

$$m_\mu \frac{d^2 x_\mu}{dt^2} = X_\mu, 1 \le \mu \le 3n. \tag{1.2}$$

Here we have $\mu = 3(i-1)+j, 1 \le i \le n, 1 \le j \le 3$, and there are really only n masses, m_i, and not $3n$, with the identification

$$m_{3(i-1)+1} = m_{3(i-1)+2} = m_{3i+3} \equiv m_i.$$

Sometimes the particles are not entirely free but are subject to some constraints. Then we need less than $3n$ coordinates to specify the configuration of the system. For example, in a diatomic molecule with fixed separation between atoms, we need to specify the coordinates of one atom and two angles to specify the orientation of the line joining the two atoms. Thus, only 5 independent coordinates are required in this case. In the case of a rigid body too, with fixed inter-particle distances, we require 5 independent coordinates to specify the position and orientation of the body. In fact, a diatomic molecule with a fixed interatomic distance is a rigid body of 2 particles. Let there be $N \le 3n$ independent coordinates; then $1 \le \mu \le N$ in (1.2) and we need N equations.

1.2.2 Kinetic energy

The kinetic energy of the system would be

$$T = \frac{1}{2} \sum_{i=1}^{n} m_i(\dot{x}_i^2 + \dot{y}_i^2 + \dot{z}_i^2) \equiv \frac{1}{2} \sum_{\mu=1}^{N} m_\mu \dot{x}_\mu^2,$$

with suitable identification of masses. Then,

$$\frac{\partial T}{\partial \dot{x}_\mu} = m_\mu \dot{x}_\mu = p_\mu,$$

where p_μ is the μth linear momentum, $1 \le \mu \le N$. Thus,

$$\frac{d}{dt}\left(\frac{\partial T}{\partial \dot{x}_\mu}\right) = \dot{p}_\mu = m_\mu \ddot{x}_\mu. \tag{1.3}$$

In a conservative system it is possible to define a potential function $V(x_1, x_2, ..., x_N)$ such that $X_\mu = -\partial V/\partial x_\mu$; then the equations of motion become

$$\frac{d}{dt}\left(\frac{\partial T}{\partial \dot{x}_\mu}\right) = -\frac{\partial V}{\partial x_\mu}. \tag{1.4}$$

Here we have two functions, T and V. It would be convenient if we could express the equations of motion in terms of one function only. Lagrange used $L = T - V$ and Hamilton used $H = T + V$. Thus we get Lagrangian and Hamiltonian forms of equations of motion.

1.2.3 Lagrangian equations

Since T does not contain x_j, $\partial T/\partial x_j = 0$. Further, with conservative forces, V is independent of $\dot{x}_j$. Therefore, defining the Lagrangian function as

$$L \equiv L(x_j, \dot{x}_j) = T - V,$$

we get

$$\frac{\partial T}{\partial \dot{x}_j} = \frac{\partial L}{\partial \dot{x}_j}, \quad -\frac{\partial V}{\partial x_j} = \frac{\partial L}{\partial x_j}. \tag{1.5}$$

Therefore the equations of motion become

$$\frac{d}{dt}\left(\frac{\partial L}{\partial \dot{x}_j}\right) - \frac{\partial L}{\partial x_j} = 0. \tag{1.6}$$

This is the Lagrangian form of equations of motion. There are N equations of the second order as indicated before.

1.2.4   Hamilton's equations

The total energy of the system of particles is

$$H = T + V = 2T - L,$$

where L is the *Lagrangian* introduced above. But

$$2T = \sum_{j=1}^{N} m_j \dot{x}_j^2 = \sum_j p_j \dot{x}_j,$$

where the momenta p_j are given by

$$p_j = \frac{\partial T}{\partial \dot{x}_j} = \frac{\partial L}{\partial \dot{x}_j}. \tag{1.7}$$

Therefore the Hamiltonian function is defined as

$$H = \sum_j p_j \dot{x}_j - L, \tag{1.8}$$

with $L = T - V$. Thus,

$$L = \sum_j \frac{p_j^2}{2m_j} - V,$$

$$H = \sum_j \frac{p_j^2}{m_j} - L = \sum_j \frac{p_j^2}{2m} + V.$$

The function $H \equiv H(x_j, p_j)$ is called the *Hamiltonian*. It gives

$$\frac{\partial H}{\partial p_j} = \dot{x}_j, \quad \frac{\partial H}{\partial x_j} = -\frac{\partial L}{\partial x_j} = -\frac{d}{dt}\left(\frac{\partial L}{\partial \dot{x}_j}\right) = -\dot{p}_j. \tag{1.9}$$

These may be written concisely as

$$\dot{x}_j = \frac{\partial H}{\partial p_j}, \quad \dot{p}_j = -\frac{\partial H}{\partial x_j}. \tag{1.10}$$

This is the Hamiltonian form of equations of motion which contains $2N$ equations of the first order. Hamiltonian equations are also known as canonical equations, and x_j, p_j are called *conjugate canonical variables*. They play an important role in quantum mechanics.

It can be seen that $H = E$ (total energy) gives

$$\frac{\partial H}{\partial t} = \dot{E}, \quad \frac{\partial H}{\partial E} = 1 = \dot{t}.$$

Hence E and t are also canonically conjugate variables. The Lagrangian and Hamiltonian ways of describing a dynamical system are essentially equivalent. One can pass from one to the other via (1.8).

1.3 Generalised Quantities

1.3.1 Generalised coordinates

Let $q_j \equiv q_j(x_1, x_2, \cdots, x_N)$ be N functions of coordinates such that the inverse transformation $x_k \equiv x_k(q_1, q_2, \cdots, q_N)$ is unique. Then q_j are called *generalised coordinates*. They need not have dimensions of length (assuming that x_k have dimensions of length).

For example, consider a particle with Cartesian coordinates x, y, z, in three dimensions. The spherical coordinates (also called spherical polar coordinates) r, θ, ϕ form a set of generalised coordinates; see Fig. 1.1.

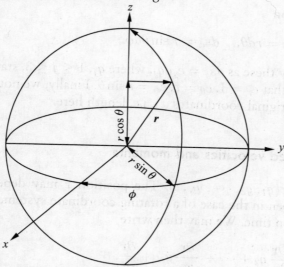

Fig. 1.1   *The spherical polar coordinate system*

They are related to each other by

$$r = (x^2 + y^2 + z^2)^{1/2}, \quad \theta = \tan^{-1} \frac{(x^2 + y^2)^{1/2}}{z}, \quad \phi = \tan^{-1} \frac{y}{x}; \qquad (1.11a)$$

$$x = r \sin\theta \cos\phi, \quad y = r \sin\theta \sin\phi, \quad z = r \cos\theta. \qquad (1.11b)$$

If x, y, z are lengths, r has the dimensions of length while θ and ϕ are dimensionless.

Their infinitesimal increments can be worked out to be

$$dx = dr \sin\theta \cos\phi + r \cos\theta \cos\phi \, d\theta - r \sin\theta \sin\phi \, d\phi,$$

$$dy = dr \sin\theta \sin\phi + r \cos\theta \sin\phi \, d\theta + r \sin\theta \cos\phi \, d\phi,$$

$$dz = dr \cos\theta - r \sin\theta d\theta.$$

Thus an element of 'length' is given by

$$ds^2 = dx^2 + dy^2 + dz^2$$

$$= dr^2 + r^2 d\theta^2 + r^2 \sin^2\theta \, d\phi^2.$$

We could write this in the form

$$ds^2 = ds_r^2 + ds_\theta^2 + ds_\phi^2,$$

with the identification

$$ds_r = dr, \quad ds_\theta = r d\theta, \quad ds_\phi = r \sin\theta d\phi.$$

If we further identify these as $ds_j = e_j dq_j$, where q_j, $1 \le j \le 3$, stand for the spherical coordinates, we see that $e_r = 1, e_\theta = r, e_\phi = r \sin\theta$. Finally, we note that $e_j dq_j$ has the dimensions of the original coordinates x_j, i.e, length here.

1.3.2   Generalised velocities and momenta

We can write $\mathbf{r} \equiv \mathbf{r}(q_1, q_2, \cdots, q_N, t)$. The position $\mathbf{r}$ may depend intrinsically on time, as would happen in the case of a rotating coordinate system, and also indirectly, because q_i depend on time. We may then write

$$\dot{\mathbf{r}} = \frac{\partial \mathbf{r}}{\partial q_1} \dot{q}_1 + \frac{\partial \mathbf{r}}{\partial q_2} \dot{q}_2 + \cdots + \frac{\partial \mathbf{r}}{\partial q_N} \dot{q}_N + \frac{\partial \mathbf{r}}{\partial t}. \qquad (1.12)$$

The quantities $\dot{q}_i$ are called generalised components of velocities. As before, they need not be the physical dimensions of velocity, though $e_j \dot{q}_j$ will have the dimensions of velocity.

Again, in spherical coordinate system, we have

$$\dot{\mathbf{r}} = \dot{r}\mathbf{u}_r + r\dot{\theta}\mathbf{u}_\theta + r\sin\theta\dot{\phi}\mathbf{u}_\phi, \qquad (1.13)$$

where $\mathbf{u}_r, \mathbf{u}_\theta, \mathbf{u}_\phi$ are unit vectors in directions of increasing r, θ, ϕ. Here $\dot{r}, \dot{\theta}, \dot{\phi}$ are generalised velocities, though it is $\dot{r}, r\dot{\theta}, r\sin\theta\dot{\phi}$ which will have dimensions of velocity. In general, we can write $\dot{\mathbf{r}} \equiv \dot{\mathbf{r}}(q_j, \dot{q}_j, t)$.

Since linear momentum is given by $\mathbf{p} = m\dot{\mathbf{r}}$, and $\mathbf{p} \equiv (p_x, p_y, p_z) \equiv (p_r, p_\theta, p_\phi)$, we can treat p_r, p_θ, p_ϕ as spherical components of generalised momentum. From (1.13) we see that

$$p_r = m\dot{r}, p_\theta = mr\dot{\theta}, p_\phi = mr\sin\theta\dot{\phi}. \qquad (1.14)$$

1.3.3 Work done and generalised forces

When forces $\mathbf{F}_i$ act on particles i and produce displacements $\mathbf{dr}_i$ in them, the work done is

$$\delta W = \sum_{i=1}^{n} \mathbf{F}_i . \mathbf{dr}_i = \sum_{j=1}^{N} X_j dx_j,$$

the latter sum being over the generalised coordinates. This can further be written as

$$\delta W = \sum_{i=1}^{N} \sum_{j=1}^{N} X_i \frac{\partial x_i}{\partial q_j} dq_j + \sum_{j=1}^{N} X_i \frac{\partial x_i}{\partial t} dt.$$

Let

$$Q_j = \sum_{i=1}^{N} X_i \frac{\partial x_i}{\partial q_j}.$$

Then Q_j are called the generalised forces. We see that

$$Q_j = -\sum_{i=1}^{N} \frac{\partial V}{\partial x_i} \frac{\partial x_i}{\partial q_j} = -\frac{\partial V}{\partial q_j}.$$

Thus the generalised forces can be derived from the potential V, but they may not have the same dimensions of force. Since

$$\delta W = \sum_i Q_j dq_j + \sum_i \mathbf{F}_i \cdot \frac{\partial \mathbf{r}_i}{\partial t},$$

we see that $Q_j dq_j$ have dimensions of work. Since $e_j dq_j$ have dimensions of length, we find that Q_j/e_j must have dimensions of force.

In spherical coordinates, $V \equiv V(r, \theta, \phi)$, and

$$a_r = -\frac{\partial V}{\partial r}, \quad a_\theta = -\frac{\partial V}{\partial \theta}, \quad a_\phi = -\frac{\partial V}{\partial \phi}.$$

The actual forces in $\mathbf{u_r}, \mathbf{u_\theta}, \mathbf{u_\phi}$ directions are given by

$$-\frac{\partial V}{\partial r}, \quad -\frac{1}{r}\frac{\partial V}{\partial \theta}, \quad -\frac{1}{r \sin \theta}\frac{\partial V}{\partial \phi}.$$

1.3.4 Lagrangian and Hamiltonian

We see that the kinetic energy of the system is given by

$$T \equiv T(q_j, \dot{q}_j, t) = \frac{1}{2}\sum_{i=1}^{n} m_i \mid \dot{\mathbf{r}}_i \mid^2 .$$

With the Lagrangian $L \equiv L(q_j, \dot{q}_j, t) = T - V$, the canonically conjugate momenta are given by $p_j = \partial L/\partial \dot{q}_j$, and the Hamiltonian is given by an equation similar to (1.8), that is

$$H = \sum p_j \dot{q}_j - L(q_j, \dot{q}_j, t).$$

Again, consider the motion of a particle in spherical coordinates. Then we have

$$T = \frac{1}{2}m(\dot{r}^2 + r^2\dot{\theta}^2 + r^2 \sin^2 \theta \dot{\phi}^2).$$

This gives the Lagrangian and the generalised momenta as

$$L = \frac{1}{2}m(\dot{r}^2 + r\dot{\theta}^2 + r^2 \sin^2 \theta \dot{\phi}^2) - V(r, \theta, \phi),$$

$$p_r = m\dot{r}, p_\theta = mr^2\dot{\theta}, p_\phi = mr^2 \sin^2 \theta \dot{\phi}. \tag{1.15}$$

We can write the Lagrangian in terms of q_j, p_j, rather than in terms of $q_j, \dot{q}_j$ in the form

$$L = \frac{1}{2m}\left(p_r^2 + \frac{p_\theta^2}{r^2} + \frac{p_\phi^2}{r^2 \sin^2 \theta}\right) - V(r, \theta, \phi).$$

The Hamiltonian can now be obtained as

$$H = p_r \dot{r} + p_\theta \dot{\theta} + p_\phi \dot{\phi} - L = \frac{1}{2m}\left(p_r^2 + \frac{p_\theta^2}{r^2} + \frac{p_\phi^2}{r^2 \sin^2 \theta}\right) + V(r, \theta, \phi). \qquad (1.16)$$

1.4 Lagrangian and Hamiltonian Equations in Generalised Coordinates

1.4.1 Lagrange's form

Since $Q_j = -\partial V/\partial q_j$ and $V \equiv V(q_j)$, we get

$$\frac{\partial V}{\partial \dot{q}_j} = 0, \quad \frac{d}{dt}\left(\frac{\partial V}{\partial \dot{q}_j}\right) = 0.$$

Hence

$$Q_j = \frac{d}{dt}\left(\frac{\partial V}{\partial \dot{q}_j}\right) - \frac{\partial V}{\partial q_j}. \qquad (1.17)$$

In nonconservative fields where Q_j depends on $\dot{q}_j$ also, (1.17) is taken as the definition of the generalised forces Q_j.

Now $T = \frac{1}{2}\sum\limits_{i=1}^{n} m_i \mid \dot{\mathbf{r}}_i \mid^2$.

Therefore, $\quad \dfrac{\partial T}{\partial q_j} = \sum\limits_{i} m_i \dot{\mathbf{r}}_i \cdot \dfrac{\partial \dot{\mathbf{r}}_i}{\partial q_j}$

$$\frac{\partial T}{\partial \dot{q}_j} = \sum_{i} m_i \dot{\mathbf{r}}_i \cdot \frac{\partial \dot{\mathbf{r}}_i}{\partial \dot{q}_j} = \sum_{i} m_i \dot{\mathbf{r}}_i \cdot \frac{\partial \mathbf{r}_i}{\partial q_j}.$$

Hence

$$\frac{d}{dt}\left(\frac{\partial T}{\partial \dot{q}_j}\right) - \frac{\partial T}{\partial q_j} = \sum_{i}\left[\frac{d}{dt}\left(m_i \dot{\mathbf{r}}_i \frac{\partial \mathbf{r}_i}{\partial q_j}\right) - m_i \dot{\mathbf{r}}_i \frac{\partial \dot{\mathbf{r}}_i}{\partial q_i}\right]$$

$$= \sum_{i}\left\{m_i \ddot{\mathbf{r}}_i \cdot \frac{\partial \mathbf{r}_i}{\partial q_j} + m_i \dot{\mathbf{r}}_i \cdot \left[\frac{d}{dt}\left(\frac{\partial \mathbf{r}_i}{\partial q_j}\right) - \frac{\partial \dot{\mathbf{r}}_i}{\partial q_j}\right]\right\}. \qquad (1.18)$$

Further, a little algebra shows that

$$\frac{d}{dt}\left(\frac{\partial \mathbf{r}_i}{\partial q_j}\right) = \sum_k \frac{\partial^2 \mathbf{r}_i}{\partial q_k \partial q_j}\dot{q}_k + \frac{\partial^2 \mathbf{r}_i}{\partial t \partial q_j} = \frac{\partial \dot{\mathbf{r}}_i}{\partial q_j},$$

so the second term in (1.18) vanishes and we have

$$\frac{d}{dt}\left(\frac{\partial T}{\partial \dot{q}_j}\right) - \frac{\partial T}{\partial q_j} = \sum_i m_i \ddot{\mathbf{r}}_i \cdot \frac{\partial \mathbf{r}_i}{\partial q_j} = Q_j. \tag{1.19}$$

Comparing (1.17) and (1.19), we get

$$\frac{d}{dt}\left(\frac{\partial L}{\partial \dot{q}_j}\right) - \frac{\partial L}{\partial q_j} = 0. \tag{1.20}$$

These are Lagrange's equations in generalised coordinates, one for each set of generalised coordinate and velocity.

1.4.2 Hamiltonian form

Since the Hamiltonian in generalised coordinates is given, in analogy with (1.8), by $H(p_j, q_j, t) \equiv \sum_j p_j \dot{q}_j - L$, we get

$$dH = \sum_j \left\{ dp_j \dot{q}_j + p_j d\dot{q}_j - \frac{\partial L}{\partial q_j}dq_j - \frac{\partial L}{\partial \dot{q}_j}d\dot{q}_j - \frac{\partial L}{\partial t}dt \right\}$$

$$= \sum_j (\dot{q}_j dp_j - \dot{p}_j dq_j) - \frac{\partial L}{\partial t}dt$$

from Lagrange's equations (1.7) and (1.9). Hence,

$$\frac{\partial H}{\partial p_j} = \dot{q}_j, \qquad \frac{\partial H}{\partial q_j} = -\dot{p}_j. \tag{1.21}$$

These are the Hamiltonian equations of motion in generalised coordinates and momenta.

1.4.3 Hamilton's equation in spherical coordinates

We wish to develop, and verify, Hamilton's equations in spherical coordinates. This is really a special case of the basics developed so far. We shall do it through a guided exercise.

▶ Guided Exercise 1.1

Derive Hamilton's equations in spherical coordinates from the first principles.

Hints

(a) Like the unit Cartesian vectors $\mathbf{i}, \mathbf{j}, \mathbf{k}$, define unit spherical vectors, $\mathbf{u_r}, \mathbf{u}_\theta, \mathbf{u}_\phi$, at a point $\mathbf{r} \equiv (r, \theta, \phi)$, $\mathbf{u_r}$ in the direction of increasing r, $\mathbf{u}_\theta$ in the direction of increasing θ and $\mathbf{u}_\phi$ in the direction of increasing ϕ. Note that, like the Cartesian unit vectors, the spherical unit vectors are orthogonal to each other[2].

(b) Express $\mathbf{i}, \mathbf{j}, \mathbf{k}$ in terms of $\mathbf{u_r}, \mathbf{u}_\theta, \mathbf{u}_\phi$ and vice versa. For example,

$$\mathbf{i} = \sin\theta\cos\phi\,\mathbf{u_r} + \cos\theta\cos\phi\,\mathbf{u}_\theta - \sin\phi\,\mathbf{u}_\phi, \tag{1.22a}$$

with similar equations for $\mathbf{j}, \mathbf{k}$ and

$$\mathbf{u}_r = \sin\theta\cos\phi\,\mathbf{i} + \sin\theta\sin\phi\,\mathbf{j} + \cos\theta\,\mathbf{k}, \tag{1.22b}$$

with similar expressions for $\mathbf{u}_\theta$ and $\mathbf{u}_\phi$.

(c) Note that Cartesian components of a vector $\mathbf{v} \equiv (v_x, v_y, v_z)$ and its spherical components (v_r, v_θ, v_ϕ) will be similarly related to each other through Eqs.(1.11a) and (1.11b).

(d) Noting that the Cartesian components of acceleration are $(\ddot{x}, \ddot{y}, \ddot{z})$, write the spherical components of acceleration, (a_r, a_θ, a_ϕ).

(e) Differentiating (1.11b) twice with respect to time and substituting in the equations obtained in step (d), obtain

$$a_r = \ddot{r} - r\,\dot{\theta}^2 - r\sin^2\theta\,\dot{\phi}^2,$$
$$a_\theta = r\,\ddot{\theta} + 2\dot{r}\dot{\theta} - r\sin\theta\cos\phi\,\dot{\phi}^2 r,$$
$$a_\phi = r\sin^2\theta\,\ddot{\phi} + 2\sin^2\theta\,\dot{r}\dot{\phi} + 2r\sin\theta\cos\theta\,\dot{\theta}\,\dot{\phi}. \tag{1.23}$$

(f) Taking the Hamiltonian in spherical coordinates from (1.16) with momentum components given by (1.15), obtain the desired Hamilton's equations in spherical coordinates:

$$\frac{\partial H}{\partial p_r} = \dot{r}, \frac{\partial H}{\partial p_\theta} = \dot{\theta}, \frac{\partial H}{\partial p_\phi} = \dot{\phi}, \tag{1.24a}$$

---

[2]Such systems in which unit vectors in the directions of increasing coordinates are orthogonal to each other are called *orthogonal coordinate systems*.

$$\frac{\partial H}{\partial r} = -\dot{p}_r, \quad \frac{\partial H}{\partial \theta} = -\dot{p}_\theta, \quad \frac{\partial H}{\partial \phi} = -\dot{p}_\phi, \tag{1.24b}$$

◀

1.4.4   A free particle in a rotating reference frame

Consider a Cartesian coordinate system (ξ, η, ζ) rotating with an angular velocity ω around the z axis. Then

$$\xi = x \cos \omega t + y \sin \omega t, \quad \eta = -x \sin \omega t + y \cos \omega t, \quad \zeta = z,$$

and conversely,

$$x = \xi \cos \omega t - \eta \sin \omega t, \quad y = \xi \sin \omega t + \eta \cos \omega t, \quad z = \zeta.$$

So the kinetic energy $T = m(\dot{x}^2 + \dot{y}^2 + \dot{z}^2)/2$ comes out to be

$$T = \frac{1}{2}m\{\dot{\xi}^2 + \dot{\eta}^2 + \dot{\zeta}^2 + \omega^2(\xi^2 + \eta^2) - 2\omega(\dot{\xi}\eta - \xi\dot{\eta})\}, \tag{1.25}$$

and, with $V = 0$ for a free particle, the Lagrangian becomes $L = T$, which is given in (1.25) above. Then ,

$$\frac{\partial L}{\partial \dot{\xi}} = m(\dot{\xi} - \omega\eta), \quad \frac{\partial L}{\partial \dot{\eta}} = m(\dot{\eta} + \omega\xi), \quad \frac{\partial L}{\partial \dot{\zeta}} = m\dot{\zeta},$$

and

$$\frac{\partial L}{\partial \xi} = m\omega^2\xi, \quad \frac{\partial L}{\partial \eta} = m\omega^2\eta, \quad \frac{\partial L}{\partial \zeta} = 0.$$

So the Lagrangian equations of motion give

$$m\ddot{\xi} = m\omega(\dot{\eta} + \omega\xi), \quad m\ddot{\eta} = m\omega(-\dot{\xi} + \omega\eta), m\ddot{\zeta} = 0.$$

Thus we see that the force can be divided into two parts:

$\mathbf{Q_1}$, with components $m\ddot{\xi} = m\omega^2\xi, m\ddot{\eta} = m\omega^2\eta, m\ddot{\zeta} = 0$ and

$\mathbf{Q_2}$ with components $m\ddot{\xi} = m\omega\dot{\eta}, \; m\ddot{\eta} = -m\omega\dot{\xi}, \; m\ddot{\zeta} = 0.$

If we put $\rho = \xi\mathbf{i} + \eta\mathbf{j}$ and $\boldsymbol{\omega} = \omega\mathbf{k}$, we see that

(i) $\mathbf{Q_1} = m\omega^2\rho$, which is known as *centrifugal force*, and

(ii) $\mathbf{Q_2} = m(\dot{\rho} \times \boldsymbol{\omega})$, which is known *Coriolis force*.

1.5 Principle of Least Action

It will be appropriate to introduce the various spaces we will have to deal with.

1.5.1 Generalised spaces

The q-space

If we consider the generalised coordinates as coordinates in an N-dimensional space, each point X in this q-space will represent a geometrical configuration of the n-particle system; see Fig. 1.2. Similarly, each curve like AB, CD or EF in Fig. 1.2 passing through X will be a possible course of events which can be marked by the time parameter t. Initial values of coordinates q_j^0 and initial velocities $\dot{q}_j^0$ will determine the actual course of events. Since $\dot{q}_j$ are arbitrary, an infinite number of trajectories can start from or intersect at the same point in the q-space.

The q–p space

We can think of a q–p space of $6n$ dimensions, known as the *phase space*; see Fig. 1.3. Every point in this space represents a dynamical state of the system. All subsequent or prior dynamical states are completely determined by the initial state through the equations of motion. Hence one and only one curve can pass through any point in the phase space. A change in the initial conditions changes the starting point and the trajectory passing through it is entirely different. Thus two curves representing two different paths of the system cannot cross each other. Points on each curve can be marked by the time parameter t. Closed curves in phase-space represent periodic motion.

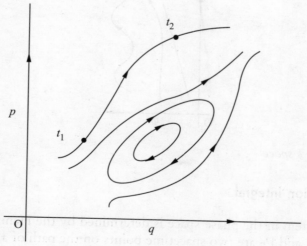

Fig. 1.2   *The coordinate space of an n-body system*

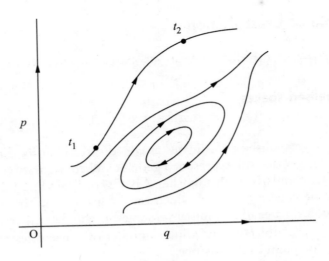

Fig. 1.3 *The phase space and the trajectories in it*

The phase-space with time axis

We can generalise the space still further by adding a time axis, giving rise to a $6n + 1$ dimensional space; see Fig. 1.4. In this space too the trajectories cannot cross each other. Further they must move monotonically in the time axis. Periodic motion will be represented by a wavy line.

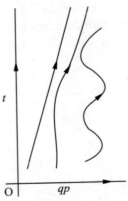

Fig. 1.4 *The qp–t space*

1.5.2 The action integral

The path of a system in the phase space is determined by the Hamilton's principle of least action. If P_1 and P_2 are two spacetime points on the path of a system, the action in going from P_1 to P_2 is given by the integral

$$A = \int_{P_1}^{P_2} L dt = \int_{P_1}^{P_2} \{p_j dq_j - H dt\}, \tag{1.26}$$

which is a functional of the path connecting the two points P_1 and P_2. A is called the *action integral* in spacetime. The principle of least action states that the necessary and sufficient condition for defining an actual path R between the points P_1 and P_2 is that $\delta A = 0$ in the neighbourhood of the path R; see Fig. 1.5(a). We shall now discuss its proof.

Let R be the actual path the system takes while R′ be a slightly different path in the neighbourhood of R, both connecting P_1 with P_2. Then the variation in the action integral over R and R′ is

$$\delta A = \int_{P_1}^{P_2} \text{along } R' - \int_{P_1}^{P_2} \text{along } R$$

$$= \int_{P_1}^{P_2} \sum \left\{ \delta p_j dq_j + p_j \delta(dq_j) - \frac{\partial H}{\partial q_j} \delta q_j dt - \frac{\partial H}{\partial p_j} \delta p_j dt - \frac{\partial H}{\partial t} \delta t dt - H \delta(dt) \right\}.$$

(a)

(b)

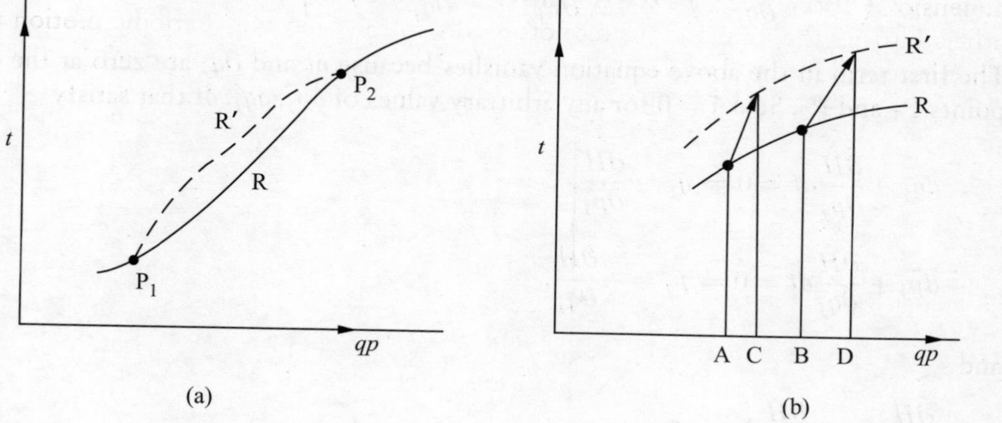

Fig. 1.5 *(a) Two possible paths between two fixed end points in qp–t space; (b) variation of the path between two fixed end points in qp–t space.*

Here δ represents the difference between two adjacent paths while d is the increment on one path itself. Now, from Fig. 1.5 (b), we see that

$$\delta(dq_j) = \text{CD} - \text{AB} = \text{BD} - \text{AC} = d(\delta q_j).$$

Similarly,

$$\delta(dt) = d(\delta t),$$

which leads to

$$p_j \delta(dq_j) = p_j d(\delta q_j) = d(p_j \delta q_j) - dp_j \delta q_j, \quad \text{and}$$

$$H\delta(dt) = H d(\delta t)$$

$$= d(H\delta t) - dH\delta t$$

$$= d(H\delta t) - \left(\frac{\partial H}{\partial p_j} dp_j + \frac{\partial H}{\partial q_j} dq_j + \frac{\partial H}{\partial t} dt \right) \delta t.$$

Therefore,

$$\delta A = \int_{P_1}^{P_2} d \left\{ \sum_j p_j \delta q_j - H\delta t \right\} + \int_{P_1}^{P_2} \left(\sum \left\{ \left(dq_j - \frac{\partial H}{\partial p_j} dt \right) \delta p_j \right. \right.$$

$$\left. \left. - \left(dp_j + \frac{\partial H}{\partial q_j} dt \right) \delta q_j + \left(\frac{\partial H}{\partial p_j} dp_j + \frac{\partial H}{\partial q_j} dq_j \right) \delta t \right\} \right). \tag{1.27}$$

The first term in the above equation vanishes because δt and ∂q_j are zero at the end points P_1 and P_2. So $\delta A = 0$ for any arbitrary values of δp_j, δq_j, δt that satisfy

$$dq_j - \frac{\partial H}{\partial p_j} dt = 0 \Rightarrow \dot{q}_j = \frac{\partial H}{\partial p_j},$$

$$dp_j + \frac{\partial H}{\partial q_j} dt = 0 \Rightarrow \dot{p}_j = -\frac{\partial H}{\partial q_j},$$

and

$$\frac{\partial H}{\partial p_j} dp_j + \frac{\partial H}{\partial q_j} dq_j = 0.$$

The first two of the above conditions give Hamiltonian equations, while the third one does not give anything new. This shows that if $\delta A = 0$, the Hamiltonian equations are valid at every point on the path.

   On the other hand, if Hamiltonian's equations are valid all along a path, then (1.27) shows that $\delta A = 0$, and thus A is an extremum for an actual path. This completes the proof.

1.5.3   Fermat's principle

For photons, $V = 0$, so that

$$L = K = H = h\nu. \tag{1.28}$$

Therefore, the action integral between two points on the path of a photon becomes

$$A = h\nu \int_{P_1}^{P_2} dt = h\nu \int_{P_1}^{P_2} \frac{ds}{v}, \tag{1.29}$$

where ds is the displacement along the path and v the velocity of the photon.

This shows that the path of light (photon) is determined by the condition that $\int ds/v$ is extremum. This is Fermat's principle from which follow the laws of geometrical optics—rectilinear propagation, laws of reflection, laws of refraction, etc.

1.6   Poisson Brackets

In the further development of classical Hamiltonian mechanics, it is convenient to define Poisson brackets. If $F(q_i, p_i, t)$ and $G(q_i, p_i, t)$ are two functions of the generalised coordinates and momenta and time, then their *Poisson bracket* (PB) is defined as

$$\{F, G\} = \sum_{i=1}^{n} \left(\frac{\partial F}{\partial q_i} \frac{\partial G}{\partial p_i} - \frac{\partial F}{\partial p_i} \frac{\partial G}{\partial q_i} \right). \tag{1.30}$$

The following properties of Poisson brackets then easily follow from this definition:

$$\{F, F\} = 0, \quad \{F, G\} = -\{G, F\}, \quad \{F, c\} = 0,$$

where c is independent of q_i, p_i, though it may depend on t. If E is yet another function of q_i, p_i, t, then

$$\{E + F, G\} = \{E, G\} + \{F, G\},$$

$$\{E, FG\} = \{E, F\}G + \{E, G\}F.$$

As a special case, the Poisson brackets between the coordinates and momenta are themselves found to be

$$\{q_i, q_j\} = 0, \quad \{p_i, p_j\} = 0.$$

But,

$$\{q_i, p_j\} = \sum_{r=1}^{n} \left(\frac{\partial q_i}{\partial q_r} \frac{\partial p_j}{\partial p_r} - \frac{\partial q_i}{\partial p_r} \frac{\partial p_j}{\partial q_r} \right)$$

$$= \sum_{r=1}^{n} \delta_{ir} \frac{\partial p_j}{\partial p_r} = \frac{\partial p_j}{\partial p_i}$$

$$= \delta_{ij},$$

where,

$$\delta_{ij} = \begin{cases} 1 & \text{if} \quad i = j \\ 0 & \text{if} \quad i \neq j \end{cases}$$

is the Kronecker delta function.

1.6.1 Equation of motion

The Poisson bracket defined above is useful in recasting the Hamiltonian equations of motion using Poisson brackets. For any function $F(q_i, p_i, t)$, we have

$$\frac{dF}{dt} = \frac{\partial F}{\partial t} + \sum_{i=1}^{n} \left(\frac{\partial F}{\partial q_i} \dot{q}_i + \frac{\partial F}{\partial p_i} \dot{p}_i \right). \tag{1.31}$$

Using Hamilton's equations, Eqs.(1.10), the above equation becomes

$$\frac{dF}{dt} = \frac{\partial F}{\partial t} + \sum_{i=1}^{n} \left(\frac{\partial F}{\partial q_i} \frac{\partial H}{\partial p_i} - \frac{\partial F}{\partial p_i} \frac{\partial H}{\partial q_i} \right) = \frac{\partial F}{\partial t} + \{F, H\}. \tag{1.32}$$

This equation can be considered as the fundamental equation of motion for the entity F. The justification for this comes from the fact that if we successively put $F = q_i, p_i$ and H, equation (1.32), with $\frac{\partial q_i}{\partial t} = \frac{\partial p_i}{\partial t} = 0$ gives

$$\frac{dq_i}{dt} = \sum_{r=1}^{n} \left\{ \frac{\partial q_i}{\partial q_r} \frac{\partial H}{\partial p_r} - \frac{\partial q_i}{\partial p_r} \frac{\partial H}{\partial q_i} \right\}$$

$$\Rightarrow \dot{q}_i = \frac{\partial H}{\partial p_i}, \frac{dp_i}{dt} = -\frac{\partial H}{\partial q_i}, \frac{dH}{dt} = \frac{\partial H}{\partial t} = 0. \tag{1.33}$$

Thus H is a constant of motion. Further, (1.32) shows that $\Delta dF/dt = 0$ if $\Delta \{F, H\} = -\partial F/\partial t$. So F will be a constant of motion if $\{F, H\} = 0$.

1.6.2 Poisson brackets for combined variables

Let us define new variables by[3]

$$w_{2r-1} = q_r, \quad w_{2r} = p_r, \quad 1 \leq r \leq n.$$

Then it is easy to see that

$$\{w_{2r-1}, w_{2s-1}\} = \{q_r, q_s\} = 0,$$

$$\{w_{2r}, w_{2s}\} = \{p_r, p_s\} = 0,$$

$$\{w_{2r-1}, w_{2s}\} = -\{w_{2r}, w_{2s-1}\} = \delta_{rs}.$$

This can further be put in a compact form as shown below. Let us define, for $n = 1$,

$$\epsilon_{\mu\nu} = \begin{vmatrix} \{q_1, q_1\} & \{q_1, p_1\} \\ \{p_1, q_1\} & \{p_1, q_1\} \end{vmatrix} = \begin{bmatrix} 0 & 1 \\ -1 & 0 \end{bmatrix}, \tag{1.34}$$

where $\mu, \nu = 1, 2$. Further, $\epsilon_{\mu\nu} = 1$ or -1 according as whether (μ, ν) is an even or odd permutation of $(1, 2)$, and $\epsilon_{\mu\nu} = 0$ otherwise. For general n with $1 \leq \mu, \nu \leq 2n$, we define

$$\epsilon_{\mu\nu} = \{w_\mu, w_\nu\}; \tag{1.35a}$$

we write $\epsilon_{\mu\nu}$ as a matrix of order $2n$ in which the matrix of (1.34) is repeated n time along the principal diagonal, i.e.,

$$\epsilon_{\mu\nu} = \begin{bmatrix} 0 & 1 \\ -1 & 0 \\ & & 0 & 1 \\ & & -1 & 0 \\ & & & & 0 & 1 \\ & & & & -1 & 0 \\ & & & & & & 0 & 1 \\ & & & & & & -1 & 0 \end{bmatrix}. \tag{1.35b}$$

[3]See Mukunda, N. (1974)

Let us define the reciprocal (inverse) matrix $\epsilon^{\mu\nu}$ such that

$$\epsilon^{\mu\nu}\epsilon_{\mu\nu} = \delta_\alpha^\nu,$$

where δ_α^ν is the Kronecker delta. Thus for $n = 1$, $\epsilon^{\mu\nu}$ will be

$$\epsilon^{\mu\nu} = \begin{bmatrix} 0 & -1 \\ 1 & 0 \end{bmatrix},$$

and for general n, it will take the form similar to (1.35b) with interchange of 1 and -1.

If $f(w)$ and $g(w)$ are any two functions, where w stands for the set of generalised coordinates and momenta, n each, then their Poisson bracket can be expressed as

$$\{f(w), g(w)\} = \sum_{\mu,\nu=1}^2 n\epsilon_{\mu\nu} \frac{\partial f(w)}{\partial w_\mu} \frac{\partial g(w)}{\partial w_\nu}.$$

1.7 Contact Transformation

Since the generalised coordinates q_j and the generalised momenta p_j occur at the same footing and are not essentially different, we can consider all of them as fundamental and make a more general transformation. Thus, we consider two sets of functions

$$Q_j \equiv Q_j(q_k, p_k, t), P_j \equiv P_j(q_k, p_k, t), \tag{1.36}$$

so that the Hamiltonian $H(q_j, p_j, t)$ is transformed in to $K(Q_j, P_j, t)$. In general, such a transformation will not preserve the Hamiltonian form of equations of motion. But if it does, so that

$$\dot{Q}_j = \frac{\partial K}{\partial P_j}, \dot{P}_j = \frac{\partial K}{\partial Q_j}, \tag{1.37}$$

then such a transformation is called a *contact transformation*. We shall see later why it is called by this name.

The advantage of a contact transformation is that if we can make K independent of Q_j or P_j, then $\dot{P}_j = 0$ or $\dot{Q}_j = 0$, and the corresponding Q_j or P_j will be constants of motion. This simplifies the problem.

Various contact transformations can be obtained from generating functions $S \equiv S(q_j, p_j, Q_j, P_j, t)$. Out of the $4n$ variables q_j, p_j, Q_j, P_j, only $2n$ are independent. So S is really a function of $2n + 1$ variables including t. There is a theorem due to Caratheodory[4] according to which there are 2^n different ways of making a contact transformation. We choose one group of $2n$ variables x_r ($1 \leq r \leq n$) from old q_j, p_j

---

[4]See Mukunda, N.(1974)

and X_r $(1 \leq r \leq n)$ from the new Q_j, P_j in a particular way and put the rest in another group x'_r $(1 \leq r \leq n)$ and X'_r $(1 \leq r \leq n)$. Then it is possible to find a function $S(x_r, X_s)$ such that $dS(x_r, X_s) = \epsilon^{\mu\nu}(x'_\mu dx_\nu - X'_\mu dX_\nu)$ is a perfect differential. Then,

$$x'_\lambda = -\epsilon_{\lambda\mu} \frac{\partial S(x_r, X_s)}{\partial x_\mu},$$

$$X'_\lambda = \epsilon_{\lambda\mu} \frac{\partial S(x_r, X_s)}{\partial X_\mu}. \tag{1.38}$$

We shall consider four simple ways of choosing the two sets out of four in which all x are from one set and all X from another set.

1.7.1 Some contact transformations

(i) Let us choose $S \equiv S(q_j, Q_j, t)$. We can obtain a contact transformation if we make the difference of $L dt$, that is

$$dS = \left(\sum p_j dq_j - H dt \right) - \left(\sum P_j dQ_j - K dt \right),$$

a perfect differential. In that case, $S = \int_{P_1}^{P_2} dS$ would be independent of path and $\delta S = 0$. Now

$$\delta S = \delta \int_{P_1}^{P_2} \left(\sum p_j dq_j - H dt \right) - \delta \int_{P_1}^{P_2} \left(\sum P_j dQ_j - K dt \right)$$

$$\equiv \delta A - \delta B,$$

where

$$A = \int_{P_1}^{P_2} \left(\sum p_j dq_j - H dt \right), \quad B = \int_{P_1}^{P_2} \left(\sum P_j dQ_j - K dt \right). \tag{1.39}$$

We note from Section 1.5.2 that $\delta A = 0$ near an actual path in the (q, p) phase space and $\delta B = 0$ near an actual path in the (Q, P) phase space, which are different. Thus, P_j and Q_j satisfy Hamilton's equations (1.37).

If the above S is to be a perfect differential, then the old and the new variables are related by

$$\frac{\partial S}{\partial q_j} = p_j, \quad \frac{\partial S}{\partial Q_j} = P_j \text{ with } \frac{\partial S}{\partial t} = K - H. \tag{1.40}$$

Thus, knowing S, we can obtain the new variables.

As an example, consider $S = \sum q_i Q_i$. Then

$$\frac{\partial S}{\partial q_j} = Q_i = p_i, \quad \frac{\partial S}{\partial Q_i} = q_i = -P_i,$$

where we have used (1.40). Thus the old coordinates and momenta interchange their roles. This is an example of contact transformation.

(ii) Let us choose $S \equiv S(q_j, P_j, t)$. Then we have to make

$$dS = \left(\sum p_j \, dq_j - H \, dt\right) - \left(\sum P_j \, dQ_j - K \, dt\right)$$

a perfect differential. But $P_j \, dQ_j = d(P_j Q_j) - Q_j \, dP_j$, giving

$$dS = \left(\sum p_j \, dq_j - H \, dt\right) - d \sum P_j Q_j + \left(\sum Q_j \, dP_j + K \, dt\right).$$

Since $d \sum P_j Q_j$ is already a perfect differential, we should make

$$dS = \left(\sum p_j \, dq_j - H \, dt\right) + \left(\sum Q_j \, dP_j + K \, dt\right)$$

a perfect differential. This requires that

$$\frac{\partial S}{\partial q_j} = p_j, \quad \frac{\partial S}{\partial P_j} = Q_j, \quad \frac{\partial S}{\partial t} = K - H. \tag{1.41}$$

If we now consider choosing $S = \sum q_i P_i$, we get $P_i = p_i$, $q_i = Q_i$, which is an identity transformation.

(iii) For $S \equiv S(p_j, P_j, t)$, a similar treatment as before with a simple choice such as $S = \sum p_i P_i$ gives

$$\frac{\partial S}{\partial p_j} = -q_j, \quad \frac{\partial S}{\partial P_j} = Q_j, \quad \frac{\partial S}{\partial t} = K - H,$$

leading to $Q_j = p_j$, $P_j = -q_j$.

(iv) Finally, if we choose $S = S(p_j, Q_j, t)$, with the choice $S = \sum p_i Q_i$, we arrive at

$$\frac{\partial S}{\partial p_j} = -q_j, \quad \frac{\partial S}{\partial Q_j} = -P_j, \quad \frac{\partial S}{\partial t} = K - H,$$

which leads to $Q_j = -q_j$, $P_j = -p_j$.

1.8 Hamilton–Jacobi Equation

With suitable transformation of the $2n$ variables q_i, p_i, Hamilton's equations can be put in another suitable form which makes it easier to spot the constants of motion. A contact transformation would be useful if it makes the new variables (Q_j, P_j) constant, which will be the case if

$$\dot{P}_j = -\frac{\partial K}{\partial Q_j} = 0, \quad \dot{Q}_j = \frac{\partial K}{\partial P_j} = 0. \tag{1.42}$$

If any of the above equations is satisfied for any j, then K is independent of the corresponding partner canonical coordinate.

As a particular case, suppose $K = 0$. Then from (1.40), we have $H + \frac{\partial S}{\partial T} = 0$. If $S \equiv S(q, Q, t)$ or $S(q, p, t)$, then (1.40) or (1.41) shows that $p_j = \frac{\partial S}{\partial q_j}$. Here we have dropped the subscript for brevity so that a symbol such as q stands for the whole set of coordinates q_j. This gives

$$\frac{\partial S}{\partial t} + H\left(q, \frac{\partial S}{\partial q}, t\right) = 0. \tag{1.43}$$

This is known as Hamilton–Jacobi equation. In solving it for $S(q, Q, t)$, additive constants of integration are not required. On obtaining $S(q, Q, t)$, we get constants of integration Q_j, P_j.

Considering $S \equiv S(p, Q, t)$ or $S(p, P, t)$ so that $q_j = -\partial S/\partial p_j$, we get

$$\frac{\partial S}{\partial t} + H\left(-\frac{\partial S}{\partial p}, p, t\right) = 0, \tag{1.44}$$

which is another form of Hamilton–Jacobi equation.

1.8.1 Nature of contact transformation

Let P_1 be a fixed point and P a variable point on the actual path R in Fig. 1.5 (a). Then the change in the action integral will be

$$dA = d\int_{P_1}^{P} (pdq - Hdt)$$

$$= pdq - Hdt - p^{(1)}dq^{(1)} + H^{(1)}dt^{(1)}, \tag{1.45}$$

where the superscript (1) indicates initial values at time $t^{(1)}$ at point P_1 and pdq stands for $\sum p_k dq_k$. Thus A is a function of (q, p, t) which has $2n$ constants of integration,

$p^{(1)}$ and $q^{(1)}$. Taking them as initial and final coordinates, $q^{(1)}$ and q, instead of $q^{(1)}$ and $p^{(1)}$, we get from (1.45),

$$\frac{\partial A}{\partial q_k} = p_k, \quad \frac{\partial A}{\partial q_k^{(1)}} = p_k^{(1)}, \quad \frac{\partial A}{\partial t} = -H.$$

This gives

$$\frac{\partial A}{\partial t} + H\left(q, \frac{\partial A}{\partial q}, t\right) = 0,$$

showing that A satisfies the Hamilton–Jacobi equation. So A is a contact transformations which has $q^{(1)}$ and $p^{(1)}$ as constants of motion.

In other words, the actual path is a succession of infinitesimal contact transformations; hence the name contact transformation. It is the basis of considering the effect of perturbations on the orbits of planets in terms of the variation of orbital parameters.

1.8.2   Some applications of Hamilton–Jacobi equation

It would be appropriate to discuss some applications of Hamilton–Jacobi formulation to appreciate its utility. We discuss here some simple cases such as a free particle, a particle falling under gravity, and a one-dimensional simple harmonic oscillator.

Free particle
In this case, we have

$$V = 0, \quad T = \frac{1}{2}mv^2 = \frac{1}{2}m(\dot{x}^2 + \dot{y}^2 + \dot{z}^2).$$

Then we have $L = T - V = T$, so that

$$p_x = \frac{\partial L}{\partial \dot{x}} = m\dot{x}, \quad p_y = m\dot{y}, \quad p_z = m\dot{z}.$$

This gives

$$L = \frac{1}{2m}(p_x^2 + p_y^2 + p_z^2)$$

and

$$H = \sum_i p_i q_i - L = \frac{1}{2m}(p_x^2 + p_y^2 + p_z^2).$$

The Hamilton–Jacobi equation becomes

$$\frac{\partial S}{\partial t} + \frac{1}{2m}\left[\left(\frac{\partial S}{\partial x}\right)^2 + \left(\frac{\partial S}{\partial y}\right)^2 + \left(\frac{\partial S}{\partial z}\right)^2\right] = 0.$$

Let us choose

$$S = T(t) + X(x) + Y(y) + Z(z).$$

Then we have

$$\frac{\partial T}{\partial t} + \frac{1}{2m}\left[\left(\frac{\partial X}{\partial x}\right)^2 + \left(\frac{\partial Y}{\partial y}\right)^2 + \left(\frac{\partial Z}{\partial z}\right)^2\right] = 0.$$

In order to separate the time part, we put

$$\left(\frac{\partial X}{\partial x}\right)^2 + \left(\frac{\partial Y}{\partial y}\right)^2 + \left(\frac{\partial Z}{\partial z}\right)^2 = V_0^2,$$

which, together with the above equations gives

$$T(t) = -\frac{V_0^2}{2m}t.$$

Now, we let

$$\frac{\partial X}{\partial x} = Q_x, \quad \frac{\partial Y}{\partial y} = Q_y, \quad \frac{\partial Z}{\partial z} = Q_z,$$

so that

$$V_0^2 = Q_x^2 + Q_y^2 + Q_z^2,$$

and $X(x) = Q_x x, \; Y(y) = Q_y y, \; Z(z) = Q_z z,$

giving $\Delta S = Q_x x + Q_y y + Q_z z - \dfrac{t}{2m}(Q_x^2 + Q_y^2 + Q_z^2).$

We then get

$$p_x = \frac{\partial S}{\partial x} = Q_x = mv_x^0,$$

$$p_y = \frac{\partial S}{\partial y} = Q_y = mv_y^0,$$

$$p_z = \frac{\partial S}{\partial z} = Q_z = mv_z^0,$$

where we have introduced three initial velocity components $\Delta v_x^0, v_y^0$ and v_z^0. Therefore, the three components of momenta are constant. In turn, this implies that the three components of velocity are also constant. In other words, the particle moves with a uniform velocity.

Also, we have

$$-P_x = \frac{\partial S}{\partial Q_x} = x - \frac{Q_x}{m}t = x - v_x^0 t, \quad \text{etc.}$$

This gives

$$P_x = -x_0, \quad P_y = -y_0, \quad P_z = -z_0,$$

where $\Delta x_0, y_0, z_0$ are the three initial value of the coordinates at $t = 0$. These three values provide the other three constants of motion.

The final solution of the problem is therefore

$$x = x_0 + v_x^0 t, \quad y = y_0 + v_y^0 t, \quad z = z_0 + v_z^0 t,$$

which indicates a straight-line motion.

Particle falling under gravity

Consider a particle of mass m falling freely under gravity (g = acceleration due to gravity). Then, taking x as the coordinate measured upward from a suitable arbitrary ground level, we have $V = mgx$. Therefore,

$$L = \tfrac{1}{2}m\dot{x}^2 - mgx,$$

giving

$$p_x = \frac{\partial L}{\partial \dot{x}} = m\dot{x}.$$

Then

$$H = p_x \dot{x} - L = \frac{1}{2}m\dot{x}^2 + mgx = \frac{p_x^2}{2m} + mgx.$$

Then the Hamilton–Jacobi equation becomes

$$\frac{\partial S}{\partial t} + \frac{1}{2m}\left(\frac{\partial S}{\partial x}\right)^2 + mgx = 0. \tag{1.46}$$

Let $S = T(t) + X(x)$ so that (1.46) reduces to

$$\frac{dT}{dt} + \frac{1}{2m}\left(\frac{dX}{dx}\right)^2 + mgx = 0. \tag{1.47}$$

In order to separate the x and t parts, let us put $dT/dt = -A$, a constant, so the (1.47) gives

$$T = -At, \quad \frac{dX}{dt} = [2m(A - mgx)]^{1/2},$$

which then gives

$$X = (2m)^{1/2}\int_x^h (A - mgx)^{1/2}dx,$$

where h is some arbitrary level (see Fig. 1.6) given by

$$h = A/mg.$$

This then gives

$$S = (2m)^{1/2}\int_x^h (A - mgx)^{1/2}dx - At.$$

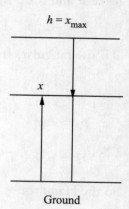

Fig. 1.6   *A particle falling under gravity*

This yields

$$\frac{\partial S}{\partial A} \equiv -B = -t + \left(\frac{m}{2}\right)^{1/2}\int_x^h (h - mgx)^{-1/2}dx.$$

If $t = T$ at $x = h$, then $B = T$, which gives, on integrating,

$$x = h - \frac{1}{2}g(t - T)^2.$$

We also find that

$$p_x = \frac{\partial S}{\partial x} = [2m(h - mgx)]^{1/2},$$

giving $g(h - x) = \frac{1}{2}\dot{x}^2$.

This can be written in another form:

$$\frac{1}{2}m\dot{x}^2 + mgx = mgh = E,$$

where E is the total energy, the sum of kinetic and potential energies, and is a constant of motion. It also turns out, from (1.47) and equation following it, that

$$A = E, \quad h = E/mg. \tag{1.48}$$

One-dimensional simple harmonic oscillator

For a one-dimensional simple harmonic oscillator of mass m the acceleration $\ddot{x}$, the force F and the potential energy $V(x)$ are given by

$$\ddot{x} = -\omega^2 x, \quad F(x) = -m\omega^2 x, \quad V(x) = m\omega^2 x^2/2,$$

ω is the (angular) frequency of oscillations. Then,

$$L = \frac{1}{2}m\dot{x}^2 - \frac{m\omega^2 x^2}{2},$$

$$p_x = m\dot{x}$$

$$H = p_x\dot{x} - L = \frac{p_x^2}{2m} + \frac{m\omega^2 x^2}{2}.$$

So the Hamilton–Jacobi equation becomes

$$\frac{\partial S}{\partial t} + \frac{1}{2m}\left(\frac{\partial S}{\partial x}\right)^2 + \frac{m\omega^2 x^2}{2} = 0. \tag{1.49}$$

Let us put

$$S = T(t) + X(x), \tag{1.50}$$

so that

$$\frac{dT}{dt} + \frac{1}{2m}\left(\frac{dX}{dx}\right)^2 + \frac{m\omega^2 x^2}{2} = 0.$$

To separate the space and time coordinates, we put

$$\frac{dT}{dt} = -Q,$$

a constant, so as to get

$$\frac{dX}{dx} = m\omega\left[\frac{2Q}{m\omega^2} - x^2\right]^{1/2}.$$

With these, (1.50) becomes

$$S = -Qt + m\omega\int\left(\frac{2Q}{m\omega^2} - x^2\right)^{1/2} dx.$$

This gives

$$P = -\frac{\partial S}{\partial Q} = t - \frac{1}{\omega} \sin^{-1} \frac{x}{(2Q/m\omega^2)^{1/2}},$$

or

$$x = \left(\frac{2Q}{m\omega^2}\right)^{1/2} \sin \omega(t - P). \tag{1.51}$$

This represents a periodic motion of frequency $\omega/2\pi$, with an amplitude

$$\Delta x_{max} = (2Q/m\omega^2)^{1/2},$$

and phase P at $t = 0$. We can also consider a three-dimensional oscillator with potential given by

$$V = \frac{1}{2} \left(\omega_1^2 x^2 + \omega_2^2 y^2 + \omega_3^2 z^2 \right) \quad \text{per unit mass.}$$

We leave this as an exercise.

▶Guided Exercise 1.2

Show that all the three applications discussed in this section satisfy the condition $K = 0$.

Hints The three applications discussed in this section refer to a free particle, a particle falling under gravity and a one-dimensional harmonic oscillator. In each case, proceed along the suggested steps.

(a) Write down x, y, z and p_x, p_y, p_z in terms of the new coordinate Q_x, Q_y, Q_z and P_x, P_y, P_z.

(b) Write the Lagrangian and the Hamiltonian in terms of the new coordinates. Take S from the respective section, Section 1.9.2 and Section 1.9.3.

(c) Show that $K = H + \partial S/\partial t = 0$ in each case.

◀

1.9 The Two-Body Problem

The two-body problem in classical mechanics is one of the few exactly solvable problems. It is also the first step beyond the point mass of Newtonian mechanics. This is why it holds a prominent place in the teaching of physics. It turns out that many real systems like the masses in a dumb-bell (neglecting the mass of the light rod), the earth-moon system, binary stars (which outnumber other stars in the universe), diatomic molecules (in quantum mechanics), etc., can be treated as two-body systems. In this section, we discuss their kinematics and dynamics, concluding the section with orbital parameters of a planet.

1.9.1   Hamilton–Jacobi equation for two-body system

Consider two masses m_1 and m_2 with coordinates $r_1 \equiv (x_1, y_1, z_1)$ and $r_2 \equiv (x_2, y_2, z_2)$, respectively, with central potential $V(r)$ between them. It could be the gravitational potential $V(r) = -Gm_1m_2/r$, where $r = [(x_2 - x_1)^2 + (y_2 - y_1)^2 + (z_2 - z_1)^2]^{1/2}$ is the distance between them. Then the system will be described by the Hamiltonian

$$H = \frac{1}{2}m_1(\dot{x}_1{}^2 + \dot{y}_1{}^2 + \dot{z}_1{}^2) + \frac{1}{2}m_2(\dot{x}_2{}^2 + \dot{y}_2{}^2 + \dot{z}_2{}^2) + V(r)$$

$$= \frac{1}{2m_1}(p_{1x}^2 + p_{1y}^2 + p_{1z}^2) + \frac{1}{2m_2}(p_{2x}^2 + p_{2y}^2 + p_{2z}^2) + V(r). \qquad (1.52)$$

The Hamilton–Jacobi equation is

$$\frac{1}{2m_1}\left[\left(\frac{\partial S}{\partial x_1}\right)^2 + \left(\frac{\partial S}{\partial y_1}\right)^2 + \left(\frac{\partial S}{\partial z_1}\right)^2\right]$$

$$+ \frac{1}{2m_2}\left[\left(\frac{\partial S}{\partial x_2}\right)^2 + \left(\frac{\partial S}{\partial y_2}\right)^2 + \left(\frac{\partial S}{\partial z_2}\right)^2\right] + \frac{\partial S}{\partial t} + V(r) = 0. \qquad (1.53)$$

This equation cannot be solved by the method of separation of variables. But it can be decoupled by choosing two suitable new coordinates. This is what we shall do in the next section.

1.9.2   Separation of variables

Let us consider the relative coordinate

$$\mathbf{r} = \mathbf{r} - \mathbf{r_1}, \qquad (1.54a)$$

which is the position of particle 2 with respect to particle 1, and the position $\mathbf{R}$ of the centre of mass of the two bodies given by

$$\mathbf{R} = \frac{m_1\mathbf{r_1} + m_2\mathbf{r_2}}{m_1 + m_2}, \qquad (1.54b)$$

as the new variables. Then we have

$$\mathbf{r_1} = \mathbf{R} - \frac{m_2\mathbf{r}}{m_1 + m_2}, \quad \mathbf{r_2} = \mathbf{R} + \frac{m_1\mathbf{r}}{m_1 + m_2}.$$

Then the kinetic energy of the two particles can be recast in a nice form as

$$\frac{1}{2}m_1\dot{\mathbf{r}}_1^2 + \frac{1}{2}m_2\dot{\mathbf{r}}_2^2 = \frac{1}{2}M\dot{\mathbf{R}}^2 + \frac{1}{2}\mu\dot{\mathbf{r}}^2,$$

where

$$M = m_1 + m_2, \quad \mu = \frac{m_1 m_2}{m_1 + m_2}, \tag{1.55}$$

are the total mass and the reduced mass of the two particles.

The Lagrangian can be recast using (1.54) as

$$L = \frac{1}{2}M\dot{R}^2 + \frac{1}{2}\mu\dot{r}^2 + V(r). \tag{1.56}$$

Corresponding to the six components of **r** and **R**, we have the six momentum components of **p** and **P** which can be found from the Lagrangian equations. Using (1.56), they are seen to be

$$p_x = \mu\dot{x}, p_y = \mu\dot{y}, p_z = \mu\dot{z},$$

$$P_X = M\dot{X}, P_Y = M\dot{Y}, P_Z = M\dot{Z}.$$

Then the Lagrangian and Hamiltonian can be written in the form

$$L = \frac{1}{2M}(P_X^2 + P_Y^2 + P_Z^2) + \frac{1}{2\mu}(p_x^2 + p_y^2 + p_z^2) - V(r),$$

$$H = \frac{1}{2M}(P_X^2 + P_Y^2 + P_Z^2) + \frac{1}{2\mu}(p_x^2 + p_y^2 + p_z^2) + V(r).$$

The Hamilton–Jacobi equation then becomes

$$\frac{\partial S'}{\partial t} + \frac{1}{2M}\left[\left(\frac{\partial S'}{\partial X}\right)^2 + \left(\frac{\partial S'}{\partial Y}\right)^2 + \left(\frac{\partial S'}{\partial Z}\right)^2\right]$$

$$+ \frac{1}{2\mu}\left[\left(\frac{\partial S'}{\partial x}\right)^2 + \left(\frac{\partial S'}{\partial y}\right)^2 + \left(\frac{\partial S'}{\partial z}\right)^2\right] + V(r) = 0.$$

So let

$$S' = \sigma(\mathbf{R}, t) + S(\mathbf{r}, t),$$

so that we have

$$\frac{\partial \sigma}{\partial t} + \frac{1}{2M}\left[\left(\frac{\partial \sigma}{\partial X}\right)^2 + \left(\frac{\partial \sigma}{\partial Y}\right)^2 + \left(\frac{\partial \sigma}{\partial Z}\right)^2\right] + \frac{\partial S}{\partial t}$$

$$+\frac{1}{2\mu}\left[\left(\frac{\partial \sigma}{\partial x}\right)^2 + \left(\frac{\partial \sigma}{\partial y}\right)^2 + \left(\frac{\partial \sigma}{\partial z}\right)^2\right] + V(r) = 0.$$

Now the equation can be split into two parts,

$$\frac{\partial \sigma}{\partial t} + \frac{1}{2M}\nabla_{\mathbf{R}}^2 \sigma = 0, \tag{1.57a}$$

$$\frac{\partial S}{\partial t} + \frac{1}{2\mu}\nabla_{\mathbf{r}}^2 S + V(r) = 0, \tag{1.57b}$$

where $\nabla_{\mathbf{R}}$ and $\nabla_{\mathbf{r}}$ are the Laplacians with respect to $\mathbf{R}$ and $\mathbf{r}$, respectively.

Equation (1.57a) is that for a free particle of mass M, and shows that the centre of mass moves with a uniform velocity through space, while (1.57b) is the equation of relative motion of the two bodies. Due to the appearance of the central force term $V(r)$ in the latter equation, it cannot be solved in Cartesian coordinates. However, it is obvious that the variables (the three components of $\mathbf{r}$) could be separated in spherical polar coordinates.

1.9.3   Solution in spherical coordinates

In spherical coordinates $\mathbf{r} \equiv (r, \theta, \phi)$, the Laplacian is given by

$$\nabla_{\mathbf{r}}^2 = p_r^2 + \frac{p_\theta^2}{r^2} + \frac{p_\phi^2}{r^2 \sin^2 \theta}.$$

Using this in the Hamilton–Jacobi equation (1.57b) and letting

$$S = T(t) + R(r) + U(\theta) + F(\phi),$$

we get

$$\frac{dT}{dt} + \frac{1}{2\mu}\left[\left(\frac{dR}{dr}\right)^2 + \frac{1}{r^2}\left(\frac{dU}{d\theta}\right)^2 + \frac{1}{r^2 \sin^2 \theta}\left(\frac{dF}{d\phi}\right)^2\right] + V(r) = 0.$$

The time variable is immediately separated out by letting $dT/dt = -Q_1$, a constant, so that $T(t) = -Q_1 t$ and

$$\left(\frac{dR}{dr}\right)^2 + \frac{1}{r^2}\left(\frac{dU}{d\theta}\right)^2 + \frac{1}{r^2\sin^2\theta}\left(\frac{dF}{d\phi}\right)^2 + 2\mu V(r) = 2\mu Q_1.$$

The next variable to be separated, as usual, is ϕ. This leads to the two equations

$$\frac{dF}{d\phi} = Q_2 \Rightarrow F(\phi) = Q_2\phi, \tag{1.58a}$$

and

$$r^2\left(\frac{dR}{dr}\right)^2 + 2\mu r^2 V(r) - 2\mu Q_1 r^2 = -\left(\frac{dU}{d\theta}\right)^2 - \frac{Q_2^2}{\sin^2\theta}. \tag{1.58b}$$

In (1.58b), r and θ are also separated, so that putting each side equal to a constant Q_3^2, we get the equations

$$\left(\frac{dU}{d\theta}\right)^2 = Q_3^2 - \frac{Q_2^2}{\sin^2\theta},$$

$$\left(\frac{dR}{dr}\right)^2 = 2\mu Q_1 - 2\mu V(r) - \frac{Q_3^2}{r^2} \equiv f(r),$$

which defines $f(r)$. Let r_1 be the smaller root of $f(r) = 0$. Then we can put

$$R(r) = \int_{r_1}^{r}\left[2\mu Q_1 - 2\mu V(r) - \frac{Q_3^2}{r^2}\right]^{1/2} dr. \tag{1.59a}$$

Similarly, starting the U integral from $\theta = \pi/2$ where $z = 0$, we can write

$$U(\theta) = \int_{\pi/2}^{\theta}\left[Q_3^2 - \frac{Q_2^2}{\sin^2\theta}\right]^{1/2} d\theta, \tag{1.59b}$$

which results in

$$S = -Q_1 t + Q_2\phi + \int_{\pi/2}^{\theta}(Q_3^2 - \frac{Q_2^2}{\sin^2\theta})^{1/2}d\theta + \int_{r_1}^{r}\left[2\mu Q_1 - 2\mu V(r) - \frac{Q_3^2}{r^2}\right]^{1/2} dr.$$

Then $P_i = -\partial S/\partial Q_i$ gives

(i) $P_1 = t - \int_{r_1}^{r} \frac{\mu dr}{[2\mu Q_1 - 2\mu V(r) - Q_3^2/r^2]^{1/2}} + f^{1/2}(r_1)\frac{\partial r_1}{\partial Q_1}.$

But as $f(r_1) = 0$, the integral vanishes at r_1, and we get $P_1 = t(r_1)$.

(ii) $P_2 = -\phi + \int_{\pi/2}^{\theta} \dfrac{Q_2 d\theta}{\sin^2\theta (Q_3^2 - Q_2^2/\sin^2\theta)^{1/2}} = -\phi(\theta = \pi/2).$

(iii) $P_3 = -\int_{\pi/2}^{\theta} \dfrac{Q_3 d\theta}{(Q_3^2 - Q_2^2/\sin^2\theta)^{1/2}} + \int_{r_1}^{r} \dfrac{(Q_3/r^2)dr}{f^{1/2}(r)} - f^{1/2}(r_1)\dfrac{\partial r_1}{\partial Q_3}$

which gives

$$P_3 = \int_{-\pi/2}^{\theta(r_1)} \frac{Q_3 d\theta}{(Q_3^2 - Q_2^2/\sin^2\theta)^{1/2}}.$$

1.9.4   Orbital elements of a planet

We shall consider the sun–planet system. To specify the dynamical state of the planet in its orbit around the sun at any moment, we require three components of position and three components of velocity. These can be recast into a set of six new independent parameters which are more meaningful and convenient. They are the semi major axis a of the ellipse, eccentricity e, longitude Ω of the ascending node N where the planet crosses the plane of the ecliptic from south to north (see Fig. 1.7), the inclination i of the orbit with the ecliptic plane, angle ω of the perihelion P from the ascending node, and T_P, the time of the (last) perihelion passage of the planet. It should be clear that the first five parameters characterise the elliptical orbit completely in space and the last one helps in locating the planet in its orbit at any moment. Let us see how these parameters are related to P_i and Q_i.

(i)   Taking x-axis in the direction of vernal equinox V (Fig. 1.7), y-axis at right angles in the plane of the orbit and z-axis at right angles to it, the node N occurs at $\theta = \pi/2$, so $P_2 = -\Omega$.

(ii)   In the integral for $U(\theta)$, equation (1.59), the term under the radical must be non-negative. Therefore θ has a minimum value given by

$$\sin\theta_{\min} = Q_2/Q_3.$$

This is reached when the planet reaches its maximum northern latitude $i = \theta_{\min}$. Hence

$$i = \cos^{-1}(Q_2/Q_3).$$

(iii)   From the spherical triangle NPQ of Fig. 1.7, we have

$$\cos\theta(r_1) = \sin\omega \sin i.$$

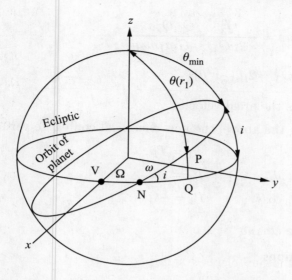

Fig. 1.7 *Orientation of the orbit of a planet with respect to the ecliptic; V − vernal equinox, N − node, P − perihelion.*

Then

$$P_3 = -\int_{\pi/2}^{\theta(r_1)} \frac{Q_3 d\theta}{(Q_3^2 - Q_2^2/\sin^2\theta)^{1/2}}$$
$$= \pi/2 - \cos^{-1}(\sin\omega) = -\omega,$$

giving

$$P_3 = -\omega.$$

(iv) Now for $Q_1 < 0$, both roots of $f(r)$ are positive and we have an elliptic orbit with

$$r_1 + r_2 = -\frac{2G\mu m_1 m_2}{2\mu Q_1} = -\frac{Gm_1 m_2}{Q_1},$$

$$r_1 r_2 = -Q_3^2/2\mu Q_1.$$

As $r_1 = a(1-e)$ and $r_2 = a(1+e)$, we have

$$2a = -\frac{Gm_1 m_2}{Q_1}$$

$$\Rightarrow Q_1 = \frac{Gm_1 m_2}{2a} = E, \quad \text{which is the total energy of the planet.}$$

(v) We also have

$$r_1 r_2 = a^2(1-e^2) = ap$$

$$\Rightarrow ap = -\frac{Q_3^2}{2\mu Q_1} = \frac{Q_3^2 a}{Gm_1 m_2 \mu}$$

$$\Rightarrow Q_3 = 2\mu.(\frac{1}{2}r^2\dot{\theta}),$$

where $\frac{1}{2}r^2\dot{\theta}$ is the areal velocity.

Summarising the above, the new canonical conjugate variables are

$$Q_1 = E, \qquad\qquad P_1 = T_P,$$
$$Q_3 = \mu r^2 \dot{\theta}, \qquad\qquad P_3 = -\omega,$$
$$Q_2 = Q_3 \cos i, \qquad\quad P_2 = -\Omega, \qquad\qquad\qquad (1.60)$$

1.9.5 Perturbations

There would be gravitational perturbation to an orbit due to other planets. In such a case, we write

$$H = H_0 + R,$$

where H_0 is the unperturbed potential and R is the perturbing potential. The perturbation method is useful provided the perturbation has a small effect, $R \ll H_0$. The solution of H_0 will be a good starting point for solving the perturbed equation. Then Q_i, P_i will be approximately constant. Therefore we transform to these variables and solve the complete equation $H = H_0 + R$ by expanding Q_i, P_i into series. Thus,

$$\dot{Q}_i = \left(\frac{\partial H}{\partial P_i}\right) = \frac{\partial H_0}{\partial P_i} + \frac{\partial R}{\partial P_i}.$$

But

$$\frac{\partial H_0}{\partial P_i} = \dot{Q}_{i0} = 0,$$

giving

$$\dot{Q}_i = \partial R / \partial P_i. \qquad\qquad\qquad (1.61a)$$

Similarly,

$$\dot{P}_i = -\frac{\partial H_0}{\partial Q_i} = \frac{\partial H_0}{\partial Q_i} - \frac{\partial R}{\partial Q_i}.$$

Again,

$$\frac{\partial H_0}{\partial Q_i} = -\dot{P}_{i0} = 0,$$

giving

$$\dot{P}_i = -\partial R/\partial Q_i. \tag{1.61b}$$

Equations (1.61) give the variables Q_i, P_i as functions of time. This is the principle of obtaining the variation of the orbital parameters of planets.

1.10 Virial Theorem

1.10.1 The virial and the virial theorem

Consider a system of n particles described by the equations of motion

$$\mathbf{F}_i = m_i \ddot{\mathbf{r}}_i.$$

Then we have

$$\mathbf{r}_i.\mathbf{F}_i = m_i \mathbf{r}_i.\ddot{\mathbf{r}}_i$$

$$= m_i \left[\frac{d}{dt}(\mathbf{r}_i.\dot{\mathbf{r}}_i) - \Delta \mid \dot{\mathbf{r}}_i \mid^2 \right].$$

Let the force $\mathbf{F}_i$ on the ith particle be split into two parts

$$\mathbf{F}_i = \sum_{j \neq i} \mathbf{F}_{ji} + \mathbf{G}_i,$$

where $\mathbf{F}_{ji}$ are mutual internal forces and $\mathbf{G}_i$ are external forces. Then the above expression becomes

$$m_i \left[\frac{1}{2}\frac{d^2}{dt^2} \mid \mathbf{r}_i \mid^2 - \mid \dot{\mathbf{r}}_i \mid^2 \right] = \mathbf{r}_i. \left[\sum_{j \neq i} \mathbf{F}_{ji} + \mathbf{G}_i \right].$$

Summing over all the particles, we get

$$\frac{1}{2}\frac{d^2}{dt^2} \sum_i m_i \mid \mathbf{r}_i \mid^2 - \sum_i m_i \mid \dot{\mathbf{r}}_i \mid^2 = \sum_i \sum_{j \neq i} \mathbf{r}_i.\mathbf{F}_{ji} + \sum_i \mathbf{r}_i.\mathbf{G}_i. \tag{1.62}$$

The two terms on the right hand side are respectively called the *internal virial* and the *external virial*.

As $\mathbf{F}_{ji} = -\mathbf{F}_{ij}$ by Newton's third law of motion, the internal virial can be written as

$$\sum_i \sum_{j \neq i} \mathbf{r}_i.\mathbf{F}_{ji} = \sum_{ij} \mathbf{F}_{ji}.(\mathbf{r}_i - \mathbf{r}_j),$$

each pair being counted only once. Let

$$I = \sum_i m_i \mid \mathbf{r}_i \mid^2 = \sum_i m_i(x_i^2 + y_i^2 + z_i^2) = \frac{1}{2}(I_x + I_y + I_z),$$

where I_x, I_y, I_z are the moments of inertia of the body about the x, y, z axes, respectively. Also, we have

$$\frac{1}{2}\sum_i m_i \mid \dot{\mathbf{r}}_i \mid^2 = K.$$

With these, (1.62) becomes

$$\frac{1}{2}\frac{d^2 I}{dt^2} = 2K + \sum_{ij} \mathbf{F}_{ji}.(\mathbf{r}_i - \mathbf{r}_j) + \sum_i \mathbf{r}_i.\mathbf{G}_i.$$

If the internal forces obey the inverse square law (eg., gravitation, electrostatic), then

$$\mathbf{F}_{ji} = C_{ij}\frac{\mathbf{r}_i - \Delta\mathbf{r}_j}{\mid \mathbf{r}_i - \Delta\mathbf{r}_j \mid^3}$$

$$\Rightarrow \mathbf{F}_{ji}.\Delta(\mathbf{r}_i - \Delta\mathbf{r}_j) = \frac{C_{ij}}{\mid \mathbf{r}_i - \Delta\mathbf{r}_j \mid} \equiv V_{ji}, \tag{1.63}$$

which defines V_{ji}. Then the internal virial of (1.62) becomes

$$\sum_{ij} \mathbf{F}_{ji}.(\mathbf{r}_i - \mathbf{r}_j) = \sum_{ij} V_{ji} \equiv V,$$

which is the potential energy. This finally gives

$$\frac{1}{2}\frac{d^2 I}{dt^2} = 2K + V + \sum_i \mathbf{r}_i.\mathbf{G}_i. \tag{1.64}$$

This is known as the *virial theorem*.

If the external force is in the form of pressure $P(\mathbf{r})$, it will act only at the boundary, where it will be given by $-P\mathbf{n}d\sigma$, where $\mathbf{n}$ is the unit outward normal at the element of surface area $d\sigma$. Then,

$$\sum_i \mathbf{r}_i.\mathbf{G}_i = -\oint_S \mathbf{r}.P\mathbf{n}d\sigma,$$

where we have a closed surface integral over the surface S. If P is constant over the surface, the external virial becomes $-P \oint_S \mathbf{r}.n d\sigma$. Using Gauss theorem and noting that $\nabla.\mathbf{r} = 3$, this becomes

$$\sum \mathbf{r}_i.\mathbf{G}_i = -3Pv,$$

where v is the volume inside the closed surface. Then (1.63) takes the form

$$\frac{1}{2}\frac{d^2 I}{dt^2} = 2K + V - 3Pv.$$

Further, in a steady or quasi-steady state, we will have $d^2 I/dt^2 = 0$, which requires

$$2K + V = 3Pv,$$

or taking the energies K and V per unit volumes,

$$2K + V = 3P. \tag{1.65}$$

1.10.2 Applications

We shall now consider some applications of the virial theorem.

Perfect gas

Although we shall deal with a perfect gas in Chapter 4, we may point out a few aspects here. Since there are assumed to be no inter-particle interactions in a perfect gas, we have potential energy $V = 0$. Since the gas is in a steady state and the pressure is constant everywhere, (1.64), leads to $P = 2K/3$. But $K = 3NkT/2$, where N is the number of particles in volume V (not to be confused with potential) and k is the Boltzmann constant. If V_m is the molar volume and N_A the Avogadro number, we have $PV_m = N_A kT = RT$, where $R = N_A k$ is the gas constant.

No external force

If there is no external force, then $P = 0$ and we have $2K + V = 0$. If $U = K + V$ is the total internal energy, we get $U + K = 0$ or $U = -K = V/2$.

Since K is always positive, for steady state, we must have $U < 0$ and so also $V < 0$. This situation arises, for example, in the two-body problem. If the potential energy of the two particles, in accordance with (1.63), is $V = C_{12}/r$, where r is the separation between the two particles, then $U = -C_{12}/2r$ and $V = -C_{12}/r$, giving $K = C_{12}/2r$. Here C_{12} would be $Gm_1 m_2$ for gravity and $q_1 q_2$ for electrostatic forces. Here m_i and q_i, $i = 1, 2$, are the masses and charges of the two particles.

Masses of star clusters

If we consider a star or a star cluster as an isolated body in steady state and in equilibrium, we can use $2K + V = 0$. If M is the total mass of the cluster and $\bar{v}$ is the average speed of a star in it, then we have $2K = M\bar{v}^2$. If we assume that the cluster is in a polytrop[5] of index n, then

$$V = -\frac{3}{5-n}\frac{GM^2}{R}, \tag{1.66}$$

where R is the radius of the cluster. Therefore,

$$M\bar{v}^2 = \frac{3}{5-n}\frac{GM^2}{R} \Rightarrow M = \Delta\frac{5-n}{3}\frac{R\bar{v}^2}{G}.$$

We shall show how we can estimate masses of cluster with the above equation in the following cases:

(a) Galactic cluster: In this case

$$\sqrt{(\bar{v}^2)} = 0.25 \text{ km/s}, \ R = 6 \text{ parsecs}, \ n = 0. \text{ Therefore, } M/M_\odot \approx 140.$$

(b) Globular cluster: In this case, we have

$$\sqrt{(\bar{v}^2)} = 15 \text{ km/s}, \ R = 15 \text{ parsecs}, \ n = 3, \text{ which gives } M/M_\odot \approx 5 \times 10^5.$$

Minimum self-gravitating mass

An average star is in a steady state, where its mass tends to decrease its radius and its core temperature and pressure tend to increase it. In the absence of external forces, the situation would be governed by (1.63). In the steady state I would be constant and $dI/dt = 0$ and $d^2I/dt^2 = 0$, so that we would have $2K + V = 0$. Contraction or collapse would occur if I were already maximum or $\frac{1}{2}d^2I/dt^2 = 2K + V < 0$. Therefore, for self-gravitation to occur, we must have

$$2K < -V. \tag{1.67}$$

Consider a uniform gaseous sphere of mass M and radius R. Then, with $n = 0$, (1.65) becomes

$$V = -\frac{3}{5}\frac{GM^2}{R}.$$

If T is the temperature and μ the molecular weight of the gas, the total number of particles in the galaxy becomes $N = M/\mu m_H$, where m_H is the mass of the hydrogen atom. As the kinetic energy per particle is $3kT/2$, we get

---

[5]See K D Abhyankar (2001) for details

$$K = \frac{3}{2}NkT = \frac{3MkT}{2\mu m_H}.$$

Therefore for collapse, we must have, in accordance with (1.67),

$$M > \frac{5kTR}{G\mu m_H}.$$

But we also have $M = (4/3)\pi R^3 \rho$, where ρ is the mass density given by $\rho = n\mu m_H$, n being the number density. Then the criterion finally becomes

$$M^2 > \frac{375k^3T^3}{4\pi n G^3 \mu^4 m_H^4}.$$

We can get simple estimates from the following considerations. Interstellar medium consists of neutral hydrogen atoms with a density of 1 atom per cm^3 at a temperature of 100K. So taking $\mu = 1, n = 1/\text{cm}^3$, T = 100K and M$_\odot$ = 2×10^{33}gm, we see that for gravitational collapse to occur, we must have $M/M_\odot > 10^5$. Embedded in the interstellar medium we find clouds of molecular hydrogen at a temperature of 10K with a density of 10 to 1000 molecules per cc. Putting these numbers, the criterion for this situation becomes $M/M_\odot > 1$ to 100.

Problems

1.1. Verify the Hamiltonian equation for cylindrical coordinates ρ, ϕ, z.

1.2. Write the Hamiltonian for n gravitationally interacting particles, and obtain the Hamilton-Jacobi equations for the system.

1.3. Derive the following ten constants of integration for n gravitationally interacting particles:

(i) Three initial coordinates of the centre of mass.

(ii) Three components of uniform velocity of the centre of mass.

(iii) Three components of angular momentum about the three principal axes passing through the centre of mass.

(iv) The total energy of the system.

1.4. (a) Prove the gravitational equivalent of Gauss law of electrostatics. [Hint: If $\rho(r')$ is the mass density in space, define

$$\mathbf{g}(\mathbf{r}) = -\frac{G\rho(r')(\mathbf{r} - \mathbf{r}')}{\mid \mathbf{r} - \mathbf{r}' \mid^3}$$

as the gravitational field at $\mathbf{r}$. Then the gravitational flux leaving a closed surface S is $\phi = \oint_S \mathbf{g}(\mathbf{r}) \cdot \mathbf{ds}$ and show that it is equal to $-4\pi G m_S$, where m_S is the mass inside S.]

(b) Hence prove that the gravitational field due to a spherically symmetric body of mass M at a point outside it is the same as that due to a point mass at the centre.

1.5. Show that the laws of reflection and refraction for light follow from Fermat's principle. (Note: Velocity of light in a medium of refractive index μ is c/μ.)

1.6. Solve the problem of three-dimensional oscillator mentioned at the end of Section 1.9(c).

1.7. Consider the restricted problem of three bodies in which two larger masses M_1 and M_2 move in circular orbits around their common centre of mass and an infinitesimal mass m moves in the gravitational field of the large masses. The motion of the bodies takes place in the x–y plane with angular velocity ω. Consider the system of coordinates ξ, η, ζ rotating with angular velocity ω around the z-axis (which is also the ζ-axis). Derive the generalised coordinates, velocities and forces in that system and obtain an expression for the (gravitational) potential Ω.

1.8. A child is drowning in a rectangular swimming pool at some point. A lifeguard sitting outside notices it and wants to reach the child in the shortest possible time. The guard must run some distance on the ground in a straight line at a known speed and swim some distance in water at a lower speed. Show that the guard must enter water at a point where the angles outside and inside water satisfy Snell's law (of refraction).

2

Special and General Theories of Relativity

IN this chapter we discuss the theories of relativity—special and general. It is said that the time was ripe around 1905 for the development of special theory, and somebody or the other would have been lead to it sooner or later. But this was not the case with the general theory of relativity which is a unique contribution of Einstein. We start with the background for the special theory in early 1905 with Maxwell's equations of electrodynamics, the appearance of a constant velocity in it, the consequent search for a unique frame of reference (the ether), and the postulates of special theory of relativity boldly put forward by Einstein. Then we deal with Lorentz transformation and their kinematical applications. We introduce the four-dimensional Minkowski space and the concepts of mass, energy, momentum, force and acceleration in it. We go on to the general theory of relativity and some of its significant consequences, including gravitational lensing.

2.1 Background

2.1.1 Classical mechanics

We have seen in Chapter 1 that classical mechanics is based on Newton's laws of motion. We restate the second law which reads

$$\mathbf{f} = \frac{d\mathbf{p}}{dt} = \frac{d}{dt}(m\mathbf{v}) = m\frac{d\mathbf{v}}{dt} = m\frac{d^2\mathbf{r}}{dt^2} = m\mathbf{a}. \tag{2.1}$$

Here $\mathbf{f}$ is the force and $\mathbf{r}$, $\mathbf{v}$, $\mathbf{p}$ and $\mathbf{a}$ are positions, velocity, momentum and acceleration, respectively. In the absence of any net external force, $\mathbf{p}$ and hence $\mathbf{v}$ is a constant, which is a part of the first law of motion. The spacetime coordinate systems in which the first law is valid are known as *inertial frames*.

In Newtonian mechanics, length, time and mass are taken as absolute quantities which do not depend on the frame of reference, i.e., the chosen coordinate system. The space itself is assumed to be isotropic and Euclidean in which the space interval ds is given by $ds^2 = dx^2 + dy^2 + dz^2$, where dx, dy, dz are the Cartesian components of ds.

Under a Galilean transformation, which represents a transformation between two frames of reference where S' $= (x', y', z', t')$ and S $= (x, y.z, t)$, the first of which is moving uniformly with respect to the second, say in the x direction with velocity v, we have

$$x' = x - vt, y' = y, z' = z, t' = t. \tag{2.2}$$

Then

$$u_x' = \frac{dx'}{dt'} = \frac{dx}{dt} - v = u_x - v, u_y' = u_y, u_z' = u_z. \tag{2.3}$$

Here $\mathbf{u}' = (u_x', u_y', u_z')$ is the velocity of a body in frame S' and $\mathbf{u} = (u_x, u_y, u_z)$ is the velocity of the body in the frame S. Eqs. (2.3) contain the vector law of relative velocities. $\mathbf{u}' = \mathbf{u} - \mathbf{v}$, where the relative velocity $\mathbf{v}$ in an arbitrary direction. Differentiating Eqs. (2.3) once more we get

$$a_x' = \frac{du_x'}{dt'} = \frac{du_x}{dt} = a_x, a_y' = a_y, a_z' = a_z. \tag{2.4}$$

Thus acceleration, and hence force, is invariant under Galilean transformation.

Thus Newton's laws are valid and invariant with respect to a Galilean transformation. One can easily play a game of table tennis on a stationary ship as well on a ship moving with a uniform velocity with respect to the former. All systems moving uniformly with respect to any one inertial frame of reference are themselves inertial and the laws of motion are identical in them. This is the principle of relativity. The choice of a particular (approximate) inertial system depends upon circumstances, e.g. the earth for laboratory experiments, fixed stars for the solar system and external galaxies for stars in the Milky Way.

In accelerated frames of reference, Newton's laws remain valid only by the introduction of some fictitious forces known as inertial forces. Consider a system for which $x' = x - gt^2/2, y' = y, z' = z$, where g is the acceleration of the body under consideration along the x-axis. The acceleration in this system is given by $a_x' = a_x - g, a_y' = a_y, a_z' = a_z$ or $\mathbf{a}' = \mathbf{a} + \mathbf{g}$, where the acceleration $\mathbf{g}$ could be in an arbitrary direction. Then the equation of motion $\mathbf{f} = m\mathbf{a}$ becomes $\mathbf{f} = m\mathbf{a}' + m\mathbf{g}$ or $\mathbf{f} - m\mathbf{g} = m\mathbf{a}$. Thus we have to introduce a fictitious force $-m\mathbf{g}$ to satisfy the equation of motion. Here $-m\mathbf{g}$ represents the inertial force in the accelerated frame of reference. In the case of earth which represents a rotating frame of reference, we have to introduce two fictitious forces (see Section 1.4.4): (i) a centrifugal force $m\omega^2\mathbf{r}$, and (ii) a Coriolis force $m\mathbf{r} \times \boldsymbol{\omega}$, where $\mathbf{r}$ is the position and $\boldsymbol{\omega}$ is the angular velocity.

It may be mentioned here that the electromagnetic phenomenon is not invariant under Galilean transformation. A charge at rest produces only electrostatic field and no magnetic field. But a moving charge exhibits both electric and magnetic fields. Hence we expect trouble in applying Gallilean transformation to the electric phenomenon.

2.1.2 Maxwell's equations and ether

Maxwell's equations for electromagnetic fields in vacuum contain constants ϵ_0 and μ_0, respectively the permittivity and permeability of free space. It so happens that $(\epsilon_0\mu_0)^{-1/2}$ has the physical dimensions of velocity. It was soon recognised that its numerical value comes to be 3×10^{10} cm/s, which is the speed of light in vacuum. Since this factor appears in Maxwell's wave equation for electromagnetic fields, it was suggested that light consists of electromagnetic waves. From experience, it was known well before Maxwell that waves normally require a medium for transmission. So it was assumed that empty space is filled with a hypothetical substance called *ether* which, however, required very strange properties. As only solids can transmit transverse waves, ether was considered to be a solid. Further, since velocities of transverse and longitudinal waves are respectively given by $v_t = \sqrt{\eta/\rho}$ and $v_l = \sqrt{(k + 4\eta/3)/\rho}$, where k is the bulk modulus and η the shear modulus, ether was attributed with a positive shear modulus and negative bulk modulus so that $k + 4\eta/3 = 0$, and in spite of its solidity, ether would not offer any resistance to the bodies moving through it.

It was commented in Section 1.1.1 that Newton's laws of motion did not contain velocity and therefore it was not possible to measure the velocity of an inertial observer with respect to any fixed universal frame of reference such as the ether. Thus, while Newtonian mechanics suggested that there is no preferred frame of reference, Maxwell's electromagnetism suggested that there should be a unique frame of reference in which ether can be assumed to be at rest and in which light travels with a constant speed of 3×10^{10} cm/s. All bodies move in the backdrop of this ether. It should be possible to determine this preferred coordinate system by observing the velocity of light in different directions.

It was also commented earlier that Newton's equations are invariant under Galilean transformations but Maxwell's equations of electromagnetism are not. Lorentz and Fitzgerald in 1900 and Poincare around 1905 tried to find out the most general transformations under which Maxwell's equations would be invariant. This formed the basis for Einstein's special theory of relativity.

2.1.3 Aberration of light

Consider observing a star with a telescope; see Fig. 2.1. Let OAC be the direction of an incoming light ray and v the velocity of the earth (telescope) with respect to *ether* in the direction BC. Then we have to tilt the telescope in the direction BA so that the photons entering the objective along OA will also enter the eyepiece, which in the meantime has moved from B to C through a distance vdt. If θ the true angle of the ray

OAC with the direction of motion and θ' the apparent angle ABC, we have

$$\frac{\sin\theta'}{\sin(\theta-\theta')} = \frac{cdt}{vdt} = \frac{c}{v}$$

$$\Rightarrow \sin(\theta-\theta') = \frac{v}{c}\sin\theta'.$$

If we take v to be the speed of the earth in its orbit around the sun, then v is about 30 km/s, and $v/c \simeq 10^{-5} \ll 1$. Therefore, $\Delta\theta = \theta - \theta' \simeq (v/c)\sin\theta$.

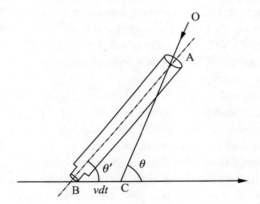

Fig. 2.1   *Aberration of light in air*

This is the phenomenon of aberration which was discovered by Bradley in 1728. He found that during the course of the year all stars in the sky appear to trace out an ellipse with semimajor axis $a = 20.5$" and semiminor axis $b = a\sin\beta$, where β is the celestial latitude of the star. For stars on the ecliptic, $\beta = 0$, and the ellipse stratches out into a straight line of length $2a$, while for stars near the celestial poles, $\beta = \pi/2$, the ellipse becomes a circle of radius a. In fact, a represents the maximum value of $\Delta\theta$ at $\theta = \pi/2$, i.e., $a = v/c$. The phenomenon of aberration proves that the earth moves round the sun in a nearly circular orbit with a velocity of 30 km/s. However, it does not measure the velocity of the sun with respect to ether.

Now the velocity of light in water is c/n, where n is the refractive index of water. Hence, if the telescope tube is filled with water, the value of aberration constant a' should come out larger, equal to nv/c. In 1871 Airy performed such an experiment and found no change in the aberration, i.e. $a' = a$. This observation can be explained by assuming that a medium like water drags ether with it by a fraction f of v. In Fig. 2.2, we show a telescope filled with water. Here AC is the path of light ray in air and AD that in water so that

$$\frac{\sin\psi}{\sin\psi'} = n,$$

where ψ and ψ' are the respective angles made by CA and DA with the telescope axis. If v' is the velocity of ether with respect to water, then, as before,

$$\sin(\phi' - \phi) = \sin\psi' = \frac{nv'}{c}\sin\phi.$$

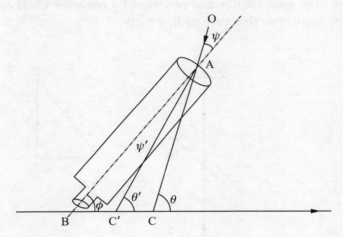

Fig. 2.2   *Aberration of light in a water filled telescope*

But

$$\sin(\theta - \phi) = \sin\psi = (v/c)\sin\phi.$$

From these equations we arrive at

$$v' = v/n^2.$$

Hence the dragging coefficient is

$$f = \frac{v - v'}{v} = 1 - \frac{1}{n^2}.$$

f is known as Fresnel dragging coefficient after Fresnel, who explained Arago's 1810 experiments which showed that the refractive index of a glass prism is unaltered by placing it in front of a telescope and pointing the telescope in different directions. The formula for dragging coefficient was verified in laboratory by Fizeau in 1851.

For vacuum, $n = 1$ and for air $n \simeq 1$, hence air does not drag ether with it. So one should be able to measure the ether current past the earth by an appropriate experiment in the laboratory.

2.1.4   Michelson–Morley experiment

In this experiment in 1887, Michelson and Morley used a Michelson interferometer with equal arms; see Fig. 2.3. Here a light beam from source S is split into two beams by the beam splitter B and sent to two mirrors M_1 and M_2 placed at equal distance l in directions x and y. The plate C compensates for the extra path in beam SM_2 (because the beam splits at the front surface of the beam splitter B). After reflection the signals are combined by the beam splitter and observed by observer O. If t_1 and t_2 are the times required by the beams to return to B, we have

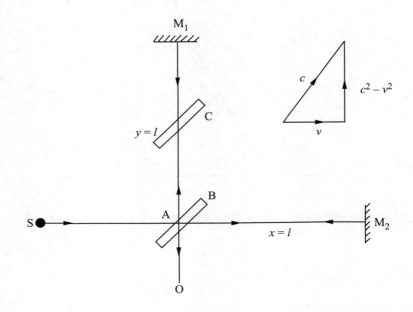

Fig. 2.3   *Michelson–Morley experiment*

$$t_1 = \frac{2l}{\sqrt{c^2 - v^2}} = \frac{2l}{c}(1 - \beta^2)^{-1/2} \simeq \frac{2l}{c}\left(1 + \frac{1}{2}\beta^2\right),$$

$$t_2 = \frac{l}{c - v} + \frac{l}{c + v} = \frac{2l}{c}(1 - \beta^2)^{-1} \simeq \frac{2l}{c}(1 + \beta^2),$$

where $\beta = v/c$. Note that the instrument is moving with velocity v with respect to the ether in the direction SM_2. Then,

$$\Delta t = t_2 - t_1 = lv^2/c^3.$$

This gives a path difference $\Delta x = c\Delta t = lv^2/c^2$ and a fringe shift $lv^2/\lambda c^2$, where λ is the wavelength of light. A rotation through $\pi/2$ would give a fring shift of double the amount, $\Delta N = 2lv^2/\lambda c^2$. Michelson and Morley used a system of mirrors to increase l to 11 m. With $\lambda = 5100 \, \overset{\circ}{A}$ and $v = 30$ km/s, one gets $\Delta N = 0.37$, a fringe shift which was easily observable in the experiment. Michelson and Morley, however, did not find any fringe shift by rotating the apparatus so that SM$_2$ was one way or the other. The result was null throughout the year, which should have taken into account the motion of the earth and the sun as well. It showed that the velocity of light c was the same for all observers, and the existence of ether became doubtful.

Reference frames and phenomena

A reference frame is a set of coordinates (three for our purpose) to specify a point and a set of clocks placed around everywhere to record the time of an event. Events and phenomena take place in spacetime; they do *not* take place in a reference frame. When an event occurs, any number of inertial observers can record it and specity it by giving the three coordinates of where the event occurred and the time at which it occurred, as measured by each observer. In fact, this set of four values, three space coordinates and a time coordinate, is called an event in spacetime. A phenomenon is a succession of events. A phenomenon similarly takes place in spacetime, not in a reference frame, and it can be recorded and measured by any number of observers using their respective spacetime coordinate systems.

2.2 Lorentz Transformation

2.2.1 Einstein's postulates

Michelson–Morley experiment can be explained if we assume that the arm of the interferometer in the direction of motion of the earth is contracted by a factor $(1 - \beta^2)^{1/2}$. Then,

$$t_2 = \frac{(2l/c)(1-\beta)^{1/2}}{1-\beta^2} = t_1.$$

This is called the Lorentz–Fitzgerald contraction. It would cause double refraction in transparent media; experiments by Rayleigh (1902) and Brace(1904) showed no effect of this kind. But the Lorentz–Fitzgerald contraction follows from Lorentz transformation which is discussed below. Lorentz transformation also keeps the Maxwell's equations invariant as shown by Poincare in 1905. Therefore, it was felt that we should abandon Galilean transformation and accept Lorentz transformation for moving observers. Einstein gave a justification for this by proposing his special theory of relativity.

Einestein in 1905 based his theory on the following two postulates:

1. All the laws of physics (mechanics, electromagnetism, etc.) should have the same form i.e., they should be invariant in all inertial frames of reference moving uniformly with respect to each other. In this way he extended the relativity of inertial frames from mechanics to all physics.

2. Velocity of light c is an absolute constant. It does not depend upon the motion of the source or the observer. This postulate, which was based on the measurement of the speed of light, the result of the Michelson–Morley experiment and observations of double stars, explains the result of Michelson–Morley experiment in a natural way.

2.2.2 Derivation of Lorentz transformation

Consider two inertial frames with their origins O and O′ and respective Cartesian coordinates x, y, z, t and x', y', z', t', such that the respective space axes are parallel to each other. Let the frame O′ be moving with respect to O with uniform velocity v in the x-direction; see Fig. 2.4. Let us try to find the most general linear transformation between the two sets of coordinates, satisfying Einstein's postulates. Let us start with

$$x' = a_x x + a_y y + a_z z + a_t t,$$

$$y' = b_x x + b_y y + b_z z + b_t t,$$

$$z' = c_x x + c_y y + c_z z + c_t t,$$

$$t' = d_x x + d_y y + d_z z + d_t t. \tag{2.5}$$

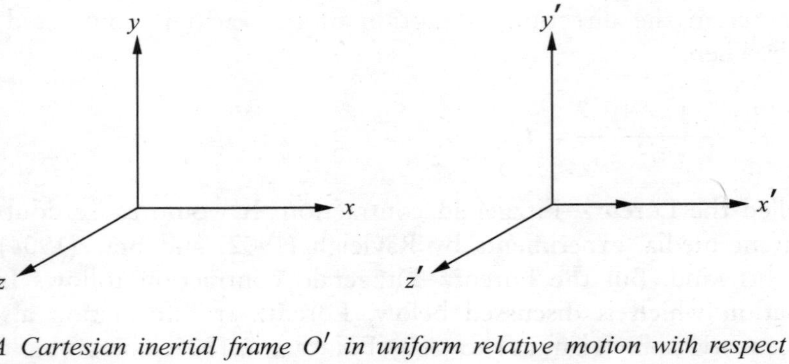

Fig. 2.4 *A Cartesian inertial frame O′ in uniform relative motion with respect to another Cartesian inertial frame O*

By appropriate arguments (see Problem 2.13) and a comparison with Galilean transformation, 12 of the above 16 coefficients can be determined and we can write (2.5) in the form

$$x' = px + qt, \ \ t' = rx + st, \ \ y' = y, \ \ z' = z. \tag{2.6}$$

Let the origins O and O' coincide at time $t = 0$ and let a pulse of light be emitted at the origin at time $t = 0$. Since (i) the pulse of light will be propogated as a spherical wave in both systems, and (ii) the velocity of light is a constant, c, in both the systems, we have, for any spacetime point on the wavefront,

$$x'^2 + y'^2 + z'^2 - c^2 t'^2 = x^2 + y^2 + z^2 - c^2 t^2$$

$$\Rightarrow x'^2 - ct'^2 = x^2 - c^2 t^2. \tag{2.7}$$

Now we shall give a complete solution through a guided exercise.

▶ Guided Exercise 2.1

Obtain the most general transformation from x, t to x', t' of Equations (2.6) subject to (2.7).

Hints: (a) Substitute equations (2.6) into (2.7) and equate coefficients of x^2, xt and t^2 to obtain

$$p^2 - c^2 r^2 = 1, \ \ pq - c^2 rs = 0, \ \ q^2 - c^2 s^2 = -c^2. \tag{2.8}$$

(b) Since the origins O and O' coincide at $t = 0$, you can write

$$x = vt \text{ when } x' = 0 \text{ for all } t.$$

Substitute this in the first of the equations of (2.6) to get

$$q = -pv. \tag{2.9}$$

(c) Substitute (2.9) into the last two of the equations of (2.8) to get

$$p^2 v - c^2 rs = 0, \ \ p^2 v^2 - c^2 s^2 = -c^2.$$

It is convenient to define

$$\beta = v/c \ \ \gamma = (1 - \beta^2)^{-1/2}. \tag{2.10}$$

From the above, note that

$$\beta^2 \gamma^2 = \gamma^2 - 1. \tag{2.11}$$

(d) Eliminate s between the above two equations and use the first of the equations of (2.8) for c^2r^2 to obtain

$$p = \pm(1 - v^2/c^2)^{-1/2}. \tag{2.12}$$

(e) With (2.9) and the other equations, now obtain

$$p = \pm\gamma, \quad q = \mp\gamma v, \quad r = \mp\gamma v/c^2. \tag{2.13}$$

(f) Remember that Galilean transformation is $x' = x - vt, t' = t$ and that Lorentz transformation of (2.6) must reduce to these for $v \ll c$. This decides the signs of p, q, s. The second equation of (2.8) decides the sign of r. Hence you will finally arrive at the Lorentz transformation

$$x' = \gamma(x - vt), \quad y' = y, \quad z' = z, \quad t' = \gamma(t - vx/c^2). \tag{2.14}$$

Note that these transformations can also be written as

$$x' = \gamma(x - \beta ct), ct' = \gamma(ct - \beta x). \tag{2.15}$$

Note the symmetry between x and ct and x' and ct' in the above.   ◀

However, it is important to note that this is not how Lorentz obtained his famous transformations. Lorentz obtained these transformations from different arguments in 1900 while Einstein gave his postualtes of special relativity in 1905. At the same time, it may be noted that in hindsight Lorentz transformations can be obtained *ab initio* from Einstein's postulates.

2.2.3   Lorentz transformations as imaginary rotation

From (2.11), it may be noted that the limits of various variables are

$$0 \leq v < c, \quad 0 \leq \beta < 1, \quad 1 \leq \gamma < \infty. \tag{2.16}$$

These limits and the algebraic form of (2.11) and (2.12) allow us to choose

$$\gamma = \cosh \xi, \quad \beta = \tanh \xi, \quad \beta\gamma = \sinh \xi. \tag{2.17}$$

The parameter ξ is obviously determined by the relative velocity v.

The complete Lorentz transformation can then be written in a matrix form, using (2.17), as

$$
\begin{bmatrix} x' \\ y' \\ z' \\ ct' \end{bmatrix} = \begin{bmatrix} \cosh\xi & 0 & 0 & -\sinh\xi \\ 0 & 1 & 0 & 0 \\ 0 & 0 & 1 & 0 \\ -\sinh\xi & 0 & 0 & \cosh\xi \end{bmatrix} \begin{bmatrix} x \\ y \\ z \\ ct \end{bmatrix}.
\tag{2.18}
$$

This reminds one of the orthogonal rotation matrix involving trignometric functions. Noting that $\cos ix = \cosh x$ and $\sin ix = i\sinh x$, it is often said that Lorentz transformation, for relative velocity along the x axis, represents a rotation in the $x - t$ plane through an imaginary angle.

Dropping the coordinates y' and z' for motion in the x-direction, the Lorentz transformation matrix in the (x, t) space can be written as

$$
A(\xi) = \begin{bmatrix} \cosh\xi & -\sinh\xi \\ -\sinh\xi & \cosh\xi \end{bmatrix}.
\tag{2.19}
$$

2.2.4 General Lorentz transformation

In the previous section we obtained the Lorentz transformation between two inertial frames moving relative to each other with a uniform velocity in the x-direction. Let us now consider the case when reference frame O$'$ moves with a uniform velocity $\mathbf{v}$ in an arbitrary direction with respect to O.

Note that the choice of Cartesian axes in each frame is arbitrary and a matter of convention with the objective of simplicity. Let us, therefore, choose another set of Cartesian axes (X, Y, Z) in the reference frame O such that X is along the relative velocity $\mathbf{v}$, with Y and Z suitably chosen. Let us also choose Cartesian axes (X', Y', Z') in the reference frame O$'$ such that the respective axes in frames O and O$'$ are parallel to each other.

The new axes X, Y, Z in frame O can be obtained from the old system (x, y, z) by a rotation throught ϕ about the z-axis, followed by a rotation through θ around the new y-axis.

Let us split the position vector $\mathbf{r}$ (of a particle under study) into $\mathbf{r}_\parallel$ and $\mathbf{r}_\perp$, respectively parallel and normal to the relative velocity $\mathbf{v}$. Then,

$$\mathbf{r}_\parallel = X\mathbf{i}, \quad \mathbf{r}_\perp = Y\mathbf{j} + Z\mathbf{k}.$$

Equations (2.18) show that the coordinates parallel to the ralative velocity undergo Lorentz transformation while coordinates normal to it remain unchanged. Therefore, we can write, in analogy with (2.18),

$$\mathbf{r}'_\parallel = \gamma(\mathbf{r}_\parallel - \mathbf{v}t), \quad \mathbf{r}'_\perp = \mathbf{r}_\perp,
\tag{2.20}$$

where the primed variables refer to similar quantities measured by the moving frame O'. Then the position vector becomes,

$$\mathbf{r}' = \mathbf{r}_\parallel' + \mathbf{r}_\perp' = \gamma(\mathbf{r}_\parallel - \mathbf{v}t) + \mathbf{r}_\perp$$

$$= \mathbf{r} + (\gamma - 1)\mathbf{r}_\parallel - \gamma t\mathbf{v}$$

$$= \mathbf{r} + (\gamma - 1)\frac{(\mathbf{r}.\boldsymbol{\beta})\boldsymbol{\beta}}{\beta^2} - \gamma\mathbf{v}t,$$

and

$$ct' = \gamma(ct - \mathbf{r}.\boldsymbol{\beta}). \tag{2.21}$$

2.3 Kinematic Applications

The Lorentz transformation between two sets of space-time coordinates on two inertial frames of reference moving relative to each other with a uniform velocity is a complete departure from Galilean transformation. The latter is based on the assumption that time flows uniformly for all observers. Lorentz transformation, on the other hand, suggested that space and time are intricately connected with each other. There was no evidence in favour of Lorentz transformation except the theoretical fact that it explained the negative result of the Michelson–Morley experiment.

Soon it was realised that Lorentz transformation affected kinetmatics of spacetime to a great extent. Here we discuss some of these aspects.

2.3.1 Length contraction and time dilatation

For low relative velocities, $v \to 0$ (more correctly $v/c << 1$),

$$\beta \to 0, \gamma \to 1.$$

The Lorentz transformation of (2.14) then reduces to the Galilean transformation

$$x' = x - vt, \quad y' = y, \quad z' = z, \quad t' = t. \tag{2.22}$$

The inverse Lorentz transformation is obtained from (2.14) by changing the sign of β and interchanging the primed and the unprimed coordinates.

$$x = \gamma(x' + ct'), \quad y = y', \quad z = z',$$

$$ct = \gamma(ct' + \beta x'). \tag{2.23}$$

Differentiating (2.23), we get

$$dx = \gamma(dx' + \beta c\, dt'), \quad dy = dy', \quad dz = dz',$$
$$dt = \gamma(dt' + \beta dx'/c).$$

$$(2.24)$$

Now if O' measures a rod of length $dx' = l_0$ kept parallel to the x'-axis, by noting its ends simultaneously, that is, with $dt' = 0$, we have

$$dx = \gamma l_0, \quad dt = \gamma\beta l_0/c.$$

$$(2.25)$$

Thus O concludes that the measurements of the two ends were not made simultaneously. Subtracting the distance moved during time dt, O' gets for the length of the rod

$$l = dx - v\,dt = l_0/\gamma.$$

$$(2.26)$$

So O' concludes that the moving rod is contracted by a factor $1/\gamma$. This is the Lorentz–Fitzgerald contraction.

The time (interval) measured by a clock stationary relative to an observer is called the *proper time*, and is generally denoted by τ. If the observer O' has a clock at a fixed position in his frame, say $x' = 0$, and notes the time at two different moments, then $dx' = 0$ and the time interval between those two moments for O' will be $dt' = \tau$. For O, these two ticks of the clock have not occurred at the same place, but at spatial separation of $dx = \gamma vt$, and therefore, $dt = \gamma\tau > \tau$. This means that the moving clock runs slower as measured by O.

The relativistic effects are mutual and symmetrical. If O' moves relative to O with a velocity v, O' finds that O moves relative to O' with the same velocity in the opposite direction. Therefore, just as O finds lengths in O' contracted by a factor γ and time intervals in O' to be dilated by γ, O' finds lengths of objects in O to be contracted and time intervals in O to be dilated by the same factor γ.

2.3.2 Relativistic law of addition of velocities

Consider a body moving in O' with a velocity u' along the x-axis, for the sake of convenience in the same direction in which O' moves with respect to O with a velocity v. What will be the velocity of the body as measured by O?

According to Galilean transformation, it would simply have been $u = u' + v$. But now, differentiating (2.24), we find that

$$u_x = \frac{dx}{dt} = \frac{u_x' + v}{1 + vu_x'/c^2},$$

$$u_y = \frac{dy}{dt} = \frac{u_y'}{\gamma(1 + vu_x'/c^2)}, \quad u_z = \frac{dz}{dt} = \frac{u_z'}{\gamma(1 + vu_x'/c^2)}.$$

$$(2.27)$$

Here we find that not only the component parallel to the relative velocity changes, transverse velocity components also change. This is because although transverse distances do not change, the time interval changes in the two reference frames, affecting all components of velocity. The inverse transformation would be

$$u_x' = \frac{u_x - v}{1 - vu_x/c^2},$$

$$u_y' = \frac{u_y}{\gamma(1 - vu_x/c^2)}, \quad u_z' = \frac{u_z}{\gamma(1 - vu_x/c^2)}. \tag{2.28}$$

In vector form, for arbitrary directions of motion of the body and the reference frames, we have to split the velocities in components parallel and perpendicular to the relative velocity $\mathbf{v}$. Thus, if a body moves with a velocity $\mathbf{u}$ as observed by O, its velocity $\mathbf{u}'$ measured by O' moving relative to O with a uniform velocity $\mathbf{v}$ will be, in the notation introduced in Section 2.2.3,

$$u_\parallel' = \frac{(\mathbf{u} - \mathbf{v})_\parallel}{1 - \mathbf{v}.\mathbf{u}/c^2}, \quad \mathbf{u}_\perp' = \frac{\mathbf{u}_\perp}{\gamma(1 - \mathbf{v}.\mathbf{u}/c^2)}. \tag{2.29}$$

These can further be combined, as before, into

$$\mathbf{u}' = [\mathbf{u} + (\gamma - 1)\frac{\mathbf{u}.\boldsymbol{\beta}}{\beta^2}\boldsymbol{\beta} - \gamma\mathbf{v}]/\{\gamma(1 - \mathbf{v}.\mathbf{u}/c^2)\}. \tag{2.30}$$

2.3.3 Transformation of acceleration

An observer in an inertial frame O finds a body moving with an acceleration a_x, the acceleration being in the x-direction for convenience. Then an observer in another inertial frame O' moving relative to O with a velocity v along the x-direction will measure the acceleration of the body to be

$$a_x' = \frac{du_x'}{dt'} = \frac{a_x}{\gamma^3(1 - vu_x/c^2)^3},$$

$$a_y' = \frac{1}{\gamma^2(1 - vu_x/c^2)^2}\{a_y + \frac{vu_x/c^2}{1 - vu_x/c^2}a_x\},$$

$$a_z' = \frac{1}{\gamma^2(1 - vu_x/c^2)^2}\{a_z + \frac{vu_x/c^2}{1 - vu_x/c^2}a_x\}. \tag{2.31}$$

This can be put in a vector form, for arbitrary direction of acceleration $\mathbf{a}$ and velocity $\mathbf{u}$ with respect to the relative velocity $\mathbf{v}$, as

$$\mathbf{a}' = \frac{\gamma\mathbf{a}(1 - \mathbf{u}.\mathbf{v}/c^2) - (\gamma - 1)(\mathbf{a}.\boldsymbol{\beta})\boldsymbol{\beta}/\beta^2 + \gamma(\mathbf{a}.\mathbf{v})\mathbf{u}/c^2}{\gamma^3(1 - \mathbf{u}.\mathbf{v}/c^2)^3}. \tag{2.32}$$

2.3.4 Proper acceleration

Consider a system O' moving with a uniform velocity in the x-direction with respect to another system O. Consider a body moving in the same direction, with a variable velocity. Let the instantaneous velocities of the body measured by the two systems be u' and u, respectively. They would be related to each other by the first of equations (2.28). Differentiating it, we find that the velocity increments would be related by

$$du' = du(1 - v^2/c^2)/(1 - u'v/c^2)^2.$$

Then the acceleration of the body, a and a', measured by the two frames, would be related by

$$a' = du'/dt' = a\gamma^3(1 - u'v/c^2)^2. \tag{2.33}$$

We define proper acceleration α as the acceleration when initial velocity is zero, $u' = 0$. In that case, $u = v, \gamma = \gamma(u)$, and $\alpha = a\gamma^3(u)$.

Proper acceleration α is not Lorentz invariant. But if it is so, then

$$\gamma^3(u)\frac{du}{dt} = \frac{d}{dt}[\gamma(u)u] = \alpha. \tag{2.34}$$

Conversely, if (2.34) is satisfied, then integrating the above, we find $\gamma(u)u = \alpha t + k$, where k is a constant. If the body starts accelerating from rest, $u = 0$ at $t = 0$, then $k = 0$ and we have $\gamma(u)u = \alpha t$. This can be rewritten in the form

$$u^2 = \alpha^2 t^2(1 - u^2/c^2)$$

$$\Rightarrow u = \frac{\alpha t}{(1 + \alpha^2 t^2/c^2)^{1/2}}. \tag{2.35}$$

Integrating this, we get

$$x = \int \frac{ct\,dt}{(1 + \alpha^2 t^2/c^2)^{1/2}},$$

which yields

$$x^2 - c^2 t^2 = c^4/\alpha^2 = X^2, \tag{2.36}$$

which defines X. This is a hyperbola as against the classical solution $x = \alpha t^2/2 +$ constant, which is a parabola. X is the value of x at $t = 0$, which is the vertex of the

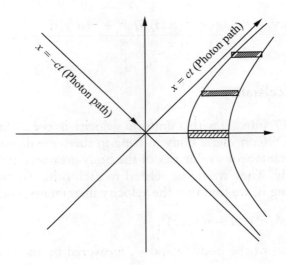

Fig. 2.5 *Motion of a particle with uniform acceleration*

hyperbola as shown in Fig. 2.5. When $X = 0$, we have $x = ct$, which represents a photon.

Figure 2.5 shows the trajectories of (2.36) in the $x - ct$ plane. Here AB represents a rigid rod of certain length placed in the x-direction. As it accelerates to higher velocities, it gets contracted, and with passage of time its length tends to zero as shown in Fig. 2.5.

It may be noted that acceleration is *absolute*. This means that if the acceleration of a body is zero (nonzero) in one inertial frame, then it is zero (nonzero) in every inertial frame.

2.4 Minkowski Space

The four-dimensional space with three space coordinates and one time coordinate which needs to be introduced in special theory of relativity is called *Minkowsky space*. Now we wish to extend other concepts of 3-D Newtonian space to Minkowski space, which should be valid in all inertial frames of reference.

2.4.1 Scalars and vectors in Minkowski space

In 3-D Newtonian mechanics, we classify physical quantities as scalars, vectors and tensors, depending on the number of their components and their transformation properties from one Galilean frame to another. Thus a scalar is a quantity that can be described by only one number which remains invariant for all Galilean observers (stationary relative to each other). Mass, length, volume, time (intervals), energy, pressure, electric charge, are such Galilean scalars. A vector quantity can be specified in a frame of reference by giving three components which should transform to another Galilean frame by specified laws of vector transformation. Velocity, momentum, force, acceleration, element of area, electric and magnetic fields, are such vectors. Then we have tensors.[6] A tensor of rank r in a d-dimensional space is specified by d^r components which must transform from one frame to another by specified laws. It is clear that a scalar is a tensor of rank zero and a vector is a tensor of rank 1.

As has been seen so far, quantities like length and time are no longer scalars in 4-D Minkowski space because they do not remain invariant under Lorentz transformations (between two inertial frames moving with a uniform relative velocity with respect to each other). Speed of electromagnetic radiation in vacuum, c, electric charge, and action (defined in Chapter 1) are among the quantities which remain invariant for all inertial observers, and are therefore *world-scalars*. A world-scalar is of course also a Galilean scalar because Galilean transformations are a subset of Lorentz transformations.

Coming to vectors, the least we need is to add a fourth component. But merely having four components is not enough; they must satisfy Lorentz transformations. We shall now develop these concepts.

Under Lorentz transformations, $x^2 + y^2 + z^2 - c^2t^2$ is invariant in all inertial frames of reference moving with uniform velocity relative to each other. So we take x, y, z and ict as coordinates in Minkowski space. A point in Minkowski space is called an *event*, specified by the 'event vector' r_μ $(1 \leq \mu \leq 4)$ having components x, y, z, ict. It has a 'length' equal to $(x^2 + y^2 + z^2 - c^2t^2)^{1/2}$. We represent $r_\mu = (x, y, z, ict) = (\mathbf{r}, ict)$, where $\mathbf{r}$ is the three-dimensional Newtonian vector. Such vectors are called *four-vectors* or *world-vectors* and they satisfy Lorentz transformation, which, in matrix form, can be written in analogy with (2.18) as

$$\begin{bmatrix} x' \\ y' \\ z' \\ ict' \end{bmatrix} = \begin{bmatrix} \gamma & 0 & 0 & i\beta\gamma \\ 0 & 1 & 0 & 0 \\ 0 & 0 & 1 & 0 \\ -i\beta\gamma & 0 & 0 & \gamma \end{bmatrix} \begin{bmatrix} x \\ y \\ z \\ ict \end{bmatrix}. \tag{2.37}$$

Let $x_\nu (1 \leq \nu \leq 4)$ represent a four-vector as measured in one frame of reference and $x_\mu' (1 \leq \mu \leq 4)$ the same four-vector as measured in another inertial frame. Then the components would be related by

$$x_\mu' = a_{\mu\nu} x_\nu, \tag{2.38}$$

where Einstein's summation convention is assumed over repeated indices. The matrix $a_{\mu\nu}$ of coefficients is a tensor of rank two in the Minkowski space. In the particular

---

[6]See Joshi, A. W. (2005).

case of relative velocity along x-direction, the transformation would reduce to that of (2.18).

It can be verified that $a_{\mu\nu}$ is an orthogonal matrix, that is

$$a_{\mu\nu}a_{\nu\lambda} = \delta_{\mu\lambda}. \tag{2.39}$$

The inverse transformation is given by

$$x_\lambda = a_{\lambda\mu}x_\mu \tag{2.40}$$

The matrix occurring in (2.40) is the transpose or inverse of that occurring in (2.38).

If $A = (A_1, A_2, A_3, iA_4)$ is a four-vector, its space coordinates (A_1, A_2, A_3) form a Galilean vector **A** and its time component A_4 is a Galilean scalar. We may mention a few other four-vectors in special relativity. Some of these are:

Velocity four-vector:

$$U_\mu = (\gamma \mathbf{u}, i\gamma c), \tag{2.41a}$$

where **u** is the Galilean velocity of a body;

Energy-momentum four-vector:

$$p_\mu = (m\mathbf{u}, imc^2), \tag{2.41b}$$

where m is the relativistic mass;

Four-wave vector: $k_\mu = (\mathbf{k}, i\,\omega/c)$, $\tag{2.41c}$

where **k** is the Galilean wave vector of a wave and ω its frequency.

In each of these cases, it can be verified that, for the four-vector A_μ, $A_1{}^2 + A_2{}^2 + A_3{}^2 - A_4{}^2$ is a world-scalar.

2.4.2   Intervals in Minkowski space

Using (2.24), it can be seen that[7]

$$dx^2 + dy^2 + dz^2 - c^2 dt^2 = dx'^2 + dy'^2 + dz'^2 - c^2 dt'^2.$$

Thus the quantity

$$ds^2 = d\sigma^2 - c^2 dt^2, \tag{2.42a}$$

---

[7]With coordinates (x, y, z, ct), the metric is 1, 1, 1, -1.

where

$$d\sigma^2 = dx^2 + dy^2 + dz^2, \tag{2.42b}$$

is invariant under Lorentz transformations. ds is called the *interval* between the two events P $= (x, y, z, t)$ and Q $= (x + dx, y + dy, z + dz, t + dt)$. It is the 'length' of the four-vector $(d\sigma, icdt)$ connecting the two events.

Consider the two events: emission of a photon at the event P $= (x, y, z, t)$ and its absorption at Q $= (x + dx, y + dy, z + dz, t + dt)$, the two space points being separated by vacuum. Then the speed of the photon is $d\sigma/dt = c$ and hence we would have

(i) $$ds^2 = \left[\frac{1}{c^2} \left(\frac{d\sigma}{dt} \right)^2 - 1 \right] c^2 dt^2 = 0. \tag{2.43a}$$

On the other hand, if a material particle is emitted at P and reaches Q, then the time elapsed would have to be larger than that given by (2.43a) because the particle velocity must be smaller than c. Thus, if

(ii) $$ds^2 = d\sigma^2 - c^2 dt^2 = \left[\frac{1}{c^2} \left(\frac{d\sigma}{dt} \right)^2 - 1 \right] c^2 dt^2 < 0, \tag{2.43b}$$

it is called a *time-like interval*. It can be covered by moving with a velocity $v = d\sigma/dt < c$. Actually, it represents $dt = d\tau$, the proper time interval of a particle when $d\sigma/dt = 0$, as would happen if the reference frame is moving with the body.

Finally, if,

(iii) $$ds^2 = d\sigma^2 - c^2 dt^2 = \left[\frac{1}{c^2} \left(\frac{d\sigma}{dt} \right)^2 - 1 \right] c^2 dt^2 > 0, \tag{2.43c}$$

it is called a *space-like interval*, because it cannot be covered by any velocity $v < c$.

The first case of (2.43a) represents the path of a light ray. Figure 2.6 represents Minkowski space on a two-dimensional (σ, t) plane. Taking $c = 1$, the lines b_1 and b_2 with a slope[8] $\pi/4$ represent light rays. Time-like intervals have slopes greater than $\pi/4$ while space-like intervals have slopes less than $\pi/4$. Then with respect to the origin O, all *past* and *future* events which can be covered with velocities smaller than c lie in the shaded cone-shaped region known as the *light cone*. The unshaded region represent events which cannot be reached by actual travel with a velocity $v < c$. They are called *elsewhere*.

---

[8]Negative sign has been disregarded here; only magnitude of the slope is considered.

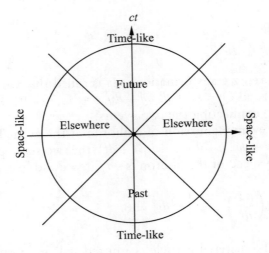

Fig. 2.6 *Minkowsky space*

We can now talk about space-like and time-like four-vectors. For example, $r_\mu = (\mathbf{r}, ict)$ is a space-like four vector with 'length' $(x^2 + y^2 + z^2 - c^2 t^2)^{1/2}$ and $(r_\mu = ct, i\mathbf{r})$ is a time-like four-vector with 'length' $(c^2 t^2 - x^2 - y^2 - z^2)^{1/2}$. Here $\mathbf{r}$ is the three-dimensional Galilean vector.

2.4.3 Operations with scalars and vectors

We shall now develop the algebra and calculus of four-vectors. We describe here various operations with four-vectors.

Scalar product

Like vectors in three-dimensional space, we define scalar product of two four-vectors $p_\mu = (p_1, p_2, p_3, ip_4) = (\mathbf{p}, ip_4)$ and $q_\mu = (q_1, q_2, q_3, iq_4) = (\mathbf{q}, iq_4)$, where $\mathbf{p}$ and $\mathbf{q}$ are Galilean vectors, as

$$p.q = p_1 q_1 + p_2 q_2 + p_3 q_3 - p_4 q_4. \tag{2.44}$$

Note that the symbol for the scalar product of two four-vectors is a simple dot placed between the two vectors.

It can also be seen that the scalar product of two four-vectors, as defined in (2.44), is invariant under Lorentz transformations as can be readily seen. If p' and q' are the two four-vectors in the primed reference frame, then

$$p'.q' = \sum_\mu p_\mu' q_\mu' = \sum_{\mu, \nu, \alpha} a_{\mu\nu} p_\nu a_{\mu\alpha} q_\alpha$$

$$= \sum_{\nu,\alpha} \delta_{\nu\alpha} p_\nu q_\alpha = \sum_\alpha p_\alpha q_\alpha = p.q. \tag{2.45}$$

The scalar product of a four-vector with itself gives the square of its norm or of 'length', under this Minkowski metric. Thus we can say,

$$p = (p_\mu p_\mu)^{1/2} = (p_1{}^2 + p_2{}^2 + p_3{}^2 - p_4{}^2)^{1/2}. \tag{2.46}$$

Gradient

Like the four-vectors, we can also have four-vector operators. Thus we define the four-vector gradient operator having components which are differentiations with respect to the coordinates (x, y, z, ict) as

$$\partial_\mu = \left(\frac{\partial}{\partial x}, \frac{\partial}{\partial y}, \frac{\partial}{\partial z}, \frac{1}{ic} \frac{\partial}{\partial t} \right). \tag{2.47}$$

Then the four-vector gradient of Lorentz scalar function $\phi(\mathbf{r}, t)$ and the divergence of a four-vector $b(\mathbf{r}, t)$ become

$$\partial_\mu \phi = \left(\frac{\partial \phi}{\partial x}, \frac{\partial \phi}{\partial y}, \frac{\partial \phi}{\partial z}, -\frac{i}{c} \frac{\partial \phi}{\partial t} \right),$$

$$\partial_\mu b_\mu = \frac{\partial b_1}{\partial x} + \frac{\partial b_2}{\partial y} + \frac{\partial b_3}{\partial z} - \frac{1}{c} \frac{\partial b_4}{\partial t}.$$

Also the d'Alembertian $\square^2$, which is the equivalent of three-dimensional Laplacian ∇^2, becomes

$$\square^2 = \frac{\partial^2}{\partial x^2} + \frac{\partial^2}{\partial y^2} + \frac{\partial^2}{\partial z^2} - \frac{1}{c^2} \frac{\partial^2}{\partial t^2} = \nabla^2 - \frac{1}{c^2} \frac{\partial^2}{\partial t^2}, \tag{2.48}$$

where ∇ stands for the three dimensional gradient symbol.

Vector product

In three-dimensional space, we define the vector product or cross-product of two vectors $\mathbf{p}$ and $\mathbf{q}$ as a vector with the three Cartesian components $p_2 q_3 - p_3 q_2$, $p_3 q_1 - p_1 q_3$ and $p_1 q_2 - p_2 q_1$. This is because these three components transform from one coordinate system to another in accordance with Galilean transformation of a vector.

This vector product is a very special operation in three-dimensional Newtonian space. This can be seen by looking at it another way, by asking this question: Can we create (a) a scalar, (b) a vector from a bilinear combination of two vectors? For (a), the answer is yes; a combination such as $\sum p_i q_i$ is a scalar in a space of any dimensions. But for (b), it turns out that it is possible to find n bilinear combinations of n-dimensional vectors p and q only for specific values of n which includes $n = 3$. For other values of

n it is not possible to define a vector product of two vectors. For the same reason, it is not possible to define the curl operator in Minkowsky space.

2.5 Relativistic Mechanics

In this section we shall deal with what happens to mass in relativity and how we can keep account of conservation of mass, momentum, and energy. In Newtonian mechanics, mass and energy are scalars while momentum and force are vectors. We shall see what happens to them under Lorentz transformations and how, if at all, we can generate four-vectors from any of them.

2.5.1 Transformation of mass

Consider the phenomenon of inelastic collision of two bodies travelling in the x-direction, coelasting into a single body and continuing to travel in the same direction as shown in Fig. 2.7. An observer O' in the inertial frame Σ' measures both the masses to be m moving in opposite direction with velocities u' and $-u'$. Since the masses are equal and the velocities are equal and opposite, they collide and the combined mass $2m$ comes to rest in Σ'. Frame Σ' is moving relative to another inertial frame Σ with a velocity v in the x-direction. Observer O measures the masses to be m_1 and m_2, moving respectively with velocities u_1 and u_2 before collision and the combined mass M moving with a velocity v.

It is clear that in the frame Σ', momentum is conserved. If it is to be conserved in the frame Σ too, we must have

$$m_1 + m_2 = M, \quad m_1 u_1 + m_2 u_2 = Mv. \tag{2.49}$$

The velocities would be related by

$$u_1 = \frac{u' + v}{1 + u'v/c^2}, \quad u_2 = \frac{-u' + v}{1 - u'v/c^2}. \tag{2.50}$$

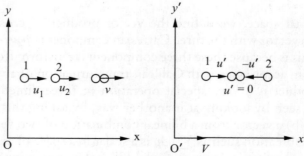

Fig. 2.7 *Relative description of collision of two masses in two inertial frames of references*

Substituting these in (2.49) and working out the algebra, we find that

$$m_1/m_2 = (1 + u'v/c^2)/(1 - u'v/c^2).$$

Using the result of Problem 2.4, this can be seen to be

$$m_1/m_2 = [(1 - u_2^2/c^2)/(1 - u_1^2/c^2)]^{1/2}.$$

We may write this in the form

$$m_1 = m_0/(1 - u_1^2/c^2)^{1/2}, \quad = m_0/(1 - u_2^2/c^2)^{1/2},$$

or generally,

$$m = \frac{m_0}{(1 - u^2/c^2)^{1/2}} = \gamma(u)m_0. \tag{2.51}$$

This equation shows two things. If a mass is at rest in a reference frame, $u = 0$, $\gamma(u) = 1$, and $m = m_0$. This is the mass of a body measured when it is at rest in a reference frame and is called its *rest mass*. Secondly, for any nonzero velocity, $\gamma(u) > 1$, which shows that the mass of a moving body appears to be greater than its rest mass. As the speed of the moving body increases and approaches c, its measured mass approaches infinity. As a consequence, it is clear that no material particle can travel with a speed c. A material particle is one which has a nonzero rest mass. On the other hand, particles travelling with speed c must have zero rest mass.

2.5.2 Conservation of mass and momentum

Consider some masses m_i moving with various velocities. For example, they could be molecules in a gas in a steady state or particles of a rigid body which is in motion. Let $\mathbf{u_i}$ be the velocity of mass m_i, with Cartesian components u_{ix}, u_{iy} and u_{iz} in the frame O. These particles may collide with each other and change their velocities. But as long as the velocities are small and non-relativistic with respect to O, O will find that Σm, Σmu_x, Σmu_y and Σmu_z are conserved in time with respect to the processes occurring in O. Here we have dropped the symbol i, so that for example Σmu_x stands for $\Sigma m_i u_{ix}$.

Consider another inertial frame O' and assume that it is moving with a velocity v with respect to O in the x-direction. Since v need not be small compared to c, all the above speeds of particles in gas and rigid body could be relativistic. Then it can be shown that if Σm and $\Sigma m\mathbf{u}$ are conserved for O, then $\Sigma m'$ and $\Sigma m'\mathbf{u}'$ are also conserved for any other inertial frame. Even if Σm is not equal to $\Sigma m'$, each of them is a constant in their respective frames of reference.

Using (2.24), (2.28) and the result of Problem 2.4, we have

$$\sum m' = \sum \gamma(u')m_0 \; = \sum \gamma(u)\gamma(v)(1 - vu_x/c^2)m_0 \; = \gamma(v)\sum m - \gamma(v)vmu_x/c^2.$$

Since each term on the right-hand side is separately conserved by assumption, it follows that $\sum m'$ is conserved. Next, consider

$$\sum m'u_x' = \sum \gamma(u')m_0u_x'$$

$$= \sum \gamma(u)\gamma(v)m_0(u_x - v)$$

$$= \gamma(v)\left[\sum mu_x - v\sum m\right],$$

showing that the quantity on the left-hand side is conserved. Next, consider

$$\sum m'u_y' = \sum \gamma(u')m_0u_y'$$

$$= \sum \gamma(u)m_0u_y,$$

which once again shows that the quantity on the left-hand side is conserved. Since the relative velocity between O and O$'$ is along x-direction, y and z are interchangeable; it follows that $\sum m'u_z'$ is also conserved.

2.5.3   Kinetic energy

If $\mathbf{f}$ is the force acting on a mass m and $\mathbf{dl}$ its displacement in time dt, then the change in the kinetic energy of the mass will be equal to the work done by the force, which is

$$\frac{dK}{dt} = \mathbf{f}.\frac{\mathbf{dl}}{dt} = \mathbf{f}.\mathbf{u},$$

where $\mathbf{u}$ is the velocity of the body after time dt. But

$$\mathbf{f} = \frac{d}{dt}(m\mathbf{u}) = m\frac{d\mathbf{u}}{dt} + \frac{dm}{dt}\mathbf{u},$$

which gives

$$\frac{dK}{dt} = \frac{dm}{dt}\mathbf{u}.\mathbf{u} + m\frac{d\mathbf{u}}{dt}.\mathbf{u}. \tag{2.52}$$

Using (2.51), we see that

$$\frac{dm}{dt} = \frac{m_0u\gamma^3(u)}{c^2}\frac{du}{dt}. \tag{2.53}$$

This gives

$$\frac{dK}{dt} = m_0 u \gamma^3(u) \frac{du}{dt},$$

which together with (2.53) gives

$$\frac{dK}{dt} = c^2 \frac{dm}{dt}.$$

We could say that the work done per unit time on the mass m under the action of a force is

$$dK = c^2 dm,$$

giving, for the kinetic energy of the mass,

$$K = mc^2 + \text{constant}.$$

When $u = 0$, $m = m_0$ and the mass is at rest. We take this as the zero of kinetic energy, giving the constant in the above equation to be $m_0 c^2$. This finally gives

$$K = mc^2 - m_0 c^2. \tag{2.54}$$

Equation (2.54) indicates equivalence of mass and energy, and we call $m_0 c^2$ the *rest mass energy* or just rest energy. The total energy is given by

$$E = K + m_0 c^2 = mc^2. \tag{2.55}$$

The validity of this equation has been amply demonstrated in modern particle accelerators and in the interiors of stars.

2.5.4 Energy–momentum four-vector

Consider a particle of rest mass m_0 travelling with a speed u in a certain inertial frame of reference. The mass–energy formula can be written with the use of (2.12) as

$$E^2 = m^2 c^4 = m_0{}^2 \gamma^2(u) c^4 = m_0{}^2 c^4 + m^2 u^2 c^2$$

$$= m_0{}^2 c^4 + p^2 c^2$$

$$\Rightarrow E = c(p^2 + m_0{}^2 c^2)^{1/2}, \tag{2.56a}$$

where

$$p = mu = m_0 \gamma u \qquad\qquad (2.56b)$$

is the relativistic momentum. Equation (2.56a) gives the total relativistic energy of the particle, consisting of the rest energy and the kinetic energy. The kinetic energy is then given by

$$K = c(p^2 + m_0{}^2 c^2)^{1/2} - m_0 c^2.$$

Differentiating E of (2.56a) with respect to p, we get

$$dE/dp = pc^2/E = p/m = u.$$

It may be noted that this holds in classical mechanics too. From (2.56a), we can further see that

$$E^2/c^2 - p^2 = m_0{}^2 c^2, \qquad\qquad (2.57)$$

which is a Lorentz invariant. It may be noted that in these equations $\mathbf{p}$ and $\mathbf{u}$ could be three-dimensional vectors.

A comparison of (2.57) with (2.42a) suggests that the left hand side of the former may contain components of a four-vector. It would thus be appropriate to work out the Lorentz transformations of E and p_x, p_y, p_z. As before, suppose an observer in a reference frame O finds a particle of rest mass m_0 moving with a speed u_x along the x-axis. Its mass as observed by O would be $\gamma(u_x)m_0$. Let another observer O' moving with respect to O with a velocity v in the x-direction also measure the mass, energy and momentum of the particle. Its velocity u'_x measured by O' would be obtained by relativistic addition of velocities u_x and v. The mass measured by O' would be $m' = \gamma(u_x{}')m_0$. On using the result of Problem 2.4, we can write

$$E' = m'c^2 = \gamma(v)mc^2(1 - vu_x/c^2) = \gamma(v)(E - vp_x).$$

Also,

$$p_x{}' = m'u_x{}' = \gamma(v)\gamma(u_x)m_0(u_x - v) = \gamma(v)\left(p_x - vE/c^2\right), \qquad\qquad (2.58)$$

$$p_y{}' = p_y, \quad p_z{}' = p_z.$$

This suggests that $(E/c, i\mathbf{p})$ is a time-like four-vector. It is known as momentum four-vector of 'length' $m_0 c$. Since $E = mc^2$ and $\mathbf{p} = m\mathbf{u}$, we get on dividing by m_0 the velocity four-vector $(\gamma c, i\gamma\mathbf{u})$. It has the 'length' $\gamma(v)(c^2 - v^2)^{1/2} = c$, which is Lorentz invariant.

Differentiating the momentum four-vector with respect to invariant proper time interval $d\tau$, we get the force four-vector, with components.

$$F_\mu = \left(\frac{1}{c} \frac{dE}{d\tau}, i \frac{d\mathbf{p}}{d\tau} \right).$$

Note that here $d\tau = [dt^2 - (dx^2 + dy^2 + dz^2)/c^2]^{1/2} = dt/\gamma$. Noting further that $E = \mathbf{F}.\mathbf{r}$, $d\mathbf{r}/dt = \mathbf{u}$ and $d\mathbf{p}/dt = \mathbf{F}$, where $\mathbf{u}$ is the Newtonian velocity of the body and $\mathbf{F}$ the Newtonian force acting on the body, we get the four-vector force as

$$F_\mu = (\gamma \mathbf{F}.\mathbf{u}/c, i\gamma \mathbf{F}). \tag{2.59}$$

2.5.5 Transformation of four-force

We shall now obtain the transformation of the components of (Galilean) force vector for the case when the system O' moves with respect to O with a uniform velocity v along the x-direction. On using Eqs. (2.14), (2.55) and (2.58), we have

$$F_x' = \frac{dp_x'}{dt'} = \frac{\gamma(dp_x - vdm)}{\gamma(dt - vdx/c^2)} = \frac{dp_x/dt - vdm/dt}{(1 - vu_x/c^2)}.$$

Now

$$\frac{dm}{dt} = \frac{dE}{c^2 dt} = \frac{d(\mathbf{F}.\mathbf{r})}{c^2 dt} = \frac{1}{c^2}\mathbf{F}.\mathbf{u}.$$

This gives

$$F_x' = \frac{(F_x - v\mathbf{F}.\mathbf{u}/c^2)}{1 - vu_x/c^2}. \tag{2.60a}$$

It can be similarly seen that

$$F_y' = \frac{dp_y'}{dt'} = \frac{F_y}{\gamma(1 - vu_x/c^2)}, \quad F_z' = \frac{F_z}{\gamma(1 - vu_x/c^2)}. \tag{2.60b}$$

Vector ally we can write the above equations for a general direction of the force and the particle velocity in the form

$$\mathbf{F}_\parallel' = \frac{[\mathbf{F} - (\mathbf{v}/c^2)\mathbf{F}.\mathbf{u}]_\parallel}{1 - \mathbf{v}.\mathbf{u}/c^2}, \quad \mathbf{F}_\perp' = \frac{[\mathbf{F} - (\mathbf{v}/c^2)\mathbf{F}.\mathbf{u}]_\perp}{\gamma(1 - \mathbf{v}.\mathbf{u}/c^2)}. \tag{2.61}$$

2.5.6 Photons

Photons travel with the speed c in free space. To avoid the photon mass from tending to infinity, it is assigned a zero rest mass, $m_0 = 0$. But in accordance with the work of Planck, Einstein and several others, a photon can be assigned energy, frequency, wavelength and momentum, related to each other by $E = h\nu$, $p = h\nu/c$, where ν is the frequency. These can be used for deriving relativistic formulas for aberration and for Doppler effect.

Consider two inertial systems O and O′, where the latter moves with a uniform velocity v along the common x-axis. Consider a beam of photons coming at an angle α with the x-axis as observed by O and angle α' as observed by O′; see Fig. 2.8. The two observers measure the frequencies and wavelengths of a photon to be ν, λ and ν', λ', respectively. Then we have, for its energy and components of momenta,

$$E = h\nu, \quad p_x = -(h\nu/c)\cos\alpha, \quad p_y = -(h\nu/c)\sin\alpha;$$

$$E' = h\nu', \quad p_x' = -(h\nu'/c)\cos\alpha', \quad p_y' = -(h\nu'/c)\sin\alpha'.$$

Using the transformation of energy from (2.58), we get

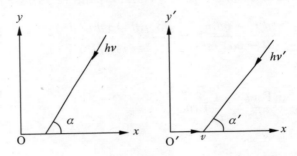

Fig. 2.8   *Relativistic aberration and Doppler shift*

$$\nu' = \gamma\nu\,(1 + \beta\cos\alpha)\,, \quad \nu = \nu'/\left[\gamma\,(1 + \beta\cos\alpha)\right];$$

$$\lambda = \lambda'\gamma\,(1 + \beta\cos\alpha)\,, \quad \lambda' = \lambda/\left[\gamma\,(1 + \beta\cos\alpha)\right]. \tag{2.62}$$

For $\alpha = 0$, a photon approaching along the x-axis, we have

$$\lambda = \lambda'\left(\frac{1+\beta}{1-\beta}\right)^{1/2}\,, \quad \lambda' = \lambda\left(\frac{1-\beta}{1+\beta}\right)^{1/2}\,,$$

which is the *longitudinal Doppler effect*. Also, for $\alpha = \pi/2$, a photon approaching in a plane normal to the x-axis, we get

$$\lambda = \gamma(v)\lambda'\,, \quad \lambda' = \lambda/\gamma(v),$$

which is the *transverse Doppler effect*.

Transverse Doppler effect is a relativistic phenomenon related to the slowing down of moving clocks. For the longitudinal Doppler effect, where $v \ll c$, we get

$$\lambda' = \lambda(1 - \beta) \Rightarrow \Delta\lambda/\lambda = (\lambda' - \lambda)/\lambda = \beta,$$

which is the classical result.

Also, we see that

$$p_x' = -\frac{h\nu'}{c}\cos\alpha' = -\frac{\gamma h\nu}{c}(\cos\alpha + \beta),$$

$$p_y' = \frac{h\nu'}{c}\sin\alpha' = -\frac{h\nu}{c}\sin\alpha,$$

which results in

$$\tan\alpha' = \frac{\tan\alpha}{\gamma(1 + \beta\sec\alpha)}, \tag{2.63}$$

which is the relativistic formula for aberration. As $\beta \to 1$ and $\alpha \to \pi/2$, α' becomes very small and we can write

$$\alpha' \to 1/\beta\gamma \to 1/\gamma. \tag{2.64}$$

This result is quite important for synchrotron radiation which is produced by the motion of relativistic electrons in the presence of a magnetic field.

2.5.7 Relativistic Lagrangian

Consider a free particle of rest mass m_0, moving with a velocity $\mathbf{u}$ as observed by an inertial observer. For relativistic processes, the action integral is given by

$$A = \int_{\tau_1}^{\tau_2} \gamma L d\tau,$$

where L is the Lagrangian and τ the proper time, with the two values τ_1 and τ_2 at two points of the path. Since A and $d\tau$ are world-scalars, it follows that γL must also be a relativistic scalar. For a free particle as above, the only parameters on which the Lagrangian can depend on are its rest mass m_0 and its four-velocity $U_\mu = (\gamma(u)\mathbf{u}, i\gamma(u)c)$. Dimensional arguments suggest that m_0 should occur in first power and U_μ in second power in L. Also, the only way to obtain a world-scalar from a world-vector U_μ is to take the scalar product of U_μ with itself. Therefore, the only candidate appropriate for γL is

$$\gamma L = km_0 U_\mu U_\mu = -km_0 c^2,$$

where k is a constant. Conventionally, k is taken to be 1, so that we have

$$L = -m_0 c^2 / \gamma(u) = -m_0 (1 - u^2/c^2)^{1/2} c^2. \tag{2.65}$$

Now we can get the canonical momentum and the Hamiltonian. We find that

$$p_x = \frac{\partial L}{\partial \dot{x}} = \frac{\partial L}{\partial u_x} = -m_0 c^2 \frac{(-u\gamma(u))}{c^2} \cdot \frac{u_x}{u}$$

$$= m u_x,$$

with similar expressions for p_y and dp_z. This gives the relativistic momentum of the particle as $\mathbf{p} = m\mathbf{u}$. This allows us to write

$$L = -mc^2 + p^2/m. \tag{2.66}$$

Then $H = \mathbf{p}.\mathbf{u} - L = mc^2.$

This can also be written as

$$H = (m^2 c^4)^{1/2} = (m_0{}^2 \gamma^2(u) c^4)^{1/2}$$

$$= c[m_0{}^2 \gamma^2 c^2 (1 - \frac{u^2}{c^2} + \frac{u^2}{c^2})]^{1/2}$$

$$= c(p^2 + m_0{}^2 c^2)^{1/2}. \tag{2.67}$$

This gives for the kinetic energy

$$K = H - m_0 c^2 = c(p^2 + m_0{}^2 c^2)^{1/2} - m_0 c^2,$$

as we have already seen.

2.6   Elements of General Theory of Relativity

2.6 .1   Equivalence of gravity and acceleration

General theory of relativity arose in an attempt to extend the principle of relativity to accelerated frames. The clue came from the observed equivalence of gravitational and inertial masses. The gravitational force on a body near the surface of the earth is $GM_E m_g / R_E{}^2$, where M_E is the inertial mass of the earth, R_E its radius and m_g the gravitational mass of the body. If m_i is the inertial mass of the body, we have

$$\frac{GM_E m_g}{R_E{}^2} = m_i g,$$

where g is the acceleration due to gravity near the surface of the earth. Galileo's experiments from the tower of Pisa showed that g has the same value for all bodies. Hence $m_i \propto m_g$, and the two can be made identical by choosing appropriate units.

The equivalence of gravitational and inertial masses can be understood if we note that an accelerated frame simulates a gravitational field. If we consider a closed lift far removed from all other bodies and give it a constant acceleration g in the upward direction as indicated in Fig. 2.9 (a), all experiments carried out within the lift will indicate the presence of a downward directed gravitational field. For example, a projectile thrown (or a photon shot) horizontally will take a parabolic path, bending downward. Since no forces are acting, we are dealing with an inertial mass, but the same mass appears to experience a gravitational field. Similarly, we can cancel a gravitational field locally by putting ourselves inside a freely falling lift (Fig. 2.9(b)). Now, inside the lift all bodies will appear to obey Newton's first law corresponding to absence of a net force, e.g. projectiles will move in straight lines. Thus the gravitational field is equivalent to our accelerated frame of reference, at least locally. But if the lift is made big, the gravitational field can be perceived by convergence of the paths of freely falling bodies towards the centre of the earth; see Fig. 2.9 (c). This indicates that the spacetime around a gravitational body is not Euclidean as assumed in the Minkowski space of special relativity. We say that gravitating masses produce a curvature in the spacetime continuum.

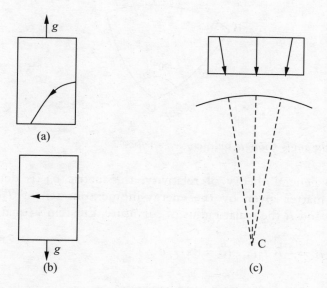

Fig. 2.9 *(a) Acceleration, (b) a freely falling lift, and (c) field in a big lift.*

2.6.2 Einstein's equation

In the Minkowski space, we have the Euclidean metric

$$ds^2 = c^2 dt^2 - (dx^2 + dy^2 + dz^2).$$

But in general, we can write for the metric[9]

$$ds^2 = \sum_{i,k} g_{ik} dx^i dx^k, \qquad (2.68)$$

where i, k run over 1, 2, 3, 4, corresponding to time and the three space coordinates. The tensor of rank 2, g_{ik}, is known as the *metric tensor* of the four-dimensional space. It tells us about the structure of the space, particularly about its curvature. If $g_{ik} = 0$ for $i \neq k$, it is called a Euclidean space. A non-Euclidean space can always be approximated by a Euclidean space in the neighbourhood of any point. For example, consider the spherical triangle ABC in Fig 2.10, with sides a, b, c. Here angle $c = \pi/2$, so $\cos c = \cos a \cos b - \sin a \sin b \cos c = \cos a \cos b$. For small a, b, c (in units of radian or as length divided by radius of the sphere), we can expand the trigonometric functions and find that $c^2 = a^2 + b^2$, which is true for Euclidean space.

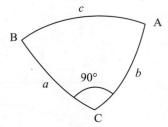

Fig. 2.10 *A right-angle triangle on a curved surface*

According to general theory of relativity, the metric g_{ik} is determined by the distribution of matter given by the energy-momentum tensor T_{ik}. If R_{ik} is the curvature tensor and R the scalar radius of curvature, Einstein's equation is written as

$$R_{ik} - \frac{1}{2} g_{ik} R = -K T_{ik}, \quad K = 8\pi G/c^4.$$

A solution of this equation gives g_{ik} and hence ds^2.

The paths of test particles (tiny point masses) are given by geodesics which make $\int ds$ between two spacetime points an extremum. If $\int ds$ is positive, it is called a

---

[9]See Joshi, A. W. (2004)

time-like geodesic, while if it is negative it is a space-like geodesic. If $\int ds = 0$, it is called a null geodesic which represents the path of a photon.

Out of the several solutions of Einstein's equation, we shall consider the following two solutions.

Schwarzschild solution

For a spherical mass distribution like the sun, we have

$$ds^2 = (1 - 2GM/rc^2)dt^2 - \frac{1}{c^2}\left\{\frac{dr^2}{(1 - 2GM/rc^2)} + r^2 d\theta^2 + r^2\sin^2\theta d\phi^2\right\}. \quad (2.69)$$

Since $-GM/r = \phi(r)$, the gravitational potential at a distance r from the centre of the star, we see that $1 - 2GM/rc^2$ is also written as $1 + 2\phi/c^2$. This solution is valid outside the mass. In empty space (with $M = 0$), we get back the Minkowski metric

$$ds^2 = dt^2 - \frac{1}{c^2}(dr^2 + r^2 d\theta^2 + r^2\sin^2\theta d\phi^2). \quad (2.70a)$$

Robertson–Walker metric

For a uniform distribution of matter like what we can obtain by spreading all the matter in the universe uniformly throughout the space, we have the metric

$$ds^2 = dt^2 - \frac{R^2(t)}{c^2}\frac{dr^2 + r^2 d\theta^2 + r^2\sin^2\theta d\phi^2}{1 + kr^4/4}, \quad (2.70b)$$

where $R(t)$ is the constant curvature of the whole space and $k=0, \pm 1$ in different situations. When $k=0$, we have a flat Euclidean universe; $k=1$ indicates an elliptical (closed, finite) universe; $k = -1$ indicates a hyperbolic or open, infinite universe.

2.6.3   Tests of general theory of relativity and other effects

The special theory of relativity had immediate relevance in laboratory conditions where it could be tested. But the general theory of relativity was amendable to verification only on a much larger scale of the solar system and the universe of stars and galaxies. Some of these tests are discussed here.

Advance of perihelion of Mercury

From (2.69) one finds that gravitational attraction goes as $1/r^{2+n}$ close to the central body where n is a small fraction. It leads to a slow advance of the line of apsides for planets given by

$$\Delta\omega = \frac{6\pi GM_{\mathrm{S}}}{c^2 a(1-e^2)} \quad \text{radians/revolution,} \tag{2.71}$$

where M_{S} is the mass of the sun. For Mercury, $a = 0.387\,\mathrm{AU} = 5.79 \times 10^7$ km, $e = 0.2056$, which gives $\Delta\omega = 5.02 \times 10^{-7}$rad/rev. With the period of Mercury being $87.064d$, this comes out to be $42''.9$ per century. In fact, a discrepancy of $(42.3'' \pm 1'')$ per century was earlier found for Mercury even after taking into account the effect of perturbations due to the other planets. The explanation of this discrepancy was a great triumph for the general theory of relativity, which thus proves to be a small but definite improvement over the Newtonian theory of gravitation. Recently, the binary pulsar PSR 1915+16, which has a very intense gravitational field, was found to have an advance of perihelion of $4°$/year in agreement with the calculation made on the basis of the general theory of relativity.

Bending of light rays

If a photon of energy $h\nu$ is considered to be equivalent to a mass of $h\nu/c^2$, even according to the special theory of relativity, it should have a hyperbolic orbit around the sun. It predicts a deflection of $2GM_{\mathrm{S}}/c^2 R_{\mathrm{S}}^2 = 0''.87$ for light rays grazing the solar limb, where R_{S} is the radius of the sun. But the general theory of relativity, which requires a gravitational attraction varying as $1/r^{2+n}$, predicts twice this value, that is $4GM_{\mathrm{S}}/c^2 R_{\mathrm{S}} = 1.75''$. Observations of stars near the solar limb at the time of the total solar eclipse of 1919 by Eddington confirmed the bigger value, and consequently the correctness of the general theory of relativity. It led to the concept and use of gravitational lensing as a probe of the universe discussed in Section 2.7.

Gravitational red-shift

Consider a spectral line of frequency ν and wavelength λ_{S} emitted by an atom from the surface of a star of mass M_{S} and radius R_{S}, having a surface potential $\phi_{\mathrm{S}} = -GM_{\mathrm{S}}/R_{\mathrm{S}}$. Now, the coefficient of dt in (2.69) gives the gravitational time dilatation factor. Then the frequency of the same spectral line ν_{E} emitted on earth's surface is related to ν_{S} by

$$\frac{\nu_{\mathrm{S}}}{\nu_{\mathrm{E}}} = \sqrt{\frac{1 - 2GM_{\mathrm{S}}/R_{\mathrm{S}}c^2}{1 - 2GM_{\mathrm{E}}/R_{\mathrm{E}}c^2}} = \sqrt{\frac{1 + 2\phi_{\mathrm{S}}/c^2}{1 + 2\phi_{\mathrm{E}}/c^2}} \simeq 1 + \frac{\phi_{\mathrm{S}} - \phi_{\mathrm{E}}}{c^2}. \tag{2.72}$$

It may be noted that once the photon leaves the star there is no change in its frequency except by Doppler effect. It will, however, change its direction in varying gravitational field due to the phenomenon of gravitational bending.

From (2.72) we see that

$$(\nu_{\mathrm{S}} - \nu_{\mathrm{E}})/\nu_{\mathrm{E}} = (\phi_{\mathrm{s}} - \phi_{\mathrm{E}})/c^2$$

$$\Rightarrow \frac{\lambda_S - \lambda_E}{\lambda_E} = \frac{\phi_E - \phi_S}{c^2}. \tag{2.73}$$

As GM/R for stars is very much larger than that for the earth, $\lambda_S - \lambda_E > \lambda_E$. This, is known as *gravitational red-shift*.

For a photon of wavelength $\lambda = 6000$ Å emitted from the surface of the sun, we get $\Delta\lambda = 0.012$ Å, which is small. But for a white dwarf of the same mass of the sun, and radius 50 times smaller than that of the sun, we get $\Delta\lambda = 0.6$ Å. This shift has been verified in the case of Serius B.

Rates of atomic clocks

Nowadays, we can test the results for gravitational redshift more accurately by measuring the rates of atomic clocks in satellites. We have to consider two effects—relative motion using special theory of relativity and gravitational effect using general theory of relativity. The special relativity effect gives the Doppler effect

$$\nu_1' = \nu(1 - \beta^2)^{1/2} \simeq \nu(1 - \beta^2/2) \quad \Rightarrow \Delta\nu_1/\nu = -\beta^2/2.$$

We are concerned here only with transverse Doppler effect.

If a satellite is orbiting the earth in a circular orbit of radius r, with the earth's mass being M_E and its radius R_E, we have,

$$v^2 = \frac{GM_E}{r} = g\frac{R_E^2}{r}, \quad g = \frac{GM_E}{R_E^2}.$$

This gives,

$$\delta\nu_1/\nu = -\frac{gR_E}{2c^2}\frac{R_E}{r}.$$

The general relativistic effect gives a shift

$$\frac{\delta\nu_2}{\nu} = \frac{\delta\phi}{c^2},$$

where $\delta\phi = -GM_E\left(\frac{1}{r} - \frac{1}{R_E}\right)$, i.e., the potential difference between the satelite and the earth's surface. We have replaced here $\delta\nu_2/\nu'$ by $\Delta\nu_2/\nu$, which from (2.72) and (2.73), is correct to first order in $\delta\phi/c^2$. With the above $\delta\phi$ in the context of the satellite, we finally get

$$\frac{\delta\nu_2}{\nu} = \frac{gR_E}{c^2}(1 - \frac{R_E}{r}). \tag{2.74}$$

Combining these two effects, we get the total shift in frequency of the clock on the satellite as compared to a similar one on the earth:

$$\frac{\delta \nu}{\nu} = g \frac{R_E}{c^2} \left(1 - \frac{3R_E}{2r} \right). \tag{2.75}$$

This shows that at $r = 3R_E/2$, the satellite clock shows no shift. This occurs for a satelite orbiting the earth with a radius of about 9600 km.

For a satellite placed in an orbit below this, say at a height of 500 km, $r \simeq$ 7000 km, $\delta\nu < 0$, and the satelite clock will run slow. However, due to the low gravity on the earth and the appearance of c^2 in the denominator, the shift is extremely small, of the order of 10^{-10}. This phenomenon is observed in global positioning systems (GPS).

Black holes

Schwarzschild solution, ((2.69) of Einstein's equation), has a singularity at $R_S = 2GM/c^2$, which is known as Schwarzschild radius. It can be seen that at this distance from a star of mass M the escape velocity becomes equal to c. Hence no particle, not even a photon, can escape from a distance below R_S. As a consequence, if the radius of a star shrinks to a value below R_S, nothing can escape it. That is why such a configuration is known as a black hole. The Schwarzschild radius for the sun is only 3 km. If the sun shrinks to a radius below this, it will become a black hole.

But the sun and ordinary sun-like stars cannot become black holes. During the course of evolution of a star, after exhausting the nuclear fuel, it shrinks into a compact object. There are three kinds of compact objects. There are *white dwarfs* with densities of the order of 10^5 to 10^6 gm/cm^3 and radii of the order of 1000–10000 km and an upper mass limit of $1.4M_S$ (1.4 times the solar mass which called is the *Chandrashekhar limit*). Many white dwarfs are known in the sky. These are *Neutron stars* with densities like 10^{14} gm/cm^3 and radii of the order of 10 to 20 km. They have a maximum mass of 2 to 3 times solar mass. Hundreds of neutron stars have now been discovered as pulsars. Finally, there are black holes which are more massive objects; in fact one is supposed to be present in the X-ray source Cygnus X-1. X-rays are produced when the material from a companion star of about 12 solar masses falls into the black hole. Very massive black holes of 10^5 to 10^6 solar masses have been found at the centre of active galaxies including the Milky Way.

The expanding universe

The Robertson–Walker metric of (2.70) has curvature parameter $R(t)$. An increasing $R(t)$ (decreasing curvature) represents an expanding universe. Observations of galaxies show that they are receding and their rate of recession increases with distance. This shows that the universe is expanding and it corresponds to one solution of the equation of general theory of relativity. Astronomers are trying to find out the value of k which will tell us whether the universe is bounded and finite or open and infinite. Discussion of this cosmological question is out of the scope of this book and the reader is referred to Narlikar (1998).

2.6.4   Significance of Einstein's work

The ideas of Einstein, scientist *par excellence* of the twentieth century, produced lasting and revolutionary effects in science in general and physics and astronomy in particular. Here the special and general theories of relativity have to be considered on different footing. While the former was a modification of Galilean relativity necessitated by the process of measurement, in the light of Maxwell's equations and Michelson-Morley experiment, the latter was an entirely new concept regarding nature of gravity. While the former had immediate relevance to laboratory conditions, the latter applied to larger domains of the solar system and the universe of stars and galaxies.

The most important consequence of special theory of relativity pertains to the basic concepts of space, time, mass and energy. Firstly, space and time not only lost their absoluteness but they also merged into the common spacetime continuum of Minkowski space. However, space and time behaved differently in the transformation from one frame to another, and one had to distinguish between space-like and time-like intervals. Secondly as a consequence of merging space and time, it was necessary to abolish the distinction between mass and energy through the equation $E = mc^2$. One had to combine the laws of conservation of mass and energy into a single law of conservation of their sum. Thirdly, the special theory of relativity which is applicable to inertial frames having uniform relative motion paved the way for a search for laws which are valid in accelerated frames of reference.

Now, acceleration was most often met in gravitational fields, hence for these the equations appropriate for accelerated motion were required to replace Newton's law of gravitation. The equivalence of a gravitational field and an accelerated frame of reference became a hypothesis of Einstein's general theory of relativity. This equivalence was established through the curvature of spacetime continuum in the presence of mass.

Here we can draw attention to the important difference between Newton and Einstein. Newton's law of gravity is beautifully simple but its application to actual situations is extremely complicated by virtue of the existence of infinite number of particles. So one has to integrate the effects of all of them and take limits. It is for this reason that Newton had to invent *de novo*, the mathematical discipline of fluxions, now known as differential calculus. On the other hand, Einstein was fortunate because he already had available to him a fully developed Riemannian differential geometry which forms the very backbone of his laws of general relativity. Even so, as someone has said, while anyone could have discovered the special theory of relativity, Einstein alone was responsible for the invention of the general theory.

2.6.5   Twin paradox

We shall end this section with a discussion of this famous paradox. Let X and Y be two identical twins. Suppose X remains fixed at a space point and Y moves out with uniform speed v to a certain distance and returns back with the same speed. Let t_X be the total time measured by X and t_Y that by Y. Now, according to X, $t_X = \gamma t_Y$, which implies that $t_Y < t_X$. So we will conclude that when Y returns to join X, he will be younger. In the same way Y will think that it was X who moved away and then came back, and so X should be younger. This paradox can be resolved by an appeal to general theory of relativity.

Let OABCDE be the trajectory of Y in the x-t plane of X as shown in Fig. 2.11. In the short section OA, the twin Y accelerates from rest to velocity v with respect to X, then moves with uniform velocity in section AB, decelerates to reverse the velocity in section BC, again moves with uniform velocity in the opposite direction on CD, and finally decelerates to zero velocity in the short section DE to join X. Let t_1 to t_5 be the time intervals in the sections as measured by X and as shown in Fig. 2.11, and τ_1 to τ_5 those measured by Y. By symmetry, $t_4 = t_2$ and $\tau_4 = \tau_2$.

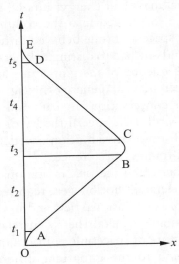

Fig. 2.11   *Path of twin Y in the frame of reference of twin X*

Now, according to the general theory of relativity, Y experiences a gravitational field in sections OA, BC and DE, which multiplies X's time intervals by a factor $(1 + (\phi_Y - \phi_X)/c^2)$, where the ϕ's are gravitational potentials seen by them. X and Y are close to one another in sections OA and DE so that this factor is unity in those intervals. The velocity is $v=0$ or very low in those sections so that $t_1 = \tau_1$ and $t_5 = \tau_5$. The velocity v is constant in sections AB and CD. So Y concludes that $t_2 + t_4 = 2t_2 = 2\tau_2/\gamma$, that is,

Warping of spacetime near the earth

The general theory of relativity (GTR) suggests that spacetime around a massive object is curved, leading to effects like bending of light and precession of the orbit of a body around the larger mass. Both these effects have been experimentally verified in the case of the sun. Now these effects are being verified for the much weaker gravitational field in the near-earth spacetime.

The earth-orbiting observatory GRAVITY PROBE B (GP-B) devoted to testing the GTR near the earth has measured the warping of spacetime with a precision of 1%. The satellite GP-B carries four gyroscopes onboard, although only three are essential. Each gyroscope is a smooth sphere of the size of a ping-pong ball whose radius does not very by more than 10 nm over the entire surface. They are electrostatically held in a small case in a vacuum of 10^{-12} torr and can be put into rotation at 4000 rpm. They are covered with niobium and maintained at a temperature of a few kelvin. Thus the balls are rotating superconductors, and as such develop a tiny magnetic moment which serves to signify the orientation of the sphere. A precession of the axis was observed in early 2007 and measured to be 6.6 arcseconds per year, which is close to that predicted by the GTR. The experimental team hopes to improve the precession to 0.01% by December 2007.

$$t_2 + t_4 \simeq 2\tau_2(1 - \beta^2/2) \tag{2.76}$$

to second order in β if v is not too large.

In the interval BC, Y experiences acceleration $g = -2v/\tau_3$. So the potential difference between Y and X will be

$$\Delta\phi = (-2v/\tau_3)x' = (-2v/\tau_3)(-v\tau_2) = 2v^2\tau_2/\tau_3.$$

Hence

$$t_3 = \tau_3(1 + 2\beta^2\tau_2/\tau_3). \tag{2.77}$$

Combining the above two equations, we get

$$t_2 + t_3 + t_4 = \tau_2(2 + \beta^2) + \tau_3. \tag{2.78}$$

Now, according to X, $t_3 = \tau_3$ as v is small in this interval while

$$t_2 + t_4 = 2t_2 = 2\gamma\tau_2 = 2\tau_2(1 + \beta^2/2).$$

This gives the same result for $t_2 + t_3 + t_4$ as (2.78). Thus the interpretations of both X and Y agree with one another and there is no paradox. It may be noted that it is Y who uses rockets to accelerate, so Y experiences a gravitational field while X always remains inertial. The result of all this is that Y remains younger but that Y thinks that X should have remained younger. If both of them know the general theory of relativity, their time measurements will agree with each other and both of them will agree that there is no paradox in that Y is younger than X.

2.7 Gravitational Lensing

In the section we discuss a fascinating phenomenon of current interest called gravitational lensing.

2.7.1 History

As we saw earlier, Einstein's prediction about the bending of light rays around a gravitating body, based on his general theory of relativity, was brilliantly confirmed by Eddington during the total solar eclipse of 1919. That the sun or any other star can act as a lens to focus light was first pointed out by Oliver Lodge in 1919 with the following argument.

The deflection of light passing near a star of mass M is given by

$$\psi = 4GM/rc^2 = -4\phi/c^2$$

or $\psi = K/r$, where $K = 4GM/c^2$. $\qquad\qquad\qquad\qquad$ (2.79)

Here, r is the distance from the centre of the star to the passage of the ray, that is, the impact parameter of scattering. Accordingly, parallel rays in a cylinder of radius r about the star will get focussed at the vertex of a cone with semiangle given by ψ. Figure 2.12 shows the scheme. Rings A, B and C projected on the sky plane passing near the sun S will be focussed at points A′, B′, C′, respectively. In this way a nebula behind the sun will be focussed along a line. The observer's position along the line will decide the angular size of the observed ring. If the distance from the centre of the sun to the observer is L, we have

$$\phi = r/L.$$

With (2.79), this gives

$$L = r^2/K.$$

In the sun's case the minimum meaningful r is its radius ($= 7 \times 10^8$ m); also K for the sun is 6×10^3 m. So the minimum distance from the sun at which a ring due to gravitational focussing can be seen comes to about 550 A.U. (1 A.U. = sun-earth distance = 1.5×10^{11} m.) Thus there is no chance for us to observe gravitational focussing by the sun.

In the mid-1930's Einstein showed that if one star is occulted by another, the distant source star will be seen as a ring around the intervening star. Around the same time, it was also shown that galaxies too can act as gravitational lenses for objects behind them. The first example of a gravitational lens was provided in 1979 where light from a distant quasar passed through an intervening galaxy. Several examples have since been reported.

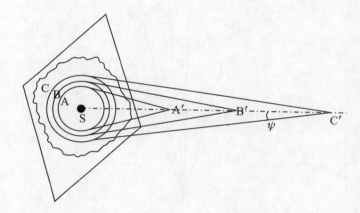

Fig. 2.12   *Imaging of a nebula by the sun*

2.7.2   Formation of images by a stellar lens

Figure 2.13 shows an observer O, a star S, and a lensing star L near the ray SO. Let AB and CD be the sky planes, normal to the line OL, passing through S and L, respectively. Consider a point I in the plane CD at a distance r from L such that the ray SI gravitationally bends due to star L and just reaches the observer O. Let θ be the bending angle (at a distance r from L) as shown.

We extend lines OL and OI to meet AB at E and F, respectively. Let the distance OS of the source star be d_S and that of the lensing star OL be d_L. Note that OE is almost equal to OS. Rays from S going beyond the distance r from star L will miss the observer and will meet the line EO beyond O. Let the angle LOI be α and angle LOS be β. Also, let ES=x and EL=d_{SL}.

Noting that all the angles α, β and θ are small, we may write

$$x = \beta d_S, \mathrm{EF} = \alpha d_S, r = \alpha d_L,$$

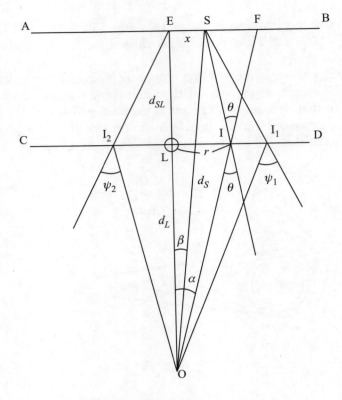

Figure 2.13 *Formation of two images by a stellar lens*

giving

$$\text{SF} = (\alpha - \beta)d_S.$$

In the triangle SIF, we may approximately consider

$$\text{SI} \simeq d_S - d_L$$

so that

$$\text{SF} \simeq (d_S - d_L)\theta\,.$$

This therefore gives

$$\theta = \frac{(\alpha - \beta)d_S}{d_S - d_L}\,. \tag{2.80}$$

The bending angle ψ, from (2.79), is given by

$$\psi = \frac{K}{r} = \frac{K}{\alpha d_L},$$

where $K = 4GM/c^2$ for star L. Equating the above two expressions and rearranging, we get

$$\alpha^2 - \beta\alpha - K\frac{d_S - d_L}{d_S d_L} = 0 \qquad (2.81)$$

Equation (2.81) has two solutions

$$\alpha_{1.2} = \beta/2 \pm [(\beta/2)^2 + \alpha_0{}^2]^{1/2},$$

where $\quad \alpha_0{}^2 = K\dfrac{d_S - d_L}{d_S d_L} = K\left(\dfrac{1}{d_L} - \dfrac{1}{d_S}\right).$

It is to be noted that the star S is not seen at an angle β from L but appears to be shifted and moved away in the sky from the lensing star.

It is convenient to denote the above radical by

$$\alpha = [(\beta/2)^2 + \alpha_0{}^2]^{1/2}, \qquad (2.82a)$$

so that the two solutions can be written as

$$\alpha_1 = \alpha + \beta/2, \, \alpha_2 = -\alpha + \beta/2. \qquad (2.82b)$$

Note that $\alpha > \beta/2$, indicating that $\alpha_2 < 0$. This means that the star S will produce two images at I_1 and I_2 in the sky plane. I_1 is on the same side of L as the star S actually is, while I_2 is on the other side of L. This can be understood by referring to Fig 2.14. Ray SI_1, after bending around L reaches observer O, and O sees the star of an angle α_1 from L instead of the actual angle β. On the other hand, a ray SI_2 passing around the star from the other side at a distance close enough to L may also, after bending reach O and produce another image at an angle α_2 from L on the other side.

2.7.3   Sizes of the images

Figure 2.14 shows the configuration of Fig. 2.13 in the sky plane. L is the centre of the lensing star and S that of the source star. Let the angular radius of S and L be u and v, respectively. The angular separation LS is β, as before. Let M be mid-point of LS so that LM$=\beta/2$. Then consider a circle with centre at M and radius α of (2.74a). Draw tangents to the star S from L and let them meet the periphery of the circle at Q_1 and P_1 on the sides of the actual star and at Q_2 and P_2 on the other side.

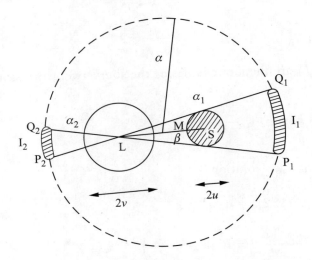

Fig. 2.14   *Extension of the images in the sky plane*

Then the linear size of the images I_1, I_2 along the circle can be easily calculated from the geometry of similar triangles in Fig. 2.14. The angular size of the two images is seen to be

$$I_1 = u(1 + 2\alpha/\beta), \quad I_2 = u(1 - 2\alpha/\beta). \tag{2.83a}$$

The images will also have a thickness in the radial direction. This can be seen from the fact that while β is the angular separation between the two stars, the angular separation of the left (closer to the star L) and right (farther from the star L) edges of S will be $\beta - u$ and $\beta + u$, respectively. A simple calculation then shows that the thickness of the two images would be

$$\Delta\alpha_1 = (u/2)(1 + \beta/2\alpha), \Delta\alpha_2 = (u/2)(1 - \beta/2\alpha). \tag{2.83b}$$

These equations show that the images I_1 on the side of the actual star (right in Fig. 2.15) is bigger and thicker than the other images I_2.

2.7.4   Luminosity of the images

It is well known that in any image formation the intensity of light remains constant, ignoring attenuation. Hence the luminosities of the source and the two images will be proportional to the solid angles subtended by them. The solid angle for the source star is $\Delta\Omega_S = \pi u^2$. The solid angles subtended by the two images can be found from their extensions in the radial and angular directions. They turn out to be, respectively for I_1 and I_2,

$$\Delta\Omega_1 = \frac{\pi u^2}{4}\left(\frac{\beta}{2\alpha} + \frac{2\alpha}{\beta} + 2\right), \quad \Delta\Omega_2 = \frac{\pi u^2}{4}\left(\frac{\beta}{2\alpha} + \frac{2\alpha}{\beta} - 2\right).$$

Thus if F_S, F_1, F_2 are the luminosities of the source and the two images, respectively, then

$$F_1 = \frac{\Delta\Omega_1}{\Delta\Omega_S}F_S, \quad F_2 = \frac{\Delta\Omega_2}{\Delta\Omega_S}F_S.$$

We show in Table 2.1 the fractional intensities of the two images as a function of the parameter β/α. It is interesting to consider two limiting cases, (i) $\beta \to 0$, that is, the source star and the lensing star are close to each other in the sky, and (ii) $\beta >> \alpha$, that is, the source star is far away from the lensing star. We now discuss the two cases.

Table 2.1 *Luminosities of images formed by a lensing star*

β/α	F_1/F_S	F_2/F_S	$(F_1 + F_2)/F_S$
8.0	1.000	0.000	1.000
2.0	1.030	0.030	1.060
1.0	1.171	0.171	1.342
0.5	1.591	0.591	2.182
0.2	3.037	2.037	5.074
0.1	5.519	4.519	10.034
0.0	ring	ring	$2\alpha/u$

(i) $\beta << \alpha$:   In this case the source star overlaps or is behind ($\beta = 0$) the lensing star. The images are almost equidistant from the lensing star, on either side, and both have a width of $\simeq u/2$. $\beta = 0$ would produce a circular image around the lens star. When the source star is exactly behind the lens star, light coming from it toward the lens star and passing at an appropriate distance from it bends around and reaches the observer. Thus the star produces a circular image around the lens star. In a sense, the two images I_1 and I_2 merge with a single circular image of width u. This is the *Einstein ring*.

(ii) $\beta >> \alpha$:   This implies that the source star is far away (in sky) from the lensing star. Image I_2 goes on becoming fainter and image I_1 merges with the star S itself. Light from star S going round the star L from the other side cannot bend and reach the observer. A sequence of these images corresponding to parameters in Table 2.1 is shown in Fig. 2.15.

This is what happens when one star passes behind another star, although it must be noted that the probability of such an occurrence in our galaxy is quite small. Such events have been recently observed as sudden increase in the brightness of the star. They are interpreted as passing of a brown dwarf or a planet in front of another star.

The theory of gravitational lensing finds its best applications in the explanation of the variation of the brightness of the images of a quasar formed by a foreground galaxy. We shall next consider galaxies as gravitational lenses.

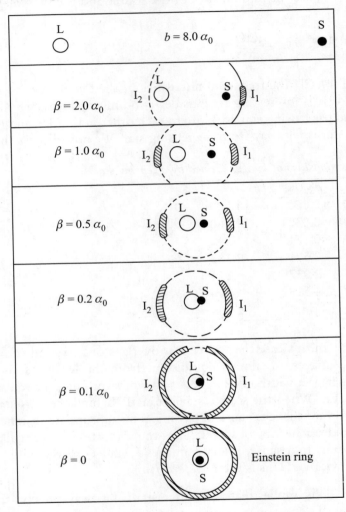

Fig. 2.15   *Sequential development of gravitational images during the occultation of one star by another*

2.7.5 A galaxy as a gravitational lens

A galaxy is a massive body containing billions of stars. It has a large gravitational potential which can deflect the light of sources like quasars that lie behind them and produce gravitational lensing. A typical galaxy like our Milky Way has a mass of about 10^{12} solar masses and a radius of about 20 kpc. So it can produce a deflection of about $2''$. But a galaxy cannot be considered to be a point-like object like a star. Firstly, its potential is more complicated than that due to a simple inverse square law of force. Secondly, the distances between the stars in it are so large that light can easily pass through the body of the galaxy with only a small amount of extinction in the thin interstellar medium of gas and dust.

Now, a galaxy is a three-dimensional entity, but we may consider only its projection on the sky plane as a thin lens. This is because the diameter of a galaxy is much smaller than inter-galactic distances. We assume that the projection has circular symmetry, though it may not be so in practice. Thus we shall consider the deflection ψ to be a function of the distance of the path of a light ray from the centre; thus $\psi \equiv \psi(r) \equiv \psi(x, y)$.

The potential $\phi(r)$ or the deflection $\psi(r) = -4\pi\phi(x, y)/c^2$ for this case (apart from the sign) are shown in Fig. 2.16. It is seen that at the centre, the deflection is zero, then it increases linearly with r near the centre, as for ordinary convex lens. In the outer region, it again decreases with increasing r, as in the case of a star with inverse square law force. We can represent it approximately as

$$\psi(r) = \frac{K_1 r}{1 + K_2 r^2}, \tag{2.84}$$

where K_1, K_2 are constants. It turns out that this gives three solutions for r, the distance at which a light ray must pass in order to reach the observer. One of the images is formed by a ray passing through the central part while the other two from either side of the intervening galaxy, as seen earlier.

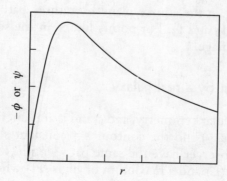

Fig. 2.16 *Gravitational potential $\phi(r)$ or reflection $\psi(r)$ of a ray as function of distance from the centre of a galaxy with circular symmetry*

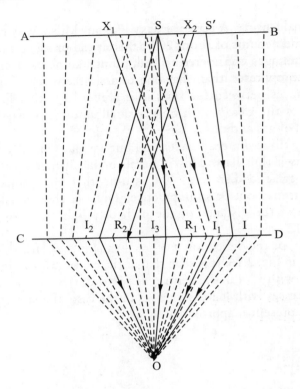

Fig. 2.17 *Ray paths for lense in the plane CD*

The paths of the rays from source plane AB, passing through the lens plane CD, and reaching the observer O are shown in Fig. 2.17. Here the central region R_1R_2 behaves like a convex lens while the outer region behaves like a star. So the rays from a source S within the region X_1X_2 in the source plane have three paths to reach O, and thus produce three images I_1, I_2 and I_3. For points like S' in the source plane outside X_1, X_2, we can see only one image.

2.7.6 Image formation by a real galaxy

In this case there is no circular symmetry, and ϕ and ψ are functions of x, y. Figure 2.18 projects the source at $s(x, y)$ and the contour $s =$ constant shows points where rays from the source to the observer have the same path length. These contours normally show a minimum or low (L) and a maximum or high (H) where $\delta s/\delta x = \delta s/\delta y = 0$. For $\delta^2 s/\delta x^2$ and $\delta^2 s/\delta y^2$, there are two familiar choices — both positive (corresponding to minimum s) and both positive (corresponding to maximum s). In addition, we will also have a saddle point P where the second derivative of s is negative in one direction and positive in another direction (1 and 2 in Fig. 2.18, respectively). We will thus see three images of the source corresponding to L, H and P. If there is an additional low or high point, then mathematics demands that there will be one more saddle point too. Thus the number of images will always be odd, though some of them may merge into each other due to the low resolution of the observing equipment.

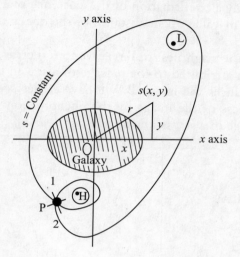

Fig. 2.18 *A galaxy as a gravitational lens: L—low point, H—high point, P—saddle point, 1 and 2 — directions along which the second derivative of s is negative and positive, respectively.*

2.7.7 Examples of gravitational lensing

Several phenomena have been observed in the recent years which are ascribed to gravitational lensing. We describe here a few of them. Fig. 2.19 shows four cases which we discuss below.

Quasar Q957+561

This binary quasar was discovered by Walsh, Carswell and Weynmann in the 1970s. The two components are seen on either side of a galaxy. The red shift of both the components (A and B in Fig. 2.19 (a)) of the quasar is $z = 1.406$ while that of the galaxy G is $z = 0.36$. Since z is a measure of distance in cosmology, they argued that the duplicity in the quasar was the result of gravitational lensing of a single quasar by the intervening galaxy. The existence of identical jets in both images confirmed this guess. But there was one difficulty. We have seen that the deflection produced by a galaxy is of the order of $2''$, but the separation between the two components of the quasar here was seen to be $6''$. This discrepancy is resolved when we consider that the lens galaxy is a member of a cluster of galaxies. Now a cluster of galaxies is about 100 times in mass and about 10 times in size as compared to a single galaxy. So the cluster can produce a deflection of 20″. Hence the observed separation of 6″ is the combined effect of the lens galaxy and the cluster to which it belongs.

Since quasars are known to vary in brightness with time, the two components were monitored and it was found that the variation of B was delayed with respect to A by 1.1 year. This was not only a confirmation of the existence of a gravitational lens, but it also gave an independent value of the distance of the quasar and an improved value of Hubble's constant.

It was also noted that the brightness of B relative to A increased by 2% over a period of 8 years. This has been attributed to the passage of a star in the galaxy through the path of light which produces the image B. We have earlier seen that such a passage can enhance the brightness of the image by the phenomenon of Einstein ring. Such brightening by the passage of a star is known as *microlensing*.

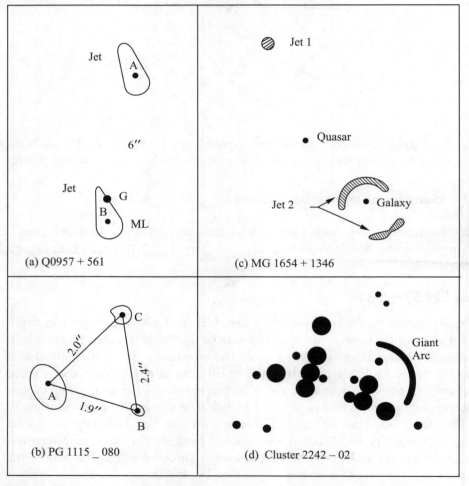

Fig. 2.19   *Four examples of images formed by gravitational lensing detected in recent decades*

Quasar PG1115+080

This triple quasar seen in Fig. 2.19 (b) was discovered by Weynmann in 1980. As all the three components have an identical red shift of $z=1.723$, they are considered to be the images of a single quasar. The elongation of component A suggests that it may itself be double. However, the lens galaxy of this system is not seen either in optical photographs or radio maps.

Quasar MG1654+1346

Quasars are usually accompanied by jets on either side. Here in Fig. 2.19 (c), we see one jet but the other jet is hidden by a galaxy, which shows it in the form of an Einstein ring around the lens galaxy.

Cluster 2242+02

In Fig. 2.19(d), we see a cluster of galaxies containing a giant arc which is supposed to be the image of a distant galaxy produced by gravitational lensing due to the cluster.

These and such other phenomena are considered to be a confirmation of gravitational lensing, though one often fails to see the requisite number of images as required by the theory. Anyway, it is certain that we can visualise galaxies and clusters of galaxies as lenses enabling us to probe the universe at large distances. In addition to giving an improved value of Hubble's constant, they tell us about the density of matter in the universe which is crucial for all cosmological theories.

Problems

2.1. Show that the phenomena of length contraction and time dilatation are relative. This means, that with reference to Section 2.3(a), the observer O$'$ finds lengths in the frame O to be contracted and clocks in O running slow by the same factor $\gamma = (1 - \beta^2)^{-1/2}$.

2.2. Taking $c=1$, represent the following events or processes graphically in both $x - t$ and $x'-t'$ planes of O and O$'$: $x = 0$, $t = 0$, $x = x_0$, $t = t_0$, $x = vt$, $x = \pm ct$, $t = vx$, and $x' = 0$, $t' = 0$, $x' = x'_0$, $t = t'_0$, $x' = -vt'$, $x' = \pm ct'$, $t' = -vx$.

2.3. Represent graphically the events $x'=0$, $t' = \alpha$ and $x' = \xi$, $t'=0$, in the $x - t$ plane for different values of v, where α and ξ are particular values of the variables.

2.4. An observer in an inertial frame O sees a body moving with a uniform velocity **u** in an arbitrary direction. Consider another inertial observer O$'$ in another inertial frame which is moving with respect to O with a velocity v in the x-direction. O$'$ finds the velocity of the body to be **u$'$**. Show that

$$\gamma(u') = \gamma(v)\gamma(u)(1 - vu_x/c^2,$$

and similarly,

$$\gamma(u) = \gamma(v)\gamma(u')(1 + vu_x/c^2.$$

[**Hint:** Start with

$$1 - \frac{u'^2}{c^2} = 1 - \frac{u_x'^2 + u_y'^2 + u_z'^2}{c^2}$$

and use the transformation of velocity components.]

2.5. Find the velocity at which the kinetic energy of a relativistic particle is equal to its rest mass energy.

2.6. A small particle of mass m is attracted towards a large mass M, and directly falls towards it from infinity. Its velocity which is $v = (GM/r)^{1/2}$ will exceed velocity of light at small r. How does the particle avoid this in special theory of relativity?

2.7. Prove that (a) $\gamma v = c(\gamma^2 - 1)^{1/2}$, (b) $c^2 d\gamma = \gamma^3 v dv$ and (c) $d(\gamma v) = \gamma^3 dv$.

2.8. Draw a graph of γ versus v. (a) At what speed is the γ-factor equal to 2 10 100? (b) What is the γ-factor at a speed of 0.95 0.99 0.999 of the speed of light?

2.9. Discuss operations with time-like four vectors.

2.10. Show that, when the velocity of a body in an inertial frame of reference tends to c, it also tends to c in any other inertial frame of reference.

2.11. Show that (2.63) leads to $\sin \alpha' = \dfrac{\sin \alpha}{\gamma(1 + \beta \cos \alpha)}$.

2.12. An observer in a reference frame O records two events with spacetime coordinates $(0, 0, 0, 0)$ and $(2ct, 0, 0, t)$ where $t > 0$. Find the inertial frames in which (a) the two events are simultaneous, (b) the second event precedes the first event by t and (c) the events occur at the same point.

2.13. A body A moves with a uniform velocity v in the x-direction as seen by an inertial observer O. Another body B moves with a uniform velocity w in the x'-direction, which is parallel to x-direction as recorded by an inertial observer O' fixed with the body A. Using the Lorentz transformation matrix of (2.37) successively for v and w, obtain the relativistic formula for addition of velocities.

2.14. Write down the 4×4 general linear Lorentz transformation between x', y', z', t' and x, y, z, t with the respective space axes parallel to each other. Argue that if the former frame is moving with a velocity v with respect to the latter along the x axis, then x' and t' should not depend on y and z. Similarly, argue that y' should not contain x, z and t' and z' should not contain x, y and t. Finally argue that the coefficient of y in y' and of z in z' must be unity. (This determines 12 of the transformationcoefficients and reduces it to a 2×2 transformation discussed in Guided Exercise 2.1.)

2.15. Three astronauts are moving towards celestial north pole with velocities $0.1c$, $0.5c$ and $0.95c$, respectively. Draw diagrams to show how they will see the northern and southern hemispheres of the sky due to the effect of aberration. What will be seen when $v \rightarrow c$?

3

Classical Theory of Radiation

W E begin this chapter by introducing Maxwell's equations of electrodynamics and electromagnetic waves (EMW). We then proceed with electromagnetic radiation (EMR) by a molecule. We revisit the harmonic oscillator and consider the case when it is an oscillating charge. The properties of the medium play an important role in the transmission of EMR, and we highlight its essential aspects. We then go on to discuss the relativistic transformation of electromagnetic quantities such as fields, potentials, and charge and current densities. This chapter ends with a discussion of scattering of EMR by small particles, considering different cases of particle size smaller and larger than the wavelength, thus covering the size range from nanometers to millimeters, and when the particle is a conductor or a nonconductor.

3.1 Maxwell's Equations of Electrodynamics

3.1.1 The beginnings of electromagnetism

Electric and magnetic phenomena have been observed critically over the past four centuries. Various empirical laws have been formulated in the beginning, which include Coulomb's (1785) law of electrostatics, Biot–Sawart and Ampere's laws (1820) for magnetic field due to a current, and Faraday's law (1821) of induction. They were also quantitatively formulated. For a long time, electricity and magnetism were considered separate phenomena. Then with Faraday's work, it was realized that they are linked with each other. It turns out that static, time-independent, electric and magnetic phenomena are independent of each other but that a time-dependent electric field produces a magnetic field and a time-dependent magnetic field produces an electric field.

These laws were analysed for their validity or otherwise for time-dependent electromagnetic quantities. In static situations, there are two laws which give the electric field in terms of charges and two laws which give the magnetic field in terms of currents[10]. Maxwell (1873) worked out the necessary modifications and gave his four equations which cover *all* classical electromagnetic phenomena. Here 'classical' also covers all bulk relativistic phenomena; it excludes microscopic, quantum phenomena.

Coulomb's law of electrostatics states that two charges q_1 and q_2 at $\mathbf{r}_1$ and $\mathbf{r}_2$, respectively, act on each other with a force proportional to the product of the charges and inversely proportional to the square of the distance between them. Further, the force depends upon the intervening medium, the force being inversely proportional to the dielectric constant of the medium. Taking the proportionality constant to be unity[11], the force due to charge q_2 on charge q_1 is given by

$$\mathbf{F}_{12} = \frac{q_1 q_2}{\epsilon r_{12}^3} \mathbf{r}_{12},\qquad(3.1)$$

where $\mathbf{r}_{12} = \mathbf{r}_1 - \mathbf{r}_2$ is the vector from $\mathbf{r}_2$ to $\mathbf{r}_1$, and ϵ the *dielectric constant* of the medium; $\epsilon = 1$ for vacuum and $\epsilon > 1$ for any other medium. As the algebraic sign indicates, two like charges repel each other while unlike charges attract each other.

The electrostatic force acting on a charge at a point, divided by the charge, in the limit when the charge tends to zero, is called the electric field at that point. Thus the electric field $\mathbf{E}(\mathbf{r})$ at a point $\mathbf{r}$ is

$$\mathbf{E}(\mathbf{r}) = \lim_{q \to 0} \frac{\mathbf{F}(\mathbf{r})}{q},$$

where $\mathbf{F}(\mathbf{r})$ is the force acting on a charge q placed at $\mathbf{r}$ due to surrounding electric charges. This description is often simplified by saying that electric field at a point is the force acting on a small unit positive charge (or test charge) kept at that point.

It is convenient, though not necessary, to represent the electric field in a region by electric field lines. An electric field line originates at a positive charge (or at infinity) and terminates at a negative charge (or at infinity). The tangent to the field line at a point indicates the direction of the electric field at that point. Thus a field line is a well-defined concept. However, it must be emphasised that their number, or number density is an ill-defined concept because one can draw a field line passing through any given point. As such the number of lines crossing any area is infinite (equal to the number of points in that area)!

It has not been possible to isolate magnetic monopoles, and the smallest unit of source of magnetic field is a magnetic dipole. There is a theoretical conjecture of their existence, possibly at extremely high energies. If we assume the existence of magnetic monopoles, we need two types for the overall neutrality. These two types

---

[10]Since charges and currents produce electromagnetic fields, they are called *sources* in this context.

[11]We are using CGS-Gaussian units throughout this book; see Box in this section.

are conventionally called *north* and *south*, and we might say they are like the positive and the negative in the context of electric charges. We will then find that the force law between magnetic monopoles is exactly like the Coulomb's law, (3.1), with charges q_i replaced by magnetic monopoles m_i and ϵ replaced by $1/\mu$, where μ is the permittivity of the medium. Again $\mu = 1$ for vacuum. We could also define the magnetic induction at a point $\mathbf{r}$ as

$$\mathbf{B}(\mathbf{r}) = \lim_{m \to 0} \frac{\mathbf{F}(\mathbf{r})}{m},$$

where $\mathbf{F}(\mathbf{r})$ is the force on a north pole m placed at $\mathbf{r}$ due to magnetic monopoles around it. We could write the magnetic force due to a magnetic monopole m_2 at $\mathbf{r}_2$ on a monopole m_1 at $\mathbf{r}_1$ in exact analogy with (3.1) as

$$\mathbf{F}_{12} = \frac{\mu m_1 m_2}{r_{12}^3} \mathbf{r}_{12}. \tag{3.2}$$

The electric field due to a charge q placed at the origin in vacuum in the space around it is derivable from the electrostatic potential

$$\varphi_e(\mathbf{r}) = \frac{q}{r},$$

where $\mathbf{r}$ is the position of the observation point with respect to the origin. This allows us to write the electric field at $\mathbf{r}$ as $\mathbf{E}(\mathbf{r}) = -\nabla \varphi_e(\mathbf{r})$. In a similar manner, we could define the magnetostatic potential in vacuum as $\varphi_m(\mathbf{r}) = m/r$, with $\mathbf{B}(\mathbf{r}) = -\nabla \varphi_m(\mathbf{r})$. This is possible because electromagnetic force is a conservative force.

Consider an electric field $\mathbf{E}$ or magnetic induction $\mathbf{B}$ in vacuum and a material medium placed in it. The external fields polarise the medium due to the presence of microscopic electric charges and magnetic dipoles. In the medium, it is convenient to define the electric displacement $\mathbf{D} = \epsilon \mathbf{E}$, and magnetic field $\mathbf{H} = \mathbf{B}/\mu$. The field $\mathbf{D}$ was introduced by Maxwell and he called it the *displacement current*. Both these relations are point relations, which means they are valid for every point $\mathbf{r}$ and every moment of time t.

It is found that the total electric charge is conserved and the total magnetic charge is zero. Also, the movement of electric charge produces electric current. When charge moves in a medium, we define the current density $\mathbf{j}$ as the charge flowing per unit area normal to the direction of charge flow, per unit time. It is a vector and is a function of $\mathbf{r}$ and t. Current i is defined as the charge flowing per unit time, say between points A and B, irrespective of the path it follows. Current i is a scalar and depends on the two points A and B in question.

Thus consider a battery. In one situation we connect its terminals to the ends of a thin wire. The charge flowing from one end to another per unit time is the current flowing through the wire. In another situation, we connect the terminals to some two points A and B of a material of arbitrary shape. Again some charge will flow from A to B through the material, but now the elementary charges could follow widely different paths through the material between A and B. Here it is convenient to talk of current density, because it tells us about the distribution of charge flow in the material. Of course, the total charge flowing between A and B per unit time is again the current.

We further define the charge density $\rho(\mathbf{r}, t)$ as the amount of charge per unit volume at $\mathbf{r}$ at time t. Thus $\rho d^3 r$ would be the charge contained in an elementary volume $d^3 r$ around $\mathbf{r}$. It may be seen that if $\mathbf{v}$ is the velocity of charge flow at a point $\mathbf{r}$ where the charge density is $\rho(\mathbf{r})$, then the current density there is $\mathbf{J}(\mathbf{r}) = \rho(\mathbf{r})\mathbf{v}$.

An electric charge experiences a force due to electric and magnetic fields. The electromagnetic force on a charge q moving with a velocity $\mathbf{v}$ in the presence of fields $\mathbf{E}$ and $\mathbf{B}$, known as *Lorentz force*, is

$$\mathbf{F} = q(\mathbf{E} + \mathbf{v} \times \mathbf{B}/c). \tag{3.3}$$

If there is a current passing through an isotropic medium in the presence of electric field only, it is found that $\mathbf{J}$ is proportional to the applied electric field $\mathbf{E}$ for small fields. If we write this relationship as $\mathbf{J} = \sigma \mathbf{E}$, then σ is called the (electrical) *conductivity* of the medium. In an anisotropic medium $\mathbf{J}$ is no longer parallel to $\mathbf{E}$, and σ becomes a tensor.

Electric currents give rise to magnetic field. Ampere's experiments led to his law which states that a static current i produces a magnetic field $\mathbf{B}$ in the surrounding region such that the line integral $\oint_C \mathbf{B} \cdot \mathbf{dl}$ around a closed loop enclosing the current equals $4\pi i$. In general, if we take any closed loop C in a region of static magnetic field which is produced by some current distribution, then Ampere's law can be stated as

$$\oint \mathbf{B} \cdot \mathbf{dl} = 4\pi i_C, \tag{3.4}$$

where i_C is the current passing through the loop C. Here $\mathbf{dl}$ is an element of length along the loop and $\mathbf{B}$ the magnetic field at that point. The line integral is taken around the current in the sense specified by the right hand screw rule.

About units and dimensions in electromagnetism

It has become a fashion worldwide to use SI units in all areas of physics. In this book, however, we have used CGS-Gaussian units because, for one, we believe they are simpler. We would much prefer to write the potential seen by an electron in a hydrogen-like atom in vacuum as $-Ze^2/r$ than $-Ze^2/(4\pi\epsilon_0 r)$. One takes the permittivity and permeability of free space in these units as 1. Second, dimensions of physical quantities are more important in physics than units. Third, in physics we use all kinds of conventional units such as angstrom, electron-volt, Rydberg, atomic mass unit, light-year, parsec and others, which have little connection with either of the two systems of units. They are used for convenience and depend on the context. Fourth, scientists prefer to write even algebraic equations in reduced units, and use phrases like 'Taking $\hbar = 1$'. Thus they would begin the Schrodinger equation with $-\nabla^2\psi$ rather than with $-\hbar^2\nabla^2\psi/2m$. And finally, the great classic book by J. D. Jackson, which we have closely used all our lives, uses CGS-Gaussian units. As a simple example, suppose you want to calculate something elementary like the electrostatic potential energy between two charges separated by a certain distance. You may begin with any system of units and come out with the same answer in eV.

With all this, the important message to be impressed upon the reader is: *Let physics not suffer because of the debate about the proper choice of units.* For example, it is important to understand that charge-squared upon distance leads to energy, with or without ϵ_0, as the case may be. Similarly, current-squared upon distance leads to force per unit length, with or without μ_0 and other constants like c and 4π. The product of electric field and magnetic field leads to energy flux, that is energy per unit time per unit area or energy density, again with or without factors of c, etc. The essential point is that bilinear products of electromagnetic quantities with each other or their squares are related to mechanical quantities.

For dimensional check, students are generally accustomed to reducing all quantities to $L-M-T$. One need not do that. Leave force as force (F), energy as energy (E), momentum as momentum (p), angular momentum as angular momentum (L), velocity as velocity (v), etc. Remember the connection between them. For example, energy is $F \times r$ (distance), power is E/T, $M \times a$ (acceleration) is F, v^2/r as well as $\omega^2 r$ is a, and these multiplied by M is F, mv is p while mvr is L, etc.

In the case of electromagnetic quantities, we deal with **E** (electric field), **B** (magnetic induction), q (charge), I (current), ρ, σ, λ (volume, surface and linear charge densities, respectively), **J** (current density), **K** (surface current density), **A** (vector potential), ϕ (or V, scalar potential), **D** (electric displacement), **p** (electric dipole moment), **P** (polarization), $\boldsymbol{\mu}$ (or **m**, magnetic dipole moment), **M** (magnetization), **H** (magnetic field), α (polarizability), χ (susceptibility), etc. Remember the relationships among them. Apart from the appropriate factors of ϵ_0, μ_0, and c, q/r is ϕ, q/r^2 is E, q/T is I, J is I/L^2 or q/TL^2, K is I/L, ρ is q/L^3, σ is q/L^2, B is E/c (that is E/velocity, in SI units) and also I/r (Biot–Savart law), and A is $B \times r$. Magnetic moment μ is $I \times$ area, χ is P/D (electric) or M/H (magnetic), and α is p/E. Note that the relation between E and ϕ is dimensionally similar to that between B and A. P is p/r^3 and M is m/r^3.

Finally, as said above, the square of an electromagnetic quantity or a bilinear product of two such quantities is a mechanical quantity. This provides the connection between electromagnetic and mechanical quantities. Thus q^2 is Fr^2 (Coulomb's law), whereas q^2/r is energy. $E \times B$ is energy/area/time (Poynting vector). If we multiply and divide this by length, we see that it is also equal to energy density $\times$ velocity. Similarly, $q\phi$ and μH are energy, whereas I^2 is force and E^2 is force/area or pressure.

However, after all this, there is a word of caution. There is a certain justification for using SI system as against the CGS-Gaussian system, i.e., the concept of polarization, whether of a medium or vacuum. (SI system was introduced before the concept of vacuum polarization.) This has been the rationale behind the SI system. One must realise that two quantities may have the same physical dimensions and yet may be quite different physically. Even if **E, B, P, D, M, H** have the same physical dimensions in CGS-Gaussian system, all of them are different physical quantities. Whereas the **E**-field is defined through the force acting on a unit change at a point, the field produced by a charge at a point in space is the **D**-field because the intervening medium plays a role here even if it is vacuum. Similarly, the **B**-field is defined through the force acting on a current; the field produced at a point due to a current is the **H**-field. This distinction is emphasised and brought out in the SI system. After all, good old Jackson also seems to have changed to it. Even so, the fact remains that the CGS-Gaussian system is easier for numerical calculations and most scientists are comfortable with it. In practice, the need for using units arises only in the final step when we wish to calculate numerical quantities. So long as we do algebra, it hardly matters which system of units we use.

3.1.2 Electrostatics and Gauss law

Consider an electric charge q in vacuum in volume V bounded by the closed surface S. The field produced by it at a surface element dS is $\mathbf{E} = (q/r^2)\hat{\mathbf{r}}$, where r is the distance between the charge and the surface element and $\hat{\mathbf{r}}$ a unit vector along it; see Fig. 3.1. The total electric flux leaving the volume V is given by

$$\varphi_e = \oint_S \mathbf{E} \cdot \mathbf{n} dS = \oint_S \frac{q}{r^2}\hat{\mathbf{r}} \cdot \mathbf{n} dS.$$

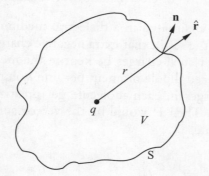

Fig. 3.1 *An arbitrary volume V and its boundary surface S for Gauss's law*

But $\hat{\mathbf{r}} \cdot \mathbf{n} dS/r^2$ is the solid angle $d\Omega$ subtended by the surface element dS at the charge. Hence

$$\varphi_e = \oint_S q d\Omega = 4\pi q. \tag{3.5}$$

Instead of a single charge, if we have a charge distribution $\rho(\mathbf{r}')$ in the space, each element of charge $\rho d^3 r'$ at r' will contribute to the flux leaving the volume V. Hence the total flux due to the entire charge distribution becomes

$$\varphi_e = 4\pi \int_V \rho d^3 r' = 4\pi Q_V, \tag{3.6}$$

where Q_V is the charge contained in the volume V. Thus we have finally

$$\varphi_e = \oint_S \mathbf{E} \cdot \mathbf{n} dS = 4\pi \int_V \rho(r') d^3 r' = 4\pi Q_V. \tag{3.7}$$

This is known as *Gauss law of electrostatics*. Stated in words, it says that the total electric flux leaving any closed surface S enclosing an arbitrary volume V in vacuum is equal to 4π times the total charge contained in the volume.

Using the Gauss divergence theorem, we may write

$$\oint_S \mathbf{E} \cdot \mathbf{n} dS = \int_V \nabla \cdot \mathbf{E} d^3 r. \tag{3.8}$$

Since (3.8) and (3.9) are valid for an arbitrary volume V bounded by a closed surface S, it follows that

$$\nabla \cdot \mathbf{E}(\mathbf{r}) = 4\pi \rho(\mathbf{r}). \tag{3.9}$$

This is Maxwell's first equation of electrostatics for vacuum.

Now consider a charge q situated in a dielectric medium of dielectric constant ϵ, which gets polarized with the result that extra negative charge builds up in the volume V as shown in Fig. 3.2. Let such charges be kept in the medium with charge density $\rho(\mathbf{r})$ and let $\mathbf{P}$ be the induced dipole moment per unit volume, called *polarization*. The positive and negative charges in each molecule get polarized, and they produce an electric dipole moment $\mathbf{p}$. Then $\mathbf{P}$ would be the vector sum of moments $\mathbf{p}_i$ of each molecule over a unit volume,

$$\mathbf{P} = \sum_i N\mathbf{p}_i,$$

where N is the number of molecules per unit volume.

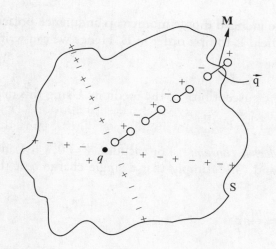

Fig. 3.2 *Polarization due to a charge in a dielectric medium*

The charge q or the charges making up $\rho(\mathbf{r})$ are called *free charges* whereas the charges in the induced dipole moments, going right upto the boundary surface S, are called *bound charges*. A free charge placed in a medium induces an equal and opposite charge around it. The total opposite induced charge in the volume V will be given by

$$\oint_S \mathbf{n} \cdot \sum_i N\mathbf{p}_i dS = \oint_S \mathbf{n} \cdot \mathbf{P} dS = \int_V \boldsymbol{\nabla} \cdot \mathbf{P} d^3 r,$$

where we have used Gauss divergence theorem for $\mathbf{P}$. Therefore, the induced charge density will be $-\boldsymbol{\nabla} \cdot \mathbf{P}$. Since there is already the free charge density $\rho(\mathbf{r})$ in the medium, the total charge density will be $\rho - \boldsymbol{\nabla} \cdot \mathbf{P}$. Then (3.10) becomes

$$\boldsymbol{\nabla} \cdot \mathbf{E} = 4\pi(\rho - \boldsymbol{\nabla} \cdot \mathbf{P}).$$

This can be written as

$$\boldsymbol{\nabla} \cdot (\mathbf{E} + 4\pi\mathbf{P}) = 4\pi\rho.$$

If we define

$$\mathbf{D} = \mathbf{E} + 4\pi\mathbf{P}, \tag{3.10}$$

we get

$$\boldsymbol{\nabla} \cdot \mathbf{D} = 4\pi\rho. \tag{3.11}$$

This is the *first Maxwell's equation for a medium.*

It is found that the induced dipole moment **p** and hence polarization **P** is proportional to the electric field **E** to first order in **E**. Hence we can write

$$\mathbf{P} = \chi_e \mathbf{E},$$

where χ_e is the electric susceptibity of the medium. Using this in (3.11), we see that

$$\mathbf{D} = \mathbf{E}(1 + 4\pi\chi_e) \equiv \epsilon\mathbf{E}, \tag{3.12}$$

which defines the *dielectric constant* ϵ. For vacuum, $\chi = 0, \epsilon = 1$ and $\mathbf{D} = \mathbf{E}$.

Equations (3.12) and (3.13) imply that a single charge q at the origin produces a field at **r** equal to

$$\mathbf{D} = \frac{q}{r^2}\hat{\mathbf{r}}, \quad \mathbf{E} = \frac{q}{\epsilon r^2}\hat{\mathbf{r}} \tag{3.13}$$

and, for a charge distribution $\rho(\mathbf{r})$,

$$\nabla \cdot \mathbf{E} = 4\pi\rho/\epsilon$$

The Coulomb potential can be written in this case as $\varphi(\mathbf{r}) = -q/\epsilon r$ for a single charge q, and as

$$\varphi(\mathbf{r}) = \int_V \frac{\rho(\mathbf{r}')d^3r'}{\epsilon|\mathbf{r} - \mathbf{r}'|}$$

for a charge distribution.

3.1.3 Magnetostatics

Here $\mathbf{B}, 1/\mu$ and $\mathbf{H}$ take the place of $\mathbf{E}, \epsilon$ and $\mathbf{D}$, respectively, and **P** is replaced by the induced magnetization **M**, which is the induced magnetic dipole moment per unit volume. Thus we have

$$\mathbf{B} = \mu\mathbf{H} = \mathbf{H} + 4\pi\mathbf{M}.$$

Then, corresponding to (3.12) we would have an equation for **B** which would involve magnetic monopole density in space. But since magnetic monopoles have not been found to exist, we have

$$\nabla \cdot \mathbf{B} = 0. \tag{3.14}$$

This is the *Maxwell's second equation*.

3.1.4   Faraday's law of induction

Consider an open surface S bound by a closed loop L. Then any changes in the electric field in loop L and magnetic field passing through S should conform to the law of conservation of energy. Consider an element of area $\mathbf{dS}$ on the surface S and let the magnetic field at $\mathbf{dS}$ be $\mathbf{B}$, as shown in Fig. 3.3. If $\mathbf{n}$ is a unit normal at $\mathbf{dS}$, we have $\mathbf{dS} = dS\mathbf{n}$, and the magnetic flux passing through the surface S is $\varphi_m = \oint_S \mathbf{B} \cdot \mathbf{dS} = \oint_S \mathbf{B} \cdot \mathbf{n} dS$.

Faraday found that a change in the magnetic flux φ_m passing through any closed conducting loop L gives rise to an electromotive force $\mathcal{E}$ around the loop. His experiments further showed that $\mathcal{E}$ was proportional to the rate of change of φ_m. Thus we could write Faraday's law of induction as

$$\mathcal{E} = -\frac{1}{c}\frac{d\varphi_m}{dt}. \tag{3.15}$$

Here the constant of proportionality has been taken as $-1/c$. The negative sign can be seen as a consequence of Lenz's law; experiments show that any change in magnetic flux through the loop L produces a current in a conducting wire kept along the loop L which in turn produces a magnetic field which opposes the original change in flux. The factor $1/c$ is consistent with the CGS-Gaussian system of units and agrees numerically with experiments.

When the loop L has a conducting wire along it, the electromotive force $\mathcal{E}$ gives rise to a current. If $\mathbf{F}$ is the force acting on a charge q at a point on the loop, then $\mathcal{E}$ is equal to the work done in taking a unit charge around the loop once in the anticlockwise sense. Therefore $\mathcal{E}$ is given in terms of the electromagnetic fields in the region as

$$\mathcal{E} = \oint_L (\mathbf{E} + \mathbf{v} \times \mathbf{B}/c) \cdot \mathbf{dr}. \tag{3.16}$$

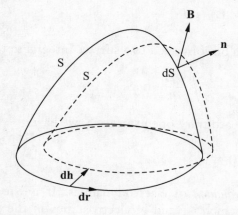

Fig. 3.3   *An open surface S bound by a loop L, being displaced by an amount* $\mathbf{dh}$

The magnetic flux through the surface S can change because the electromagnetic fields may change with time or because the surface S may move and change orientation. Then the total time rate of change of the magnetic flux is given by

$$\frac{d\varphi_m}{dt} = \frac{\partial \varphi_m}{\partial t} + \frac{\delta \varphi_m}{\delta t},$$

where $\partial/\partial t$ indicates the change due to a change in fields and $\delta/\delta t$ the change due to a change in the position of S. Then Faraday's law can be written as

$$\frac{1}{c} \int_S \frac{\partial \mathbf{B}}{\partial t} \cdot \mathbf{n} dS + \frac{1}{c} \frac{\delta}{\delta t} \int_S \mathbf{B} \cdot \mathbf{n} dS = - \oint_L (\mathbf{E} + \mathbf{v} \times \mathbf{B}/c) \cdot \mathbf{dr}. \tag{3.17}$$

Let now the loop L and surface S move, and let **dh** be the displacement of an element **dr** of the loop. Then $\mathbf{n} dS = \mathbf{dh} \times \mathbf{dr}$. The change in the magnetic flux due to the displacement of surface S is

$$\delta \int_S \mathbf{B} \cdot \mathbf{n} dS = \delta \oint_L \mathbf{B} \cdot (\mathbf{dh} \times \mathbf{dr}) = \delta \oint_L (\mathbf{B} \times \mathbf{dh}) \cdot \mathbf{dr},$$

so that

$$\frac{1}{c} \frac{\delta}{\delta t} \int_S \mathbf{B} \cdot \mathbf{n} dS = \oint_L (\mathbf{B} \times \mathbf{v}) \cdot \mathbf{dr} = - \oint_L (\mathbf{v} \times \mathbf{B}) \cdot \mathbf{dr}.$$

With this, (3.15) becomes

$$\oint_L \mathbf{E} \cdot \mathbf{dr} = -\frac{1}{c} \int_S \frac{\partial \mathbf{B}}{\partial t} \cdot \mathbf{n} dS.$$

Using Stokes theorem of vector calculus,

$$\oint_L \mathbf{E} \cdot \mathbf{dr} = \int_S (\boldsymbol{\nabla} \times \mathbf{E}) \cdot \mathbf{n} dS,$$

the line integral of **E** can be converted to a surface integral in the form

$$\int_S (\boldsymbol{\nabla} \times \mathbf{E}) \cdot \mathbf{n} dS = -\frac{1}{c} \int_S \frac{\partial \mathbf{B}}{\partial t} \cdot \mathbf{n} dS. \tag{3.18}$$

Since this is true for any arbitrary open surface S, we have

$$\boldsymbol{\nabla} \times \mathbf{E} + \frac{1}{c} \frac{\partial \mathbf{B}}{\partial t} = \mathbf{0}. \tag{3.19}$$

This is *Maxwell's third equation* of electrodynamics. It is the modified form of the equation $\boldsymbol{\nabla} \times \mathbf{E} = \mathbf{0}$, which is valid in electrostatics in the case of time-dependent fields.

3.1.5 Ampere's law

Once again, we consider an open surface S bounded by a loop L in a region where charges are flowing, giving rise to electromagnetic fields. Let **J** be the current density; see Fig. 3.4. Then the total current passing through loop L, which is also the current passing through S, is given by

$$I_{\mathrm{L}} = \int_{\mathrm{S}} \mathbf{J} \cdot \mathbf{n} dS.$$

Ampere performed several experiments and found that when the current density, and hence current I_{L} through loop L are constant, independent of time, the line integral of the tangential component of the magnetic field **B** along the loop L is given by

$$\oint_{\mathrm{L}} \mathbf{B} \cdot \mathbf{dr} = \frac{4\pi}{c} I_{\mathrm{L}}.$$

Using Stokes theorem, the line integral can be written as

$$\oint_{\mathrm{L}} \mathbf{B} \cdot \mathbf{dr} = \int_{\mathrm{S}} (\boldsymbol{\nabla} \times \mathbf{B}) \cdot \mathbf{n} dS.$$

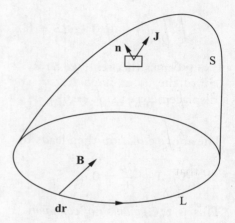

Fig. 3.4 *An open surface S bound by a loop L for Ampere's law.*

Hence we have

$$\int_{\mathrm{S}} (\boldsymbol{\nabla} \times \mathbf{B}) \cdot \mathbf{n} dS = \frac{4\pi}{c} \int_{\mathrm{S}} \mathbf{J} \cdot \mathbf{n} dS.$$

Since this is true for any arbitrary surface S, we have

$$\boldsymbol{\nabla} \times \mathbf{B} = \frac{4\pi}{c} \mathbf{J}. \tag{3.20}$$

This is true only for static currents and fields. Since

$$\boldsymbol{\nabla} \cdot (\boldsymbol{\nabla} \times \mathbf{B}) = 0,$$

we also find that

$$\boldsymbol{\nabla} \cdot \mathbf{J} = 0$$

in static situations. But the situation changes when charges and currents vary in time.

Consider an arbitrary volume V bound by a closed surface S in vacuum as shown in Fig. 3.5. Charges may be flowing in and out of the volume V. But since charge must be conserved in any arbitrary volume V at any time, the rate of increase of charge in volume V must be equal to the net inflow of charge through the surface S. If ρ is the charge density, then we must have

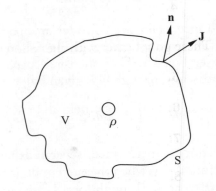

$$\int_V \frac{\partial \rho}{\partial t} d^3 r + \oint_S \mathbf{J} \cdot \mathbf{n} dS = 0. \qquad (3.21)$$

Using Gauss theorem, we have

$$\oint_S \mathbf{J} \cdot \mathbf{n} dS = \int_V \mathbf{\nabla} \cdot \mathbf{J} d^3 r.$$

The above equation then leads to

Fig. 3.5   *An arbitrary volume V bound by S for charge conservation*

$$\mathbf{\nabla} \cdot \mathbf{J} + \frac{\partial \rho}{\partial t} = 0. \qquad (3.22)$$

This is the *equation of continuity*. It represents conservation of charge in the case of time-dependent sources.

Using the first equation of Maxwell, (3.10), we can write the above equation as

$$\mathbf{\nabla} \cdot \mathbf{J} + \frac{1}{4\pi} \frac{\partial}{\partial t} \mathbf{\nabla} \cdot \mathbf{E} = 0 \Rightarrow \mathbf{\nabla} \cdot \left(\mathbf{J} + \frac{1}{4\pi} \frac{\partial \mathbf{E}}{\partial t} \right) = 0.$$

This shows that we must replace $\mathbf{J}$ by $\mathbf{J} + (1/4\pi)\partial \mathbf{E}/\partial t$ in the case of time-dependent sources in vacuum. With this identification, (3.20) becomes

$$\mathbf{\nabla} \times \mathbf{B} = \frac{4\pi}{c} \mathbf{J} + \frac{1}{c} \frac{\partial \mathbf{E}}{\partial t}. \qquad (3.23)$$

This is the *fourth equation of Maxwell*. In analogy with the current density $\mathbf{J}$, Maxwell called the new term $\partial \mathbf{E}/4\pi \partial t$, the *displacement current* (density).

We finally summarise the four Maxwell's equations of electrodynamics in vacuum as:

1.   Gauss law: $\mathbf{\nabla} \cdot \mathbf{E} = 4\pi \rho$ \hfill (3.24a)

2.   Absence of magnetic monopole: $\mathbf{\nabla} \cdot \mathbf{B} = 0$ \hfill (3.24b)

3.   Faraday's law: $\mathbf{\nabla} \times \mathbf{E} = -\frac{1}{c} \frac{\partial \mathbf{B}}{\partial t}$ \hfill (3.24c)

4.   Ampere-Maxwell's law: $\nabla \times \mathbf{B} = \dfrac{4\pi}{c}\left(\mathbf{J} + \dfrac{1}{4\pi}\dfrac{\partial \mathbf{E}}{\partial t}\right)$ (3.24d)

These go together with the following four laws:

5.   Electric field and displacement: $\mathbf{D} = \mathbf{E} + 4\pi\mathbf{P} = \epsilon\mathbf{E}$ (3.25a)

6.   Magnetic field and induction: $\mathbf{B} = \mathbf{H} + 4\pi\mathbf{M} = \mu\mathbf{H}$ (3.25b)

7.   Lorentz's force: $\mathbf{F} = q(\mathbf{E} + \mathbf{v} + \mathbf{B}/c)$ (3.25c)

8.   Equation of continuity: $\nabla \cdot \mathbf{J} + \dfrac{\partial \rho}{\partial t} = 0.$ (3.25d)

In a ponderable medium, Maxwell equations take the form

$$\nabla \cdot \mathbf{D} = 4\pi\rho, \tag{3.26a}$$

$$\nabla \cdot \mathbf{B} = 0 \tag{3.26b}$$

$$\nabla \times \mathbf{E} = -\frac{1}{c}\frac{\partial \mathbf{B}}{\partial t} \tag{3.26c}$$

$$\nabla \times \mathbf{H} = \frac{4\pi}{c}(\mathbf{J} + \frac{1}{4\pi}\frac{\partial \mathbf{D}}{\partial t}) \tag{3.26d}$$

3.2 Electromagnetic Waves

Maxwell's formulation of electrodynamics immediately lead to electromagnetic waves, propagating in vacuum as well as in dielectric media. In this section we will briefly discuss how Maxwell's formalism leads to a wave equation for the **E** and **B** fields followed by the nature of electromagnetic waves.

3.2.1 Uniform medium with no sources

In this case, we have $\rho = 0, \mathbf{J} = \mathbf{0}$ so that Maxwell's equations take the form

$$\nabla \cdot \mathbf{D} = 0, \qquad\qquad \nabla \cdot \mathbf{B} = 0,$$

$$\nabla \times \mathbf{E} = -\frac{1}{c}\frac{\partial \mathbf{B}}{\partial t}, \qquad \nabla \times \mathbf{H} = \frac{1}{c}\frac{\partial \mathbf{D}}{\partial t}. \tag{3.27}$$

Taking the curl of the third of the above equations, we get

$$\nabla \times (\nabla \times \mathbf{E}) = -\frac{1}{c}\frac{\partial}{\partial t}(\nabla \times \mathbf{B}) = -\frac{\mu}{c^2}\frac{\partial^2 \mathbf{D}}{\partial t^2}, \qquad (3.28a)$$

or

$$\nabla \times (\nabla \times \mathbf{D}) = -\frac{\mu\epsilon}{c^2}\frac{\partial^2 \mathbf{D}}{\partial t^2}.$$

Since $\nabla \times (\nabla \times \mathbf{D}) = \nabla(\nabla \cdot \mathbf{D}) - \nabla^2 \mathbf{D}$, this gives

$$\nabla^2 \mathbf{D} = \frac{\epsilon\mu}{c^2}\frac{\partial^2 \mathbf{D}}{\partial t^2}. \qquad (3.28b)$$

Doing a similar operation on the fourth equation of (3.24b), we find that $\mathbf{H}$ also satisfies an identical equation

$$\nabla^2 \mathbf{H} = \frac{\epsilon\mu}{c^2}\frac{\partial^2 \mathbf{H}}{\partial t^2}. \qquad (3.28c)$$

Equations (3.28) are homogeneous wave equations and have solutions of the type $\mathbf{F}(\mathbf{r} \cdot \mathbf{n} \mp vt)$ and $\mathbf{G}(r \mp vt)/r$, where $v = c/(\epsilon\mu)^{1/2}$ is the speed of the wave and $\mathbf{n}$ a unit vector in the direction of propagation. Let $\mathbf{n} = (p, q, r)$ so that p, q, r are the direction cosines of $\mathbf{n}$. Then it can indeed be verified that these functions $\mathbf{F}$ and $\mathbf{G}$ satisfy (3.28).

The function $\mathbf{F}(\mathbf{r} \cdot \mathbf{n} \mp vt)$ represents a plane wave moving with velocity v in the direction $\pm\mathbf{n}$, while the function $\mathbf{G}(r \mp vt)/r$ represents a spherical wave going out or in, with a velocity v and an amplitude which varies as $1/r$.

Thus the solutions of Maxwell's equations represent electromagnetic waves travelling with velocity v. In vacuum, $\epsilon = \mu = 1$ and $v = c = 3 \times 10^{10}$ cm/s, which is the speed of light. This leads to a strong belief that light is an electromagnetic wave and $(\epsilon\mu)^{1/2}$ represents the refractive index of the medium. Soon it was found that the spectrum of electromagnetic radiation extends all the way from γ rays, through X-rays, UV radiation, visible light, to IR, microwave and radio waves.

3.2.2 Nature of electromagnetic waves

Let us consider a plane electromagnetic wave travelling in a source-free medium in the positive x direction with electric and magnetic fields given by

$$\mathbf{E} = \mathbf{f}(x - vt), \qquad \mathbf{H} = \mathbf{g}(x - vt).$$

Since $\nabla \cdot \mathbf{E} = 0$ and $\nabla \cdot \mathbf{H} = 0$, we find that

$$\partial E_x/\partial x = 0, \qquad \partial H_x/\partial x = 0.$$

Thus E_x and H_x must be constant. Since we are interested in wave propagation, we are not concerned with constant quantities. We could as well put $E_x = H_x = 0$ and look for the other components. Thus, electromagnetic waves are transverse waves.

Let the transverse components of **E** be

$$E_y = f_y(x - vt), E_z = f_z(x - vt).$$

Then (3.24c) with (3.25b) gives

$$\frac{\partial H_y}{\partial t} = \frac{c}{\mu} f'_z(x - vt), \ \frac{\partial H_z}{\partial t} = -\frac{c}{\mu} f'_y(x - vt),$$

where the primes denote differentiation with respect to the argument. Note that for any function $\psi(u)$ where $u = x - vt$, we have

$$\partial \psi(u)/\partial x = \partial \psi(u)/\partial u = -\partial \psi(u)/v \partial t.$$

Using this the above equations become

$$H_y = -\frac{c}{v\mu} f_z(x - vt), \ H_z = \frac{c}{v\mu} f_y(x - vt). \tag{3.29}$$

If **i**, **j**, **k** are Cartesian unit vectors, then in the present case we have **n** = **i**. Also, from the above

$$\mathbf{E} \times \mathbf{H} = \mathbf{i}(E_y H_z - E_z H_y) = \mathbf{i}\frac{c}{v\mu}(f_y^2 + f_z^2).$$

This shows that $\mathbf{E} \times \mathbf{H}$ is in the direction of propagation. Also $\mathbf{E} \cdot \mathbf{H}$ can be seen to vanish, showing that **E** and **H** are normal to each other. Finally, it can be seen that

$$\mathbf{n} \times \mathbf{E} = \frac{v\mu}{c}\mathbf{H},$$

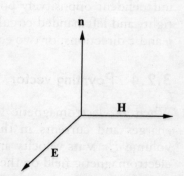

Fig. 3.6 *A right-handed Cartesian system for* **E**, **H** *and* **n**

showing that **E**, **H**, **n** form a right-handed system as shown in Fig. (3.6). To see the relation between **E** and **H** in a medium, we have

$$E^2 = f_y^2 + f_z^2 = \frac{v^2\mu^2}{c^2}(H_y^2 + H_z^2) = \frac{\mu}{\epsilon}H^2,$$

or $\ \ \epsilon E^2 = \mu H^2.$

3.2.3 Polarization

In terms of frequency ω, let the components of the electric field be

$$E_y = Ae^{i\omega(x/v-t)}, \quad E_z = Be^{i[\omega(x/v-t)+\delta]},$$

where A and B are arbitrary. In general the phase difference δ between the two components is also arbitrary. In other words, we have an elliptically polarized wave in the general case, as shown in Fig. 3.7. If $0 < \delta < \pi$, we have right-handed polarization, while if $\pi < \delta < 2\pi$, we have left-handed polarization. If $\delta = 0$ or π, we have *linearly polarized* or *plane-polarized* wave. If $A = B$, we have a circularly polarized wave.

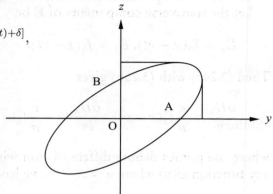

Fig. 3.7 *Elliptical polarization of electromagnetic radiation*

Natural radiation is unpolarized because it consists of many independent variations with different phase differences. On averaging, they can be represented by two independent oppositvely polarized components of equal intensity, that is, two equal right- and left-handed circularly polarized beams, two equal beams plane polarized in y and z directions, or two equal right- and left-handed elliptically polarized beams.

3.2.4 Poynting vector

When an electromagnetic wave travels through a medium, it exerts a force on the charges and currents in the medium. Since ρd^3r is the charge in the element of volume d^3r, $\mathbf{v}$ its velocity and $\rho\mathbf{v} = \mathbf{J}(r,t)$ the current density, the work done by the electromagnetic field on the charge ρd^3r for unit time is given by

$$\frac{dW}{dt} = \rho\mathbf{F}\cdot\mathbf{v}d^3r = \left[\mathbf{E} + \frac{\mathbf{v}\times\mathbf{B}}{c}\right]\cdot\mathbf{v}d^3r$$

$$= \mathbf{J}\cdot\mathbf{E}d^3r.$$

The work done on an arbitrary volume V bounded by the closed surface S would then be

$$\frac{dW}{dt} = \int_V \mathbf{J}\cdot\mathbf{E}d^3r.$$

It is convenient to eliminate **J** in favour of **H** using the Maxwell equation connecting them. This gives

$$\frac{dW}{dt} = \frac{c}{4\pi} \int_V \mathbf{E} \cdot \left[\boldsymbol{\nabla} \times \mathbf{H} - \frac{\epsilon}{c} \frac{\partial \mathbf{E}}{\partial t} \right] d^3r.$$

Using the vector identity

$$\boldsymbol{\nabla} \cdot (\mathbf{E} \times \mathbf{H}) = \mathbf{H} \cdot (\boldsymbol{\nabla} \times \mathbf{E}) - \mathbf{E} \cdot (\boldsymbol{\nabla} \times \mathbf{H}),$$

and the Maxwell equation (3.24c) for $\boldsymbol{\nabla} \times \mathbf{E}$, this can be written as

$$\frac{dW}{dt} = -\frac{c}{4\pi} \int_V \left[\boldsymbol{\nabla} \cdot (\mathbf{E} \times \mathbf{H}) - \frac{\partial}{\partial t} \frac{\epsilon E^2 + \mu H^2}{8\pi} \right] d^3r.$$

Finally, using Gauss's theorem for the first term above, we get

$$\frac{dW}{dt} = -\frac{c}{4\pi} \oint_S (\mathbf{E} \times \mathbf{H}) \cdot \mathbf{n} dS - \frac{\partial}{\partial t} \int_V \frac{\epsilon E^2 + \mu H^2}{8\pi} d^3r.$$

Note here that **n** is the unit outward normal to the surface S. Since **n** is outward to surface S, with negative sign, the first term denotes the gain of energy per unit time through the surface S. The second term in the above equation is the loss of energy per unit time in the volume V.

We therefore define the *Poynting vector* by

$$\mathbf{S} = \frac{c}{4\pi} (\mathbf{E} \times \mathbf{H}). \tag{3.30a}$$

It represents energy flow per unit time per unit cross-sectional area. It points in the direction of propagation of the electromagnetic wave. We also define

$$U = (\epsilon E^2 + \mu H^2)/8\pi \tag{3.30b}$$

as the energy density. Since $\epsilon E^2 = \mu H^2$, it could also be written as $U = \epsilon E^2/4\pi = \mu H^2/4\pi$.

3.2.5 Momentum associated with electromagnetic wave

We shall proceed to obtain the momentum and pressure associated with electromagnetic radiation through a guided exercise.

▶ Guided Exercise 3.1

Obtain the momentum and pressure associated with electromagnetic radiation.

Hints

(a) Let $\mathbf{M}d^3r$ be the momentum associated with a volume element d^3r through which electromagnetic waves are passing. Then write

$$\frac{d\mathbf{M}}{dt} = \int_V \mathbf{F}d^3r = \int_V \rho(\mathbf{E} + \mathbf{v} \times \mathbf{B}/c)d^3r.$$

(b) Use $\rho\mathbf{v} = \mathbf{J}, \mathbf{B} = \mu\mathbf{H}$ and the appropriate Maxwell's equations to show that

$$\frac{d\mathbf{M}}{dt} = \frac{1}{4\pi} \int_V \left[\epsilon(\nabla \cdot \mathbf{E})\mathbf{E} + \mu(\mathbf{H} \cdot \nabla)\mathbf{H} - \mu(\nabla\mathbf{H}) \cdot \mathbf{H} - \frac{\epsilon\mu}{c}\frac{\partial}{\partial t}(\mathbf{E} \times \mathbf{H}) \right.$$

$$\left. + \frac{\epsilon\mu}{c}\mathbf{E} \times \frac{\partial\mathbf{H}}{\partial t} \right] d^3r.$$

(c) Using Maxwell's equation (3.24b) and the vector identity

$$(\nabla \times \mathbf{E}) \times \mathbf{E} = (\mathbf{E} \cdot \nabla)\mathbf{E} - \frac{1}{2}\nabla E^2,$$

put the rate of momentum transfer in the form

$$\frac{d\mathbf{M}}{dt} = \frac{1}{4\pi} \int_V [\epsilon(\nabla \cdot \mathbf{E} + \mathbf{E} \cdot \nabla)\mathbf{E} + \mu(\nabla \cdot \mathbf{H} + \mathbf{H} \cdot \nabla)\mathbf{H}$$

$$- \frac{1}{2}\nabla(\mu H^2 + \epsilon E^2) - \frac{\epsilon\mu}{c}\frac{\partial}{\partial t}(\mathbf{E} \times \mathbf{H})]d^3r.$$

(d) Using Gauss theorem of vector calculus and noting that $\oint_S \mathbf{n}dS = 0$, write this finally in the form

$$\frac{d\mathbf{M}}{dt} = \oint_S \frac{\epsilon\mathbf{E}(\mathbf{E} \cdot \mathbf{n} + \mu\mathbf{H}(\mathbf{H} \cdot \mathbf{n})}{4\pi}dS - \frac{\partial}{\partial t}\int_V \frac{\epsilon\mu}{4\pi c}(\mathbf{E} \times \mathbf{H})d^3r. \qquad (3.31)$$

Note that the first term contains the tensor

$$T = \frac{\epsilon\mathbf{E}\mathbf{E} + \mu\mathbf{H}\mathbf{H}}{4\pi}. \qquad (3.32a)$$

This second rank tensor is called the *momentum flux tensor*. The expression in the second term is the *momentum density vector* and is related to the Poynting vector. We denote it by

$$\rho_M = \frac{\epsilon\mu}{4\pi c}\mathbf{E} \times \mathbf{H}. \qquad (3.32b)$$

(e) In analogy with mechanics, define *radiation pressure* p_r as the momentum transfer per unit area normal to it per unit time so that p_r is given by

$$p_r = \rho_M v = \frac{1}{c}\mathbf{S}, \qquad (3.32c)$$

where we have used $v = c/(\epsilon\mu)^{1/2}$. ◀

3.3 Electromagnetic Radiation by a Molecule

A molecule contains positive and negative charges which may be in motion, so computation of electric and magnetic forces exerted by them will be quite complicated. In the theory of gravitation, when we have to deal with forces due to some distribution of mass, it is convenient to use potentials. In the same way in the electromagnetic theory too, it would be convenient to talk of potentials instead of electric and magnetic fields.

3.3.1 Electromagnetic potentials

One of Maxwell's equations, (3.24b), suggests that we can write **B** as the curl of some vector field, say **A**. Thus

$$\mathbf{B} = \boldsymbol{\nabla} \times \mathbf{A}; \tag{3.33}$$

this quantity **A** is called the *vector potential*. Like any other potential, its choice is arbitrary. In the present case,

$$\mathbf{A}' = \mathbf{A} + \boldsymbol{\nabla}\psi,$$

where ψ is an arbitrary scalar field, also leads to the same **B**. Using the above equation in (3.24c), it follows that

$$\boldsymbol{\nabla} \times \mathbf{E} = -\frac{\mu}{c}\boldsymbol{\nabla} \times \frac{\partial \mathbf{A}}{\partial t} \Rightarrow \boldsymbol{\nabla} \times \left(\mathbf{E} + \frac{\mu}{c}\frac{\partial \mathbf{A}}{\partial t}\right) = 0. \tag{3.34}$$

This implies that the bracketed quantity in the above equation can be written as the curl of an arbitrary scalar, say φ. This gives

$$\mathbf{E} = -\boldsymbol{\nabla}\varphi - \frac{1}{c}\frac{\partial \mathbf{A}}{\partial t}. \tag{3.35}$$

This quantity φ is called the *scalar potential*. It is arbitrary to the extent that $\varphi' = \varphi + \chi$ and χ should be such that this, together with (3.35), should yield the same **E**. **A** and φ together are called *electromagnetic potentials*. Equations (3.35) and (3.33) express the **E** and **B** fields in terms of **A** and φ. In several situations it is convenient to work with electromagnetic potentials than electromagnetic fields.

With suitable manipulations of the above equations, we can eliminate **A** and φ alternately and obtain an equation containing only one of the potentials. To this end, we notice that (3.35) and (3.24a) give

$$\boldsymbol{\nabla} \cdot \mathbf{E} = -\nabla^2 \varphi - \frac{\mu}{c}\frac{\partial}{\partial t}(\boldsymbol{\nabla} \cdot \mathbf{A}) = \frac{4\pi\rho}{\epsilon}. \tag{3.36}$$

Also, (3.24d) with (3.33) gives

$$\nabla \times \mathbf{H} - \frac{\epsilon}{c}\frac{\partial \mathbf{E}}{\partial t} = \nabla \times (\nabla \times \mathbf{A}) + \frac{\epsilon}{c}\nabla\frac{\partial \varphi}{\partial t} + \frac{\epsilon\mu}{c^2}\frac{\partial^2 \mathbf{A}}{\partial t^2} = \frac{4\pi}{c}\mathbf{J}. \tag{3.37}$$

Using $\nabla \times (\nabla \times \mathbf{A}) = \nabla(\nabla \cdot \mathbf{A}) - \nabla^2 \mathbf{A}$, these equations for $\mathbf{A}$ and φ can be written in the form

$$\nabla^2\varphi + \frac{\mu}{c}\frac{\partial}{\partial t}(\nabla \cdot \mathbf{A}) = -\frac{4\pi\rho}{\epsilon},$$

$$\nabla(\nabla \cdot \mathbf{A}) - \nabla^2 \mathbf{A} + \frac{\epsilon}{c}\nabla\frac{\partial \varphi}{\partial t} + \frac{\epsilon\mu}{c^2}\frac{\partial^2 \mathbf{A}}{\partial t^2} = \frac{4\pi\mathbf{J}}{c}. \tag{3.38}$$

We have seen that $\mathbf{A}$ is arbitrary upto the addition of the gradient of a scalar field and the arbitrariness in φ should cancel with that in $\mathbf{A}$ such as to yield the same $\mathbf{E}$ field. Let ψ be an arbitrary function and let $\chi = -\mu\partial\psi/c\partial t$. Then it can be seen that the new potentials

$$\varphi' = \varphi + \frac{\mu}{c}\frac{\partial \psi}{\partial t}, \quad \mathbf{A}' = \mathbf{A} + \nabla\psi \tag{3.39}$$

yields the same electromagnetic fields as φ and $\mathbf{A}$ do. Thus all the arbitrariness is transferred to the single function ψ.

Both (3.38) for φ and $\mathbf{A}$ are coupled equations. Two choices for ψ are particularly convenient. These are discussed below.

(i) *Coulomb gauge:* If we choose

$$\nabla \cdot \mathbf{A} = 0,$$

then we get $\nabla^2\psi = \nabla \cdot \mathbf{A}'$ and $\nabla^2\varphi = -4\pi\rho/\epsilon$. This is the Poisson equation for electrostatic potential and thus φ is simply the Coulomb potential which gives $\mathbf{E} = -\nabla\varphi$. Further, the second of equations of (3.38) reduces to

$$\nabla^2 \mathbf{A} - \frac{\epsilon\mu}{c^2}\frac{\partial^2 \mathbf{A}}{\partial t^2} = -\left(\frac{4\pi}{c}\mathbf{J} + \frac{\epsilon}{c}\frac{\partial \mathbf{E}}{\partial t}\right).$$

(ii) *Lorentz gauge:* If we put

$$\nabla \cdot \mathbf{A} + \frac{\epsilon}{c}\frac{\partial \varphi}{\partial t} = 0,$$

then (3.38) reduce to

$$\Box^2\varphi \equiv \nabla^2\varphi - \frac{\epsilon\mu}{c^2}\frac{\partial^2 \varphi}{\partial t^2} = -\frac{4\pi\rho}{\epsilon},$$

$$\Box^2 \mathbf{A} \equiv \nabla^2 \mathbf{A} - \frac{\epsilon\mu}{c^2}\frac{\partial^2 \mathbf{A}}{\partial t^2} = -\frac{4\pi \mathbf{J}}{c}, \tag{3.40}$$

where $\Box^2$ is the d'Alembertian operator. This gauge involves the Lorentz invariant four-vector $(\mathbf{A}, \varphi)$ and is hence called the Lorentz gauge. Both equations of (3.40) are identical inhomogeneous wave equations. In fact, $\Box^2$ is a Lorentz invariant scalar operator. It is a generalisation of the three-dimensional Laplacian operator.

Since ρ and $\mathbf{J}$ are related by the equation of continuity, (3.24d), $\mathbf{A}$ and φ must also be related and derivable from a single quantity. Let us define $\mathbf{Z}$ such that

$$\frac{\partial \mathbf{Z}}{\partial t} = \mathbf{J}. \tag{3.41a}$$

Then the equation of continuity gives us

$$\rho = -\nabla \cdot \mathbf{Z}. \tag{3.41b}$$

These two equations together determine $\mathbf{Z}$.

Let us now define a *superpotential* $\mathbf{\Pi}$ which satisfies

$$\nabla^2 \mathbf{\Pi} - \frac{\epsilon\mu}{c^2}\frac{\partial^2 \mathbf{\Pi}}{\partial t^2} = -4\pi \mathbf{Z}. \tag{3.42}$$

This is inhomogeneous wave equation. Then $\mathbf{A}$ and φ can be derived from $\mathbf{\Pi}$ through

$$\mathbf{A} = \frac{1}{c}\frac{\partial \mathbf{\Pi}}{\partial t}, \quad \varphi = -\frac{\nabla \cdot \mathbf{\Pi}}{\epsilon}. \tag{3.43}$$

3.3.2 Application to atoms and molecules

We may treat an atom or a molecule as a finite continuous time-dependent charge and current distribution. We wish to calculate the radiation at an observation point X far away from the molecule. Let us choose some origin O within the molecule and let vector OX = $\mathbf{r}$. Consider a source point P in the molecule and let vector OP= $\mathbf{r}'$. Also, $\mathbf{r} - \mathbf{r}' = \mathbf{R}$ which is the vector from P to X; see Fig. 3.8. Our strategy will be as follows.

(i) Obtain $\mathbf{Z}$ from ρ and $\mathbf{J}$ from (3.41).

(ii) Solve (3.42) to obtain $\mathbf{\Pi}$.

(iii) Obtain $\mathbf{A}$ and φ by using (3.43).

(iv) Obtain $\mathbf{E}$ and $\mathbf{B}$ by using (3.35) and (3.33).

(v) Calculate energy flux (Poynting vector)

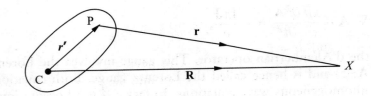

Fig. 3.8 *The geometry of the molecule, the source point P, and the observation point X*

Suppose a disturbance starts from point O at time t. It will reach X at time $t + r/v$. Similarly, disturbances from the other regions of the molecule will also travel to the point X. Since we are interested in finding the effect at X at time $t + r/v$, we must consider the disturbance which started from P at time $t = t + r/v - R/v$. Then the solution of (3.42) is given by (see Appendix A3.1)

$$\mathbf{\Pi}(r, t + r/v) = \int_V \frac{[\mathbf{Z}]_{t'}}{R} d^3 r'$$

$$+ \frac{1}{4\pi} \oint_S \left[\frac{\mathbf{\Pi} \cos \theta}{R^2} + \frac{1}{R^2} \mathbf{n}.\nabla \mathbf{\Pi} + \frac{\cos \theta}{vR} \frac{\partial \mathbf{\Pi}}{\partial t} \right]_{t'} \cdot d\mathbf{S},$$

where V is the volume containing the sources and S is its boundary surface. If the observation point is taken far away from the source, $r \gg r'$, $R \gg r'$, then the quantity in square brackets in the above equation can be made to tend to zero. Hence

$$\mathbf{\Pi}(\mathbf{r}, t + r/v) = \int_V \frac{[\mathbf{Z}]_{t'}}{R} d^3 r'. \tag{3.44a}$$

$[\mathbf{Z}]_{t'}/R$ is the *retarded potential*. It gives $[\mathbf{A}]_{t'}$ and $[\varphi]_{t'}$, the retarded vector and scalar potentials. They, in turn, are called *Lienard–Wischart potentials*.

We may write

$$\mathbf{Z}(t') = \mathbf{Z}(t) + (t' - t)\dot{\mathbf{Z}}(t),$$

to first order in $t' - t$. But

$$t' - t = (r - R)/v, \quad \text{and} \quad R^2 = r^2 + r'^2 - 2\mathbf{r} \cdot \mathbf{r}'.$$

This gives

$$R = r \left[1 + \frac{r'^2}{R^2} - \frac{2\mathbf{R} \cdot \mathbf{r}'}{r^2} \right]^{-1/2} \approx r \left(1 - \frac{\mathbf{r} \cdot \mathbf{r}'}{r^2} \right).$$

Thus we have

$$r - R = \frac{\mathbf{r} \cdot \mathbf{r}'}{r} \quad \text{and} \quad R = r,$$

giving

$$t' - t = \frac{\mathbf{r} \cdot \mathbf{r}'}{rv},$$

and $\mathbf{Z}(t') = \mathbf{Z}(t) + \dfrac{\mathbf{r} \cdot \mathbf{r}'}{rv}\dot{\mathbf{Z}}(t).$

Hence (3.44) gives

$$\mathbf{\Pi}(r) = \frac{1}{r}\int_V \mathbf{Z}d^3r' + \frac{1}{vr^2}\int_V (\mathbf{r} \cdot \mathbf{r}')\dot{\mathbf{Z}}d^3r'. \tag{3.44b}$$

Then

$$\mathbf{A} = \frac{1}{c}\frac{\partial \mathbf{\Pi}}{\partial t} = \frac{1}{cr}\int_V \mathbf{J}d^3r' + \frac{1}{cvr}\int_V (\mathbf{n} \cdot \mathbf{r}')\dot{\mathbf{J}}d^3r', \tag{3.45}$$

where $\mathbf{n}$ is a unit vector from the molecule to the point of observation, that is along $\mathbf{r}$.

In order to get φ from (3.43), we will need $\nabla \cdot \mathbf{\Pi}$. Noting that $\mathbf{\Pi}$ contains terms like $f(t)/R$, we work out some of its derivatives as follows. We have

$$\frac{\partial}{\partial x}\left[\frac{f(t)}{R}\right] = \frac{f'(t)}{R}\frac{\partial t}{\partial R}\frac{\partial R}{\partial x} + f(t)\frac{\partial}{\partial x}\left(\frac{1}{R}\right).$$

But $\partial R/\partial x = x/R$, and $t + R/v = $ constant gives $\partial t/\partial R = -1/v$. Therefore

$$\frac{\partial}{\partial x}\left[\frac{f(t)}{R}\right] = -\frac{x}{vR^2}f'(t),$$

neglecting the terms in $1/R^3$ and others. This will be valid for large R. This leads to

$$\varphi = \frac{\mathbf{n}}{\epsilon v} \cdot \frac{\partial \mathbf{\Pi}}{\partial t}$$

$$= \frac{\mathbf{n}}{\epsilon v R} \cdot \int_V \mathbf{J}d^3r + \frac{\mathbf{n}}{\epsilon v^2 R} \cdot \int_V \mathbf{r}'(\dot{\mathbf{J}} \cdot \mathbf{n})d^3r. \tag{3.46}$$

3.3.3 Electromagnetic radiation by a molecule

We shall work out the electric and magnetic fields due to a molecule as seen at the observation point X. The electric field can be seen to be

$$\mathbf{E} = -\frac{\mu}{c}\frac{\partial \mathbf{A}}{\partial t} - \nabla\varphi$$

$$= -\frac{\mu}{c^2 R}\int_V \dot{\mathbf{J}}d^3r - \frac{\mu\mathbf{n}.}{c^2 vR}\int_V \mathbf{r}'\,\ddot{\mathbf{J}}d^3r'$$

$$+ \frac{\mathbf{n}}{\epsilon v^2 R}\int_V \mathbf{J}\cdot\mathbf{n}d^3r + \frac{\mathbf{n}.}{\epsilon v^3 R}\int_V \mathbf{r}'(\ddot{\mathbf{J}}\cdot\mathbf{n})d^3r'$$

The magnetic field is obtained as

$$\mathbf{H} = \nabla \times \mathbf{A}$$

$$= \frac{1}{cvR}\int_V \dot{\mathbf{J}} \times \mathbf{n}d^3r + \frac{\mathbf{n}.}{cv^2 R}\int_V \mathbf{r}'(\ddot{\mathbf{J}} \times \mathbf{n})d^3r'.$$

This shows that $\mathbf{H}$ is perpendicular to $\mathbf{n}$. Also

$$\mathbf{H} \times \mathbf{n} = \frac{1}{cvR}\int_V (\dot{\mathbf{J}} \times \mathbf{n}) \times \mathbf{n}d^3r + \frac{\mathbf{n}}{cv^2 R} \cdot \int_V \mathbf{r}'(\mathbf{J} \times \mathbf{n}) \times \mathbf{n}d^3r$$

$$= (\epsilon/\mu)^{1/2}\mathbf{E}.$$

We thuse see that $\mathbf{E}, \mathbf{H}, \mathbf{n}$ form a right-handed system. It can also be verified that $\mu H^2 = \epsilon E^2$. The fields thus have all the properties of electromagnetic radiation.

The electric and magnetic fields themselves are given by

$$\mathbf{H} = \frac{(\epsilon\mu)^{1/2}}{c^2 R}\int_V \dot{\mathbf{J}} \times \mathbf{n}d^3r + \frac{\epsilon\mu}{c^3 R}\int_V (\mathbf{n}\cdot\mathbf{r}')(\ddot{\mathbf{J}} \times \mathbf{n})d^3r',$$

$$= \frac{\mu}{c^2 R}\int_V (\dot{\mathbf{J}} \times \mathbf{n}) \times \mathbf{n}d^3r + \frac{\mu(\epsilon\mu)^{1/2}}{c^3 R}\int_V (\mathbf{n}\cdot\mathbf{r}')(\ddot{\mathbf{J}} \times \mathbf{n}) \times \mathbf{n}d^3r'. \tag{3.47}$$

The first term on the right-hand side of the above equations represents *electric dipole* radiation. The second term in each contains *electric quadrupole* and *magnetic dipole* radiation. Because of the extra factor of $1/c$, the latter are normally weak compared to the electric dipole radiation and can be ignored. But when the electric dipole moment vanishes, the electric quadrupole and magnetic dipole radiations become important. A molecule emits (or absorbs) radiation when it makes a transition from one quantum state to another. When a transition takes place under the influence of electric dipole radiation, it is called an *allowed transition*. When this transition is not possible, mostly on grounds of symmetry, higher order transitions, called *forbidden transitions*, are seen.

3.3.4 Electric dipole radiation

Equation (3.47) can be put in a compact form. We start with the vector identity

$$\nabla \cdot (x\mathbf{J}) = (\mathbf{J} \cdot \nabla)x + x\nabla \cdot \mathbf{J}$$

$$= J_x - x\frac{\partial \rho}{\partial t},$$

where we have used the equation of continuity. We can write this in the form

$$J_x = \nabla \cdot (x\mathbf{J}) + x\frac{\partial \rho}{\partial t},$$

to obtain

$$\int_V J_x d^3r = \oint_S \mathbf{n} \cdot (x\mathbf{J})dS + \frac{\partial}{\partial t}\int_V x\rho d^3r.$$

Taking the surface S such that it encloses the localised current distribution, no current crosses the surface S, and the surface integral can be made to vanish. Defining

$$\int_V \mathbf{r}\rho d^3r = \mathbf{p} \qquad (3.48)$$

as the electric dipole moment of the localised charge distribution, we see finally that

$$\int_V \mathbf{J}d^3r = \dot{\mathbf{p}}.$$

Using this in (3.47), the first term in **H** and **E** can be reduced to

$$\mathbf{H} = \frac{(\epsilon\mu)^{1/2}}{c^2R}\ddot{\mathbf{p}} \times \mathbf{n},$$

$$\mathbf{E} = \frac{\mu}{c^2R}(\ddot{\mathbf{p}} \times \mathbf{n}) \times \mathbf{n}.$$

The flux of radiation is then given by

$$\mathbf{S} = \frac{c}{4\pi}(\mathbf{E} \times \mathbf{H}) = \frac{\mu}{4\pi vc^2R^2}[(\ddot{\mathbf{p}})^2 - (\mathbf{n} \cdot \ddot{\mathbf{p}})^2]\mathbf{n}.$$

If θ is the angle between **n** and $\ddot{\mathbf{p}}$, then

$$(\ddot{\mathbf{p}})^2 - (\mathbf{n} \cdot \ddot{\mathbf{p}})^2 = \ddot{\mathbf{p}}^2 \sin^2\theta$$

and the above equation reduces to

$$\mathbf{S} = \frac{\mu}{4\pi v c^2 R^2}(\ddot{\mathbf{p}})^2 \sin^2\theta \, \mathbf{n} = \frac{c}{4\pi}\left(\frac{\epsilon}{\mu}\right)^{1/2} E^2 \mathbf{n}. \tag{3.49}$$

The total power radiated in all directions will be

$$W = \frac{\mu}{4\pi v c^2 R^2}(\ddot{\mathbf{p}})^2 \int_\Omega R^2 \sin^2\theta dS$$

$$= \frac{2\mu}{3v c^2}(\ddot{\mathbf{p}})^2. \tag{3.50}$$

The total power is seen to be related to the flux as

$$\mathbf{S} \equiv \mathbf{S}(\theta) = \frac{W}{4\pi R^2} \cdot \frac{3}{2}\sin^2\theta \mathbf{n}. \tag{3.51}$$

From the middle equation of (3.49), we see that the flux radiated by the molecule depends on (i) $(\ddot{\mathbf{p}})^2$, (ii) inverse square of the distance, and (iii) $\sin^2\theta$. The angular dependence is shown in Fig. 3.9. It is seen that there is no radiation along the line of $\ddot{\mathbf{p}}$ ($\theta = 0$), while there is maximum radiation along directions in a plane perpendicular to it ($\theta = \pi/2$).

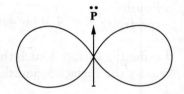

Fig. 3.9 *The radiation pattern from a molecule with dipole moment* **p**. *The instantaneous direction of* $\ddot{\mathbf{p}}$ *is taken as the z-axis. The radiation shows the* $\sin^2\theta$ *pattern.*

For an electron in vacuum, we have

$$\mathbf{p} = e\mathbf{r}, \; \mu = 1, v = c,$$

so that

$$W = \frac{2e^2}{3c^3}(\ddot{\mathbf{r}})^2,$$

$$\mathbf{S}(\theta) = \frac{e^2}{4\pi R^2 c^3}(\ddot{\mathbf{r}})^2 \sin^2\theta \, \mathbf{n},$$

$$|\mathbf{E}(\theta)| = \frac{\mu e \ddot{r} \sin\theta}{c^2 R}$$

$$|\mathbf{H}(\theta)| = \frac{\epsilon^{1/2}\mu e \ddot{r} \sin\theta}{c^2 R}. \tag{3.52}$$

The nature of polarization depends on the motion of the electron. Linear motion always gives rise to linear polarization while circular motion appears as circular, elliptic, or linear depending upon the direction of observation.

We can now calculate the power radiated per unit solid angle. Noting that the Poynting vector **S** gives the energy transmitted per unit area per unit time, the power radiated per unit solid angle is see to be

$$\frac{dW(t)}{d\Omega} \equiv \frac{dE(t)}{d\Omega dt} = R^2|\mathbf{S}| = \frac{\mu}{4\pi v c^2}(\ddot{\mathbf{p}})^2 \sin^2\theta, \tag{3.53}$$

where E represents energy[12].

3.3.5 The spectrum of radiation

As the charges in the molecule change places with passage of time, **p** is a fucntion of time. Consequently the radiation shows a spectrum of frequencies, which means it consists of waves of all frequencies with different amplitudes. We introduce the Fourier transform pair $\mathbf{p}(t)$ and $\mathbf{p}(\omega)$ representing the dipole moment and its frequency spectrum, respectively, through

$$\mathbf{p}(\omega) = \frac{1}{\sqrt{2\pi}} \int_{-\infty}^{\infty} \mathbf{p}(t)e^{i\omega t}dt,$$

$$\mathbf{p}(t) = \frac{1}{\sqrt{2\pi}} \int_{-\infty}^{\infty} \mathbf{p}(\omega)e^{-i\omega t}d\omega. \tag{3.54}$$

Since ω cannot be negative, we note that

$$\mathbf{p}(-\omega) = \frac{1}{\sqrt{2\pi}} \int_{-\infty}^{\infty} \mathbf{p}(t)e^{-i\omega t}dt = \mathbf{p}^*(\omega).$$

Then by Percival's theorem, we have

$$\int_{-\infty}^{\infty} |\mathbf{p}(t)|^2 dt = \int_{-\infty}^{\infty} |\mathbf{p}(\omega)|^2 d\omega = 2\int_{0}^{\infty} |\mathbf{p}(\omega)|^2 d\omega.$$

Since the second time derivative of dipole moment is

$$\ddot{\mathbf{p}}(t) = -\frac{1}{\sqrt{2\pi}} \int_{-\infty}^{\infty} \omega^2 \mathbf{p}(\omega)e^{-i\omega t}d\omega,$$

we have

$$\int_{0}^{\infty} |\ddot{\mathbf{p}}(t)|^2 dt = \int_{0}^{\infty} \omega^4 |\mathbf{p}(\omega)|^2 d\omega, \tag{3.55}$$

---

[12]We have used the same symbol for electric field (magnitude) and energy, and for magnetic field (magnitude) and Hamiltonian, hoping that there would be no confusion.

where we have used the integral representation of Dirac delta function,

$$\int_{-\infty}^{\infty} dt\, e^{i(\omega'-\omega)t} = 2\pi\delta(\omega'-\omega).$$

We can obtain the power emitted over all solid angles from (3.53), and it is seen that

$$W(t) = \frac{\mu}{4\pi vc^2}|\ddot{\mathbf{p}}|^2 \int_{\Omega} \sin^2\theta d\Omega$$

$$= \frac{2\mu}{3vc^2}|\ddot{\mathbf{p}}|^2. \qquad (3.56)$$

Let $W(t)$ and $W(\omega)$ be Fourier transform pairs. Then the total power radiated by the dipole over all times or over all frequencies would be

$$W_{\text{tot}} = \int_0^{\infty} W(\omega)d\omega = \int_0^{\infty} W(t)dt.$$

But

$$\int_0^{\infty} W(t)dt = \frac{2\mu}{3vc^2}\int_0^{\infty} |\ddot{\mathbf{p}}|^2 dt = \frac{2\mu}{3vc^2}\int_0^{\infty} \omega^4 |\mathbf{p}(\omega)|^2 d\omega.$$

Hence we can write

$$\int_0^{\infty} W(\omega)d\omega = \frac{2\mu}{3vc^2}\int_0^{\infty} \omega^4 |\mathbf{p}(\omega)|^2 d\omega. \qquad (3.57)$$

$W(t), W(\omega)$ and W_{tot} can be found out if we know $\mathbf{p}(t)$.

In Fourier transform spectroscopy, one observes the changes in the fringe intensity by varying the length of one arm of a Michelson interferometer and gets its spectrum by its Fourier transform. The resolution depends on the sampling interval and the total displacement of the arm. This method is more efficient for faint sources when compared with the normal spectrographs. Fourier transforms are commonly used by radio astronomers, cosmologists, geophysicists and others for tackling various problems.

3.3.6 Hertzian oscillator in vacuum

We consider a charge placed in an electromagnetic field in vacuum and set into oscillations due to it. This is the simplest case of a Hertzian oscillator. Since there are no sources, we put $\mathbf{z} = \mathbf{0}$, and $\epsilon = \mu = 1$ for vacuum. Equation (3.42) then reduces to a homogeneous wave equation

$$\nabla^2 \mathbf{\Pi} = \frac{1}{c^2} \frac{\partial^2 \mathbf{\Pi}}{\partial t^2}. \tag{3.58}$$

We further consider a spherical wave so that the fields depend only on r. Then $\mathbf{\Pi}$ would have solutions of the form

$$\mathbf{\Pi} = \left[\frac{F(r - ct)}{r} + \frac{G(r + ct)}{r} \right] \frac{\mathbf{r}}{r},$$

which is a combination of waves travelling in opposite directions. If we further restrict to an outgoing wave, with $G = 0$, we get

$$\mathbf{H} = \nabla \times \mathbf{A} = \frac{1}{c} \frac{\partial}{\partial t} \nabla \times \mathbf{\Pi},$$

$$\mathbf{E} = -\frac{1}{c^2} \frac{\partial^2 \mathbf{\Pi}}{\partial t^2} + \nabla \cdot \mathbf{\Pi}.$$

In a Hertzian oscillator, electromagnetic radiation is produced by passing an alternating current through a gap between two electrodes. We take the direction of this gap as the polar axis. Working out $\mathbf{\Pi}$ in terms of latitude δ and azimuth λ, we find that

$$\nabla \times \mathbf{\Pi} = \left(-\frac{F_\delta'}{r} + \frac{F_\delta}{r^2} \right) \frac{\boldsymbol{\lambda}}{r} + \left(\frac{F_\lambda'}{r} - \frac{F_\lambda}{r^2} \right) \frac{\boldsymbol{\delta}}{\cos \delta},$$

where $\boldsymbol{\lambda}$ and $\boldsymbol{\delta}$ are unit vectors in the direction of increasing respective coordinate and the primes denote differentiation with respect to r. Also, all the components F_r, F_δ and F_λ are functions of r but F_λ is not a function of δ and vice versa. Also, we see that

$$\nabla \cdot \mathbf{\Pi} = \frac{1}{r^2}(F_r + rF_r') - \frac{\sin \delta}{r^2 \cos \delta} F_\delta(r).$$

Hence

$$\mathbf{H} = \left(\frac{F''}{r} - \frac{F'}{r^2} \right) \mathbf{F}_1 \times \mathbf{r}_1,$$

where $\mathbf{F}_1$ and $\mathbf{r}_1$ are unit vectors corresponding to $\mathbf{F}$ and $\mathbf{r}$. Thus $\mathbf{F}_1 = (F_r, F_\lambda, F_\delta)/F$. Then we note that

$$\mathbf{F}_1 \times \mathbf{r}_1 = \frac{1}{r} \frac{F_\delta}{F} \boldsymbol{\lambda} - \frac{1}{r \cos \delta} \frac{F_\lambda}{F} \boldsymbol{\delta}.$$

For small r, H is seen to be proportional to $1/r^2$ while for large r it is proportional to $1/r$. The transition occurs at $r \simeq |F'/F''|$.

If F is taken to be $F = e^{i(kr - \omega t)}$, then

$$|F'/F''| = |1/k| = \lambda/2\pi.$$

Therefore the transition from near-field to far-field occurs at $r \simeq \lambda/2\pi$. Hence for $r \ll \lambda$, $H \propto 1/r^2$ and the energy content varies as H^2 or as r^{-4}. On the other hand, for $r \gg \lambda$, $H \propto 1/r$ and the energy content aries as r^{-2}.

3.3.7   Electromagnetic field and the relativistic Lagrangian

In an electromagnetic field, the force on a point charge q is given by the Lorentz force, (3.25c), whereas the electromagnetic fields are described by the potentials, (3.33) and (3.35). Working out a little algebra, this force can be put in the form

$$\mathbf{F} = -q\boldsymbol{\nabla}\varphi - \frac{q}{c}\frac{\partial \mathbf{A}}{\partial t} + \frac{q}{c}\boldsymbol{\nabla}(\mathbf{A} \cdot \mathbf{v}). \tag{3.59}$$

Now let us consider the potential to be

$$V = q\varphi - \frac{q}{c}\mathbf{A} \cdot \mathbf{v} \equiv V(\mathbf{r}, \dot{\mathbf{r}}). \tag{3.60}$$

Since $\dot{x} = v_x$, we see that

$$\frac{\partial V}{\partial \dot{x}} = \frac{\partial V}{\partial v_x} = -\frac{q}{c}A_x.$$

Therefore

$$\frac{d}{dt}\left(\frac{\partial V}{\partial \dot{x}}\right) = -\frac{q}{c}\frac{\partial A_x}{\partial t}.$$

We see that the components of $\mathbf{F}$ satisfy the definition of generalised force defined by (1.17). Hence we have

$$\mathbf{F} = -\frac{q}{c}\frac{\partial A_x}{\partial t} - \boldsymbol{\nabla}(q\varphi - \frac{q}{c}\mathbf{A} \cdot \mathbf{v}).$$

Thus $\mathbf{F}$ can be derived from the potential of (3.60). Then the Lagrangian of (2.65) becomes

$$L = -m_0 c^2 (1 - v^2/c^2)^{1/2} - q\varphi + (q/c)\mathbf{A} \cdot \mathbf{v}. \tag{3.61}$$

This gives the canonical momentum as

$$\mathbf{p} = m\mathbf{v} + (q/c)\mathbf{A}, \tag{3.62}$$

and hence finally the Hamiltonian

$$H = \mathbf{p} \cdot \mathbf{v} - L = mc^2 + e\varphi. \tag{3.63}$$

Writing $m = (m_0^2 + m^2\beta^2)^{1/2}$ and using (3.62) for $m\mathbf{v}$, this can be further written in the form

$$H = c[m_0^2c^2 + (\mathbf{p} - q\mathbf{A}/c)^2]^{1/2} + e\varphi. \tag{3.64}$$

For non-relativistic particles in electromagnetic fields, we have

$$L = T - V = mv^2/2 - q\varphi + (q/c)\mathbf{A} \cdot \mathbf{v}.$$

This gives

$$\mathbf{p} = m\mathbf{v} + q\mathbf{A}/c$$

and

$$H = \frac{1}{2m}(\mathbf{p} - q\mathbf{A}/c)^2 + e\varphi. \tag{3.65}$$

3.4 Harmonic Oscillator

In the earlier days of electromagnetism, a transmitting medium was considered to be a collection of similar simple harmonic oscillators. An oscillator would consist of a heavier positive charge which would be stationary, and a lighter negative charge which oscillates around it. We shall take a look at the early theory, and consider some aspects of the radiation emitted by such an oscillator.

3.4.1 Thompson's theory

If the restoring force on the oscillating charge is proportional to its displacement from the positive charge, the equation of motion of the moving charge will be

$$m\ddot{\mathbf{r}} = -k\mathbf{r},$$

giving a solution

$$\mathbf{r} = \mathbf{A}e^{i\omega_0 t}, \tag{3.66}$$

where $\omega_0 = (k/m)^{1/2}$ is the natural frequency of the oscillator.

The potential function would be $V(r) = m\omega_0^2 r^2/2$; $\mathbf{A}$ would be complex as it may contain a phase term $e^{i\delta}$. As the electron oscillates its acceleration will give rise to electromagnetic radiation vide (3.53). Emission of radiation will reduce the energy of the oscillator, which will be reflected in the lowering of its amplitude. In other words, the oscillator will be damped.

In order to study this damping process, we note that if $\mathcal{E}$ is the total energy of the oscillator, we have

$$-\frac{d\mathcal{E}}{dt} = W = \frac{2e^2}{3c^3}(\ddot{\mathbf{r}})^2 = \frac{2e^2}{3c^3}\omega_0^4 A^2. \tag{3.67}$$

Since $\mathcal{E}$ consists of kinetic and potential energy of the oscillator, we have

$$\mathcal{E} = \frac{1}{2}m\dot{\mathbf{r}}^2 + \frac{1}{2}m\omega_0^2 r^2 = mA^2\omega_0^2.$$

Comparing the time derivative of this $\mathcal{E}$ with (3.67),

$$\frac{dA^2}{dt} = -\frac{2e^2\omega_0^2}{3mc^3}A^2,$$

which gives the solution

$$A^2 = A_0^2 \exp\left(-\frac{2e^2\omega_0^2 t}{3mc^3}\right). \tag{3.68}$$

We could thus write

$$\mathcal{E} = \mathcal{E}_0 e^{-\gamma t},$$

where

$$\mathcal{E}_0 = mA_0^2\omega_0^2, \gamma = \frac{2e^2\omega_0^2}{3mc^3}. \tag{3.69}$$

The solution for $\mathbf{r}$, equation (3.66), then becomes

$$\mathbf{r} = \mathbf{A}_0 e^{-\gamma t/2 + i\omega_0 t}. \tag{3.70}$$

The factor γ is called the *natural damping constant*. Equation (3.70) represents only the first approximation; higher approximations can be otained by the method of iteration. However, the first approximation itself will be valid if $\gamma/2 \ll \omega_0$. Using (3.69), this condition reduces to

$$\omega_0 \ll 3mc^3/e^2 \Rightarrow \omega_0 \ll 4 \times 10^{23}\text{s}^{-1}.$$

Only high energy γ-rays have frequencies higher than this. Therefore our solution is valid for most of electromagnetic spectral region of interest.

3.4.2 Equation of motion for a damped oscillator

We may write (3.70) as

$$\mathbf{r} = \mathbf{A}_0 e^{i\omega t},$$

where

$$\omega^2 = \omega_0^2 + i\gamma\omega_0,$$

where we have neglected the higher order terms in γ.

The restoring force for the damped oscillator will be

$$-m\omega^2\mathbf{r} = -m\omega_0^2\mathbf{r} - i\gamma\omega_0 m\mathbf{r}.$$

This means that the damping force is

$$-im\gamma\omega_0\mathbf{r} = -i\frac{2e^2\omega_0^3}{3c^3}\mathbf{r} \simeq \frac{2e^2}{3c^3}\frac{d^3\mathbf{r}}{dt^3},$$

where again we have dropped the higher order terms in γ. Therefore the equation of motion of the damped oscillator becomes

$$m\ddot{\mathbf{r}} = -m\omega_0^2\mathbf{r} + \frac{2e^2}{3c^3}\frac{d^3\mathbf{r}}{dt^3}. \tag{3.71}$$

It can be verified that (3.70) is a solution of the above equation when higher powers of γ are dropped.

3.4.3 Profile of the emitted line

From (3.70) we see that

$$\ddot{\mathbf{r}} \simeq -A_0\omega_0^2 f(t),$$

with

$$f(t) = e^{-\gamma t/2 + i\omega_0 t}.$$

Therefore the energy emitted in an interval dt at time t, $W(t)dt$, will be given by

$$W(t)dt = \frac{2e^2}{3c^3}A_0^2\omega_0^4|f^2(t)|dt$$

$$= \frac{2e^2}{3c^3}A_0^2\omega_0^4 e^{-\gamma t}dt. \tag{3.72}$$

It can be verified, as a matter of check, that the total energy lost, which is the integral of the above from time t equal to zero to infinity, is just the total energy $m\omega_0^2 a_0^2$ of the oscillator.

The above function $W(t)$ is shown in Fig. 3.10. It can be represented as a superposition of energies in waves of all frequencies from zero to infinity. The power spectrum can be obtained from the Fourier transform of function $f(t)$. This is achieved in a laboratory by passing the radiation through a spectrograph which introduces delays in various parts of the beam so as to produce constructive interference in certain directions that are different for different frequencies.

Let $f(t)$ and $a(\omega)$ be Fourier transforms of each other, so that

$$f(t) = \frac{1}{\sqrt{2\pi}} \int_0^\infty a(\omega)e^{i\omega t}d\omega.$$

Then

$$a(\omega) = \frac{1}{\sqrt{2\pi}} \int_0^\infty e^{-[\gamma/2 - i(\omega - \omega_0)]t}dt$$

$$= \frac{1}{\sqrt{2\pi}} \frac{i}{(\omega - \omega_0) - i\gamma/2}. \tag{3.73}$$

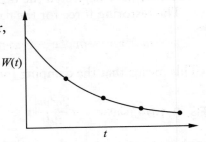

Fig. 3.10   *Power radiated by a damped harmonic oscillator as a function of time*

Therefore the power radiated at frequency ω is given by

$$I(\omega) = \frac{2e^2}{3c^3} A_0^2 \omega^4 |a(\omega)|^2.$$

Using (3.69) and (3.73), this reduces to

$$I(\omega) = \frac{\mathcal{E}_0}{\pi} \frac{\gamma/2}{(\omega - \omega_0)^2 + (\gamma/2)^2}. \tag{3.74}$$

The total power emitted, which is $\int_0^\infty I(\omega)d\omega$, indeed comes out to be $\mathcal{E}_0$, as expected.

Using frequency ν rather than ω, the power spectrum comes out to be

$$I(\nu) = I(\omega)d\omega/d\nu$$

$$= \frac{\mathcal{E}_0}{\pi} \frac{\gamma/4\pi}{(\nu - \nu_0)^2 + (\gamma/4\pi)^2}. \tag{3.75}$$

This is shown in Fig. 3.11. We see that $I(\nu)$ has a maximum of $4\mathcal{E}_0/\gamma$ at $\nu = \nu_0$. The full width at half maximum (FWHM) is seen to be $\Delta\nu = \gamma/2\pi$. In terms of wavelength,

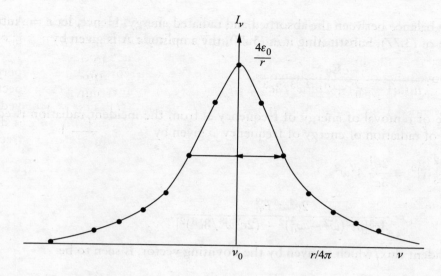

Fig. 3.11 *Spectral line produced by a damped oscillator*

we have

$$\Delta\lambda = (c/\nu_0^2)\Delta\nu = \frac{4\pi e^2}{2mc^2}.$$ (3.76)

This is the classical natural width of a line which is constant. With e and m referring to the electron, this is seen to be about $1.18 \times 10^{-4}\overset{\circ}{\text{A}}$, and is much smaller than the wavelength in most cases.

3.4.4 Scattering of radiation by an oscillator

Let electromagnetic radiation of angular frequency ω fall on the oscillator. It will exert an external force

$$\mathbf{F} = -e[\mathbf{E} + \mathbf{v} \times \mathbf{B}/c] \approx -e\mathbf{E},$$

since $|\mathbf{E}| \approx |\mathbf{B}|$ and $v \ll c$. Let $\mathbf{E} = \mathbf{E}_0 e^{i\omega t}$. Then the equation of motion of the forced, damped, harmonic oscillator, in analogy with (3.71), will be

$$-\frac{2e^2}{3c^3}\frac{d^3\mathbf{r}}{dt^3} + m\ddot{\mathbf{r}} + m\omega_0^2\mathbf{r} = -e\mathbf{E}_0 e^{i\omega t}.$$ (3.77)

As a result of the external force, the oscillator will first oscillate with its natural frequency ω_0, but soon it will be forced to oscillate with the frequency ω of the incident radiation. The oscillator will also emit radiation, but in a steady state there will be a balance between the absorbed and radiated energy. Hence, let $\mathbf{r} = \mathbf{A}e^{i\omega t}$ be a solution of (3.77). Substituting it in (3.77), the amplitude $\mathbf{A}$ is given by

$$\mathbf{A} = \frac{-e\mathbf{E}_0}{m(\omega^2 - \omega_0^2) + 2ie^2\omega^3/3c^3}. \tag{3.78}$$

The rate of removal of energy of frequency ω from the incident radiation is equal to the rate of radiation of energy of frequency ω given by

$$\frac{2e^2}{3c^3}|\ddot{\mathbf{r}}|^2 = \frac{2e^2}{3c^3}A^2\omega^2$$

$$= \frac{2e^4\omega^4 E_0^2}{3c^3[m^2(\omega^2 - \omega_0^2)^2 + (2e^2\omega^3/3c^3)^2]}.$$

The incident flux, which is given by the Poynting vector, is seen to be

$$\frac{c}{4\pi}\mathbf{E} \times \mathbf{B} = \frac{c}{4\pi}E_0^2.$$

Therefore the absorption/scattering coefficient for the oscillator will be

$$\sigma(\omega) = \frac{8\pi e^4}{3c^4} \frac{\omega^4}{m^2(\omega^2 - \omega_0^2)^2 + (2e^2\omega^3/3c^3)^2}.$$

Putting

$$\gamma = \frac{2e^2\omega_0^2}{3mc^3},$$

the above equation reduces to

$$\sigma(\omega) = \frac{8\pi e^4}{3m^2c^4} \frac{\omega^4}{(\omega^2 - \omega_0^2)^2 + (\gamma^2\omega^6/\omega_0^4)}. \tag{3.79}$$

The constant factor $\sigma_e = 8\pi e^4/3m^2c^4$ is seen to be equal to 6.6×10^{-25} cm^2. In terms of frequency ν, using $\omega = 2\pi\nu$ and $\omega_0 = 2\pi\nu_0$, the scattering cross-section comes out to be

$$\sigma(\nu) = \frac{\sigma_e\nu^4}{(\nu^2 - \nu_0^2) + (\gamma^2/4\pi^2)(\nu^6/\nu_0^4)}. \tag{3.80}$$

Since the same motion of the electron causes both absorption of radiation and its re-radiation, the frequency of both is the same. Thus we are dealing with the phenomenon of coherent scattering and not true absorption.

It is interesting to consider the following limiting cases.

(i) For $\nu \sim \nu_0$, we can write

$$\nu^2 - \nu_0^2 = (\nu + \nu_0)(\nu - \nu_0) \simeq 2\nu_0(\nu - \nu_0).$$

Hence (3.80) becomes

$$\sigma(\nu) = \frac{\sigma_0}{4} \frac{\nu_0^2}{(\nu - \nu_0)^2 + (\gamma/4\pi)^2}$$

$$= \frac{e^2}{mc} \frac{\gamma/4\pi}{(\nu - \nu_0)^2 + (\gamma/4\pi)^2}. \tag{3.81}$$

This is known as the *damping profile* of the line. The total absorption over all frequencies is seen to be

$$\int_0^\infty \sigma(\nu) d\sigma = \pi e^2/mc = 0.02 \text{ cm}^2. \tag{3.82}$$

It can be seen that (3.81) is similar to (3.75) except for the replacement of $\mathcal{E}_0$ by $\pi e^2/mc$. Thus the shapes of emission and absorption lines are similar.

(ii) When $\nu = \nu_0$, (3.81) gives

$$\sigma(\nu_0) = 3\lambda_0^2/2\pi \simeq \lambda_0^2/2.$$

(iii) For $|\nu - \nu_0| > \gamma/4\pi$, (3.80) gives

$$\sigma(\nu) = \frac{\sigma_e \nu^4}{(\nu^2 - \nu_0^2)^2}. \tag{3.83}$$

We have the following two subcases of this.

(iv) When $\nu \gg \nu_0$, we have $\sigma(\nu) = \sigma_e = $ constant. This is the Thompson scattering formula for X-rays. Since high energy photons are predominant in hot stellar interiors of massive stars, this electron scattering cross-section becomes important there.

(v) When $\nu \ll \nu_0$, we see that

$$\sigma(\nu) = \sigma_e \nu^4/\nu_0^4 = \sigma_e \lambda_0^4/\lambda^4. \tag{3.84}$$

This is the Rayleigh scattering formula for air molecules. Since blue light is scattered much more efficiently than red light, the sky looks blue during daytime and the sun looks red at sunrise or sunset.

In both Thompson and Rayleigh scattering, the scattered radiation is polarized.

3.4.5 The quantum picture

In the quantum theory of radiation to be discussed in Chapters 5 to 7, the process of radiation is quite different. However, many terms of the classical theory are retained, but they assume a new meaning. The principal modification is that the electron does not radiate even when it is accelerated provided it is in a discrete stationary state. Radiation is emitted when it jumps from a higher state of energy E_2 to a lower energy state E_1, the radiation having a frequency $\nu = (E_2 - E_1)/h$. Absorption of radiation of frequency ν raises the electron from the lower state E_1 to a higher energy state $E_2 = E_1 + h\nu$. Further, the energy levels are not sharp because they are broadened by the uncertainty principle, $\Delta E \sim \hbar/\Delta t$, where $\hbar = h/2\pi$ and Δt is the lifetime of the state. Hence the line is not sharp and we can write

$$\sigma(\nu) = \frac{\pi e^2}{mc} \frac{f}{\pi} \frac{\Gamma/4\pi}{(\nu - \nu_0)^2 + (\Gamma/4\pi)^2}. \tag{3.85}$$

Here f is called the *oscillator strength* and Γ is the damping constant given by

$$h\Gamma = \Delta E_1 + \Delta E_2 \equiv h(\Gamma_1 + \Gamma_2),$$

Γ_1 and Γ_2 being the half-widths of the two energy levels in frequency units. The total scattering cross-section is $\pi e^2 f/mc$. We still talk of the emitter as an oscillator in spite of the fact that the concept of the oscillator is discarded.

3.5 Properties of the Transmitting Medium

A transmitting medium consists of a large number of similar oscillators and we have to find their cumulative effect. We have to consider two kinds of phase lags: (i) the oscillator takes a certain amount of time to come to equilibrium with the incident radiation and begin to oscillate with the forced frequency. This is known as the *physical phase lag*. All oscillators in a homogeneous medium have the same physical phase lag, but it may vary from one medium to another. (ii) The oscillators placed at different points will receive the incident radiation at different time. Hence they will differ in phase, and this is called the *geometrical phase lag*. We shall now consider the different kinds of media.

3.5.1 Crystalline solids

A crystalline solid has a crystal structure, that is, the individual atoms are arranged in a regular geometrical pattern. Therefore the radiation scattered by various atoms will have a definite phase relation in each direction. All the waves scattered from different scattering centres interfere with each other and produce maxima and minima in various directions. The effect in any direction can be obtained by the vectorial addition of waves coming from various atoms, taking account of their phases. Generally the interference will be destructive; but when the distance between atoms is of the order of the wavelength of incident X-rays, we get constructive interference in some directions. Thus when X-rays are passed through a crystal such as mica or sodium chloride, we obtain a geometrical pattern of spots on the photographic plate placed behind the crystal as shown in Fig. 3.12. It is called *Laue pattern*, from which we can obtain information about the structure of the crystal. This is the basis of crystallographic studies used by biochemists to determine the structure of complicated molecules like the DNA and the proteins.

To obtain the intensity in any direction, we first add the amplitudes and then square the resultant. Since the sum of amplitudes is proportional to n, the number of oscillators, the intensity is proportional to n^2 in this case.

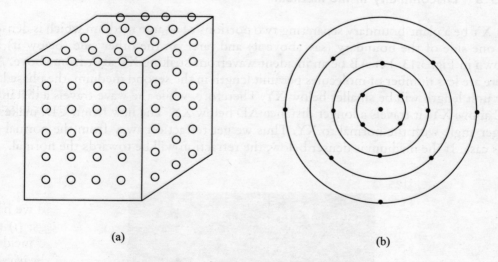

(a) (b)

Fig. 3.12 *Scattering of X-rays from a crystal: (a) the crystal and (b) schematic Laue pattern.*

3.5.2 Fluids and amorphous solids

The molecules are randomly distributed in these materials. But the number of molecules encountered in unit length of the beam is the same for all parallel rays. Therefore the total physical phase lag is the same for all of them. Particularly, in the forward direction the geometrical path length is also equal. Hence there is constructive interference in the forward direction and we get a strong beam in this direction. In other directions also the interference is not completely destructive in spite of the random arrangement of molecules. Therefore we obtain scattered radiation in all directions. Suppose the incident radiation comes along a direction l. Then the function $\varphi(l, l')$ which gives the fraction of radiation scattered into the direction l' depends upon the nature of the medium. For a harmonic oscillator it is proportional to $\sin^2 \theta$, where θ is the angle between the directions of incident and scattered radiation. The scattered energy is proportional to n, the number of oscillators. If N is the number of molecules per unit volume, the scattering cross-section is $\sigma = N\sigma_{osc}$, where σ_{osc} is the scattering cross-section of one oscillator.

If the material which scatters radiation is a gas with suspended particles like dust or water drops, called *aerosols*, their scattering property depends upon the size, shape and character of the aerosol. The radiation emitted by small particles containing many molecules (oscillators) is the subject of Mie's theory. In this case $\varphi(l, l')$ can be quite complicated with several side lobes. We shall discuss Mie scattering in Section 3.8.

3.5.3 Discontinuity in the medium

Let XY be a plane boundary separating two portions of a same medium which is denser on one side of the boundary (say above it) and rarer on the other side (below it) as shown in Fig. 3.13. Let AB be an incident wavefront striking XY at A from above. As there are less number of molecules per unit length in the second medium, the phase lag per unit length will be smaller below XY. Therefore, while the wave travels a distance BC above XY, it travels a longer distance AD below XY. The new front, CD, makes a larger angle with the normal to XY. Thus we get refraction away from the normal in this case. If the medium is denser below, the refraction will be towards the normal.

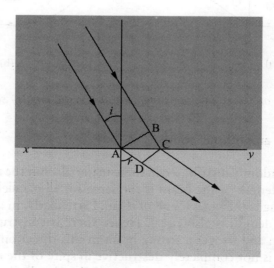

Fig. 3.13   *Refraction at the boundary between a denser and a rarer medium*

Similar effect will occur if the boundary XY separates two media whose oscillators have different phase lags. In either case, the velocity of the wave is greater in the rarer medium. The angles of incidence and refraction are related by

$$\frac{\sin i}{\sin r} = \mu = \frac{v_r}{v_i},$$

(3.86)

where v_i and v_r are the speeds in the incident and rarer media. Since the velocity of the wave is $v = c/(\epsilon\mu)^{1/2}$, the refractive index depends on the dielectric constant and magnetic susceptibility of the medium.

3.5.4 Absorption of radiation by a fluid

The displacement of a damped oscillator in the presence of incident radiation $\mathbf{E}_0 e^{i\omega t}$ was seen to be (see (3.78))

$$\mathbf{r} = \frac{-e\mathbf{E}_0 e^{i\omega t}}{m(\omega^2 - \omega_0^2) + i(2e^2/3c^3)\omega^3}.$$

Therefore the instantaneous electric dipole moment of the oscillator is

$$-e\mathbf{r} = \frac{e^2\mathbf{E}}{m(\omega^2 - \omega_0^2) + i\omega m\gamma}.$$

(3.87)

If there are N oscillators per unit volume, the dipole moment per unit volume will be given by

$$\mathbf{P} = \frac{e^2 N\mathbf{E}}{m[(\omega^2 - \omega_0^2) + i\gamma\omega]} \equiv \chi\mathbf{E},$$

(3.88)

where χ is the electric susceptibility. Then we have

$$\epsilon = 1 + 4\pi\chi = 1 + \frac{4\pi Ne^2}{m[(\omega^2 - \omega_0^2) + i\gamma\omega]}.$$

(3.89)

Consider an electromagnetic wave traveling in the x-direction and polarized in the y-direction. Then $E_z = 0$, and E_y must be of the form $E_y = f(x - vt)$. Since we are looking for a wave-like solutions, let us take

$$E_y = Ae^{i(\omega t - kx)},$$

(3.90)

where $k = 2\pi/\lambda$ is the wave vector. Then the wave equation for E_y becomes

$$\frac{\partial^2 E_y}{\partial x^2} = \frac{\epsilon}{c^2} \frac{\partial^2 E_y}{\partial t^2},$$

where we have taken $\mu = 1$, which is a very good approximation. Using E_y from (3.90) and ϵ from (3.89), we find that

$$k^2 = \frac{\omega^2 \epsilon}{c^2} = \frac{\omega^2}{c^2}(1 + 4\pi\chi). \tag{3.91}$$

Since χ is complex, k will also be complex. Therefore, let

$$\sqrt{\epsilon} = kc/\omega = n - ip/2.$$

Then (3.90) becomes

$$E_y = A \exp\left\{-\frac{in\omega}{c}\left(x - \frac{ct}{n}\right)\right\} \exp\left(-\frac{p\omega}{2e}x\right). \tag{3.92}$$

The velocity of the wave, which is the coefficient of t in the exponent, is c/n; hence we conclude that $n = \mathrm{Re}\sqrt{\epsilon}$ is the refractive index. The intensity, which is $|E_y|^2$, is seen to be

$$|E_y|^2 = |A|^2 e^{-p\omega x/c}.$$

This also suggests that $p\omega/c = 2\pi p/\lambda$ is the *volume absorption coefficient*.

Now from (3.91) and (3.92), we get

$$\frac{kc^2}{\omega^2} = (n - ip/2)^2 = 1 + 4\pi\chi. \tag{3.93}$$

Then by using (3.89), we have

$$n^2 - p^2/4 = 1 + \mathrm{Re}\left\{\frac{4\pi N e^2}{m(\omega^2 - \omega_0^2) + i\gamma\omega m}\right\},$$

$$= 1 + \frac{4\pi N e^2}{m}\frac{\omega^2 - \omega_0^2}{[(\omega^2 - \omega_0^2)^2 + \gamma^2\omega^2]}$$

$$np = -\mathrm{Im}\left\{\frac{4\pi N e^2}{m(\omega^2 - \omega_0^2) + i\gamma\omega m}\right\}$$

$$= \frac{4\pi N e^2}{m}\frac{\gamma\omega}{(\omega^2 - \omega_0^2)^2 + \gamma^2\omega^2}.$$

Since p is generally small, $n \approx 1$. Then

$$p = \frac{4\pi N e^2}{m} \frac{\gamma\omega}{(\omega^2 - \omega_0^2)^2 + \gamma^2\omega^2}.$$

Thus the volume absorption coefficient becomes

$$\frac{2\pi p}{\lambda} = \frac{8\pi^2 N e^2}{m\lambda} \frac{\gamma\omega}{(\omega^2 - \omega_0^2)^2 + \gamma^2\omega^2}. \tag{3.94}$$

This finally gives

$$n = 1 + \frac{2\pi N e^2}{m} \frac{\omega_0^2 - \omega^2}{(\omega^2 - \omega_0^2)^2 + \gamma^2\omega^2}. \tag{3.95}$$

Equations (3.94) and (3.95) are plotted in Fig. 3.14. We see that the volume absorption coefficient has a damping profile, and its value is N times the absorption coefficient per oscillator. The negative index tends to unity at small ω and $\omega_0 \to \infty$; it is above unity for $\omega < \omega_0$ and less than unity for $\omega > \omega_0$. This behaviour of the absorption coefficient and the refractive index has been experimentally verified in the laboratory.

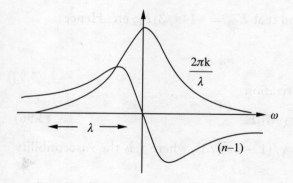

Fig. 3.14 *Variation of absorption coefficient and refractive index with frequency or wavelength*

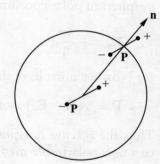

Fig. 3.15 *Self-field of an oscillator*

3.5.4 Self-field of the oscillator

In arriving at (3.94) and (3.95), we have neglected the fact that the oscillating dipole also produces a field which will cause further polarization and thus contribute to the dipole moment. Consider, for example, a sphere around a dipole centered at A, as shown in Fig. 3.15. We shall proceed further through a guided exercise.

► **Guided Exercise 3.2**

An external electrostatic field **E** is applied on a dielectric. Obtain the self field of the oscillators and the resulting polarization **P** in the dielectric. Obtain the effective relation between **E** and **P** under static equilibrium.

Hints   (a) Polarization produces a surface charge given by $\mathbf{P} \cdot \mathbf{n} d\sigma$, where **n** is a unit normal to the element of surface area $d\boldsymbol{\sigma} = \mathbf{n} d\sigma$.

(b) This charge, in turn, produces a field

$$\frac{\mathbf{P} \cdot \mathbf{n} d\sigma(-\mathbf{n})}{r^2} = -\mathbf{n}\,(\mathbf{n} \cdot \mathbf{P})d\Omega \text{ at the point A.}$$

(c) The self-field of the oscillator therefore will be obtained by a surface integral it is given by

$$\mathbf{E}_{\mathrm{S}} = -\oint_{\mathrm{S}} \mathbf{n}\,(\mathbf{n} \cdot \mathbf{P})d\Omega.$$

(d) Taking the polar axis in a certain direction and using

$$n_x = \sin\theta \cos\varphi, \quad n_y = \sin\theta \sin\varphi, \quad n_z = \cos\theta,$$

in spherical polar coordinates θ, φ, we find that $E_{\mathrm{S}x} = -(4\pi/3)P_x$, etc. Hence

$$\mathbf{E}_{\mathrm{S}} = -\frac{4\pi}{3}\mathbf{P}.$$

(e) Now we must have the effective polarization

$$\mathbf{P} = \chi(\mathbf{E} - \mathbf{E}_{\mathrm{S}}) \Rightarrow \mathbf{P} = [\chi/(1 - 4\pi\chi/3)]\mathbf{E}. \tag{3.96}$$

Thus the *effective susceptibility* reduces to $\chi/(1 - 4\pi\chi/3)$, where χ is the susceptibility of a non-polarizable medium.   ◄

We go back to (3.93) and replace χ by its effective value to get

$$(n - ip/2)^2 = 1 + \frac{4\pi\chi}{1 - 4\pi\chi/3}$$

$$= 1 + \frac{4\pi Ne^2}{m}\frac{1}{(\omega_0^2 - \omega^2 - 4\pi Ne^2/3m) + i\gamma\omega},$$

where we have taken χ from (3.89). This suggests that ω_0^2 is modified to $\omega_0^2 - 4\pi Ne^2/3m$. This then allows us to write (3.94) and (3.95) as

$$\frac{2\pi p}{\lambda} = \frac{8\pi^2 N e^2}{m\lambda} \frac{\gamma\omega}{(\omega^2 - \omega_0^2 + 4\pi N e^2/3m)^2 + \gamma^2\omega^2},$$

$$n = 1 + \frac{2\pi N e^2}{m} \frac{\omega_0^2 - \omega^2 - 4\pi N e^2/3m}{(\omega^2 - \omega_0^2 + 4\pi N e^2/3m)^2 + \gamma^2\omega^2}. \tag{3.97}$$

The shift in ω_0 is known as *pressure shift*. The maximum absorption shifts from $\omega = \omega_0$ to $\omega^2 = \omega_0^2 - 4\pi N e^2/3m$. The shift in ω_0 is given by

$$\Delta\omega = (\omega_0^2 - 4\pi N e^2/3m)^{1/2} - \omega_0 \simeq -2\pi N e^2/3m\omega_0$$

$$\Rightarrow |\Delta\omega/\omega_0| = 2\pi N e^2/3m\omega_0^2. \tag{3.98}$$

For transparent media and vapours, this pressure shift is usually small and can be neglected.

It may be noted that for a non-absorbing medium with $p = 0$, with χ replaced by effective susceptibility, we get

$$n^2 = 1 + \frac{4\pi\chi}{1 - 4\pi\chi/3} = \frac{1 + 8\pi\chi/3}{1 - 4\pi\chi/3}. \tag{3.99a}$$

The inverse relationship is found to be

$$\chi = \frac{3(n^2 - 1)}{4\pi(n^2 + 2)}. \tag{3.99b}$$

Since χ is proportional to N, the number density of oscillators, which in tern is proportional to mass density ρ, we have

$$\frac{n^2 - 1}{n^2 + 2} = K\rho \Rightarrow n^2 = \frac{1 + 2K\rho}{1 - K\rho},$$

where $K = 3\chi/4\pi N$ is a constant. This gives the relation between refractive index and density.

3.5.6 Properties of a plasma

A collection of free electrons, positive ions and neutral particles is called *plasma*. It is considered to be the fourth state of matter. Solar corona, stellar interiors, ionosphere of the earth are some examples of natural plasma. It is conjectured that about 98% of the known matter in the universe exists in the plasma state.

Since the electrons are free in a plasma, there would be no natural frequency of oscillations and no damping. Thus $\omega_0 = 0, \gamma = 0$. But if we take account of collisional effects, which now become important, it amounts to damping. Pressure shift therefore becomes important. This has to be taken into account by replacing γ by ν_{col}, the frequency of collisions. With $\mathbf{r}$ as the position vector of the electron and $\dot{\mathbf{r}}$ its velocity, the momentum lost per unit time will be $m\dot{\mathbf{r}}\nu_{col}$. This amounts to a viscous drag, suggesting an equation of motion

$$m\ddot{\mathbf{r}} = -m\dot{\mathbf{r}}\nu_{col}.$$

With $\mathbf{r} = \mathbf{A}e^{i\omega t}$ and $\dot{\mathbf{r}} = i\omega \mathbf{r}$, the damping force is seen to be equal to $-im\omega\nu_{col}\mathbf{r}$. On comparing with the damping for $-im\gamma\omega\mathbf{r}$, we see that γ is now replaced by ν_{col}. Then we have

$$n^2 - p^2/4 = 1 - \frac{4\pi N e^2}{m(\omega^2 + 4\pi N e^2/3m)},$$

$$np = \frac{4\pi N e^2}{m} \frac{\nu_{cot}\omega}{(\omega^2 + 4\pi N e^2/3m + \nu_{col}^2\omega^2)}.$$

Let us define the *plasma frequency* ω_P through

$$\omega_p^2 = 4\pi N e^2/m. \tag{3.100}$$

Then we can discuss cases depending on how ω compares with ω_p. We have the following cases.

(i) For $\omega \gg \omega_p$, we find that
$$1 \gg n^2 - p^2/4 > 0, \quad np \approx 0.$$
Therefore $p \approx 0, n \approx 1$, and we get
$$n^2 = 1 - \omega_p^2/\omega^2. \tag{3.101}$$
This is known as *Sellmeier's dispersion formula*. We further find that in this case the phase velocity is seen to be
$$\frac{c}{n} = \frac{c}{(1 - \omega_p^2/\omega^2)^{1/2}} > c.$$
Since the phase velocity is $\nu\lambda = \omega/k$, where k is the wave number, we have
$$k = (\omega^2 - \omega_p^2)^{1/2}/c \Rightarrow \omega^2 = c^2 k^2 + \omega_p^2. \tag{3.102}$$

(ii) For $\omega \ll \omega_p$, the other limiting case, we find
$$n^2 - p^2/4 \approx 1 - 3 = -2,$$
showing that p is very large.

Thus ω_p is the critical frequency for the plasma in the sense that radiation with frequencies $\omega < \omega_p$ is completely absorbed while that with $\omega > \omega_p$ is transmitted.

We also see that $n = 0$ at $\omega^2 = \omega_p^2$. Thus n is negative for $\omega \leq \omega_p$, which means that these waves are reflected back. These properties of plasma have important applications in solar and ionospheric physics.

(iii) Solar corona is the outermost atmosphere of the sun which is in the state of plasma. The electron density decreases outward from the photosphere; hence the plasma frequency also decreases outward. Consider a spherical surface through some point P in the corona where the plasma frequency is ω_p. All radiation coming out of the photosphere with $\omega > \omega_p$ will pass through this sphere, but radiations with $\omega < \omega_p$ will be absorbed. Since the plasma frequency decreases outward, along with the electron density, the radiations which can get through the inner layers of the corona can get through the entire corona and be observed on the earth. Thus observations of a certain frequency in a certain region of the corona give us the electron density in that region. Table 3.1 gives the information obtained this way.

Table 3.1 *Electron density and plasma frequency at different heights in the corona*

Height above the photosphere (in units of radius of the sun)	Electron density N_e (cm^{-3})	Plasma frequency (Hz)	Plasma wavelength
0.0086	2.5×10^{10}	10^{10}	3cm
0.056	2.5×10^{8}	10^{9}	30 cm
0.1	1.2×10^{3}	7.5×10^{7}	4m
0.4	4×10^{3}	3×10^{7}	10m
0.8	2×10^{3}	10^{7}	30m

(iv) The ionosphere of the earth consists of layers D, E, F_1, and F_2 identified by electron densities in them and their plasma frequencies. These layers are created due to ionisation of air molecule by UV radiation of the sun. Their characteristics are shown in Table 3.2. Their heights vary during day and night and also depend on other parameters.

The ionosphere plays an important role in the transmission of radio waves around the earth. In fact, it acts like a mirror for certain frequencies which then bounce back and forth between the earth's surface and ionosphere and travel round the globe. As Table 3.2 shows, frequencies below 200 kHz are absorbed by the lowest D layer. Thus it is not possible to broadcast these frequencies at all. The medium wave band extends from 0.5 MHz – 1.6 MHz. These waves are reflected from the E layer. Similarly, transmission using S_2 (3–7 MHz) and S_1 (10–20 MHz) bands is possible due to F_1 and F_2 layers, respectively.

It is also clear that terrestrial atmosphere is rather opaque to radiations of wavelength above 100 m ($\nu < 3$ MHz). If the radiation is coming from the cosmos, we can observe wavelengths $\lambda < 100$ m quite well only when $\lambda < 20$ m. Thus we can observe the solar corona upto a height of the solar radius by the help of radio telescopes placed on the earth. For better observations, we have to take the antenna above the ionosphere.

Table 3.2 *Characteristics of different layers of the ionosphere*

Layer	Height above ground (km)	Maximum electron density (cm^{-3})	Plasma frequency (MHz)	Plasma wavelength (m)
D	70-90	600	0.2	1500
E	90-150	$1.3 - 1.9 \times 10^5$	3.3-3.9	75-90
F$_1$	150-250	$2.4 - 3.6 \times 10^5$	4.4-5.4	55-70
F$_2$	> 250	$5.9 - 1.77 \times 10^5$	6.9-11.9	25-40

3.5.7 Faraday rotation

We have seen that a plasma has high damping for frequencies below the plasma frequency ω_p and allows only frequencies above ω_p to propagate. If a plasma has a magnetic field, which is the case in the interstellar medium, its component $B_{\parallel}$ in the direction of motion of the electron will make the electron gyrate with a frequency $\omega_B = eB_{\parallel}/mc$ which will cause circular polarization of the emitted radiation. Right-handed and left-handed polarization will have different frequencies, $\omega \pm \omega_B$. The corresponding wave vectors, according to (3.102), will be

$$k_{\mathrm{R,L}} = \frac{1}{c}[(\omega \pm \omega_B)^2 - \omega_p^2]^{1/2} \simeq \frac{\omega \pm \omega_B}{c}\left[1 - \frac{\omega_p^2}{2(\omega \pm \omega_B)^2}\right],$$

where the upper and lower signs respectively correspond to right and left circular pola-rization. This produces a phase difference

$$\Delta\varphi \approx \frac{\omega_p^2 \omega_B}{c\omega^2}.$$

The total phase difference along the whole path will be

$$\int \nabla\varphi ds = \frac{4\pi e^3}{m^2 c^2 \omega^2} \int NB_{\parallel} ds,$$

where ds is an element of length along the path; the electron number density N and field component $B_{\parallel}$ would vary along the path. This will cause a rotation of the plane of polarization through an angle equal to half of the above phase difference, that is

$$\Delta\theta = \frac{2\pi e^3}{m^2 c^2 \omega^2} \int N_{\parallel} ds.$$

This is known as *Faraday rotation*. The wavelengths of the two circularly polarized components can be measured and the difference between them is a measure of the total electron content in the line of sight.

3.6 Relativistic Transformation of Electromagnetic Quantities

It is clear that electromagnetic phenomena are not invariant under Galilean transformations. Suppose that there are some stationary electric charges in the frame of reference of an observer S. Then for S there will be only electric field, but no magnetic field. Another observer S′ moving with a uniform velocity relative to S will observe transitory, time-dependent currents. S′ will therefore observe electric as well as magnetic fields.

We have seen in Chapter 2 that the space part of a Lorentz vector is a Galilean vector and its time part is a Galilean scalar. But it is not possible to find Galilean scalars which, together with **E** or **B** would give a four-vector.

According to Einstein's postulate of special theory of relativity, we must formulate the theoretical description of a phenomenon such that the equations would retain the same form for all inertial observers. This makes it necessary on our part to write all equations in terms of world-scalars, world-vectors and world-tensors only. It was shown by Lorentz that Maxwell's equations of electrodynamics are invariant under what is now known as Lorentz transformations. This is best done by rewriting them entirely in terms of world-tensors.

In this section we shall show how all equations of electromagnetism can be recast using world-tensors. This will include Lorentz gauge, scalar and vector potentials, equation of continuity, and Maxwell's equations. To begin with, we note that electric charge, apart from being a Newtonian scalar, is also a Lorentz scalar. This follows from the overall charge-neutrality of the universe or any reasonably large macroscopic part of it, for all observers.

3.6.1 Charge-density–current-density four-vector

We shall work out the relativistic transformations of charge density ρ and current density **J**. Consider a set of n charges q in an element of volume dV. Consider two observers O and O′ such that O′ is moving with a uniform velocity v in the x direction with respect to O. Let the charges be moving with a velocity **u** with respect to O, and with velocity **u′** with respect to O′. Let dV_0 be the rest volume element. Then the volume elements dV and dV' as seen by the moving charges will be

$$dV = dV_0(1 - \beta^2(u))^{1/2}, \quad dV' = dV_0(1 - \beta^2(u'))^{1/2}.$$

The electric charge is a Lorentz scalar, relativistically invariant. If the number of elementary charges in volume element dV' is n', then we shall have

$$n'dV' = ndV \quad \Rightarrow \quad n' = \gamma n(1 - vu_x/c^2).$$

This gives

$$\rho' = \gamma(\rho - vJ_x/c^2),$$

because $\rho u_x = J_x$, the x-component of the current density. Then noting that $J'_x = \rho'u'_x$, we finally obtain

$$J'_x = \gamma(J_x - v\rho), \ \rho' = \gamma(\rho - vJ_x/c^2), \ J'_y = J_y, \ J'_z = J_z. \tag{3.103}$$

For arbitrary direction of relative velocity $\mathbf{v}$, these can be written as

$$J'_\parallel = \gamma(\mathbf{J} - \mathbf{v}\rho)_\parallel, \ \ \rho' = \gamma(\rho - \mathbf{v}\cdot\mathbf{J}/c^2), \ \ J'_\perp = J_\perp, \tag{3.104}$$

where the symbols $\parallel$ and $\perp$ refer to components parallel and perpendicular to $\mathbf{v}$. These equations show that J_x, J_y, J_z, ρ transform like x, y, z, ict. Therefore $J_\mu = (\mathbf{J}, ic\rho)$ is a four-vector. It is called the *four-vector current density*.

With the four-vector gradient operator defined by (2.47), the equation of continuity, (3.25d), can be written as

$$\partial_\mu J_\mu = 0, \tag{3.105}$$

which is a manifestly covariant form.

3.6.2 Transformation of electromagnetic potentials

The potentials φ and $\mathbf{A}$ satisfy the wave equation (3.40). In vacuum, we shall have $\epsilon = \mu = 1$. Since $\Box^2$ is Lorentz invariant, we can write, from (3.40),

$$\Box^2\varphi = \Box'^2\varphi = -4\pi\rho.$$

Using the appropriate equation from (3.103) and replacing v by $-v$ for inverse transformation, we can then write the above as

$$\Box'^2\varphi = -4\pi\gamma(\rho' + vJ'_x/c^2).$$

Similarly, by suitably combining (3.40) and (3.103), we can obtain

$$\Box'^2 A_x = 4\pi J_x/c = -4\pi\gamma(J'_x + \beta\rho')/c$$

and

$$\square'^2 A_y = -4\pi J'_y/c, \ \ \square'^2 A_z = -4\pi J'_z/c.$$

From these equations we then obtain the two equations

$$\square'^2 [\gamma(\varphi - \beta A_x)] = -4\pi\gamma(1 - \beta^2)\rho' = -4\pi\rho';$$

$$\square'^2 [\gamma(A_x - \beta\varphi)] = -4\pi J'_x/c.$$

Equating the right-hand sides of the above equations to $\square'^2 \varphi'$ and $\square'^2 A'_x$, respectively, we obtain the transformation of φ' and A'_x. The y and z components of $\mathbf{A}$ remain unchanged. Thus, finally we see that

$$\varphi' = \gamma(\varphi - \beta A_x), \ \ A'_x = \gamma(A_x - \beta\varphi), \ \ A'_y = A_y, A'_z = A_z. \tag{3.106}$$

Comparing this with the transformation of a four-vector $(\mathbf{r}, ict)$, we see that $(\mathbf{A}, i\varphi)$ is a four-vector in Minkowsky space.

3.6.3   Electromagnetic field-strength tensor

In Lorentz gauge, the electric field and magnetic induction are given in terms of electromagnetic potentials by (3.35) and (3.33). The x-components of these equations can be written as

$$E_x = -\frac{\partial\varphi}{\partial x} - \frac{1}{c}\frac{\partial A_x}{\partial t} = i\left(\frac{\partial(i\varphi)}{\partial x} - \frac{\partial A_x}{\partial(ict)}\right),$$

$$B_x = \frac{\partial A_z}{\partial y} - \frac{\partial A_y}{\partial z}, \tag{3.107}$$

and the other components can be similarly written.

On writing $(A_x, A_y, A_z, i\varphi)$ and (A_1, A_2, A_3, A_4) and the derivatives with respect to (x, y, z, ict) as $(\partial_1, \partial_2, \partial_3, \partial_4)$, respectively, these can be put in the form

$$E_x = i(\partial_1 A_4 - \partial_4 A_1), \ \ B_x = \partial_2 A_3 - \partial_3 A_2.$$

This prompts us to define a second rank tensor $F_{\mu\nu}$ with components

$$F_{\mu\nu} = \partial_\mu A_\nu - \partial_\nu A_\mu. \tag{3.108}$$

It is an antisymmetric tensor whose components are first derivatives of the potentials with respect to spacetime coordinates. Working out the components explicitly, we can see that

$$F_{\mu\nu} = \begin{bmatrix} 0 & B_z & -B_y & -iE_x \\ -B_z & 0 & B_x & -iE_y \\ B_y & -B_x & 0 & -iE_z \\ iE_x & iE_y & iE_z & 0 \end{bmatrix}. \tag{3.109}$$

Then the inhomogeneous Maxwell's equations I and IV (equations (3.24a) and (3.24d)) can be written as a tensor equation

$$\frac{\partial F_{\mu\nu}}{\partial x_\nu} = \frac{4\pi}{c} J_\mu, \tag{3.110}$$

where summation over repeated indices is implied and μ, ν run over x, y, z, ict. $F_{\mu\nu}$ is called the *electromagnetic field-strength tensor*. It can further be seen that the two homogeneous Maxwell equations II and III ((3.24b) and (3.24c)), can be put in the tensor form

$$\partial_\mu F_{\nu\sigma} + \partial_\nu F_{\sigma\mu} + \partial_\sigma F_{\mu\nu} = 0, \tag{3.111}$$

where μ, ν, σ are three consecutive indices from $(1, 2, 3, 4)$ taken in cyclic order. Thus (3.111) is equivalent to four equations.

3.6.4   Transformation of E and B

We use the general form of tensor transformation from one coordinate system x, y, z, ict to another x', y', z', ict'. For a second rank tensor $F_{\mu\nu}$, this can be written in the form

$$F'_{\mu\nu} = \frac{\partial x'_\alpha}{\partial x_\mu} \frac{\partial x'_\beta}{\partial x_\nu} F_{\alpha\beta}. \tag{3.112}$$

We note that the transformation factors appearing in the above equation are elements of the matrix of equation (2.37), where the rows and columns are labelled x, y, z and ict. We note that A is a Hermitian matrix. If we denote $F_{\alpha\beta}$ and $F'_{\mu\nu}$ by tensors F and F', respectively, then we can write (3.112) as

$$F' = AF\tilde{A},$$

where $\tilde{A}$ is the transpose of A, or

$$(F')_{\alpha\beta} = (A)_{\alpha\mu}(F)_{\mu\nu}(\tilde{A})_{\nu\beta}. \tag{3.113}$$

Performing this transformation and equating the resulting matrix with F' which is similar to that of (3.109) with field components replaced by primed components as seen by the moving system, we get the transformation

$$E'_x = E_x, \qquad\qquad B'_x = B_x,$$

$$E'_y = \gamma(E_y - \beta B_z), \qquad B'_y = \gamma(B_y + \beta E_z), \qquad\qquad (3.114)$$

$$E'_z = \gamma(E_z + \beta B_y), \qquad B'_z = \gamma(B_z - \beta E_y).$$

Thus we see that while components in the direction of motion remain unchanged, the components in the transverse direction get mixed up and enhanced by a factor of γ.

3.6.5 Transformation of other electromagnetic quantities

When electromagnetic fields are present in a medium which is at rest with the observer, we define the electric displacement **D** and magnetic field **H** and relate them to electric field **E** and magnetic induction **B** by

$$\mathbf{D} = \epsilon\mathbf{E}, \quad \mathbf{H} = \mathbf{B}/\mu. \qquad\qquad (3.115)$$

Also, the current density **J** is related to the electric field by the relation $\mathbf{J} = \sigma\mathbf{E}$. The three constants ϵ, μ and σ are characteristics of the medium. We also customarily define electric polarization **P** and magnetization **M** and electric and magnetic susceptibilities $\chi_e = P/E$ and $\chi_m = M/H$.

The electromagnetic field tensor was obtained in (3.109) from Maxwell's equations for **E** and **B**. Earlier, we had expressed the potentials **A** and φ as a Lorentz four-vector and put the equation of continuity in a Lorentz invariant form. Maxwell's equations for a medium in terms of **D** and **H** are similar to those for vacuum in terms of **E** and **B**. Thus we could construct an electromagnetic field strength tensor with components of **D** and **H** and write for a medium,

$$F_{\mu\nu} = \begin{bmatrix} 0 & H_z & -H_y & -D_x \\ -H_z & 0 & H_x & -iD_y \\ H_y & -H_x & 0 & -iD_z \\ iD_x & iD_y & iD_z & 0 \end{bmatrix}. \qquad\qquad (3.116)$$

Then using an argument similar to that used in Section 3.6(d), an observer moving relative to this medium with a velocity v along the x direction would measure these fields as **D'** and **H'** where

$$D'_x = D_x, \qquad\qquad H'_x = H_x,$$

$$D'_y = \gamma(D_y - \beta H_z), \qquad H'_y = \gamma(H_y + \beta D_z), \qquad\qquad (3.117)$$

$$D'_z = \gamma(D_z + \beta H_y), \qquad H'_z = \gamma(H_z - \beta D_y).$$

It is thus similar to that of $\mathbf{E}$ to $\mathbf{E}'$ and $\mathbf{B}$ to $\mathbf{B}'$.

It can be seen that electric polarization $\mathbf{P}$ and magnetization $\mathbf{M}$ also follow similar transformations. Thus the relativistic transformation of $\mathbf{P}$ to $\mathbf{P}'$ and $\mathbf{M}$ to $\mathbf{M}'$ can be obtained by replacing $\mathbf{D}$ by $\mathbf{P}$ and $\mathbf{H}$ by $\mathbf{M}$ in all the six equations of Eqs. (3.117). Thus the pairs $(\mathbf{E}, \mathbf{B})$, $(\mathbf{D}, \mathbf{H})$ and $(\mathbf{P}, \mathbf{M})$ follow similar relativistic transformations to their respective primed variables.

The constitutive relations between $\mathbf{D}$ and $\mathbf{E}$ and between $\mathbf{H}$ and $\mathbf{B}$ are, however, valid only for an observer stationary relative to the medium. Relations such as $\mathbf{D} = \epsilon\mathbf{E}$ and $\mathbf{H} = \mathbf{B}/\mu$ are valid only among fields measured by a stationary observer. The transformed fields $\mathbf{D}'$ and $\mathbf{E}'$ measured by a moving observer are *not* related in any simple manner, and the same can be said about $\mathbf{H}'$ and $\mathbf{B}'$. Thus we *cannot* write $\mathbf{D}' = \epsilon'\mathbf{E}'$ and $\mathbf{H}' = \mathbf{B}'/\mu'$. The reader can find a good discussion of this aspect in Pauli[13].

It may appear that this violates the principle of relativity. But a little thought shows that it does not. Electromagnetic phenomena are taking place in a (finite) medium, and this medium then provides a preferred frame of reference, the rest frame. The fields measured by the rest frame observer O are $\mathbf{E}, \mathbf{B}, \mathbf{D}, \mathbf{H}, \mathbf{P}, \mathbf{M}$, etc., and O assigns constants like ϵ, μ and σ. A moving observer O' measures the fields $\mathbf{E}', \mathbf{B}'$, etc. The transformed fields $\mathbf{D}', \mathbf{H}'$ contain $\mathbf{E}'$ and $\mathbf{B}'$ or $\mathbf{E}$ and $\mathbf{B}$ in an intricate mixture which depends on the relative velocity $\mathbf{v}$. Hence the relation between $\mathbf{D}', \mathbf{E}'$ or between $\mathbf{H}'$ and $\mathbf{B}'$ no longer remains simple. For the observer O at rest with the medium, the field $\mathbf{D}$ is related only to $\mathbf{E}$ (and its powers in general, and linearly under low-field approximation), and similarly, $\mathbf{H}$ is related to only $\mathbf{B}$. But, for the moving observer $\mathbf{D}'$ depends not only on $\mathbf{E}'$ but also on $\mathbf{B}'$, and $\mathbf{H}'$ depends not only of $\mathbf{B}'$ but also on $\mathbf{E}'$. In short, the constitutive relations are not relativistically invariant (see Problem 3.7).

3.7 Relativistic Electrodynamics of Moving Charges

3.7.1 Potentials due to a moving charge

As before, let the primed coordinate system move with a velocity v with respect to the unprimed coordinate system in the x-direction. Then the relevant equations for electromagnetic potentials φ' and $\mathbf{A}'$ in the presence of sources ρ' and $\mathbf{J}'$ are

$$\Box'^2\varphi' = -4\pi\rho', \quad \Box'^2\mathbf{A}' = -4\pi\mathbf{J}/c.$$

Their retarded solution is given by

$$\varphi'(\mathbf{r}', t') = \int \frac{[\rho']_{t'-r'/c}}{r'}d\tau,$$

---

[13]Pauli, W. (1958), Section 33.

$$\mathbf{A}'(\mathbf{r}', t') = \int \frac{[\mathbf{J}']_{t'-r'/c}}{r'c} d\tau.$$

For a single charge q at rest in the moving frame, the potentials at the point $\mathbf{r}'$ will be given by

$$\varphi' = q/r', \quad \mathbf{A}' = 0.$$

Then in the rest frame (unprimed coordinate system), we have, according to (3.106),

$$\varphi = \gamma(\varphi' + \beta A'_x) = \frac{\gamma q}{r'},$$

$$A_x = \gamma(A'_x + \beta\varphi') = \frac{\beta\gamma q}{r'},$$

$$A_y = A'_y = 0, \quad A_z = A'_z = 0.$$

We should now convert r' to the rest frame coordinates. Let us assume that a signal is received by the observer at origin O' at $t' = 0$ when the two frames coincide; see Fig. 3.16. The signal originated at the charge q at $\mathbf{r}'$ at time $t' = -r'/c$. Then

$$r' = -ct' = -c\gamma(t + vx/c^2) = \gamma(r - vx/c).$$

Therefore the electromagnetic potentials at O will be

$$\varphi = \frac{qc}{r - vx}, \quad \mathbf{A} = \frac{q\mathbf{v}}{cr - vx}. \tag{3.118a}$$

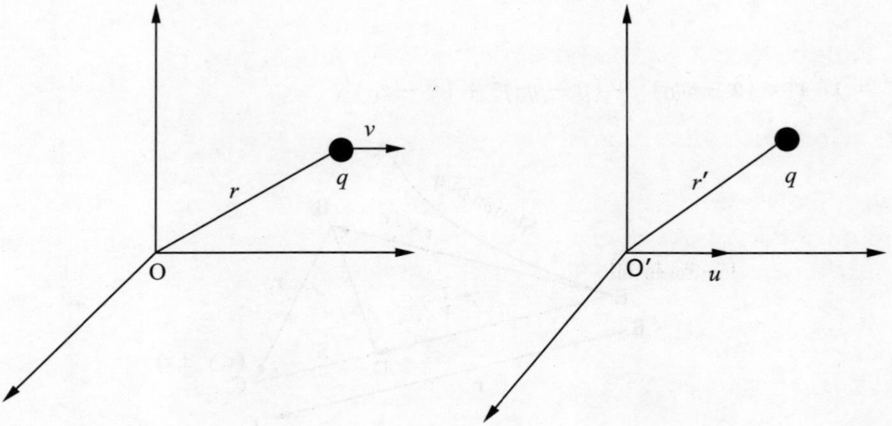

Fig. 3.16 *A stationary charge q at r' in frame O' which is moving with respect to frame O in the x-direction.*

For a general direction of motion of the relative velocity, these equations become

$$\varphi = \frac{qc}{r - \mathbf{v} \cdot \mathbf{r}}, \quad \mathbf{A} = \frac{q\mathbf{v}}{cr - \mathbf{v} \cdot \mathbf{r}}. \tag{3.118b}$$

These are called the *Lienard–Wiechart potentials* which are used for computing the radiation of a moving charge.

3.7.2   Electric and magnetic fields

The electric field $\mathbf{E}$ and magnetic induction $\mathbf{B}$ can be obtained from the potentials of (3.118a) by using (3.35) and (3.33), respectively. Here the important thing to note is that the derivatives are with respect to the point of observation, i.e., $\boldsymbol{\nabla} \equiv \boldsymbol{\nabla}_t$, but the velocity v and its derivatives relate to the time of emission so that $\mathbf{v} = \partial \mathbf{r}/\partial \tau_0$. So we have to obtain the corresponding transformation between $\boldsymbol{\nabla}$ and $\partial/\partial t$.

Let (x_0, y_0, z_0, t_0) in Fig. 3.17 be the coordinates of the emission event when the charge q was at E, and let (x, y, z, t) be the coordinates of the observer at C when the charge had moved to A in both frames of reference. Then if vector EC $= \mathbf{r}$, we have

$$\mathbf{r} = (x - x_0)\mathbf{i} + (y - y_0)\mathbf{j} + (z - z_0)\mathbf{k}, \tag{3.119a}$$

$$r = c(t - t_0). \tag{3.119b}$$

Then we see that

$$\frac{\partial \mathbf{r}}{\partial t_0} = -\frac{\partial x_0}{\partial t_0}\mathbf{i} - \frac{\partial y_0}{\partial t_0}\mathbf{j} - \frac{\partial z_0}{\partial t_0}\mathbf{k} = -\mathbf{v}$$

and

$$r^2 = \mathbf{r} \cdot \mathbf{r} = (x - x_0)^2 + (y - y_0)^2 + (z - z_0)^2.$$

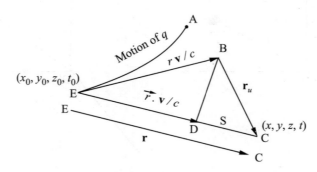

Fig. 3.17   *Electromagnetic fields at C due to a moving charge*

Then

$$2r\frac{\partial r}{\partial t_0} = 2\mathbf{r} \cdot \frac{\partial \mathbf{r}}{\partial t_0} = -2\mathbf{r} \cdot \mathbf{v},$$

and

$$\frac{\partial r}{\partial t} = c\left(1 - \frac{\partial t_0}{\partial t}\right) = \frac{\partial r}{\partial t_0}\frac{\partial t_0}{\partial t} = -\frac{\mathbf{r} \cdot \mathbf{v}}{r}\frac{\partial t_0}{\partial t}.$$

This gives

$$\frac{\partial t_0}{\partial t} = \frac{1}{1 - \mathbf{r} \cdot \mathbf{v}/rc} = \frac{rc}{cr - \mathbf{r} \cdot \mathbf{v}}. \tag{3.120}$$

Let us put

$$s = r - \mathbf{r} \cdot \mathbf{v}/c. \tag{3.121}$$

This parameter s has the following significance. Let B be the position of q at time t if it had moved with a uniform velocity $\mathbf{v}$, and let BD be the normal to EC from B. Then

$$EB = \mathbf{v}t = r\mathbf{v}/c,$$

$$ED = \mathbf{r} \cdot \mathbf{v}/c, \quad s = EC - ED.$$

Now from the second equation of (3.118a), we have

$$(\nabla r)_t = -c\nabla t_0.$$

But

$$(\nabla r)_t = (\nabla r)_{t_0} + \frac{\partial r}{\partial t_0}\nabla t_0 \Rightarrow (\nabla r)_t = (\nabla r)_{t_0} - \frac{\mathbf{r} \cdot \mathbf{v}}{r}\nabla t_0. \tag{3.122}$$

Therefore

$$-c\nabla t_0 = (\nabla r)_{t_0} - \frac{\mathbf{r} \cdot \mathbf{v}}{r}\nabla t_0.$$

Also, with $\mathbf{i}, \mathbf{j}, \mathbf{k}$ as unit Cartesian vectors,

$$(\nabla r)_{t_0} = \frac{x - x_0}{r}\mathbf{i} + \frac{y - y_0}{r}\mathbf{j} + \frac{z - z_0}{r}\mathbf{k} = \mathbf{r}/r.$$

This gives

$$\nabla t_0 = \frac{-\mathbf{r}}{cr - \mathbf{r} \cdot \mathbf{v}} = -\frac{\mathbf{r}}{cs}.$$

Then (3.118a) gives

$$(\nabla r)_t = (\nabla r)_{t_0} - \frac{\mathbf{r}}{cs}\frac{\partial r}{\partial t_0},$$

which allows us to write the operator equation

$$\nabla_t = \nabla_{t_0} - \frac{\mathbf{r}}{cs}\frac{\partial}{\partial t_0}. \tag{3.123}$$

With this background we are now ready to work out the electric and magnetic fields of the moving charge.

(i) Electric field: Using Eqs. (3.118a) and (3.120), we have

$$\mathbf{E} = -\nabla_t\phi - \frac{1}{c}\frac{\partial \mathbf{A}}{\partial t}$$

$$= \left(-\nabla_{t_0} + \frac{\mathbf{r}}{cs}\frac{\partial}{\partial t_0}\right)\frac{q}{s} - \frac{1}{c}\frac{r}{s}\frac{\partial}{\partial t_0}\left(\frac{q\mathbf{v}}{cs}\right).$$

It can be seen from (3.120) that

$$\nabla_{t_0}s = \frac{\mathbf{r}}{r} - \frac{\mathbf{v}}{c},$$

and

$$\frac{\partial s}{\partial t_0} = -\frac{\mathbf{r}\cdot\mathbf{v}}{r} + \frac{v^2}{c} - \frac{\mathbf{r}\cdot\mathbf{v}}{c}. \tag{3.124}$$

Then using (3.120) and (3.122) and defining

$$\mathbf{r}_v = \mathbf{r} - r\mathbf{v}/c, \tag{3.125}$$

we get

$$\mathbf{E} = \frac{q}{s^3}\mathbf{r}_u\left(1 - \frac{v^2}{c^2}\right) + \frac{q}{c^2 s^3}\mathbf{r}\times(\mathbf{r}_u\times\dot{\mathbf{v}}). \tag{3.126}$$

Note that the vector $\mathbf{r}_u$ defined in (3.125) is the vector BC shown in Fig. 3.17, and is different from s defined in (3.119). We also note that, for large r, the first term is proportional to $1/r^2$ so that it behaves like a quasi-static field. The second term is proportional to $1/r$, so it is radiative in nature.

(ii) Magnetic field: Using (3.118b) and (3.120), we get

$$\mathbf{H} = \boldsymbol{\nabla}_t \times \mathbf{A} = \left(\boldsymbol{\nabla}_{t_0} - \frac{\mathbf{r}}{cs} \frac{\partial}{\partial t_0} \right) \times \frac{q\mathbf{v}}{cs}$$

$$= q \left\{ -\frac{1}{cs^2} \boldsymbol{\nabla}_{t_0} s \times \mathbf{v} - \frac{\mathbf{r} \times \dot{\mathbf{v}}}{c^2 s^2} + \frac{\mathbf{r} \times \mathbf{v}}{c^2 s^3} \frac{\partial s}{\partial t_0} \right\}.$$

Then using (3.123) and (3.126), we find that

$$\mathbf{H} = \frac{q}{cs^3} \left[-\frac{\mathbf{r} \times \mathbf{v}s}{r} + \frac{\mathbf{r} \times \mathbf{v}}{c} \left(-\frac{\mathbf{r} \cdot \mathbf{v}}{r} + \frac{v^2}{c} - \frac{\mathbf{r} \cdot \dot{\mathbf{v}}}{c} \right) - \frac{\mathbf{r} \times \dot{\mathbf{v}}}{c} s \right].$$

Finally, using (3.118a), we arrive at

$$\mathbf{H} = \frac{q}{cs^3} \left[\mathbf{v} \times \mathbf{r} \left(1 - \frac{u^2}{c^2} \right) - \frac{\mathbf{r} \cdot \mathbf{v}(\mathbf{r} \times \mathbf{v})}{c^2} + \frac{\mathbf{r} \times \dot{\mathbf{v}}}{c} \left(\mathbf{r} - \frac{\mathbf{r} \cdot \mathbf{v}}{c} \right) \right]. \tag{3.127}$$

Comparing this with (3.126), we can verify that

$$\mathbf{H} = \mathbf{n} \times \mathbf{E}, \quad \mathbf{E} \times \mathbf{H} \| \mathbf{n}.$$

Thus $\mathbf{E}, \mathbf{H}$ and $\mathbf{n}$ form a right-handed orthogonal system, as seen earlier in Section 3.2.2.

3.7.3 Radiation by a moving charge

The first terms in the expressions for $\mathbf{E}$ and $\mathbf{H}$, equations (3.126) and (3.127) are proportional to $1/r^2$ and hence represent a quasi-static field. They do not give rise to radiation. The second term in each of the above equation falls off as $1/r$, and hence indicates a radiating field. Note that the $1/r^2$-term depends on the velocity of the charge but not on its acceleration whereas the $1/r$-term depends on acceleration. Thus uniform motion of charge does not produce radiation.
The flux is given by

$$\mathbf{S} = \frac{c}{4\pi} (\mathbf{E} \times \mathbf{H}) \frac{\partial t}{\partial t_0} = \frac{cs}{4\pi r} \mathbf{E} \times (\mathbf{n} \times \mathbf{E})$$

$$= \frac{q^2}{4\pi c^3 s^5 r} |\mathbf{r} \times (\mathbf{r}_u \times \dot{\mathbf{u}})|^2, \tag{3.128}$$

where we have used (3.121) and (3.126). The above equation shows that $\mathbf{S}$ vanishes in two directions EF and EG where $\mathbf{r}_u$ is parallel to $\dot{\mathbf{u}}$, as shown in Fig. 3.18, and it will have two lobes A and B. The total radiated power W will given by

$$W = \int_\Omega S r^2 d\Omega. \tag{3.129}$$

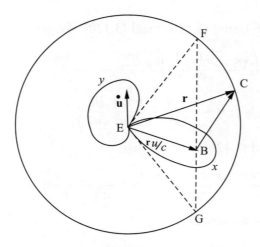

Fig. 3.18 *Angular distribution of radiation by a moving charge*

In order to evaluate W, we choose the coordinate system shown in Fig. 3.19. It would be appropriate to show all the quantities appearing in the above equations. This will help clarify the situation. We choose the z-axis along $\mathbf{u}$, and choose x-axis in such a way that $\dot{\mathbf{u}}$ is in the $x - z$ plane, making an angle θ' with the z-axis. Vector $\mathbf{r}$, as usual, has angular coordinates θ, φ. Then the various vectors occurring above will have the Cartesian components shown in Table 3.3.

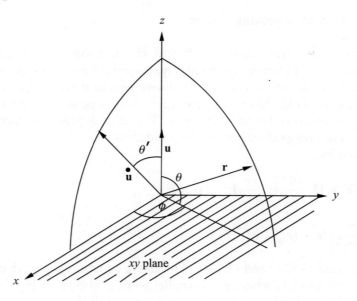

Fig. 3.19 *Orientation of* $\mathbf{r}, \mathbf{u}$ *and* $\dot{\mathbf{u}}$

Table 3.3   *Cartesian components of vectors* **u**, $\dot{\mathbf{u}}$, **r** *and* $\mathbf{r}_u$.

Vector	Component along		
	x-axis	y-axis	z-axis
u	0	0	u
$\dot{\mathbf{u}}$	$\dot{u}\sin\theta'$	0	$\dot{u}\cos\theta'$
r	$r\sin\theta\cos\varphi$	$r\sin\theta\sin\varphi$	$r\cos\theta$
$\mathbf{r}_u$	$r\sin\theta\cos\varphi$	$r\sin\theta\sin\varphi$	$r(\cos\theta - v/c)$

If vectors **r** and **v** make an angle θ between them, equation (3.121) shows that

$$s = r(1 - v\cos\theta/c).$$

Then we proceed as follows: (i) calculate $\mathbf{r}_u \times \mathbf{u}$; (ii) calculate $\mathbf{r} \times (\mathbf{r}_u \times \mathbf{u})$; (iii) calculate $|\mathbf{r} \times (\mathbf{r}_u \times \mathbf{u})|^2$; (iv) integrate over angular coordinates θ and φ; (v) note that $\dot{u}_\parallel = \dot{u}\cos\theta'$ and $\dot{u}_\perp = \dot{u}\sin\theta'$. This finally gives for the Poynting vector and power emitted,

$$\mathbf{S} = \frac{q^2\mathbf{n}}{4\pi c^3 r^2}\left[\frac{\dot{u}^2\sin^2\theta'\{(1 - u\cos\theta/c)^2 - \sin^2\theta\cos^2\varphi(1 - u^2/c^2)\} + \dot{u}^2\cos^2\theta'\sin^2\theta}{(1 - u\cos\theta/c)^5}\right],$$

(3.130)

and

$$W = \frac{2q^2\gamma^2}{3m_0^2c^3}[\gamma^4 m_0^2\dot{u}_\parallel^2 + \gamma^2 m_0^2\dot{u}_\perp^2].$$

(3.131)

Using $p = mu$, $E = mc^2$ and $m = \gamma m_0$, W can also be written as

$$W = \frac{2q^2\gamma^2}{3m_0^2c^3}\left[\left(\frac{dp}{dt}\right)^2 - \frac{1}{c^2}\left(\frac{dE}{dt}\right)^2\right].$$

(3.132)

3.7.4   Some special cases

Here we consider some special cases of radiation by a relativistic accelerated charge such as classical dipole radiation, Cerenkov radiation, Bremstrahlung and synchrotron radiation.

Classical dipole radiation   When $u \ll c$, (3.129) gives

$$\mathbf{S} = \frac{q^2\dot{u}^2\mathbf{n}}{4\pi c^3 r^2}[1 - \sin^2\theta'\sin^2\theta\cos^2\varphi - \cos^2\theta'\cos^2\theta].$$

If Θ is the angle between $\dot{\mathbf{u}}$ and $\mathbf{r}$, then from Table 3.1, we have

$$\cos^2\Theta = \sin^2\theta\sin^2\theta'\cos^2\varphi + \cos^2\theta\cos^2\theta' + 2\sin\theta\sin\theta'\cos\theta\cos\theta'\cos\varphi.$$

Therefore

$$\mathbf{S} = \frac{q^2\dot{\mathbf{u}}^2\mathbf{n}}{4\pi c^3 r^2}[\sin^2\Theta + 2\sin\theta\sin\theta'\cos\theta\cos\theta'\cos\varphi].$$

So $\mathbf{S}$ depends on φ in addition to θ as the physical situation demands. However, in certain cases the acceleration of the charge is parallel to its motion ($\theta' = 0$; charge in electric field, linear motion) or parallel to its motion ($\theta' = \pi/2$; charge in magnetic field). Then we have

$$\mathbf{S} = \frac{q^2\dot{\mathbf{u}}^2\mathbf{n}}{4\pi c^3 r^2}\sin^2\Theta.$$

In that case, with $q\dot{\mathbf{u}} = \ddot{\mathbf{p}}$, and $\mu = 1, v = c, R = r$, we recover (3.49). In other words, we get back the classical dipole radiation.

Cerenkov radiation   We have seen that when $\dot{\mathbf{u}} = 0$, we have a quasistatic field and there is no radiation. However, when a particle enters a medium with refractive index n, the velocity of light becomes $c/n < c$, and the speed u of the charge may exceed the speed of light in the medium. Then the factor $s/r = 1 - (un/c)\cos\theta$ will make the field tend to infinity in the direction $\theta = \cos^{-1}(c/un)$. Hence we will observe radiation in these directions as was found by Cerenkov in 1934. Cerenkov radiation is caused due to the sudden relative acceleration of the particle with respect to the wave velocity in crossing the boundary. This phenomenon is used for detecting the gamma rays coming from celestial objects and in many other applications.

Bremstrahlung   This kind of radiation is produced either when electrons experience retardation on hitting a target, or when they are deflected by positive charges in a plasma causing free-free transition, which is emission or absorption of a photon by a change in the kinetic energy of the electron in the presence of a positive charge.

In the first case, $\dot{\mathbf{u}}$ is parallel to $\mathbf{u}$ which leads to

$$\mathbf{S} = \frac{q^2}{4\pi c^3 r^2}\frac{\dot{\mathbf{u}}^2\sin^2\Theta}{(1 - u\cos\theta/c)^5}\,\mathbf{n} \equiv \frac{\mathbf{S}_0}{(1 - u\cos\theta/c)^5},$$

where $\mathbf{S}_0$ is the value of $\mathbf{S}$ when $u = 0$. The directional distribution of $\mathbf{S}$ is shown in Fig. 3.20. It differs from classical distribution $\mathbf{S}_0$ by the factor $(1 - u\cos\theta/c)^{-5}$.

In the case of free–free transitions, the electron moves along a hyperbola of eccentricity $e > 1$ around the positive ion of charge Z as shown in Fig. 3.21. If

p is the distance of the closest approach (when the electron is at P), then $r = p(e+1)/(1+e\cos\theta)$ and the acceleration can be written as $\dot{u} = Ze^2 mr^2$. However, near P where the acceleration is maximum, we can approximate the hyperbola by a straight line and put $r = (p^2 + u^2 t^2)^{1/2}$. Then,

$$W = \frac{2e^2}{3c^3}\dot{u}^2 = \frac{2Z^2 e^6}{3m^2 c^3}\int_{-\infty}^{\infty}\frac{dt}{(p^2 + u^2 t^2)^2}.$$

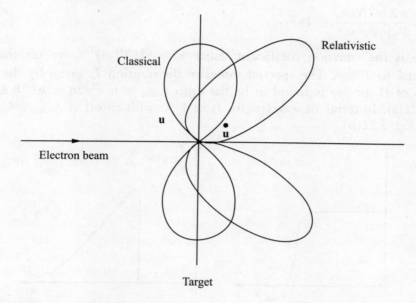

Fig. 3.20   *Bremstrahlung on hitting a target*

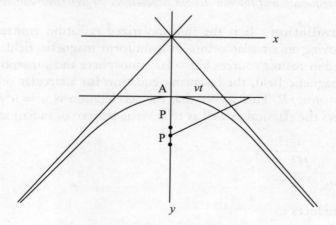

Fig. 3.21   *Free–free transition of an electron*

This integral can be evaluated easily and gives the result

$$W = \frac{\pi Z^2 e^6}{3m^2 c^3 p^3 u}.$$

If N_i is the number of ions per unit volume so that $p^3 = 1/N_i$ and N_e the electron concentration, the total radiation per unit volume will be

$$W = \frac{\pi Z^2 e^6 N_i N_e}{3m^2 c^3 u} \equiv 4\pi j_{ff},$$

where j_{ff} is the emission coefficient. Since $u \approx (2kT/m)^{1/2}$, we see that j_{ff} is proportional to $T^{-1/2}$. The spectral intensity distribution I_ν given by the Fourier transform of W or j_{ff} is found to be flat upto $\nu_{max} = mv^2/2h \approx kT/h$ as shown in Fig. 3.22(a). In terms of wavelength, $I_\lambda \propto 1/\lambda^2$ with cutoff at $\lambda_{min} = C/\nu_{max}$ as shown in Fig. 3.22(b).

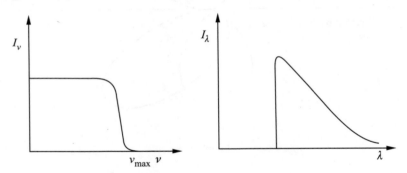

Fig. 3.22   *(a) Frequency and (b) wavelength dependence of free–free emission*

Synchrotron radiation   It is the fully polarized radiation emitted by rerlativistic electrons moving in circular orbits in a uniform magnetic field. It is found in synchrotrons and in cosmic sources like solar atmosphere and supernova remnants.

If **H** is the magnetic field, the balancing equation for a circular orbit of radius R is $e(\mathbf{u} \times \mathbf{H})/c = mu^2/R$. This shows that the acceleration is $\dot{u} = u^2/R = euH/mc$. Figure 3.23 shows the classical as well as relativistic pattern of radiation. As $m = \gamma m_0$, (3.131) gives

$$W = \frac{2}{3} \frac{e^4 \gamma^2 u^2 H^2}{m_0^2 c^5}.$$

For $u \approx c$, this reduces to

$$W = \frac{2}{3} \frac{e^4 \gamma^2 H^2}{m_0^2 c^3}.$$

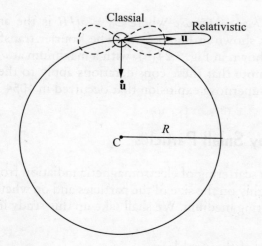

Fig. 3.23 *Emission patterns of synchrotron radiation — dotted line: classical, solid line: relativistic.*

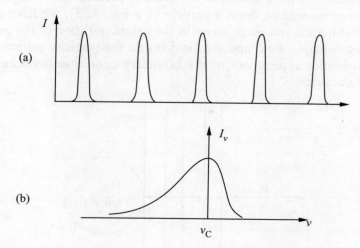

Fig. 3.24 *(a) Pulses of synchrotron radiation, (b) their frequency spectrum.*

As the energy of the electron is $E = mc^2 = \gamma m_0 c^2$, we finally get

$$W = \frac{2}{3} \frac{e^4 E^2 H^2}{m_0^4 c^5}. \tag{3.133}$$

The directional distribution of W differs from the classical distribution because of the angle transformation given by (2.63). In particular, as $u \to c$ and $\alpha \to \pi/2$, we find that $\tan \alpha' \to 1/\gamma$. Since this is small, we also have $\alpha' \to 1/\gamma$. Thus the radiation in the forward lobe is restricted to a small angle as shown in Fig. 3.23. So we see pulses of duration $\alpha'/\omega_0 = 1/\gamma\omega_0$, where $\omega_0 = u/R$ is the angular frequency and $R = cmu/eH$. This is shown in Fig. 3.24(a). The Fourier transform of this gives the frequency spectrum shown in Fig. 3.24(b), with a maximum at $\nu_c = 3eH\gamma^2/4\pi mc$.

It is interesting to note that these considerations apply to the radiation from Crab nebula, a remnant of supernova explosion that occurred in 1054 AD.

3.8 Scattering by Small Particles

We often come across scattering of electromagnetic radiation from small particles. The scattering depends highly on the size of the particles and on whether the particles make a dielectric or conducting medium. We shall take up this study in this section.

3.8.1 Formulation of the problem

Consider a particle A with a dielectric constant ϵ, permeability μ and conductivity σ. If a plane wave is incident upon a particle (see Fig. 3.25), we have to consider a combination of spherical and plane waves in the neighbourhood of the particle. In the presence of this radiation field and the field inside the particle, we have to find the transition probabilities appropriate to the boundary condition between vacuum and the medium of the particle.

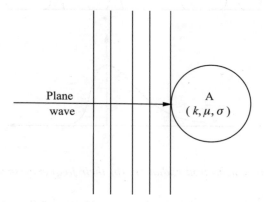

Fig. 3.25 *Plane radiation falling on a particle*

We are interested in radiation at large distances as compared to the size of the particle. We assume that there are no free charges, but there are induced charges which reside on the surface. Then the Maxwell equations for large distances are

$$\mathbf{\nabla} \times \mathbf{H} = \frac{4\pi}{c}\left(\sigma\mathbf{E} + \frac{\epsilon}{4\pi}\frac{\partial \mathbf{E}}{\partial t}\right),$$

$$\nabla \times \mathbf{E} = -\frac{\mu}{c}\frac{\partial \mathbf{H}}{\partial t},$$

$$\nabla \cdot \mathbf{D} = \nabla \cdot \mathbf{E} = 0, \nabla \cdot \mathbf{B} = \nabla \cdot \mathbf{H} = 0.$$

3.8.2 Boundary conditions

Since ϵ and μ are discontinuous at the boundary, $\nabla \cdot \mathbf{D}$ and $\nabla \cdot \mathbf{B}$ are not defined there. But we can still talk of surface integrals $\int_S \mathbf{D} \cdot \mathbf{n} dS$ and $\int_S \mathbf{B} \cdot \mathbf{n} dS$ as being zero. Similarly, $\nabla \times \mathbf{H}$ and $\nabla \times \mathbf{E}$ have no meaning on the surface. But we can talk of closed line integral such as $\oint_C \mathbf{H} \cdot d\mathbf{r}$ and $\oint_C \mathbf{E} \cdot d\mathbf{r}$ where the contour C cuts across the surface of discontinuity. Then we arrive at the following results.

Tangential component

Consider a rectangular loop cutting the boundary between two media, with its length l parallel to the boundary and the small edges cutting through it as shown in Fig. 3.26(a). Then, in the limit of small width, we have

$$\oint_C \mathbf{E} \cdot d\mathbf{r} = \int_S (\nabla \times \mathbf{E}) \cdot \mathbf{n} dS = (\mathbf{E}_1 - \mathbf{E}_2) \cdot \boldsymbol{l},$$

where C is the closed loop, S its surface and $\boldsymbol{l}$ a vector length tangential to the surface separating the two media. Since the surface integral can be made to tend to zero without limit by shrinking the loop, it follows that

$$E_{1t} = E_{2t},$$

where the subscript t denotes the tangential component. We would similarly have

$$H_{1t} = H_{2t}.$$

Thus, the tangential components of $\mathbf{E}$ and $\mathbf{H}$ are continuous at the boundary across the (uncharged) surface.

Normal component

Consider a rectangular box intersecting the boundary, with the two larges faces parallel to the boundary in two different media and the other four faces very thin and normal to the boundary; see Fig. 3.26 (b). If $d\sigma$ is the area of the surface parallel to the boundary, we get

$$\int_S \mathbf{D} \cdot \mathbf{n} d\sigma = (\mathbf{D}_1 - \mathbf{D}_2) \cdot \mathbf{n} d\sigma.$$

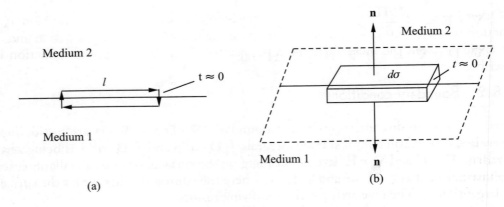

Fig. 3.26   *(a) Line and (b) surface integrals at the boundary between two media*

As the width of the box shrinks to zero, we get

$$D_{1n} = D_{2n},$$

where the subscript n represents the normal component. Similarly, we would have

$$B_{1n} = B_{2n}.$$

Thus the normal components of $\epsilon\mathbf{E}$ and $\mu\mathbf{H}$ are continuous across the boundary.

Nonetheless, the normal components of $\mathbf{E}$ and $\mathbf{H}$ and the tangential components of $\mathbf{D}$ and $\mathbf{B}$ would be discontinuous across the boundary, the discontinuity depending upon the properties of the two media. Further, the tangential component of $\mathbf{E}$ and $\mathbf{H}$ would be discontinuous if there are free charges on the boundary, and the normal component of $\mathbf{D}$ and $\mathbf{B}$ would be discontinuous if there are free currents on the boundary.

3.8.3   Absorption and scattering

We are interested in studying the general features and not in the exact solution of the problem of scattering by small particles. We wish to know (i) the absorption (or mass absorption) cross-section coefficient and (ii) scattering properties of the particles as a function of wavelength of the incident radiation and the shape and size of the particles. One of the important scattering properties is the ratio of the total scattered radiation to the total incident radiation. Suppose σ_a is the absorption cross-section and σ_s is the scattering cross-section, then the scattering efficiency or albedo is defined as

$$r = \frac{\sigma_s}{\sigma_a + \sigma_s}.$$

Since some absorption is inevitable in a real system, we have $0 < r < 1$.

Other important scattering properties are the phase function and polarization as a function of direction. If $\Phi(\theta, \varphi)$ is the direction-dependent scattering function and α the angle between the incident ray and the scattered ray, then the phase function is described by

$$g = \frac{\int \Phi(\theta, \varphi) \cos\alpha\, d\Omega}{\int \Phi(\theta, \varphi) d\Omega} \equiv \overline{\cos\alpha},$$

the weighted average of $\cos\alpha$. It is clear that $-1 \le g \le 1$; $g = 1$ or -1 implies that all radiation is scattered in the forward or backward direction. Generally, $g > 0$ for small particles which diffract most of the radiation in the forward direction and $g < 0$ for large particles which reflect most of the radiation in the backward direction; $g = 0$ signifies a balance between forward and backward scattering.

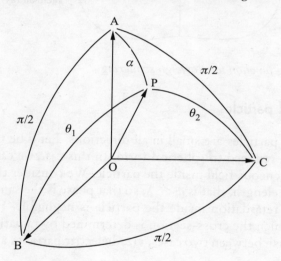

Fig. 3.27 *A beam is incident along OA, suffers dipole scattering at O and the scattered ray goes along OP.*

For example, consider the dipole scattering of electromagnetic radiation. Referring to Fig. 3.27, let an unpolarized beam of EM radiation be incident on a material along OA. Let OB and OC be normal to OA such that OA, OB, OC form a right-handed Cartesian coordinate system. The incident beam induces an electrical dipole at O and is scattered along OP. Let OP make an angle α with OA, θ_1 with OB and θ_2 with OC. Let θ be angle between OP and the dipole which will be in the plane BOC. Then the direction cosine of OP will be $\cos\theta_1, \cos\theta_2, \cos\alpha$, and we have

$$\cos^2\theta_1 + \cos^2\theta_2 + \cos^2\alpha = 1.$$

In the case of a dipole, the scattering depends upon the angle θ between the scattered ray and the direction of the dipole moment. It does not depend directly upon the angle between the scattered ray and the incident ray. But in the case of natural light where there are two states of polarization, we can get the dependence on α by averaging over the two states. Since the scattering is proportional to $\sin^2 \theta$, scattering for natural light will be proportional to $\sin^2 \theta_1 + \sin^2 \theta_2$ or to $\Phi \propto 1 + \cos^2 \alpha$. Thus for Rayleigh scattering we have the scattering proportional to $1 + \cos^2 \alpha$, which is shown in Fig. 3.28. We have to normalise Φ to unity through $\int \Phi d\Omega = 1$. Scattering patterns become very complex when $2\pi a \approx \lambda$ due to diffraction effects.

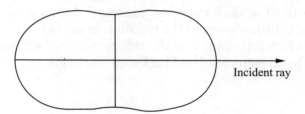

Incident ray

Fig. 3.28   *Scattering function for Rayleigh scattering*

3.8.4   Very small particles

We assume that the particles are small in all directions. Let a be the equivalent radius of the particle when reduced to spherical form. In this case we can consider radiation emitted by a homogeneous field inside the particle. We consider the size a to be much smaller than the wavelength, that is $a \ll \lambda$, so that phase is the same everywhere inside the particle. Hence retardation inside the particle is negligible. Further, we consider the dipole case in which the cross-section is determined by radiation.

We shall distinguish between two cases, viz, dielectric particles and metallic particles.

Dielectric particles

In the case of dielectrics, we are far away from those wavelengths where absorption is large. Therefore absorption coefficient is small. Hence the refractive index is real. If n is the refractive index, then (3.99) hold good. The net dipole moment $\mathbf{p}$ of the whole particle in an electric field $\mathbf{E}$ is the polarization multiplied by the volume $4\pi a^3/3$ of the particle. Thus,

$$\mathbf{p} = \chi \mathbf{E} \cdot 4\pi a^3/3 = \frac{(n^2 - 1)a^3 \mathbf{E}}{n^2 + 2}.$$

Therefore the energy radiated per unit time, from (3.50) with $\mu = 1$ and $v = c$, is

$$W = \frac{2}{3c^3}(\ddot{\mathbf{p}})^2 = \frac{2}{3c^2}\left(\frac{n^2 - 1}{n^2 + 2}\right)a^6(\ddot{\mathbf{E}})^2.$$

If $\mathbf{E}$ varies sinusoidally with an (angular) frequency ω, then $\ddot{\mathbf{E}} = \omega^2 \mathbf{E}$, and

$$W = \frac{2a^6 \omega^4 E^2}{3c^3} \left(\frac{n^2 - 1}{n^2 + 2} \right)^2.$$

As the incident flux is $cE^2/4\pi$, the equivalent cross-section, which is the ratio of the total energy scattered per unit time to the incident flux, will be

$$\sigma = \frac{8\pi a^2}{3} \left(\frac{2\pi a}{\lambda} \right)^4 \left(\frac{n^2 - 1}{n^2 + 2} \right)^2. \tag{3.134}$$

This is Rayleigh scattering of particles. In (3.134), πa^2 is just the geometrical cross-section of the particle or the target area. Since we are working in the region $a/\lambda \ll 1$ and $(n^2 - 1)/(n^2 + 2) \simeq 1$, we would have $\sigma \ll \pi a^2$, that is the scattering cross-section would be much smaller than the target area.

We can consider the mass scattering coefficient as the scattering cross-section per unit mass of the scatterer. If ρ is the mass density of the target, then the mass scattering coefficient would be given by

$$\chi = \sigma / (4\pi a^3 \rho / 3) = \frac{2}{a\rho} \left(\frac{2\pi a}{\lambda} \right)^4 \left(\frac{n^2 - 1}{n^2 + 2} \right)^2.$$

It can be noticed that the mass scattering coefficient is proportional to the target volume a^3. We also see that σ or χ is proportional to λ^{-4}. It is found that interstellar scattering is approximately proportional to λ^{-1}. Therefore Rayleight scattering does not seem applicable there.

Metallic particles

Here the dissipation of energy is mainly due to ohmic resistance, and it appears as heat in the far infrared. This is mainly a case of absorption rather than scattering.

The effective current density in this case is given by

$$\mathbf{J}_{\text{eff}} = \mathbf{J} + d\mathbf{P}/dt = \sigma \mathbf{E} + \chi d\mathbf{E}/dt$$

where σ is the conductivity. If the electric field varies as $e^{i\omega t}$, then

$$\mathbf{J}_{\text{eff}} = \sigma \mathbf{E} + i\omega \chi \mathbf{E}.$$

If n is the complex refractive index, using (3.99b) we see that the effective complex conductivity becomes

$$\sigma_{\text{eff}} = \sigma + \frac{3i\omega(n^2 - 1)}{4\pi(n^2 + 2)}.$$

The dissipation of energy per unit time per unit volume is $\mathbf{J} \cdot \mathbf{E}$. Hence the dissipation of energy in the volume of the particle of size a is

$$S = \frac{3\omega E^2}{4\pi} \mathrm{Im}\left(\frac{1-n^2}{n^2+2}\right)\frac{4\pi a^3}{3}.$$

Therefore the effective cross-section is[14]

$$\sigma = \frac{4\pi S}{cE^2} = 4\pi a^2 \left(\frac{2\pi a}{\lambda}\right) \mathrm{Im}\left(\frac{1-n^2}{2+n^2}\right).$$

Then the mass scattering cross-section comes out to be

$$\chi = \frac{3}{a\rho}\left(\frac{2\pi a}{\lambda}\right)\mathrm{Im}\left(\frac{1-n^2}{2+n^2}\right).$$

3.9 Electric Quadrupole and Magnetic Dipole Radiation

The total radiation coming out of a system can be expressed as the sum of an infinite series. The zeroth order term corresponds to a system which has no charges in it and hence has all its multipoles vanished. If the system has charges and has an electric monopole Q, the first non-vanishing term is the electric monopole term. If the system is electrically neutral as a whole, as mostly happens in natural phenomena, the dominant terms, although weaker, are the electric dipole term, followed by electric quadrupole and magnetic dipole terms, etc. We consider this case in the present section; we assume that the system has no net electric charge and also no net electric dipole moment. Writing the quadrupole radiation fields as $\mathbf{E}_2$ and $\mathbf{H}_2$ with $\mathbf{E}_2 = \mathbf{H}_2 \times \mathbf{n}$, we have, from (3.47),

$$\mathbf{E}_2 = \frac{1}{c^3 R}\int_V \mathbf{r}\cdot\mathbf{n}(\ddot{\mathbf{J}}\times\mathbf{n})\times\mathbf{n}d\tau, \tag{3.135a}$$

$$\mathbf{H}_2 = \frac{1}{c^3 R}\int_V \mathbf{r}\cdot\mathbf{n}(\ddot{\mathbf{J}}\times\mathbf{n})d\tau. \tag{3.135b}$$

To put these fields in a tractable form, let us consider

$$\int (\mathbf{n}\cdot\mathbf{r})\ddot{\mathbf{J}}d\tau = \frac{1}{2}\int [(\mathbf{n}\cdot\mathbf{r})\ddot{\mathbf{J}} + (\mathbf{n}\cdot\ddot{\mathbf{J}})\mathbf{r}]d\tau + \frac{1}{2}\int [(\mathbf{n}\cdot\mathbf{r})\ddot{\mathbf{J}} - (\mathbf{n}\cdot\ddot{\mathbf{J}})\mathbf{r}]d\tau$$

---

[14]Owing to the paucity of symbols, we use σ for conductivity as well as scattering cross-section here. But the context has been mentioned wherever there is such 'degeneracy'.

$$= \frac{1}{2} \int [(\mathbf{n} \cdot \mathbf{r})\ddot{\mathbf{J}} + (\mathbf{n} \cdot \ddot{\mathbf{J}})\mathbf{r}]d\tau - \frac{1}{2} \int \mathbf{n} \times (\mathbf{r} \times \ddot{\mathbf{J}})d\tau. \tag{3.136}$$

The orbital magnetic moment $\mathbf{m}_{\mathrm{orb}}$ of the current distribution is then given by

$$\mathbf{m}_{\mathrm{orb}} = \frac{1}{2c} \int \mathbf{r} \times \mathbf{J}(\mathbf{r})d\tau.$$

For discrete charges q_k and $\mathbf{r}_k$ moving with velocities $\mathbf{v}_k$, it will reduce to

$$\mathbf{m}_{\mathrm{orb}} = \frac{1}{2c} \sum_k \mathbf{r}_k \times (q_k \mathbf{v}_k).$$

We note that $\mathbf{r}_k \times \mathbf{v}_k$ is related to the mechanical angular momentum of the k th charge. When we disregard the factor of mass, as in $\mathbf{r}_k \times \mathbf{v}_k$, and multiply it with charge q_k, it gives the magnetic moment associated with electric currents.

We also have to add the magnetic moment due to the spin of the electron. If $s_k\hbar$ is the spin magnetic moment of the k th charge, then we can write the total magnetic moment as

$$\mathbf{m} = \frac{1}{2c} \int \mathbf{r} \times \mathbf{J}d\tau + \sum_k s_k\hbar,$$

where the summation over k, like the integral over $\mathbf{r}$, is over the entire source.

We now consider the two terms on the right-hand side of (3.136). Let us write them as

$$\mathbf{I}_1 = \frac{1}{2} \int [(\mathbf{n} \cdot \mathbf{r})\ddot{\mathbf{J}} + (\mathbf{n} \cdot \ddot{\mathbf{J}})\mathbf{r}]d\tau, \tag{3.137a}$$

$$\mathbf{I}_2 = \frac{1}{2} \int \mathbf{n} \times (\mathbf{r} \times \ddot{\mathbf{J}}(\mathbf{r}))d\tau, \tag{3.137b}$$

and note that both are vectors. Then the second integral, (3.137b), can be written as

$$I_2 = -c\mathbf{n} \times \ddot{\mathbf{m}}_{\mathrm{orb}}, \tag{3.138}$$

due to which (3.136) becomes

$$\int (\mathbf{n} \cdot \mathbf{r})\ddot{\mathbf{J}}d\tau = \mathbf{I}_1 - \mathbf{I}_2. \tag{3.139}$$

Now we will work out the first integral $\mathbf{I}_1$. To simplify it, consider a similar integral $\mathbf{I}_3$ with $\ddot{\mathbf{J}}$ replaced by $\mathbf{J}$. Then the x-component of $\mathbf{I}_3$ can be written as

$$I_{3x} = \frac{1}{2} \int [(\mathbf{n} \cdot \mathbf{r})\mathbf{J} + (\mathbf{n} \cdot \mathbf{J})\mathbf{r}]_x d\tau = \frac{1}{2} \int (\mathbf{n} \cdot \mathbf{r})J_x + (\mathbf{n} \cdot \mathbf{J})x] d\tau. \qquad (3.140)$$

Consider the vector expression $x(\mathbf{r} \cdot \mathbf{n})\mathbf{J}$ and apply Gauss divergence theorem to it over the entire source volume. The surface integral of the normal component of $\mathbf{J}$ will vanish because all the currents (and charges) are confined to a finite volume. Hence we have

$$\int \boldsymbol{\nabla} \cdot [x(\mathbf{r} \cdot \mathbf{n})\mathbf{J}] d\tau = 0. \qquad (3.141)$$

On using appropriate vector identities, this gives

$$\int [x(\mathbf{r} \cdot \mathbf{n})\boldsymbol{\nabla} \cdot \mathbf{J} d\tau + (\mathbf{r} \cdot \mathbf{n})\mathbf{J} \cdot \nabla x + x\mathbf{J} \cdot \nabla(\mathbf{r} \cdot \mathbf{n})] d\tau = 0. \qquad (3.142)$$

We note that $\mathbf{J} \cdot \nabla x = J_x, \boldsymbol{\nabla} \cdot \mathbf{J} = -\partial \rho / \partial t$, and

$$\boldsymbol{\nabla}(\mathbf{r} \cdot \mathbf{n}) = \mathbf{n}.$$

Using all these, (3.142) becomes

$$\int [-x\mathbf{r} \cdot \mathbf{n}\partial \rho / \partial t + \mathbf{r} \cdot \mathbf{n} J_x + x\mathbf{J} \cdot \mathbf{n}] d\tau = 0. \qquad (3.143)$$

Notice that the terms appearing in I_{3x} of (3.140) are contained in (3.143). We can write equations similar to (3.143) for y and z-components. Then we will find that

$$\mathbf{I}_{3i} \equiv \frac{1}{2} \int [(\mathbf{n} \cdot \mathbf{r})\mathbf{J} + (\mathbf{n} \cdot \mathbf{J})\mathbf{r}]_i d\tau = \frac{1}{2} \frac{d}{dt} \int \rho x_i x_n n_j d\tau, \qquad (3.144)$$

where x_i are components of $\mathbf{r}$ and n_i are those of $\mathbf{n}$, and summation over repeated indices is assumed. We define the *quadrupole moment* of the charge distribution as the tensor of rank two given by

$$Q_{ij} = \int \rho x_i x_j d\tau. \qquad (3.145)$$

Then (3.144) becomes

$$\frac{1}{2} \int [(\mathbf{n} \cdot \mathbf{r})\mathbf{J} + (\mathbf{n} \cdot \mathbf{J})\mathbf{r}]_i d\tau = \frac{1}{2}\dot{Q}_{ij} n_j, \qquad (3.146)$$

with summation over repeated indices. Going back to (3.137a), we see that

$$\mathbf{I}_{1i} = \frac{1}{2}\frac{d^3}{dt^3}Q_{ij}n_j. \qquad (3.147)$$

Note that $\mathbf{I}_1$ is a tensor of rank one, that is a vector. This finally gives us the quadrupole electromagnetic fields as

$$\mathbf{E}_2 = \frac{1}{2c^3 R}[(\mathbf{n} \cdot \mathbf{I}_1)\mathbf{n} - \mathbf{I}_1] + \frac{1}{c^2 R}\mathbf{n} \times \ddot{\mathbf{m}}_{\mathrm{orb}}, \quad \mathbf{H}_2 = \mathbf{n} \times \mathbf{E_2}. \tag{3.148}$$

Comparing them with the dipole fields obtained in Section 3.3(d), we see that dipole and quadrupole fields behave similarly, except for the magnitude.

If the source is electrically neutral and also has no electric dipole, then the electrical quadrupole term is the first non-vanishing term and becomes important. Although it is much weaker, it can be observed. (If the electric dipole term is present, then it becomes difficult to separate out the weak electric quadrupole contribution.) If we further drop the magnetic dipole and consider only the case of the electric quadrupole, the energy flux is seen to be

$$S = \frac{c}{4\pi}E^2 = \frac{1}{16\pi c^5 R^2}[(\mathbf{n} \cdot \mathbf{I}_1)\mathbf{n} - \mathbf{I}_1]^2 = \frac{1}{16\pi c^5 R^2}[\mathbf{I}_1^2 - (\mathbf{n} \cdot \mathbf{I}_1)^2]. \tag{3.149}$$

In general, the electric quadrupole moment tensor Q_{ij} will be non-diagonal, though symmetric. By choosing a proper coordinate system, that is, by making a principal axes transformation, it can be made a diagonal tensor with components Q_{xx}, Q_{yy}, Q_{zz}. If the unit vector $\mathbf{n}$ has components (n_x, n_y, n_z), then it can be seen from (3.147) that

$$I_1 = \frac{1}{2}\frac{d^3}{dt^3}\begin{bmatrix} Q_{xx}n_x \\ Q_{yy}n_y \\ Q_{zz}n_z \end{bmatrix}, \tag{3.150a}$$

and

$$\mathbf{n} \cdot \mathbf{I}_1 = \frac{1}{2}\frac{d^3}{dt^3}(Q_{xx}n_x^2 + Q_{yy}n_y^2 + Q_{zz}n_z^2). \tag{3.150b}$$

Then the energy flux of (3.149) becomes

$$S = \frac{1}{32\pi c^5 R^2}\frac{d^3}{dt^3}[Q_{xx}n_x^2(1-n_x)^2 + Q_{yy}n_y^2(1-n_y^2) + Q_{zz}n_z^2(1-n_z^2)$$

$$-2Q_{xx}Q_{yy}n_x^2 n_y^2 - 2Q_{yy}Q_{zz}n_y^2 n_z^2 - 2Q_{zz}Q_{xx}n_z^2 n_x^2]. \tag{3.151}$$

Appendix 3.1
Solution of Wave Equation for Superpotential

The superpotential was introduced in Section 3.3(a). We wish to find the solution of the wave equation

$$\nabla^2 \mathbf{\Pi} - \frac{1}{c^2} \frac{\partial^2 \mathbf{\Pi}}{\partial t^2} = -4\pi \mathbf{Z}. \tag{A3.1}$$

First let us consider the corresponding equation for the scalar potential φ,

$$\nabla^2 \varphi - \frac{1}{c^2} \frac{\partial^2 \varphi}{\partial t^2} = -4\pi \rho. \tag{A3.2}$$

In empty space, $\rho = 0$, and we have

$$\nabla^2 \varphi_0 - \frac{1}{c^2} \frac{\partial^2 \varphi_0}{\partial t^2} = 0. \tag{A3.3}$$

Let $\varphi_0(\mathbf{r}, t) = g(r, t)/r$; then (A3.3) reduces to

$$\frac{\partial^2 g}{\partial r^2} - \frac{1}{c^2} \frac{\partial g^2}{\partial t^2} = 0. \tag{A3.4}$$

Since the general solution of this equation is $F(r - ct)$ and $G(r + ct)$, where F and G are arbitrary functions, we have

$$\varphi_0(\mathbf{r}, t) = \frac{F(r - ct)}{r} + \frac{G(r + ct)}{r}. \tag{A3.5}$$

Here F represents an outgoing disturbance and G an incoming disturbance with speed c.

We shall consider the disturbance G arriving at a point P in volume τ bounded by a closed surface Σ. Let τ_1 be the volume of a sphere of radius ϵ centred on P, with the boundary surface Σ_1. Then returning to (A3.2) and using Gauss divergence theorem, we get

$$\int_{\tau - \tau_1} (\varphi \nabla^2 \varphi_0 - \varphi_0 \nabla^2 \varphi) d\tau = \oint_\Sigma (\varphi \nabla \varphi_0 - \varphi_0 \nabla \varphi) \cdot \mathbf{n} d\Sigma + \oint_{\Sigma_1} (\varphi \nabla \varphi_0 - \varphi_0 \nabla \varphi) \cdot \mathbf{n} d\Sigma_1,$$

where the region of integration on both sides is the annular region with the inner surface Σ_1 and the outer surface Σ, and $\mathbf{n}$ is the outward unit normal, outward for the annular region, that is inward on Σ_1 and outward on Σ. Substituting from (A3.2) and (A3.3), we get

$$\int_{\tau-\tau_1} \left(\frac{\varphi}{c^2}\frac{\partial^2 \varphi_0}{\partial t^2} - \frac{\varphi_0}{c^2}\frac{\partial^2 \varphi}{\partial t^2} + 4\pi\varphi_0\rho \right) d\tau$$

$$= \oint_{\Sigma} \left(\varphi\frac{\partial \varphi_0}{\partial r} - \varphi_0\frac{\partial \varphi}{\partial r} \right) \cos\theta \, d\Sigma - \oint_{\Sigma_1} \left(\varphi\frac{\partial \varphi_0}{\partial r} - \varphi_0\frac{\partial \varphi}{\partial r} \right) \cos\theta \, d\Sigma_1, \quad \text{(A3.6)}$$

where $d\Sigma_1 = \epsilon^2 d\Omega$ and $\cos\theta = -1$ on Σ_1. Remembering that $r = \epsilon$ on the inner surface Σ_1 and with some algebra, (A3.6) can be written as

$$\frac{1}{c^2}\int_{\tau-\tau_1} \left[\frac{\partial}{\partial t}\left(\varphi\frac{\partial \varphi_0}{\partial t} - \varphi_0\frac{\partial \varphi}{\partial t} \right) + \frac{4\pi c^2}{r}G(r+ct)\rho(t) \right] d\tau$$

$$= \frac{1}{c^2}\oint_{\Sigma} \varphi\cos\theta \left[\frac{G'(r+ct)}{r} - \frac{G(r+ct)}{r^2} - \frac{G(r+ct)}{r}\frac{\partial \varphi}{\partial r} \right] d\Sigma$$

$$- \frac{1}{c^2}\oint_{\Sigma_1} \left[\varphi\epsilon G'(\epsilon+ct) - \varphi G(\epsilon+ct) - \frac{\epsilon^2 G(\epsilon+ct)}{\epsilon}\frac{\partial \varphi}{\partial r} \right] d\Omega. \quad \text{(A3.7)}$$

As ϵ approaches zero, the second term becomes $4\pi G(ct)\varphi_{\rm P}$. Then integrating (A3.7) with respect to time from t_1 to t_2, we get

$$\frac{1}{c^2}\int_{\tau} \left\{ \varphi\frac{\partial \varphi_0}{\partial t} - \varphi_0\frac{\partial \varphi}{\partial t} \right\}_{t_1}^{t_2} + \frac{4\pi c^2}{r^2}\int_{t_1}^{t_2} G(r+ct)\rho(t)dt \Big\} dt$$

$$= \oint_{\Sigma}\int_{t_1}^{t_2} \left\{ \varphi\cos\theta \left[\frac{G'(r+ct)}{r} - \frac{G(r+ct)}{r^2} \right] - \frac{G(r+ct)}{r}\frac{\partial \varphi}{\partial r} \right\} dt \, d\Sigma$$

$$+ 4\pi\int_{t_1}^{t_2} G(ct)\varphi_{\rm P}(t)dt. \quad \text{(A3.8)}$$

Now let $G(ct)$ have a finite value at t_0 and zero at all other times, that is, we have a disturbance starting at t_0 from $\mathbf{r}$; so let

$$\int_{t_1}^{t_2} G(ct)dt = 1, \quad \text{with} \quad G(ct_1) = G(ct_2) = 0.$$

Then $\displaystyle\int_{t_1}^{t_2} G(ct)\varphi_{\rm P}(t)dt = \varphi_{\rm P}(t_0).$

Now $G(r+ct)$ will be non-vanishing only near $t = t_0 - r/c$. Then for any function $u(t)$, we have

$$\int_{t_1}^{t_2} G(r+ct)u(t)dt = u(t_0 - r/c), \quad \text{and}$$

$$\int_{t_1}^{t_2} G'(r+ct)u(t)dt = -\frac{1}{c}\frac{\partial u(t_0 - r/c)}{\partial t}.$$

Also $\varphi_0 = G(r+ct)/r$ and $\partial\varphi_0/\partial t = cG'(r+ct)/r$ vanish at t_1 and t_2. Then (A3.8), on rearranging terms, becomes

$$\varphi_P(t_0) = \int_\tau \frac{\rho(t_0 - r/c)}{r}d\tau$$

$$+\frac{1}{4\pi}\oint_\Sigma \left\{ \frac{\cos\theta}{cr}\left(\frac{\partial\varphi}{\partial t}\right) + \frac{\cos\theta}{r^2}\varphi(t_0 - r/c) + \frac{1}{r}\left(\frac{\partial\varphi}{\partial r}\right)_{t_0 - r/c} \right\}d\Sigma. \qquad \text{(A3.9)}$$

This is the solution of (A3.2). In the same way we get the solution of (A3.1) as

$$\mathbf{\Pi}(t_0) = \int_\tau \frac{\mathbf{Z}(t_0 - r/c)}{r}d\tau + \frac{1}{4\pi}\oint_\Sigma \left[\frac{\mathbf{\Pi}\cos\theta}{r^2} + \frac{\partial\mathbf{\Pi}}{\partial t}\frac{\cos\theta}{cr} + \frac{1}{r}\frac{\partial\mathbf{\Pi}}{\partial r}\right]_{t_0 - r/c}d\Sigma.$$

Problems

3.1. Show that the Hamiltonian for a non-relativistic electron in an electromagnetic field described by potentials $(\varphi, \mathbf{A})$ is

$$H = e\varphi + \frac{1}{2m}(\mathbf{p} - e\mathbf{A}/c)^2.$$

3.2. Verify that $\mathbf{r} = \mathbf{A}_0 \exp(-\gamma t/2 + i\omega_0 t)$ is a solution of (3.71) when higher order terms are neglected.

3.3. Show that $\Box^2 = \nabla^2 - \frac{1}{c^2}\frac{\partial^2}{\partial t^2}$ is invariant under Lorentz transformation. (Here ∇^2 is the 3-D Laplacian operator.)

3.4. Obtain the Lorentz transformation of $\mathbf{E}$ and $\mathbf{B}$ from that of electromagnetic potentials φ and $\mathbf{A}$, and of the spacetime gradient four-vector.

3.5. Show that $\mathbf{E}.\mathbf{B}$ and $E^2 - B^2$ are invariant under Lorentz transformations.

3.6. Discrete charges $q_i, 1 \le i \le n$, are situated at points Q_i along a straight line. Consider a field point P outside the line of charges. Let the line Q_iP make an angle θ_i with the line of charges. Show that the electric field line passing through P satisfies the equation

$$\sum_{i=1}^{n} q_i \cos\theta_i = \text{constant}.$$

3.7. Taking $\mathbf{E} \equiv \mathbf{E}(lx + my + nz - vt)$ and $\mathbf{H} \equiv \mathbf{H}(lx + my + nz - vt)$ as solutions of Maxwell's equations, show that (a) $\mathbf{E}$ and $\mathbf{H}$ are perpendicular to $\mathbf{n}$, the direction of propagation, and (b) $\mathbf{E}, \mathbf{H}$ and $\mathbf{n}$ form a right-handed orthogonal system.

3.8. Show that $F(\mathbf{r} \cdot \mathbf{n} - vt)$ indicates a plane wave and $G(r - vt)$ indicates a spherical wave. Here $\mathbf{n}$ is a fixed unit vector. (Hint: Think of wavefronts, that is, loci of points having the same phase.)

3.9. Some electromagnetic phenomena are taking place in a finite medium. An observer O stationary with rerspect to the medium, observes fields $\mathbf{E}, \mathbf{B}, \mathbf{D}, \mathbf{H}$ and relates them as $\mathbf{D} = \epsilon\mathbf{E}, \mathbf{H} = \mathbf{B}/\mu$ (with usual meanings of symbols). Another observer O' is moving with respect to O with a uniform velocity $\mathbf{v}$, and observes the same fields as $\mathbf{E}', \mathbf{B}', \mathbf{D}', \mathbf{H}'$. Obtain $\mathbf{D}'$ and $\mathbf{H}'$ in terms of $\mathbf{E}'$ and $\mathbf{B}'$. (Hint: Express $\mathbf{D}'$ in terms of $\mathbf{D}$ and $\mathbf{H}$, then express $\mathbf{D}$ and $\mathbf{H}$ in terms of $\mathbf{E}$ and $\mathbf{B}$, and $\mathbf{E}, \mathbf{B}$ back in terms of $\mathbf{E}'$ and $\mathbf{B}'$; finally do the same for $\mathbf{H}'$.)

4

Thermodynamics

THermodynamics is a phenomenological theory of great significance that was developed and perfected in the eighteenth and nineteenth centuries. It began with models and theories which failed to agree with experiments. Moreover, inconsistencies cropped up every now and then. The famous one among the failed theories was the theory of the caloric. But they were removed soon and a consistent and satisfying thermodynamics was developed. The significance of thermodynamics lies in the fact that it tries to study and predict the bulk behaviour of matter in its three phases from just external observations, without recourse to what goes on inside it. And it had to be that way because atoms as elementary constituents of matter were not known in those periods, except in a philosophical sense. In spite of this, there was a systematic development of the subject, and the internal consistency of thermodynamics remain one its amazing features.

In this chapter we start with various technical terms which are needed in this area of physics. This includes the basic thermodynamic system with the associated variables that describe it. Then follow the laws of thermodynamics, the concepts of absolute temperature and entropy, the chapter ending with the phase rule which governs the coexistence of different phases of matter in equilibrium.

4.1 Introduction

Thermodynamics is a theory of bulk matter, a macroscopic theory which does not peep into the individual constituents, viz. atoms or molecules. It would be appropriate to begin by describing some technical terms often occurring in this branch of science.

A *thermodynamic system* is an amount of matter or radiation having some volume and enclosed by a surface. The outside matter with which it can exchange energy is called the *surroundings* of the system or the *bath*. Any macroscopic system is described by a certain set of *s macroscopic thermodynamic parameters*, such as pressure p, volume V, temperature T, magnetic field $\mathbf{B}$, magnetization $\mathbf{M}$ and such others that may be required to specify the bulk system completely. A set of specific values of all these parameters is said to constitute a *s thermodynamic state* of the system. A thermodynamic system can thus have an infinite number of thermodynamic states.

4.1.1 Thermodynamic equilibrium

An isolated thermodynamic system, when left to itself for a sufficiently long time, attains a state in which

(i) there is no net transport of radiation or energy at any point in the system, i.e., we have *thermal equilibrium,*

(ii) there is no net flow of matter at any point, i.e., we have *hydrostatic* or *mechanical equilibrium,* and

(iii) there is no net change in the chemical composition of the system, i.e., we have *chemical equilibrium.*

A system is thus said to be in *thermodynamic equilibrium* when the thermodynamic state of the system does not change with time. Naturally, in such a state, all parameters of the system remain constant with respect to time. One such parameter is the *temperature* T, which remains constant in all parts of the system. That such a parameter exists is an empirical fact which is sometimes called the *zeroth law* of thermodynamics. Hydrostatic equilibrium is characterized by another quantity, viz., *pressure*, which adjusts itself to the mechanical forces acting on any portion of the system. Chemical equilibrium determines the relative abundance of various constituents giving rise to such laws as the law of mass action, phase rule, excitation and ionization equations, etc.

4.1.2 Thermodynamic coordinates

Temperature, pressure, volume, internal energy and other such attributes of the thermodynamic system are known as its thermodynamic coordinates. Some of the coordinates are proportional to the mass of the system; they are called *extensive variables*. They are conventionally denoted by capital letters. Volume V and internal energy U are examples of such parameters. On dividing these by mass of the system, we get the specific values of these quantities which are then known as *intensive variables*. An example is $v = V/M = 1/\rho$, where M is the mass of the system and ρ is the density. Intensive variables are denoted by small letters. Pressure p and temperature t are neither intensive nor extensive variables. We shall denote them by small letters.

A thermodynamic system may be *homogeneous*, in which the chemical composition and phase (i.e., solid, liquid or gas) are the same throughout the system (i.e., water, ice, etc.), or *heterogeneous*, in which there are variations of phase with position in the system. In either case, pressure p and temperature t would be constant throughout the system under equilibrium conditions.

A system may be a mixture of different chemical elements, in thermodynamic equilibrium. For example, air is a mixture of about 78% nitrogen, 21% oxygen, and trace amounts of several other gases making up the remaining 1%. There is no net transport of these components under conditions of steady state equilibrium. This means that any finite volume in the system will have the same proportion of various components and the p and t will be the same everywhere.

We can, however, define partial pressures p_i and partial volumes V_i for each component. Let N_i be the number of molecules of component i, so that the total number of molecules in the system is $N = \sum_i N_i$. Imagine that all molecules except those of component i are taken out, and only N_i molecules of component i are left in the entire volume V, the temperature being the same. Then the pressure p_i is called the *partial pressure* of component i. It is clear that the original pressure $p = \sum_i p_i$.

Obviously p_i is smaller than p. Now if we compress these N_i molecules to the pressure p, maintaining the temperature same, then the volume V_i which they occupy is the *partial volume* V_i of component i. Again it is clear that $V = \sum_i V_i$.

4.1.3   Equation of state

For every substance of a given composition, we find that p, V, t satisfy a relation

$$F(p, V, t) = 0, \tag{4.1}$$

known as the *equation of state*. In heterogeneous systems consisting of different phases, p and t are the same throughout in equilibrium, but v undergoes an abrupt change at the interface of two phases. The equation of state can be represented by a surface in the $p - V - t$ space and by its projections in the $p - V$, $p - t$ and $V - t$ planes. Here we give examples of some equations of state which are commonly followed by systems.

The ideal gas equation

It is given by

$$pv = Rt, \tag{4.2}$$

where R is a constant known as the *gas constant*. It can be derived on the basis of the kinetic theory of gases (to be discussed in Chapter 8) in which the gas is made up of weakly-interacting molecules. Most gases behave like a perfect gas at low pressures. The perfect gas equation embodies the experimentally verified Boyle's law ($pv =$ constant at constant t) and Charles law ($p \propto t$ at constant v). The gas constant has been experimentally found to be $R = 8.314 \times 10^7$ erg per degree per mole, where one gm-mole contains $N_A = 6.025 \times 10^{23}$ molecules. We can also write $pV = N_A kT$, where $k = R/N_A$ is the Boltzmann constant, equal to 1.38×10^{-16} erg/deg. If we divide the above equation by V, we get $p = nkT$, where n is the number of molecules per unit volume.

van der Waal's Equation

Van der Waal modified the ideal gas equation to

$$(p + \frac{a}{v^2})(v - b) = Rt \tag{4.3}$$

to obtain a better fit with experiments at high pressures. This equation follows by assuming that a gas molecule has finite volume and cannot be compressed beyond a limit, and that they exert attractive forces on each other. The natural gases obey this equation in an appropriate way.

Virial form of gas equation

This equation expresses the product pv in a series of inverse powers of v, and is given by

$$pv = A + \frac{B}{v} + \frac{C}{v^2} + \cdots . \tag{4.4}$$

The coefficients $A, B, C, \cdots$, are called *virial coefficients*. The virial coefficients for a perfect gas are $A = RT$, with all the rest being zero, and for a van der Waal's gas, they are $A = RT, B = RTb - a, C = RTb^2$, the rest being zero.

Beattie–Bridgemann equation

Beattie and Bridgemann suggested the equation

$$p = \frac{RT(1 - \mathcal{E})(v + B)}{v^2} - \frac{A}{v^2}$$

as a possible equation of state for all gases. Here $\mathcal{E}, A$ and B are some constants. Although the physical significance of this equation is not very clear, it is empirically found to be valid over a wide range of the variables.

From the equation of state we see that only two of the three variables p, v, t are independent. Thus, if we know two of these parameters for a certain amount of matter under equilibrium, the third one is decided by the equation of state. We can write this in the form $p \equiv p(v, t), v \equiv v(p, t)$, and $t \equiv t(p, v)$. Constant-t curves are called *isotherms*, constant-p curves are called *isobars* and constant-v curves are called *isometrics*. The corresponding changes in the state of the system are called *isothermal, isobaric* and *isometric* changes, respectively. We can also consider changes in the state of a system where no energy is added or removed from the system. They are called *adiabatic* changes; for them $\Delta Q = 0$ or $\Delta q = 0$, where ΔQ represents the total energy exchanged and Δq is the energy exchanged per unit mass.

4.1.4 Some important partial derivatives

We have remarked that since the equilibrium state of a gas is characterised by the equation of state $F(p, V, t) = $ constant, only two of the three variables are independent. We thus have six partial derivatives on the surface $F = $ constant such as $(\partial p/\partial V)_{T,F}$ etc. They are related to each other by the following relations:

$$\left(\frac{\partial p}{\partial V}\right)_{T,F} = \frac{1}{(\partial V/\partial p)_{T,F}},$$

$$\left(\frac{\partial V}{\partial T}\right)_{p,F} = \frac{1}{(\partial T/\partial V)_{p,F}},$$

$$\left(\frac{\partial T}{\partial p}\right)_{V,F} = \frac{1}{(\partial p/\partial T)_{V,F}},$$

$$\left(\frac{\partial p}{\partial V}\right)_{T,F}\left(\frac{\partial V}{\partial T}\right)_{p,F}\left(\frac{\partial T}{\partial p}\right)_{v,F} = -1. \tag{4.5}$$

These relations are in fact more generally valid, and we shall show this through a guided exercise.

▶ **Guided Exercise 4.1**   If x, y, z are quantities satisfying a functional relationship $f(x, y, z) = $ constant, prove that

(a) $$\left(\frac{\partial x}{\partial y}\right)_{z,f} = \frac{1}{(\partial y/\partial x)_{z,f}}, \tag{4.6a}$$

and other two obtained by cyclic permutation of x, y, z, and

(b) $$\left(\frac{\partial x}{\partial y}\right)_{z,f}\left(\frac{\partial y}{\partial z}\right)_{x,f}\left(\frac{\partial z}{\partial x}\right)_{y,f} = -1. \tag{4.6b}$$

(c) Also, if w is another function of x, y, z, then on the surface $f = $ constant, show that

$$\left(\frac{\partial x}{\partial y}\right)_{w,f}\left(\frac{\partial y}{\partial z}\right)_{w,f} = \left(\frac{\partial x}{\partial z}\right)_{w,f}. \tag{4.6c}$$

Hints

(a) The relation $f(x, y, z) = $ constant denotes a constant-f surface in the (x, y, z)-space. On a constant-f surface, x, y, z depend on each other as there are only two degrees of freedom. Allowing small displacements to take place on such a surface, write for small displacements dx and dy as

$$dx = \left(\frac{\partial x}{\partial y}\right)_{z,f} dy + \left(\frac{\partial x}{\partial z}\right)_{y,f} dz, \tag{4.7a}$$

$$dy = \left(\frac{\partial y}{\partial z}\right)_{x,f} dz + \left(\frac{\partial y}{\partial x}\right)_{z,f} dx. \tag{4.7b}$$

(b) Substitute (4.7b) in (4.7a) to get

$$dx = \left(\frac{\partial x}{\partial y}\right)_{z,f} \left[\left(\frac{\partial y}{\partial z}\right)_{x,f} dz + \left(\frac{\partial y}{\partial x}\right)_{z,f} dx\right] + \left(\frac{\partial x}{\partial z}\right)_{y,f} dz. \tag{4.8}$$

(c) This equation holds good for all displacements dx, dz on the constant-f surface. So equate coefficients of dx and dz separately on both sides to get

$$\left(\frac{\partial x}{\partial y}\right)_{z,f} \left(\frac{\partial y}{\partial x}\right)_{z,f} = 1, \tag{4.9a}$$

$$\left(\frac{\partial x}{\partial y}\right)_{z,f} \left(\frac{\partial y}{\partial z}\right)_{x,f} = -\left(\frac{\partial x}{\partial z}\right)_{y,f}. \tag{4.9b}$$

The above two equations are equivalent to (4.6).

(d) Note that (4.6a) is valid under specific conditions and constraints. It is known in the theory of partial differentiation that it does not hold in the general case. Since (4.6b) has been derived using (4.6a), the above remarks apply to it also.

(e) Consider (4.7a) again. Let w be another function of x, y, z. Consider constant-w contours on the constant-f surface. On such a contour, we have four variables x, y, z, w, though only two of them are independent, in terms of which the other two can be expressed. Let us treat y and w as the independent variables, and write

$$dx = \left(\frac{\partial x}{\partial y}\right)_{w,f} dy + \left(\frac{\partial x}{\partial w}\right)_{y,f} dw.$$

Putting $w=$ constant and dividing by throughout dz leads to (4.6c).

Note: There are numerous cases in physics where we come across the above situation. For example, f could denote the electrostatic potential at a point (x, y, z), $f = $ constant would be an equipotential surface, while w could be the charge density at that point. In another situation, $f(P, V, T)$ could be a function of P, V, T in thermodynamics and $f = $ constant could be the equation of state of a system; $w(P, V, T)$ could then stand for some other function such as internal energy. Finally, $f(x, y, z) = $ constant could represent the equation of a hilly terrain and $T(x, y, z) = $ constant on $f = $ constant could be the constant-temperature contours. ◀

Some typical partial derivatives and ratios are important in the context of thermodynamics of bulk matter. They lead to experimentally measurable parameters and are useful to test the theories of thermodynamics. We shall mention some of these here.

(i) Coefficient of thermal expansion

$$\beta = \frac{1}{V}\left(\frac{\partial V}{\partial t}\right)_p = \frac{1}{v}\left(\frac{\partial v}{\partial t}\right)_p \qquad (4.10a)$$

This is the fractional increase in volume with respect to temperature at constant pressure.

(ii) Compressibility:

$$K = -\frac{1}{V}\left(\frac{\partial V}{\partial p}\right)_T = -\frac{1}{v}\left(\frac{\partial v}{\partial p}\right)_T. \qquad (4.10b)$$

This indicates the fractional reduction (hence negative sign) in volume with respect to pressure at constant temperature.

(iii) Pressure coefficient:

$$\left(\frac{\partial p}{\partial T}\right)_v = -\left(\frac{\partial p}{\partial v}\right)_T\left(\frac{\partial v}{\partial T}\right)_p = \frac{\beta}{K}. \qquad (4.10c)$$

The left-hand side of this indicates the rate of change of pressure with respect to tempera-ture at constant volume. The right-hand side follows from (4.6c).

As a simple example, for the perfect gas represented by (4.2), we have

$$\left(\frac{\partial v}{\partial t}\right)_p = \frac{R}{p}, \left(\frac{\partial v}{\partial p}\right)_t = -\frac{Rt}{p^2},$$

so that

$$\beta = 1/t, \, K = 1/p, \left(\frac{\partial p}{\partial T}\right)_v = p/t. \tag{4.11}$$

4.2 Changes in Thermodynamic Systems

We must discuss how a thermodynamic system changes from one state to another. What is the work done on a system in changing its state? Does the environment do work on the system or the system does work on the surrounding? Is the change reversible or irreversible?

4.2.1 Reversible and irreversible changes

The state of a thermodynamic system can be changed by bringing it in contact with its surroundings which affect it in two ways: (i) The bath supplies energy to the system in any of the known forms such as heat, radiation, electrical or magnetic energy, chemical or radioactive energy, or surface energy. (ii) The surroundings do work on the system. Afterwards we may isolate the system so that it attains thermodynamic equilibrium again after a lapse of time. The initial and final state of the system can be represented by points A and B, respectively, in the p–v plane; see Fig. 4.1. The transition form A to B can occur in two ways.

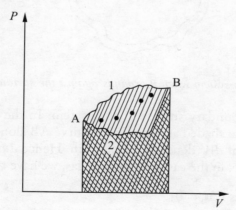

Fig. 4.1 *Reversible transitions of a system between two states A and B along two different paths*

1. If the changes occur so slowly that at every intermediate stage the system is in thermo-dynamic equilibrium, the temperature has to change very slowly. In this case we can reverse the process at every stage; hence it is called a *reversible change*.

2. The change occurs so fast that the system has no time to achieve thermodynamic equilibrium except perhaps at a few isolated instants of time. All the intermediate stages of such a *non-reversible change* cannot be shown in the *pv* diagram.

Actually all real changes are irreversible while reversibility is an idealised concept. We shall now discuss this in a little more detail.

4.2.2   Work done in a reversible change

Suppose a system experiencing pressure p at the surface element $\boldsymbol{d\sigma}$ expands by a distance $\mathbf{dr}$; see Fig. 4.2. Then the work done by the system on its surroundings in expanding through volume dV will be

$$dW = \oint_S p\,\mathbf{d\sigma} \cdot \mathbf{dr} = p\,dV, \qquad (4.12)$$

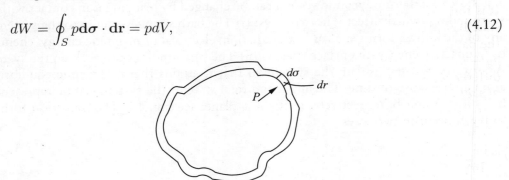

Fig. 4.2   *Work done by a system when it expands against the surroundings*

where S is the closed boundary surface of the system. In the *pv* diagram of Fig. 4.1, this is represented by the shaded area under the curve AB along the path taken by the system. It is obvious that dW depends on the path. Hence dw or dW is not a perfect differential. For example, in the case of a perfect gas, we have the following situations:

(i) In isothermal changes ($\Delta t = 0$),

$$dw_t = \int p\,dv = \int \frac{Rt}{v}\,dv = Rt\ln\frac{v_B}{v_A}.$$

(ii) In isobaric changes ($\Delta p = 0$),

$$dw_p = \int p\,dv = p(v_B - v_A).$$

(iii) In isometric changes ($\Delta v = 0$),

$$dw_v = 0.$$

(iv) In adiabatic changes $(\Delta q = 0)$, $dw_q = \int p\,dv$ can be calculated if we know the adiabatic equation of the perfect gas.

In general, any reversible change from A to B can be considered as a combination of any two of the above four processes. It is customary to use isothermal and adiabatic changes. They can be represented by a grid of curves in the pv plane which is made up of *isotherms* and *adiabats* by putting $\Delta t = 0$ and $\Delta q = 0$, respectively.

4.3 First Law of Thermodynamics

4.3.1 Experiments and the concept

It is found from experience that in taking the thermodynamic system from state A to state B along a given path, we have to supply energy dq to the system which depends on the path. Thus dq is not a perfect differential. However, one finds that the difference $dq - dw$ is independent of the path. Writing $du = dq - dw$, we say that du is the difference in the specific internal energy of the system between points B and A. Thus $u(p, v)$ is a state variable or thermodynamic coordinate. The statement

$$dq = du + dw \tag{4.13}$$

is known as the *first law of thermodynamics*.

In adiabatic changes where $dq = 0$, we have $du = -dw$. So the changes in internal energy can be measured by means of the work done on the system (note that dw is the work done by the system) to bring about the desired change. It is found that even in irreversible changes like frictional work or flow of electric current, du is always the same. Hence the first law of thermodynamics holds for all kinds of changes, both reversible and irreversible. If the work done is only in the expansion of the system against the surroundings, then we can write the first law as

$$dq = du + p\,dv. \tag{4.14}$$

4.3.2 Internal energy of an ideal gas

Experimentally it is found that when ordinary gases expand in vacuum they cool slightly because the molecules do work in separating from each other. This reduces their internal energy. However, a perfect gas is by definition a collection of non-interacting or weakly interacting particles. So their internal energy is independent of volume or pressure. Thus we can write $u \equiv u(t)$. This also follows from the kinetic theory of gases, where we get $u = 3N_A kt/2$, where k is the Boltzmann constant.

▶ **Guided Exercise 4.2**   Let f be a function of x, y, and

$$df = (3x^2 + 3y)dx + (cx + 2y)dy, \tag{4.15}$$

where c is a constant. (a) Find the value of c for which it is a perfect differential. (b) Consider the integral of df from $(0,0)$ to $(2,1)$ along the two paths (i) going along the x-axis from $(0,0)$ to $(2,0)$ and then along the y-axis upto $(2,1)$, (ii) going from $(0,0)$ to $(2,1)$ along a straight line path. Show that the integral of df along the two paths is the same for the above value of c.

Hints   (a) Write $df = M dx + N dy$ and equate $\partial M/\partial y$ with $\partial N/\partial x$ to get $c = 3$.

(b) Path (i) consists of two continuous pieces. On the first part, $y = 0$ and $0 \le x \le 2$, and on the second part, $x = 2, 0 \le y \le 1$. So we have

$$\int_{\text{Path 1}} df = \int_0^2 3x^2 dx + \int_0^1 (2c + 2y)dy = 9 + 2c. \tag{4.16}$$

(c) Along the straight line path (ii), we have $y = x/2, dy = dx/2, 0 \le x \le 2$. Thus we have

$$\int_{\text{Path 2}} df = \int_0^2 (3x^2 + 3x/2)dx + \int_0^2 (c + x)dx/2 = 12 + c. \tag{4.17}$$

The result is clear.   ◀

4.4 Specific Heats

Specific heat is the energy required to raise the temperature of a body (system) by a unit amount. It is an experimentally measurable bulk quantity. The specific heat depends on which other parameters are kept constant and which are allowed to vary.

4.4.1 General considerations

We define the specific heat at constant volume and specific that at constant pressure, respectively, by

$$c_v = \left(\frac{\partial q}{\partial t}\right)_v, \quad c_p = \left(\frac{\partial q}{\partial t}\right)_p. \tag{4.18}$$

Then considering q as a function of v and t, we can write (4.14) as

$$dq = du(v, t) + p(v, t)dv = \left(\frac{\partial u}{\partial t}\right)_v dt + \left(\frac{\partial u}{\partial v}\right)_t dv + pdv. \tag{4.19}$$

Therefore

$$c_v = \left(\frac{\partial q}{\partial t}\right)_v = \left(\frac{\partial u}{\partial t}\right)_v. \tag{4.20}$$

Again, considering q as a function of p and t, we get

$$dq = \left(\frac{\partial u}{\partial t}\right)_v dt + \left\{\left(\frac{\partial u}{\partial v}\right)_t + p\right\}\left\{\left(\frac{\partial v}{\partial p}\right)_t dp + \left(\frac{\partial v}{\partial t}\right)_p dt\right\}$$

$$= \left[c_v + \left(\frac{\partial v}{\partial t}\right)_p \left\{\left(\frac{\partial u}{\partial v}\right)_t + p\right\}\right] dt + \left(\frac{\partial u}{\partial p}\right)_t \left\{\left(\frac{\partial u}{\partial v}\right)_t + p\right\} dp.$$

Therefore,

$$c_p = \left(\frac{\partial q}{\partial t}\right)_p = c_v + \beta v \left\{\left(\frac{\partial u}{\partial v}\right)_t + p\right\}. \tag{4.21}$$

4.4.2 Application to a perfect gas

Here we have $u \equiv u(t)$ and hence $(\partial u/\partial v)_t = 0$. Also $\beta = 1/t$, so that $\beta v p = v p / t = R$. Hence

$$c_p - c_v = R. \tag{4.22a}$$

Putting $c_p/c_v = \gamma$, we get

$$c_v(\gamma - 1) = R. \tag{4.22b}$$

Now we shall discuss isotherms and adiabats for a perfect gas.

Isotherm

The perfect gas equation $pv = Rt$ (Boyle's law) is valid in this case at each stage of a transformation because the system is allowed to come to equilibrium after every infinitesimal change. Thus we have

$$\left(\frac{\partial p}{\partial v}\right)_t = -\frac{R}{v^2}. \tag{4.23}$$

Adiabats

When a transformation is carried out suddenly disallowing any exchange of heat with the surroundings, we have $dq = du + p\,dv = 0$. This gives

$$\left(\frac{\partial u}{\partial t}\right)_v dt + \frac{Rt}{v}dv = 0 \Rightarrow \frac{c_v}{t}dt + \frac{R}{v}dv = 0.$$

On integrating and using (4.22b), we get three relations in the form

$$pv^\gamma = \text{constant}, \quad tv^{\gamma-1} = \text{constant}, \quad p^{1-\gamma}t^\gamma = \text{constant}. \tag{4.24}$$

Differentiating the first one of these with respect to v, we see that

$$\left(\frac{\partial p}{\partial v}\right)_q \propto -\frac{\gamma}{v^{\gamma+1}}. \tag{4.25}$$

Comparing this with (4.23) and noting that $\gamma > 1$, we see that adiabats are steeper than isotherms, as shown in Fig. 4.3.

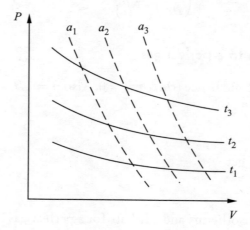

Fig. 4.3   *Some isotherms and adiabats for a perfect gas*

4.5   Second Law of Thermodynamics

Thermodynamic systems were studied extensively, among others, by Clausius and Kelvin. They formulated the second law of thermodynamics in their own ways. Here we study their thought experiments.

4.5.1 Various statements of the second law

These two statements of the law read as follows:

(i) **Clausius' statement**   It is not possible to transfer heat from a colder to a hotter body after a cycle of changes of a system without converting some work into heat.

(ii) **Kelvin–Planck statement**   It is not possible to extract heat from a single reservoir and convert it entirely into an equivalent amount of work.

The understanding of the behaviour of thermodynamic systems has not been an easy matter. Tremendous efforts have been put in by scientists and pseudo-scientists to develop a *perpetual machine*, a machine which will generate energy without putting in any of it in the first instance in any form. A clear appreciation of the impossibility of such a process has taken decades and centuries of experiments and careful and logical thought. The second law of thermodynamics is a culmination of these efforts. In a sense, it is an impossibility theorems.

The above two statements are equivalent because it can be shown that the violation of one amounts to a violation of the other. To see this, consider two reservoirs A and B at temperatures T_2 and T_1, with $T_2 > T_1$.

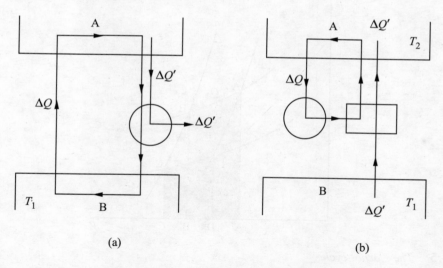

(a)

(b)

Fig. 4.4   *Proof of second law of thermodynamics. (a) Violation of Clausius' statement, (b) violation of Kelvin's statement.*

(i) Suppose Clausius' statement is violated and heat $\Delta Q > 0$ flows from B at T_1 to A at T_2 without putting in any work (see Fig. 4.4 (a)). Let us now extract heat $\Delta Q + \Delta Q'$, where $\Delta Q' > 0$, from A and operate an engine which transforms ΔQ of heat to B and converts $\Delta Q'$ into work. The net effect is to extract $\Delta Q'$ from

A and convert it into work. This is contrary to Kelvin statement, showing that a violation of Clausius statement amounts to a violation of Kelvin statement.

(ii) Suppose Kelvin statement is violated and we are able to extract ΔQ amount of heat from A and convert it into work (see Fig. 4.4 (b)). Then we can use this work to run a refrigerator which extracts $\Delta Q'$ amount of heat from B and deposits heat $\Delta Q + \Delta Q'$ into A. The net result is to transfer some heat from T_1 to $T_2 > T_1$ without doing any work. This is contrary to Clausius' statement. This shows that a violation of Kelvin statement is also a violation of Clausius' statement.

4.5.2   Carnot cycle

The process of converting heat into work is achieved by the Carnot cycle. Let t_1, t_2 be two isotherms with $t_2 > t_1$, and a_1, a_2 two adiabats of a homogeneous substance. We can use this substance in an engine which operates according to the following reversible cyclic process. Let the isotherms and adiabats intersect at points A, B, C, D, as shown in Fig. 4.5. Let AA', BB', CC' and DD' be lines parallel to the p-axis, as shown.

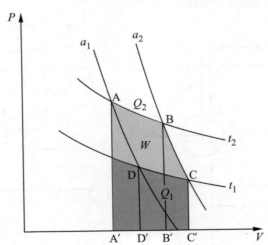

Fig. 4.5   *The Carnot cycle*

(i) The substance expands isothermally along AB, a process in which it absorbs heat Q_2 from a reservoir at temperature t_2 and does work equal to the area ABB'A' on the surroundings at temperature t_1.

(ii) It then expands adiabatically along BC in which it does work equal to the area BCC'B' without absorbing any heat. The temperature drops to t_1 in this adiabatic expansion.

(iii) Then the substance contracts isothermally, by giving out heat Q_1 to the reservoir or surrounding at temperature t_1, along the path CD. In this process, work equal to the area CC′D′D is done on the substance.

(iv) Finally, the substance contracts adiabatically along DA, a process in which work equal to the area DD′A′A is done on the substance so that the temperature again rises to t_2.

At the end of this cyclic process, known as the *Carnot cycle*, the substance has returned to its original state so that $dU = 0$. Then the first law of thermodynamics gives $dQ = dW$. In the Carnot cycle, $dQ = Q_2 - Q_1$ and dW is the net work done by the system on the surroundings, which is equal to the area ABCD. Since dW is positive, $Q_2 > Q_1$, the system absorbs more heat than it gives out. Effectively, the engine (or system) absorbs heat, converts some of it to work, and gives out the rest of it.

The efficiency of a Carnot engine E_e is defined as the work done divided by the heat absorbed at the higher temperature. Thus

$$E_e = (Q_2 - Q_1)/Q_2. \tag{4.26a}$$

The Carnot cycle is reversible, hence we can operate it counterclockwise as ADCBA. In this case the surroundings do work W on the system and transfer heat equal to $Q_2 - Q_1$ from t_1 to t_2. Thus the cycle works as a refrigerator. In this case the efficiency is the heat removed from the surroundings at t_1 divided by the amount of work done, that is,

$$E_r = Q_1/(Q_2 - Q_1). \tag{4.26b}$$

It is easy to see that $E_e \leq 1$ while E_r may have any positive value. A machine working on this principle is called a *heat engine*.

4.5.3 Efficiency of heat engine and refrigerator

Let R be a reversible Carnot cycle operating between temperatures t_1 and t_2 and doing or using work W. We can couple it to another engine or refrigerator S which requires or does the same amount of work and runs between the same two temperatures. We consider these two cases.

S acts as heat engine and R as refrigerator

Let S absorb heat Q_2' at t_2 and reject Q_1' to surroundings at t_1. Then $W = Q_2' - Q_1'$, and the efficiency of S is $E_e' = (Q_2' - Q_1')/Q_2'$. As S is coupled to R, we have $W = Q_2 - Q_1$ and $E_e = (Q_2 - Q_1)/Q_2$. Note that since W is the same, we have $Q_2' - Q_1' = Q_2 - Q_1 \Rightarrow Q_1 - Q_1' = Q_2 - Q_2'$. Now the net result of the combined action of S and R is to remove heat $Q_1 - Q_1'$ at temperature t_1 and transfer the same $(Q_2 - Q_2')$ at $t_2 > t_1$. But this is not possible according to the second law of thermodynamics. Therefore, $Q_2 - Q_2' \leq 0$ or $Q_2 \leq Q_2'$. Hence

$$E_e' = \frac{Q_2 - Q_1}{Q_2'} \leq \frac{Q_2 - Q_1}{Q_2} = E_e,$$

showing that the efficiency of S as an engine is less than or equal to that of the Carnot cycle R.

S acts as refrigerator and R as engine

In this case also, if Q_2'' and Q_1'' are the heat absorbed and heat rejected by S at t_2 and t_1, respectively, we have $Q_2'' - Q_1'' = W = Q_2 - Q_1$. The net result is to transfer $Q_1'' - Q_1 = Q_2'' - Q_2$ from lower temperature t_1 to higher temperature t_2. Again this is not possible by the second law of thermodynamics. Hence we must have $Q_1'' - Q_1 \leq 0$, or $Q_1'' \leq Q_1$, and $Q_2'' \leq Q_2$. Hence

$$E_r' = \frac{Q_1''}{Q_2'' - Q_1''} = \frac{Q_1''}{Q_2 - Q_1} \leq \frac{Q_1}{Q_2 - Q_1} = E_r,$$

showing that the efficiency of S as a refrigerator is less than or equal to that of R, the Carnot cycle.

In fact, the equality holds only if S is also a reversible cycle. In that case, $Q_2'' = Q_2'$ and $Q_1'' = Q_1'$, so that we simultaneously have $Q_2' \leq Q_2$ and $Q_2' \geq Q_2$. This is possible only if $Q_2 = Q_2'$ and $Q_1 = Q_1'$. Thus all reversible engines operating between two given temperatures are equivalent irrespective of the substance used. In particular, Q_1 and Q_2 are the same for all of them provided to do the same amount of work W.

4.6 Absolute Temperature

With the study of reversible and inreversible heat engines and with experiments on varia-tion of pressure or volume versus temperature keeping the other variable, volume or pressure, constant, it became clear that it is possible to define an absolute scale of tempera-ture which depends only on the properties of a substance and does not involve fixing arbitrary end points. We shall now discuss this logic.

4.6 .1 The logic behind the absolute temperature

From the equivalence of all reversible engines operating between temperatures t_1 and t_2, we can write

$$Q_1 = F(t_1, t_2, W), \quad Q_2 = W + F(t_1, t_2, W),$$

where F is some function of the variables indicated.

Now consider three adiabats a, b, c, as shown in Fig. 4.6. Let Q_1 be the heat given out by the substance and W the work done in the Carnot cycle ABCD and Q_1', W' the corresponding quantities in the cycle BEFG. Then we have

$$Q_1 = F(t_1, t_2, W), \quad Q_1' = F(t_1, t_2, W').$$

The cycle AEFD gives out heat equal to $Q_1 + Q_1'$ and does work equal to $W + W_1'$ so that we have

$$Q_1 + Q_1' = F(t_1, t_2, W + W'),$$

$$\Rightarrow F(t_1, t_2, W + W') = F(t_1, t_2, W) + F(t_1, t_2, W').$$

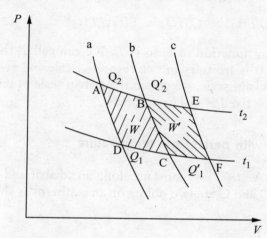

Fig. 4.6   *Concept of absolute temperature*

Putting $W' = W$, we get

$$F_1(t_1, t_2; 2W) = 2F(t_1, t_2, W) \Rightarrow F \propto W.$$

Therefore Q_1 is proportional to W, and can write

$$Q_1 = G(t_1, t_2)W \quad \text{and} \quad Q_2 = [1 + G(t_1, t_2)]W.$$

Then

$$\frac{Q_2}{Q_1} = \frac{1 + G(t_1, t_2)}{G(t_1, t_2)} \equiv H(t_1, t_2),$$

where $H(t_1, t_2)$ defined above is a universal function of t_1 and t_2. Similarly, with a third isotherm at temperature t_3 we would have

$$\frac{Q_3}{Q_1} = H(t_1, t_3), \quad \frac{Q_3}{Q_2} = H(t_2, t_3).$$

Then the fact that $(Q_3/Q_1) = (Q_3/Q_2)(Q_2/Q_1)$ gives

$$H(t_1, t_3) = H(t_1, t_2)H(t_2, t_3)$$

or

$$H(t_2, t_3) = H(t_1, t_3)/H(t_1, t_2).$$

Let t_1 be some standard temperature. Then, letting $H(t_1, t) = T(t)$, we can write

$$H(t_2, t_3) = T(t_3)/T(t_2) \Rightarrow Q_3/Q_2 = T(t_3)/T(t_2).$$

As Q is an increasing function of t, so is T. We can call T the universal absolute temperature function. It is arbitrary in two respects, scale and zero point. If we make the scale equal to the celsius scale, we obtain the Kelvin scale of temperature. The zero point of the scale can be fixed as discussed below.

4.6.2 Comparison with perfect gas temperature

For a perfect gas we have $tv^{\gamma-1} =$ constant along an adiabat and $pv = Rt =$ constant along an isotherm. If D and C are two points on an isotherm as shown in Fig. 4.5, then

$$Q_1 = \int_D^C dQ = \int_D^C p\,dv$$

$$= \int_D^C \frac{Rt_1}{v}dv = Rt_1\ln\{v(\mathrm{C})/v(\mathrm{D})\},$$

where $v(\mathrm{C})$ is the volume of the gas at the point C, etc. Similarly,

$$Q_2 = Rt_1\ln\{v(\mathrm{B})/v(\mathrm{A})\}.$$

Therefore

$$\frac{Q_2}{Q_1} = \frac{t_2}{t_1}\frac{\ln\{v(\mathrm{B})/v(\mathrm{A})\}}{\ln\{v(\mathrm{C})/v(\mathrm{D})\}}.$$

Now since A and D are on an adiabat and B and C on another adiabat, we have

$$t_2[v(A)]^{\gamma-1} = t_1[v(D)]^{\gamma-1}, t_2[v(B)]^{\gamma-1} = t_1[v(C)]^{\gamma-1}.$$

Therefore

$$v(A)/v(B) = v(D)/v(C),$$

which gives

$$Q_2/Q_1 = t_2/t_1 = T_2/T_1.$$

Therefore, the zero-point of the perfect gas scale is the same as that for Kelvin scale. Since the unit interval of the two are equal to the degree of the celsius scale, we see that the Kelvin absolute temperature is identical with the perfect gas temperature. The zero point of both is decided from the relation $T = PV/R$ and putting in the known values at $0 \circ C$. Since we have $p = 1.01325 \times 10^6$ dynes, $v = 2.2415 \times 10^4$ cm^3 and $R = 8.3144 \times 10^7$ erg. mole^{-1} deg^{-1}, we see that $0°C$ corresponds to 273 K.

4.7 Entropy

As we go from one point to another on an adiabat, what is the quantity that remains constant? It took a long time to develop a quantitative model of this concept in the eighteenth and nineteenth centuries. We shall discuss this logic here.

4.7.1 The logic behind entropy

In an isothermal change in which we go from a point on the adiabat A to a point on adiabat B, as shown in Fig. 4.7, we have

$$Q_1/Q_2 = T_1/T_2 \Rightarrow Q_1/T_1 = Q_2/T_2.$$

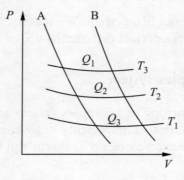

Fig. 4.7 *The concept of entropy*

If we have several isotherms between two adiabats, this suggests that

$$\frac{Q_1}{T_1} = \frac{Q_2}{T_2} = \frac{Q_3}{T_3} = \cdots = \frac{Q}{T}$$

is a function of two adiabats only. So let S_0 be a quantity for an individual adiabat and let its value for any other adiabat be given by

$$S = S_0 + Q/T, \tag{4.27}$$

where Q is calculated along any isotherm between the standard adiabat and the given adiabat, and T is the corresponding temperature. Therefore $S \equiv S(p, V)$ is a well-defined function of p and V. Hence

$$dS = \left(\frac{\partial S}{\partial p}\right)_V dp + \left(\frac{\partial S}{\partial V}\right)_p dV.$$

Considering S as a function of T and V, we can write

$$dS = \left(\frac{\partial S}{\partial T}\right)_V dT + \left(\frac{\partial S}{\partial V}\right)_T dV.$$

Along an isotherm, $dT = 0$ by definition, so $dS = dQ/T$. Also, along an adiabat, $dQ = 0$, so that $dS = 0$. So, again, $dS = dQ/T$ would hold. Since any change in the vp plane can be represented by a combination of adiabatic and isothermal changes, we can write very generally

$$dS = dQ/T. \tag{4.28}$$

Therefore

$$dS = \frac{1}{T}(dU + pdV) \Rightarrow ds = \frac{1}{T}(du + pdv). \tag{4.29}$$

This equation defines the *entropy* of a system. We see that $1/T$ is the integrating factor for dQ, which converts it into a perfect differential dS.

4.7.2   Entropy of a complex system

Here we shall consider the entropy of a complex system and then that of a perfect gas.

Let a system consist of various components designated by 1, 2, 3, etc. Then the change of heat in a transformation along an isotherm will be

$$dQ = dQ_1 + dQ_2 + dQ_3 + \cdots,$$

where dQ_i is the heat absorbed by the ith component. As T is the same for all components, we get

$$dS = dQ/T = dS_1 + dS_2 + dS_3 + \cdots.$$

Therefore entropy of a complex system is equal to the sum of the entropies of its components.

Now we come to the entropy of a perfect gas. In this case, we have

$$S = \int dS = \int (dU + pdV)/T$$

$$= m \int \frac{c_v dT + (RT/v)dv}{T}$$

$$= m[c_v \ln T + R \ln v] + \text{constant}.$$

By properly choosing the zero-point of S, we can make the constant equal to zero. Then

$$S = m[c_v \ln T + R \ln v], \quad s = [c_v \ln T + R \ln v]. \tag{4.30}$$

4.7.3 The principle of increasing entropy

According to this principle, in all natural processes the entropy of a system increases according to the applied constraints. We shall illustrate this by a few examples.

Expansion of a perfect gas in vacuum

Let V be the volume of a gas at temperature T and let m be its mass. Then $v = V/m$. Let the gas expand into vacuum of volume V', so that $V + V'$ is the final volume, and $v' = (V + V')/m$ the final specific volume; see Fig. 4.8(a). In free expansion into vacuum, the gas does not do any work, so the internal energy and temperature do not change. Then

$$dS = mR \ln \frac{V + V'}{m} - mR \ln \frac{V}{m} = mR \ln \left(1 + \frac{V'}{V}\right) > 0.$$

Thus entropy increases in the process and, in fact, reaches the maximum value in the given situation.

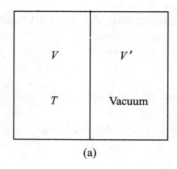 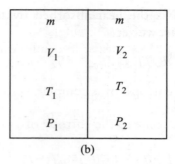

(a)   (b)

Fig. 4.8   *Principle of maximum entropy: (a) expansion in vacuum, (b) mixing of hot and cold gases.*

Mixing of hot and cold gases

Let us mix mass m of a gas at temperature T_1 and volume V_1 with an equal mass of the gas at temperature T_2 and volume V_2, originally separated by a partition as shown in Fig. 4.8(b). The initial pressures p_1 and p_2 on the two sides could be different. Hence on removing the partition between the two portions, they will do work on one another, but there will be no net work done on the system as a whole. Hence the initial energy of the system remains constant. If c_v is the specific heat of the gas, then we have

$$mc_v T_1 + mc_v T_2 = 2mc_v T \Rightarrow T = (T_1 + T_2)/2,$$

where T is the final temperature.

If S_i and S_f and the initial and final total entropies of the system, then we have

$$S_1 = m[c_v \ln(T_1 T_2) + R \ln(V_1 V_2/m^2)],$$

$$S_2 = 2mc_v \ln T + R \ln((V_1 + V_2)/2m)].$$

This gives the change in entropy to be

$$dS = m \left[c_v \ln \frac{(T_1 + T_2)^2}{4T_1 T_2} + R \ln \frac{(V_1 + V_2)^2}{4V_1 V_2} \right]. \tag{4.31}$$

It is clear that dS is a non-negative quantity, showing that the entropy increases in the process. It can be shown that equalisation of T and V maximises entropy.

4.8   The Phase Rule

Now we wish to discuss equilibrium of a system consisting of a mixture of various phases. Here phase may mean the same substance existing in solid, liquid or gaseous form, or different chemical elements capable of mixing with each other or different crystal structures of a single species. We wish to examine conditions under which different phases coexist in equilibrium, conditions for their transformation from one phase to the other, etc.

A point in the (p, v, t) space indicates a state of a single-phase system. Not every point is a state of equilibrium for the system. The points of equilibrium form a surface in the (p, v, t) space decided by the equation of state.

4.8.1 Gibbs free energy

Matter is known to exist in different phases such as solid, liquid, vapour etc. The amount of matter in different phases depends on the total mass, total volume, and temperature of the substance. The distribution can be determined by the principle of maximum entropy. Let the kth phase have mass m_k, specific volume v_k, specific energy u_k and specific entropy s_k. Then the total mass M, total volume V, total energy U and total entropy S are given by

$$M = \sum_k m_k, \quad V = \sum_k m_k v_k,$$

$$U = \sum_k m_k u_k, \quad S = \sum_k m_k s_k.$$

The extensive entropy S is a function of $3N$ variables, where N is the number of phases. We shall take m_k, v_k and u_k as the independent variables. Then for maximum entropy we have

$$dS = 0 \Rightarrow \sum_k (s_k dm_k + m_k ds_k) = 0$$

$$\Rightarrow \sum_k \left(s_k dm_k + \frac{m_k p_k}{T_k} dv_k + \frac{m_k}{T_k} du_k \right) = 0, \tag{4.32}$$

where we have used the first law of thermodynamics and the fact that $ds = dq/T$. But all the $3N$ increments dm_k, dv_k and du_k are not independent because they are governed by the conditions of constancy of total mass M, total volume V, and total energy U, that is, $dM = 0, dV = 0, dU = 0$. This gives us the three constraints:

$$\sum_k dm_k = 0, \quad \sum_k (v_k dm_k + m_k dv_k) = 0, \quad \sum_k (u_k dm_k + m_k du_k) = 0. \tag{4.33}$$

Multiplying the four equations, (4.32) and (4.33), by four independent constants, we can eliminate three out of $3N$ increments so that the remaining $3N - 3$ increments would be completely arbitrary. Thus multiplying the above four equations respectively by $\tau, \lambda, \mu, \sigma$ and adding, we get

$$\lambda \sum_k dm_k + \mu \sum_k (v_k dm_k + m_k dv_k) + \sigma \sum_k (u_k dm_k + m_k du_k)$$

$$+\tau \sum_k \left(s_k dm_k + \frac{m_k p_k}{T_k} dv_k + \frac{m_k}{T_k} du_k \right) = 0.$$

$$\Rightarrow \sum_k [dm_k(\lambda + \mu v_k + \sigma u_k + \tau s_k) + dv_k(\mu m_k + \tau m_k p_k/T_k)$$

$$+du_k(\sigma m_k + \tau m_k/T_k)] = 0. \qquad (4.34)$$

The constants $\lambda, \mu, \sigma, \tau$ are so chosen as to make the coefficients of the three increments vanish; then, since the remaining increments are arbitrary, their coefficients must be zero. Thus the coefficients of all dm_k, dv_k and du_k are zero. Therefore

$$\lambda + \mu v_k + \sigma u_k + \tau s_k = 0,$$

$$\mu m_k + \tau m_k p_k/T_k = 0,$$

$$\sigma m_k + \tau m_k/T_k = 0. \qquad (4.35)$$

Since we want to consider only the phases that are present, we put $m_k \neq 0$. This gives us

(i) $T_k = -\tau/\sigma$, independent of k. So all the phases have the same temperature, say $T = -\tau/\sigma$.

(ii) $\mu + \tau p_k/T_k = 0$, or $p_k = -\mu T_k/\tau = \mu/\sigma$, so all phases have the same pressure, $p = \mu/\sigma$.

(iii) $u_k + \tau s_k/\sigma + \mu v_k/\sigma + \lambda/\sigma = 0$ or $u_k - Ts_k + pv_k =$ constant for all phases.

The quantity

$$g_k = u_k + pv_k - Ts_k \qquad (4.36)$$

is called the specific *Gibbs free energy* or *Gibbs thermodynamic potential*. It has the same value for all phases in equilibrium. This is known as the *phase rule*. For a single-phase material, the Gibbs free energy becomes

$$G = U + pV - TS. \qquad (4.37)$$

4.8.2   Applications of phase rule

If there are p number of phases, the Gibbs free energy of each phase must be the same in steady state. This gives us $p - 1$ equations

$$g_1 = g_2 = \cdots = g_p.$$

Each g_k involves two variables p and T. Therefore the number of degrees of freedom is

$$D = 2 - (p - 1) = 3 - p.$$

Since D must be non-negative, $p \leq 3$, showing that there can be at most 3 phases with $m_k \neq 0$, which are solid, liquid, vapour, which can be in equilibrium. When $p = 3, D = 0$, so that all the variables p, v, T have each a particular fixed value. They define a state called *the triple point*.

Now consider a case when there is one degree of freedom, that is, we can have one parameter, p, v or T, at our disposal. This gives $p = 2$, that is, only two phases can coexist simultaneously. Such equilibrium states can be represented by curves in a p–T diagram of Fig. 4.9. The three curves separating two phases each meet at the triple point T. AT, the curve between solid and vapour phases, is the *sublimation* curve; BT, the curve between solid and liquid phases, is the *melting point curve* and CT, the curve separating the liquid and the vapour phases, is the *boiling point* or *vapour pressure curve*. The last curve ends in a *critical point* C where the specific volumes of the liquid and vapour phase become equal. Hence for $T > T_c$, the *critical temperature*, there is no separation between liquid and vapour and we call the substance a gas. Similarly, at some point B on the solid–liquid curve, the specific volumes of solid and liquid phases become equal. Hence for $p > p_B$, the *critical pressure*, we call the substance a *supercooled liquid*.

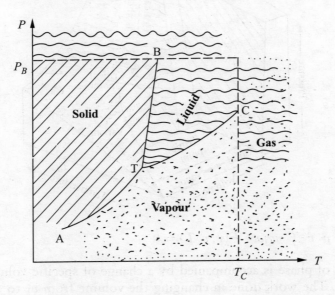

Fig. 4.9   *The phase diagram in p–T space, showing the critical points and the triple point*

4.8.3 The p–v–T surface

The equation of state $F(p, v, T) = 0$ represents a surface in the p–v–T space at each point of which the substance is in equilibrium as a whole, that is, in a steady state. It includes all the phases of the substance which occur in different regions of that surface. The surface is quite complicated and curved; see Fig. 4.10. But there are three sections, known as *ruled surfaces*, which separate the solid–liquid, liquid–vapour and solid–vapour states. Actually, the substance is found only at the boundaries of the ruled surfaces. As we change the volume more of it is transferred from the left boundary to the right boundary which represents another state. The three ruled surfaces meet at the triple point where all the three states are found together. It is characterised by coordinates p_T, v_T, T_T. The liquid–vapour surface has a maximum in pressure where the specific volumes of vapour and liquid are equal. This is the critical point and its coordinates are specified by p_c, v_c, T_c. For temperatures greater than T_c there is no distinction between vapour and liquid, and we can call this state a gas. The projection of the p–v–T surface on p–T plane was discussed earlier. The projection on the p–v plane is also shown in Fig. 4.11.

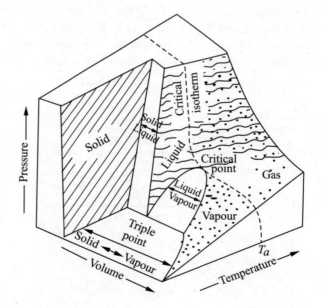

Fig. 4.10   *The p–v–T surface $F(p, v, T) = 0$*

Each change of phase is accompanied by a change of specific volume as well as in internal energy. The work done in changing the volume from v_1 to v_2 is $p(v_2 - v_1)$. Hence the specific *latent heat of transformation*, which is the sum of the work done and change in internal energy, is given by

$$l = p(v_2 - v_1) + u_2 - u_1 = (u_2 + pv_2) - (u_1 + pv_1). \tag{4.38}$$

The quantity

$$h = u + pv \tag{4.39}$$

is called the *specific enthalpy*. Hence we have $l = h_2 - h_1$.

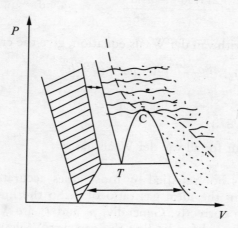

Fig. 4.11   *Phase diagram in p–v plane*

4.8.4   Critical constants for a van der Waals gas

We have discussed van der Waals equation for a real gas in section 4.1(c), equation (4.3). We shall obtain the critical constants p_c, v_c, T_c for a van der Waals gas trough a guided exercise.

▶ **Guided Exercise 4.3**   Obtain the critical constants for a van der Waals gas in terms of the parameters a and b, and show that RT_c/p_cv_c is a constant.

Hints

(a) Consider the p–v diagram for a van der Waals gas described by (4.3). Write this equation in the form

$$p = \frac{RT}{v - b} - \frac{a}{v^2}.$$

(b) Notice three things about it: (i) $p \to \infty$ as $v \to b$. (ii) p is a monotonically decreasing function of v. (iii) There is a point of inflection.

(c) The pressure, volume and temperature at the point of inflection are the critical constants of the gas.

(d) The point of inflection is given by $(\partial p/\partial v)_T = 0$ and $(\partial^2 p/\partial^2 v)_T = 0$, that is

$$\frac{RT}{(v-b)^2} = \frac{2a}{v^3}, \qquad \frac{2RT}{(v-b)^3} = \frac{6a}{v^4}.$$

(e) From these equations we get

$$b = \frac{v}{3}, \qquad a = \frac{27RTb}{8} = \frac{9RTv}{8}.$$

(f) These, together with van der Waals equation, give the critical constants

$$v_c = 3b, \quad T_c = 8a/27Rb, \quad p_c = a/27b^2.$$

(g) This shows that

$$RT_c/p_c v_c = 8/3, \tag{4.40}$$

which is a constant for all van der Waals gases. ◀

In practice, (4.40) is not satisfied by most gases accurately. Also note that this ratio differs vastly from the ideal gas ratio of 1. So the known gases do not obey van der Waals equation correctly. Generally, p_c and T_c are well determined for gases experimentally and are used for finding the van der Waals parameters a and b. The critical constants for some gases are given in Table 4.1.

Table 4.1 *Triple points and critical points for some gases*

Gas	Triple points		Critical points		
	t (^{0}C)	p (mm of Hg)	p_c (N/m^2)	v_c (m^3/kg-mole)	T_c (K)
A	-190	512			
H_2			13	0.065	33.2
H_e			2.3×10^5	0.062	5.25
N_2	-210	96.4	34	0.090	126
O_2	-218	2.0	51	0.075	154
CO_2	-56.6	3880	74	0.095	304
H_2O	0.0098	4.579	221	0.057	647

4.9 Important Thermodynamic Functions

In this section, we shall mostly summarise some important thermodynamic functions, their partial derivatives with respect to different variables, and their relationships with each other. Let us list the various parameters and functions that we have defined so far. Apart from pressure p, (specific) volume v, temperature T, we have (specific) internal energy u,

(specific) entropy s,

(specific) enthalpy $h = u + pv$,

(specific) Helmholtz free energy $f = u - Ts$,

(specific) Gibbs free energy $g = u + pv - Ts$.

They are related by the second law of thermodynamics,

$$dq = Tds = du + pdv.$$

4.9.1 Significance of thermodynamic functions

One may wonder what these different functions represent and what they do to a system. It would be pertinent to discuss some of these points.

The *internal energy* u includes the kinetic energy of the particles as well as any kind of potential — gravitational or electromagnetic — between them. It excludes any external, positional potential energy.

The infinitesimal work done (by a system on the surroundings) when a system expands by dv is pdv. In isobasic changes, $\Delta p = 0$, and we have $pdv = d(pv)$. So $dq = d(u + pv) = dh$. Thus enthalpy h is the heat required to bring about the isobasic change. Examples of change in enthalpy are latent heat of fusion, latent heat of vapourisation, dissociation energy of a molecule, etc. *When pressure is kept constant, a system likes to attain a state of minimum enthalpy.*

In an irreversible engine, $dq \leq Tds$. Using the first law for dq, we can write this as $dw \leq Tds - du$. In an isothermal change, $Tds = d(Ts)$ so that $dw \leq d(Ts - u)$, or $dw \leq -df$. Thus f is the maximum free energy that can be extracted from such an engine. We may say, in other words, a mechanically isolated system ($dw = 0$) at constant temperature attains a state of minimum Helmholtz free energy.

If we operate an irreversible engine in the presence of external pressure, we have $dq = du + dw + pdv$. Then $dq \leq Tds$ leads to $dw \leq d(Ts - u - pv)$ or $dw \leq -dg$. Thus g is the maximum work that can be extracted from such an engine. We may say that under very general conditions which allow change in pressure and temperature, a system attains a state of equilibrium when its Gibbs free energy is a minimum.

4.9.2 Maxwell's equations of thermodynamics

We shall now obtain relations among derivatives of the various functions with respect to different variables. We may rewrite the above equation as

$$du = Tds - pdv.$$

This gives

$$\left(\frac{\partial u}{\partial s}\right)_v = T, \quad \left(\frac{\partial u}{\partial v}\right)_T = -p. \tag{4.41a}$$

An increment in enthalpy then becomes

$$dh = du + pdv + vdp = Tds + vdp.$$

This equation gives the relations

$$\left(\frac{\partial h}{\partial s}\right)_p = T, = \left(\frac{\partial h}{\partial p}\right)_s = v. \tag{4.41b}$$

An increment in the Helmholtz free energy is given by

$$df = du - Tds - sdT = -pdv - sdT.$$

It follows that

$$\left(\frac{\partial f}{\partial v}\right)_T = -p, \quad \left(\frac{\partial f}{\partial T}\right)_v = -s. \tag{4.41c}$$

Finally, an increment in the Gibbs free energy is seen to be

$$dg = du + pdv + vdp - Tds - sdT = vdp - sdT.$$

This gives us the relations

$$\left(\frac{\partial g}{\partial p}\right)_T = v, \quad \left(\frac{\partial g}{\partial T}\right)_p = -s \tag{4.41d}$$

The thermodynamic potential $g(p,T)$ is an important function, perhaps the most important among the above, because we can obtain all the rest from it. For example, the first equation of (4.41d) gives us a relation connecting p, v and T, that is the equation of state. The second equation (4.41d) gives the entropy. The internal energy u, the enthalpy h, and the Helmholtz free energy can then be written in terms of Gibbs potential as

$$\begin{aligned} u &= g - pv + Ts, \\ h &= g + Ts, \\ f &= g - pv, \end{aligned} \tag{4.42}$$

which can be obtained from their definitions introduced in the beginning of this section.

Taking second derivatives of the various thermodynamic functions, we obtain more relations among them. For example, operating on the first equation of (4.41a) by $(\partial/\partial v)_s$ and on the second of these equations by $(\partial/\partial s)_v$, and noting that

$$\left(\frac{\partial}{\partial v}\right)_s \left(\frac{\partial u}{\partial s}\right)_v = \left(\frac{\partial}{\partial s}\right)_v \left(\frac{\partial u}{\partial v}\right)_s,$$

we obtain

$$\left(\frac{\partial T}{\partial v}\right)_s = -\left(\frac{\partial p}{\partial s}\right)_v.$$

Alternatively, this can also be written in the form

$$\left(\frac{\partial s}{\partial p}\right)_v = -\left(\frac{\partial v}{\partial T}\right)_s. \tag{4.43a}$$

Similarly from (4.41b), (4.41c), and (4.41d), we obtain

$$\left(\frac{\partial s}{\partial v}\right)_p = \left(\frac{\partial p}{\partial T}\right)_s, \left(\frac{\partial s}{\partial v}\right)_T = \left(\frac{\partial p}{\partial T}\right)_v, \left(\frac{\partial s}{\partial p}\right)_T = -\left(\frac{\partial v}{\partial T}\right)_p. \tag{4.43b, c, d}$$

Such equations are known as *Maxwell's equations*.

We observe that the right-hand sides of (4.43) can be expressed in terms of experimentally measurable quantities such as β, K, c_v, c_p (see (4.10a), (4.10b) and (4.18)) so they can be used for computing entropy of any substance.

4.9.3 Combinations of independent variables

Among p, v, T, we can treat any pair of variables as independent, and express other important parameters in terms of them. We shall do this in the following.

v and T as independent parameters

We may start with (4.28) and see that

$$ds = \frac{1}{T}(du + pdv)$$

$$= \frac{1}{T}\left\{\left(\frac{\partial u}{\partial T}\right)_v dT + \left(\frac{\partial u}{\partial v}\right)_T dv + pdv\right\}.$$

This gives

$$\left(\frac{\partial s}{\partial T}\right)_v = \frac{1}{T}\left(\frac{\partial u}{\partial T}\right)_v = \frac{c_v}{T}, \left(\frac{\partial s}{\partial v}\right)_T = \frac{1}{T}\left\{p + \left(\frac{\partial u}{\partial v}\right)_T\right\}. \tag{4.44}$$

At constant v, (4.29) shows that $ds = du/T$. Hence operating by $(\partial/\partial T)_v$, the second of the above equations gives

$$\frac{1}{T}\left(\frac{\partial}{\partial v}\right)_T \left(\frac{\partial u}{\partial T}\right)_v = \frac{1}{T}\left\{\left(\frac{\partial p}{\partial T}\right)_v + \frac{\partial^2 u}{\partial T \partial v}\right\} - \frac{1}{T^2}\left\{p + \left(\frac{\partial u}{\partial v}\right)_T\right\}.$$

This, together with (4.10c) shows that

$$p + \left(\frac{\partial u}{\partial v}\right)_T = \frac{T\beta}{K}. \tag{4.45}$$

Again, starting from $dq = du + pdv$, we get

$$dq = \left(\frac{\partial u}{\partial T}\right)_v dT + \left\{p + \left(\frac{\partial u}{\partial v}\right)_T\right\} dv.$$

Taking the derivatives at constant pressure, we see that

$$c_p = \left(\frac{\partial q}{\partial T}\right)_p = \left(\frac{\partial u}{\partial T}\right)_v + \left\{p + \left(\frac{\partial u}{\partial v}\right)_T\right\}\left(\frac{\partial v}{\partial T}\right)_p$$

$$\Rightarrow c_p = c_v + (T\beta/K)\beta v$$

$$\Rightarrow c_p - c_v = \beta^2 Tv/K. \tag{4.46}$$

We now wish to express the partial derivatives of u and s with respect to T and v, when the other variable is kept constant, in terms of experimental parameters. We see that $(\partial u/\partial T)_v$ is c_v, which can be taken from (4.46), while $(\partial u/\partial v)_T$ can be taken from (4.45). Then using (4.44), (4.45) and (4.46), we can express $(\partial s/\partial T)_v$ and $(\partial s/\partial v)_T$ in different ways. Thus we get

$$\left(\frac{\partial u}{\partial T}\right)_v = c_p - \frac{\beta^2 Tv}{K}, \quad \left(\frac{\partial u}{\partial v}\right)_T = \frac{T\beta}{K} - p,$$

$$\left(\frac{\partial s}{\partial T}\right)_v = \frac{c_p}{T} - \frac{\beta^2 v}{K}, \quad \left(\frac{\partial s}{\partial v}\right)_T = \frac{\beta}{K}. \tag{4.47}$$

p and T as independent parameters

Without going into details, we state that treating p and T as independent parameters, we get the following relations which express the derivatives in various forms. The important relations that we get are

$$\left(\frac{\partial u}{\partial T}\right)_p = c_p - \beta pv, \quad \left(\frac{\partial u}{\partial p}\right)_T = Kpv - \beta Tv,$$

$$\left(\frac{\partial s}{\partial T}\right)_p = \frac{c_p}{T}, \quad \left(\frac{\partial s}{\partial p}\right)_T = -\beta v. \tag{4.48}$$

p and v as independent parameters

This time, we get the partial derivatives of u and s with respect to p and v, keeping the other constant, as

$$\left(\frac{\partial u}{\partial v}\right)_p = \frac{c_p}{\beta v} - p, \quad \left(\frac{\partial u}{\partial p}\right)_v = \frac{K c_p}{\beta} - \beta T v,$$

$$\left(\frac{\partial s}{\partial v}\right)_p = \frac{c_p}{\beta v T}, \quad \left(\frac{\partial s}{\partial p}\right)_v = \frac{k c_v}{\beta T} = \frac{K c_p}{\beta T} - \beta v. \tag{4.49}$$

Partial derivatives of c_v and c_p

These specific heats are defined as the rate of change of the internal energy with respect to T at constant v and constant p, respectively. But these derivatives can have different values at different v and p, respectively. It should not be difficult to see that $c_v = (\partial u/\partial T)_v$ can be a function of v, and similarly $c_p = (\partial u/\partial T)_p$ can be a function of p. On taking the second derivatives of $(\partial s/\partial T)_v$ and $(\partial s/\partial v)_T$, and using the various above relations, we obtain

$$\left(\frac{\partial c_v}{\partial v}\right)_T = T\left\{\frac{\partial}{\partial T}\left(\frac{\beta}{K}\right)\right\}_v, \quad \left(\frac{\partial c_p}{\partial p}\right)_T = -T\left\{\frac{\partial}{\partial T}(\beta v)\right\}_p. \tag{4.50}$$

All these relations are useful in testing various models of thermodynamics, different equations of state, making connections with measured quantities, and calculating the internal energy and entropy.

4.9.4 Relations for adiabatic transformations

When there is no exchange of heat between the system and the surroundings, we have $dq = 0$ or $ds = 0$. Thus an adiabatic transformation occurs at constant entropy. There are various relations connecting the derivatives of p, v, T with respect to each other at constant entropy, with various measurable parameters. We shall prove these through a guided exercise.

► **Guided Exercise 4.4** Prove the following relations for an adiabatic transformation:

$$\left(\frac{\partial T}{\partial v}\right)_s = -\frac{(\gamma - 1)}{\beta v},$$ (4.51a)

$$\left(\frac{\partial T}{\partial p}\right)_s = \frac{K}{\beta}\frac{\gamma - 1}{\gamma},$$ (4.51b)

$$\left(\frac{\partial p}{\partial v}\right)_s = -\frac{\gamma}{Kv}.$$ (4.51c)

Hints

(a) We start with (4.14) and differentiate it using chain rule of partial differentiation with respect to pairs of variables as under:

$$dq(t, v) = \left(\frac{\partial u}{\partial T}\right)_v dT + \left[p + \left(\frac{\partial u}{\partial v}\right)_T dv\right],$$

$$dq(t, p) = c_p dT + \left[\left(\frac{\partial u}{\partial p}\right)_T + p\left(\frac{\partial v}{\partial p}\right)_T\right] dp,$$

$$dq(p, v) = \left(\frac{\partial u}{\partial p}\right)_v dp + \left[p + \left(\frac{\partial u}{\partial v}\right)_p\right] dv.$$ (4.52)

(b) In an adiabatic transformation, $dq = 0$ (constant entropy), so that the first of the above equations leads to

$$\left(\frac{\partial T}{\partial v}\right)_s = -\frac{p + (\partial u/\partial v)_T}{(\partial u/\partial T)_v} = -\frac{T\beta}{Kc_v}$$

$$= -\frac{c_p - c_v}{\beta v c_v} = -\frac{\gamma - 1}{\beta v},$$

where we have used (4.45) and (4.46). This proves (4.51a).

(c) The second equation of (4.52) leads to

$$\left(\frac{\partial T}{\partial p}\right)_s = -\frac{(\partial u/\partial p)_T + p(\partial v/\partial p)_T}{c_p} = \frac{\beta v T}{c_p}$$

$$= \frac{K}{\beta}\frac{(c_p - c_v)}{c_p} = \frac{K}{\beta}\frac{(\gamma - 1)}{\gamma},$$

where we have used (4.46) and (4.45). This proves (4.51b).

(d) The third equation of (4.52) with $dq = 0$ leads to

$$\left(\frac{\partial p}{\partial v}\right)_s = -\frac{p + (\partial u/\partial v)_p}{(\partial u/\partial p)_v} = -\frac{C_p}{\beta v} \cdot \frac{\beta}{Kc_p - \beta^2 Tv}$$

$$= -\frac{\gamma}{Kv},$$

which has been obtained by using the first two of equations (4.49) and (4.45). This proves (4.51c). ◄

4.10   Thermodynamics of Electromagnetic Radiation

The physical universe consists of matter and radiation, by which we mean electromagnetic radiation. The latter consists of waves of all frequencies. When we see an illuminated wall or a page of a book, radiation of all frequencies falls on it at all possible angles, and is also emitted from it. It would be appropriate to define some technical terms in this context.

4.10.1   Some basic definitions

Specific intensity

If dE_ν is the energy contained in the radiation of frequency between ν and $\nu + d\nu$ crossing an element of area $d\sigma$, then we define the *specific intensity* I_ν as

$$I_\nu = \frac{dE_\nu}{d\sigma \cos\theta \, d\nu \, dt \, d\Omega}, \tag{4.53}$$

where θ is the angle between the incident radiation and normal to $d\sigma$, dt is the element of time interval and $d\Omega$ is the element of solid angle into which the incident radiation spreads. Thus I_ν is the energy crossing per unit area normal to the direction of propagation per unit time per unit frequency interval per unit solid angle; see Fig. 4.12. I_ν is a function of position $\mathbf{r}$ and direction of propagation $\mathbf{l}$, that is, $I_\nu \equiv I_\nu(\mathbf{r}, \mathbf{l})$. Radiation is said to be isotropic if I_ν is independent of direction $\mathbf{l}$.

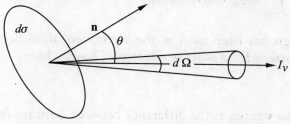

Fig. 4.12   *Radiation crossing an element of area $d\sigma$ in a direction $\mathbf{l}$ making an angle θ with the normal $\mathbf{n}$, with angular spread $d\Omega$ of incident direction.*

If we consider the total energy content at a point irrespective of frequency of radiation, we could define the *total* or *integrated intensity* as

$$I = \int I_\nu d\nu.$$

When we integrate I_ν over all the solid angle, we get the *mean intensity* as

$$J_\nu = \frac{1}{4\pi} \int I_\nu d\Omega.$$

Integrating over all frequencies, we get the *mean total intensity* as

$$J = \int J_\nu d\nu = \frac{1}{4\pi} \int I d\Omega.$$

Flux

This is taken as

$$F_\nu = \int I_\nu \cos\theta \, d\Omega, \quad F = \int F_\nu \, d\nu. \tag{4.54}$$

We can see that *flux F_ν* is the energy crossing per unit area in all directions per unit time per unit frequency interval, whereas *total flux F* is the energy crossing per unit area in all directions per unit time.

Since $d\Omega = \sin\theta \, d\theta \, d\varphi$, we have

$$F_\nu = \int_0^{2\pi} d\varphi \int_0^{\pi} d\theta I_\nu \cos\theta \sin\theta.$$

We now further define *outward flux* and *inward flux*, respectively, as

$$F_\nu^+ = \int_0^{2\pi} d\varphi \int_0^{\pi/2} d\theta I_\nu \cos\theta \sin\theta,$$

$$F_\nu^- = \int_0^{2\pi} d\varphi \int_{\pi/2}^{\pi} d\theta I_\nu (-\cos\theta) \sin\theta, \tag{4.55}$$

where a negative sign has been used in the second equation because $\cos\theta \leq 0$ for $\pi/2 \leq \theta \leq \pi$. This makes F_ν^- a positive quantity. Thus we have

$$F_\nu = F_\nu^+ - F_\nu^-,$$

that is, flux F_ν can be written as the difference between outward flux and inward flux. The above integrals can be recast into the form

$$F_\nu = 2\pi \int_{-1}^{1} I_\nu \, \mu \, d\mu,$$

$$F_\nu^+ = 2\pi \int_0^1 I_\nu \, \mu \, d\mu, \quad F_\nu^- = 2\pi \int_0^{-1} I_\nu \, \mu \, d\mu.$$

For isotropic radiation, we have $I_\nu = J_\nu$, and

$$F_\nu^+ = \pi J_\nu, \quad F_\nu^- = \pi J_\nu,$$

so that $F_\nu = F_\nu^+ - F_\nu^- = 0$.

It can be shown that flux varies as inverse square of the distance.

Energy density

It is the energy present (in vacuum) per unit volume per unit frequency interval when electromagnetic radiation is present. Referring to Fig. 4.13, consider the energy crossing a surface element $d\sigma$ in the direction **l** in a solid and $d\Omega$. After a time dt it occupies a volume $c \, dt \, d\sigma \cos\theta$, where θ is the angle between the surface normal **n** and direction **l**. So the contribution of this radiation to the energy density in volume dV is

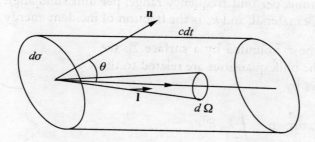

Fig. 4.13 *Volume element $d\sigma \cos\theta cdt$ for calculation of energy density.*

$$du_\nu = \frac{dE_\nu}{d\nu dV} = \frac{I_\nu d\sigma \cos\theta \, d\nu \, dt \, d\Omega}{d\nu \, c \, dt \, d\sigma . \cos\theta} = \frac{I_\nu}{c} d\Omega.$$

Integrating over all directions, we get

$$u_\nu = \frac{1}{c} \int I_\nu d\Omega = 4\pi J_\nu/c,$$

$$\Rightarrow u = \int u_\nu d\nu = 4\pi J/c. \tag{4.56}$$

It can also be seen that $F = \pi I$ in any direction.

Interaction of matter and energy

When radiation falls on a material, the material surface is characterised by *emissivity e_ν* and *absorptivity a_ν* defined by

$$e_\nu = I_\nu \text{ (at the surface)},$$

$$a_\nu = \text{energy absorped/incident energy}.$$

A body at any temperature continuously emits radiation of all frequencies in all directions. Every surface also continuously receives radiation from the surroundings. The ratio of energy absorbed to incident energy is defined as the absorptivity. The balance fraction $1 - a_\nu$ of incident energy is reflected or transmitted. If $a_\nu = 1$ for all frequencies, it represents a perfectly *black body*. In fact, a black body is defined as one which absorbs all radiation falling on it. At the other extreme, a body with $a_\nu = 0$ is called a *white body*. If a_ν is constant, independent of frequency, it is called a *grey body*. We shall denote the emissivity of a perfectly black body as E_ν, and its absorptivity as $A_\nu \equiv 1$.

The above discussion characterises a surface in terms of its absorptivity and emissivity. A bulk material as a whole may also be characterised by *emission* and *absorption coefficients*, which we shall denote by j_ν and k_ν, respectively. Thus j_ν is the energy emitted per unit time, per unit frequency range, per unit solid angle, per unit mass or volume of the bulk material, and k_ν is the fraction of incident energy absorbed per unit mass or volume.

Consider a volume τ bounded by a surface Σ; see Fig. 4.14. Then the bulk quantities are related to the surface parameters by

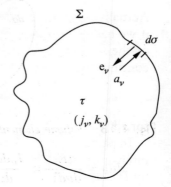

$$\int_\tau j_\nu \, d^3r = \oint_\Sigma e_\nu \, d\boldsymbol{\sigma} \cdot \mathbf{n}, \quad \int_\tau k_\nu \, d^3r$$

$$= \oint_\Sigma a_\nu \, d\boldsymbol{\sigma} \cdot \mathbf{n}, \qquad (4.57)$$

where $\mathbf{n}$ is a unit outward normal.

Fig. 4.14   *Energy balance in an enclosure*

4.10.2   Thermal radiation

Consider a cavity τ in a closed surface Σ embedded in a heat reservoir, maintained at a constant temeperature T; see Fig. 4.15. The radiation inside this cavity will be in thermal equilibrium at the given temperaure. One of the aims of thermodynamics is to derive its properties. We can proceed in the following manner.

(i) Put a speck of matter inside the enclosure. It will attain the temperature T by virtue of interaction with the radiation in the enclosure. For, if its equilibrium temperature were different from T, we could run a Carnot engine between the speck and the wall, and thus convert the energy of the surrounding reservoir into work. As this would violate the second law of thermodynamics, the equilibrium temperature of the speck must be T.

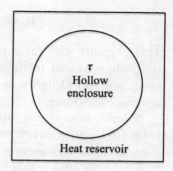

Fig. 4.15 *Thermal radiation at temperature T*

(ii) I_ν will be isotropic irrespective of the shape of the surface Σ, because if we introduce a heterogeneous body in the enclosure, it would attain the same temperature T irrespective of the orientation of its heterogeneous surface with different e_ν and a_ν.

(iii) Now consider a homogeneous body inside the enclosure, in equilibrium with the radiation there. The energy emitted by this body must be equal to the energy absorbed. Therefore, if I_ν is the energy incident on the body, then

$$\int \pi e_\nu \, d\sigma \, d\nu = \pi \int d\sigma \int a_\nu I_\nu \, d\nu$$

$$\Rightarrow \int e_\nu \, d\nu = \int a_\nu I_\nu \, d\nu.$$

Actually, we should have $e_\nu = a_\nu I_\nu$ for each frequency because if this were not so, and if we had $e_\nu > a_\nu I_\nu$ for one frequency while $e_\nu < a_\nu I_\nu$ for another, then the density of the former frequency would continuously increase while that of the latter would continuously decrease. But this does not represent equilibrium. Hence we must have

$$e_\nu = a_\nu I_\nu \Rightarrow I_\nu = e_\nu / a_\nu$$

for all frequencies for all substances.

In particular, for a black body for which $A_\nu = 1$, we would have $I_\nu = E_\nu$, that is, the intensity of radiation is equal to the emissivity of the black body. We shall represent it by B_ν, so that

$$e_\nu / a_\nu = B_\nu$$

for all substances. This is *Kirchoff's law*. Comparing it with (4.56), this also implies that $F_\nu = \pi B_\nu$.

If the body inside the enclosure is a small speck of volume dv and total surface area $d\sigma$, (4.57) suggest that

$$j_\nu dv = e_\nu d\sigma, \qquad k_\nu dv = a_\nu d\sigma$$

$$\Rightarrow e_\nu/a_\nu = j_\nu/k_\nu = B_\nu.$$

This explains why (a) a red glass piece emits green light (of higher frequency than red) when heated to incandescence, and (b) a Bunsen burner sprinkled with NaCl emits sodium D lights as well as absorbs the same radiation from the spectrum of a continuous source.

We may note that e_ν and a_ν are properties of the surface per unit area and per unit solid angle, respectively, while j_ν and k_ν are properties of the material per unit solid angle and per unit mass or volume, respectively.

Kirchhoff's law explains the formation of dark Fraunhoffer lines in stellar spectra and provides a means of identifying the constituents of stellar atmospheres.

Since the radiation inside the hollow cavity represents black body radiation, we can realise a black body by such an enclosure and observe its radiation by making a small hole in the surrounding enclosure.

4.10.3   Radiation pressure

Bartoli devised a thought experiment to show that electromagnetic radiation has pressure. We shall discuss Bartoli's thought experiment, followed by a discussion based on classical electromagnetic theory and the photon theory.

Bartoli's thermodynamic proof

Consider the radiation inside a cylinder with perfectly reflecting walls and bounded by two reservoirs at temperatures T and $T' > T$ on either side; see Fig. 4.16. The cylinder contains two perfectly reflecting diaphragms A and B near the low-temperature and high-temperature sides, respectively. Diaphragm B has a valve which opens towards the two-temperature side, as shown. We first keep the valve open in the thought experiment. Then the two portions of the cylinder divided by diaphragm A are filled with radiations at temperature T and T', respectively. We now close the valve and move diaphragm B towards A and compress the radiation between A and B, to a small volume so that the energy density there becomes larger than that above A. If we now remove diaphragm A, then two radiations will mix and their energy density will be higher than the equilibrium density appropriate for temperature T. So some radiation energy will be absorbed by the heat reservoir. The net effect is a transfer of energy from a

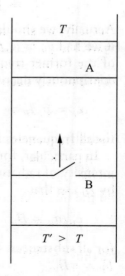

Fig. 4.16   *Bartoli's proof of radiation pressure*

reservoir at a higher temperature to one at a lower temperature. According to second law of thermodynamics, this would be possible only at the expense of some work. Hence we must have done some work in compressing the radiation as we moved the diaphragm B. This shows that radiation must exert pressure, resulting in a force, against which we had to do work.

Pressure in electromagnetic theory

Consider a parallel beam of electromagnetic radiation with fields denoted by E and H. Its momentum density is given by $E \times H/4\pi c = E^2/4\pi c$ in vacuum. Therefore radiation pressure, which is the momentum transfer per unit area per unit time, will be $P_r = E^2/4\pi c$. Consider a cylinder of cross-sectional area $d\sigma$ with its axis in the direction of E, and length $c\,dt$; see Fig. 4.17. Then the radiation contained in this cylinder must have crossed one face in time dt. Thus the momentum transfer per unit area per unit time becomes

$$P_r = \frac{E^2}{4\pi c} \cdot \frac{c\,d\sigma dt}{d\sigma dt} = \frac{E^2}{4\pi}.$$

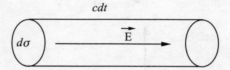

Fig. 4.17   *Rate of momentum transfer by radiation*

This gives

$$p_r = E^2/4\pi \qquad (4.58)$$

Also, the energy density in vacuum is

$$u_r = (E^2 + H^2)/8\pi = E^2/4\pi,$$

and the flux is given by

$$|\boldsymbol{F}_r| = |(c/4\pi)\,\boldsymbol{E} \times \boldsymbol{H}| = cE^2/4\pi.$$

We thus have

$$p_r = u_r = F_r/c. \qquad (4.59)$$

Photon theory

In photon theory of radiation, energy can change in multiples of $h\nu$, so that $dE_\nu = nh\nu$, and momentum p is related to energy E by $p = E/c$. Therefore momentum transfer would be, from (4.53),

$$p_\nu = \dot{E}_\nu/c = I_\nu \, d\sigma \cos\theta \, dt \, d\nu \, d\Omega / c.$$

Then for diffuse radiation absorbed by a plane surface (see Fig. 4.18) in all the frequencies and at all the angles, the normal momentum transfer per unit area per unit time would be

$$p_\nu d\nu = \frac{1}{c} \int \frac{I_\nu \, d\sigma \cos\theta \, dt \, d\nu \, d\Omega \cos\theta}{d\sigma \, dt}$$

$$= \frac{1}{c} \int I_\nu \cos^2\theta \, d\nu \, d\Omega$$

$$\Rightarrow p_\nu = \frac{1}{c} \int I_\nu \cos^2\theta \, d\Omega. \tag{4.60}$$

For isotropic radiation, I_ν is independent of θ and we get

$$p_\nu = \frac{2\pi I_\nu}{c} \int_0^\pi \cos^2\theta \sin\theta d\theta = \frac{4\pi I_\nu}{3c} = \frac{1}{3} u_\nu, \tag{4.61}$$

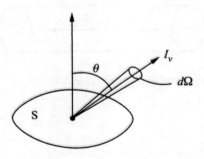

Fig. 4.18   *For calculation of radiation pressure*

which also gives $p = u/3$.

Thus photon theory shows that radiation pressure is one-third of the electromagnetic energy density. Just as u is a function of T alone, so is $p \equiv p(T)$. Finally, for a totally reflecting surface, all the normal momentum is reversed and we have $p = 2u/3$.

4.10.4   Thermal radiation as a perfect gas

Since radiation has pressure and obeys the perfect gas law $pv = $ constant or $p/u(T)$ = constant, we can use other perfect gas equations for radiation. We shall discuss here isothermal and adiabatic expansions, entropy and Gibbs free energy for radiation, followed by the case of a mixture of radiation and a perfect gas of matter in an enclosure.

Isothermal expansion

From the second law of thermodynamics, we could write the entropy of radiation in an enclosure of volume V at temperature T as

$$dS = (dU + pdV)/T.$$

Since $U = uV$, we get

$$dS = (udV + Vdu + pdV)/T.$$

Then using $p = u/3$, the entropy becomes

$$dS = (4udV/3 + Vdu)/T.$$

Therefore

$$(\partial S/\partial V)_u = 4u/3T, (\partial S/\partial u)_V = V/T \qquad (4.62)$$

This gives

$$\frac{\partial^2 S}{\partial u \partial V} = \frac{4}{3T} - \frac{4u}{3T^2}\frac{dT}{du},$$

while

$$\frac{\partial^2 S}{\partial V \partial u} = \frac{1}{T},$$

leading to

$$\frac{1}{T} = \frac{4}{3T} - \frac{4u}{3T^2}\frac{dT}{du}$$

$$\Rightarrow \frac{du}{u} = \frac{4dT}{T} \Rightarrow u = aT^4, \qquad (4.63)$$

where a is a constant.

Then as $u = 4\pi I/c$ (see (4.56)) and $F = \pi I$ is the flux emitted by the surface in solid angle π, we get

$$F = cu/4 = acT^4/4.$$

Thus $F = \sigma T^4$, with $\sigma = ac/4$. $\qquad (4.64)$

Adiabatic expansion

With $dS = 0$, the radiation satisfies the equation

$$u\,dV + V\,du + u\,dV/3 = 0,$$

leading to

$$\frac{4}{3}\frac{dV}{V} + \frac{du}{u} = 0$$

$$\Rightarrow uV^{4/3} = \text{constant}, \; pV^{4/3} = \text{constant}. \tag{4.65}$$

Thus $\gamma = 4/3$ for radiation.

Entropy

From the first equations of (4.62) and (4.63), we find that

$$S = \frac{4}{3}aT^3V + f(T),$$

where $f(T)$ is a function of T above. Now

$$\left(\frac{\partial S}{\partial T}\right)_V = \left(\frac{\partial S}{\partial u}\right)_V \frac{du}{dT} = 4aVT^2.$$

This leads to $S = \dfrac{4}{3}aT^3V + g(V)$,

where $g(V)$ is a function of V alone. Since both the expressions for entropy must be identical, we must have $f(T) = g(V) = 0$, so that we get

$$S = 4aT^3V/3 \tag{4.66}$$

for radiation.

Gibbs free energy

Using $G = U + pV - TS$, with $p = u/3$ and using (4.63) and (4.66), we see that $G = 0$ for radiation.

Radiation plus perfect matter gas

In this case, the pressure, the total internal energy and the total entropy would be

$$p = \frac{RTm}{V} + \frac{aT^4}{3} = \frac{RT}{v} + \frac{aT^4}{3},$$

$$U = c_v T_m + a T^4 V,$$

$$S = m\{c_v \ln T + R \ln v\} + \frac{4}{3} a T^3 V.$$

Then Gibbs potential can be see to be

$$G = m[c_p T - c_p T \ln T + RT \ln (T/v)]$$

or $\quad G = m[c_p T - c_p T \ln T + RT \ln(p - aT^4)/R].$ $\hfill (4.67)$

4.11 Spectrum of Thermal Radiation

Several careful experiments carried out in the later part of the nineteenth century revealed the nature of the black body spectrum. The variation of the intensity with various parameters was firmly established. It is well known that this gave birth to the first ideas of a quantum of radiation. But before that some experimental laws were established and corresponding thermodynamic theories were put forward. We discuss some of these in this section.

4.11.1 Wien's law

In the discussion of Kirchoff's law in Section 4.10 (b), the relation $B_\nu = e_\nu / a_\nu$ was introduced as the emissivity of a black body at frequency ν. We can obtain some information about $B_\nu(T)$ from thermodynamic considerations. We can do so by performing the following thought experiment.

Consider a perfectly reflecting spherical cavity at temperature T; see Fig. 4.19. Make the radiation inside it black by putting a small speck of black matter in it. Then remove the speck and allow the cavity to expand adiabatically with a velocity v of the walls. After some time the volume will change from V to V' and the nature of the radiation may change.

We can show that the radiation remains black but at a different temperature $T' < T$, since the radiation density is smaller. If the radiation were not black, it will have excess of radiation over the equilibrium value at some wavelength λ_1 and deficit at λ_2. Hence on introducing materials which absorb and emit these wavelengths respectively, the one absorbing λ_1 will gain in temperature and the other absorbing λ_2 will cool by radiation. Then we can run a heat engine between the two and extract work by extrating heat from the radiation field. This violates the second law of thermodynamics. Hence the radiation will remain black during adiabatic expansion.

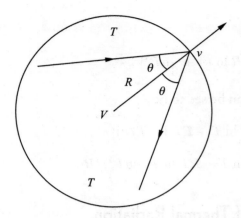

Fig. 4.19   *Thought experiment for Wien's law*

During expansion, the frequency of a photon changes at the time of reflection from the walls. If θ is the angle made by the photon path with the normal to the wall, we will have, at each reflection,

$$\frac{d\nu}{\nu} = -\frac{2v\cos\theta}{c}.$$

If R is the radius of the spherical cavity, the number of reflections in time dt would be $c\,dt/2R\cos\theta$. Therefore the frequency change in time dt would be

$$\frac{d\nu}{\nu} = -\frac{v\,dt}{R} = -\frac{dR}{R}.$$

When we relate this to the volume of the cavity, we have

$$d\nu/\nu = -dV/3V.$$

But we have seen earlier that $uV^{4/3}$ is a constant for radiation, and $u \propto T^4$. Thus we have

$$TV^{1/3} = \text{constant}.$$

This relates the temperature of radiation to volume in adiabatic expansion. It further leads to

$$\frac{dT}{T} + \frac{1}{3}\frac{dV}{V} = 0 \Rightarrow \frac{d\nu}{\nu} = \frac{dT}{T} \Rightarrow \lambda T = \text{constant}. \tag{4.68}$$

This tells us how the wavelength of each photon increases as the temperature falls during adiabatic expansion.

Now we shall obtain the dependence of u_ν on ν and T. During an adiabatic change,

$$\delta U = -p\delta V$$

for each frequency interval. Therefore

$$\delta(u_\nu \, d\nu V) = -\frac{1}{3} u_\nu \, d\nu \, \delta V.$$

This can be written as

$$\delta u_\nu \, d\nu V + \delta(d\nu) u_\nu V + u_\nu \, d\nu \delta V = -\frac{1}{3} u_\nu \, d\nu \, \delta V.$$

After rearranging terms, this leads to

$$\frac{4}{3} \frac{\delta V}{V} + \frac{\delta u_\nu}{u_\nu} + \frac{\delta(d\nu)}{d\nu} = 0. \tag{4.69}$$

Since $d\nu = \nu' - \nu$ and $\nu' \approx \nu$, we have $\delta(d\nu) = \delta\nu' - \delta\nu$, giving

$$\frac{\delta\nu}{\nu} = \frac{\delta\nu'}{\nu'} = -\frac{vdt}{R} = -\frac{dR}{R} = -\frac{1}{3}\frac{dV}{V}.$$

Therefore,

$$\delta(d\nu) = -\frac{vdt}{R}(\nu' - \nu) = \frac{\delta\nu}{\nu}d\nu.$$

Using this in (4.69) and working out a little algebra, we get

$$\frac{\delta u_\nu}{\nu} = 3\frac{\delta\nu}{\nu}$$

$$\Rightarrow u_\nu \propto \nu^3.$$

Note here that the constant of proportionality, in accordance with (4.68), must be a function of ν/T. Denoting it by $\varphi(\nu/T)$, we get

$$u_\nu = \nu^3 \varphi(\nu/T). \tag{4.70a}$$

This is *Wien's law*.

In terms of wavelength, we have

$$\nu = \frac{c}{\lambda} \Rightarrow d\nu = -\frac{c}{\lambda^2}d\lambda.$$

Since the energy content in the frequency interval $d\nu$ must be the same as the content in the wavelength interval $d\lambda$, and it must be non-negative, we have $|u_\nu d\nu| = |u_\lambda d\lambda|$. This therefore gives the wavelength spectrum of energy density as

$$u_\lambda = u_\nu \left| \frac{d\nu}{d\lambda} \right| = u_\nu \frac{c}{\lambda^2}.$$

Using (4.70a), we then have

$$u_\lambda = f(\lambda T)/\lambda^5 = T^5 F(\lambda T), \tag{4.70b}$$

where $f(\lambda T)$ and $F(\lambda T)$ are new constants, both functions of λT.

All the three functions φ, f and F appearing in (4.70) are universal functions which cannot be determined by thermodynamic arguments. Even so, it is amazing to note the understanding of the properties of radiation which can be obtained from thermodynamics alone.

Further, we can find the emissivity $B_\nu(T)$ or $B_\lambda(T)$ for all temperatures if we know it for one T. For this, we note that the specific intensity I_ν (or I_λ) introduced in (4.53) is just B_λ for a black body. Therefore we can write the first of equation (4.56) in the form

$$u_\lambda = 4\pi B_\lambda/c.$$

This then shows that

$$\frac{B_\lambda(T)}{B_{\lambda'}(T')} = \frac{u_\lambda(T)}{u_{\lambda'}(T')} = \frac{T^5}{T'^5} \text{ with } \frac{\lambda'}{\lambda} = \frac{T}{T'}. \tag{4.71}$$

Thus if we double the temperature, $T' = 2T$, then we scale down the λ-axis by a factor of 2 and scale up the u_λ-axis by a factor of 2^5. In other words, the radiation density at a wavelength λ' at a temperature $T' = 2T$ will be 2^5 times the density at a wavelength $\lambda'/2$ at the temperature T. Figure 4.20 shows B_λ at two temperatures.

The function B_λ will have a maximum at $dB_\lambda/d\lambda = 0$. This can be put as $dF(\lambda T)/d\lambda = 0$ or as $T dF(\lambda T)/d(\lambda T) = 0$. Note that T is independent of λ. The very fact that we have been able to put F as a function of λT rather than of λ and T separately, suggests that at the maximum of F with respect to λ, $\lambda_m T$ will be a constant. Thus we can write

$$\lambda_m T = b, \tag{4.72}$$

where b is a constant. Experiments have shown that this constant has a value $b = 2.9$ mm K. This is known as Wien's displacement law.

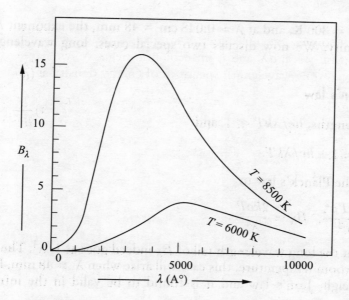

Fig. 4.20 *Function $B_\lambda(T)$ versus λ at two different tempeatures, showing Wien's displacement law*

4.11.2 Planck's law

The functions $\varphi(\nu/T)$ and $f(\lambda T)$ and hence $B_\lambda(T)$ can be determined only after introdu-cing quantum ideas and quantum statistics. We shall do so in chapter 8. Here we quote the result, that is Planck's law, according to which

$$B_\nu(T) = \frac{2h\nu^3}{c^2} \frac{1}{e^{h\nu/kT} - 1},$$

$$B_\lambda(T) = \frac{2hc^2}{\lambda^5} \frac{1}{e^{hc/\lambda kT} - 1}. \tag{4.73}$$

Here h is the Planck's constant, which has a value $h = 6.625 \times 10^{-27}$ erg s, and k is the Boltzmann constant, $k = 1.38 \times 10^{-16}$ erg K^{-1}.

Before Planck proposed the above law on the basis of his new quantum idea, the black body spectrum had been obtained experimentally. There was no theory which could reproduce the complete spectrum, but there were a few attempts which could fit with a part of the spectrum. Two partially successful attempts were made by Rayleigh–Jean and Wien. We shall discuss these developments, which appear as special cases of Planck's law.

We note that the exponent in the factor in the denominator in (4.73) is $hc/\lambda kT$. Here h, c and k are constants, and it turns out that $hc/k = 1.44$ cm.K. Thus at room temperature $T = 300$ K, and at $\lambda = 0.048$ cm $= 48$ mm, the exponent $hc/\lambda kT$ would be equal to unity. We now discuss two special cases, long wavelengths and short wavelengths.

Rayleigh–Jean's law

At long wavelengths, $hc/\lambda kT \ll 1$, and

$$e^{hc/\lambda kT} \approx 1 + hc/\lambda kT.$$

This reduces the Planck's law to

$$B_\nu = \frac{2kT\nu^2}{c^2}, \quad B_\lambda = \frac{2kcT}{\lambda^4}. \tag{4.74}$$

It is found that the long wavelength tail of B_λ indeed goes as $1/\lambda^4$. The above estimate shows that at room temperature, this case will arise when $\lambda \gg 48$ mm. Equation (4.74) is called Rayleigh–Jean's law and it is found to be valid in the infrared and radio wavelengths.

Wien's approximation

At short wavelengths and high frequencies, on the other hand, $hc/\lambda kT \gg 1$, and we can write

$$B_\nu = \frac{2h\nu^3}{c^2}e^{-h\nu/kT}, \quad B_\lambda = \frac{2hc^2}{\lambda^5}e^{-hc/\lambda kT}. \tag{4.75}$$

This is Wien's law, which is found to be good for ultraviolet, X-ray and γ-ray regions at room temperature.

Let us denote

$$2hc^2 = c_1 = 1.19 \times 10^{-5} \text{ erg cm}^2/\text{s}, \quad hc/k = c_2 = 1.44 \text{ cm K}.$$

Then we can write

$$B_\lambda = \frac{c_1}{\lambda^5}\frac{1}{e^{-c_2/kT} - 1}.$$

We shall now discuss some outcomes of Planck's black body spectrum, which are particularly of experimental interest.

Stephen–Boltzmann law

To obtain the total energy radiated by a black body at a temperature T, we integrate B_λ of (4.73) to obtain

$$B = \int_0^\infty B_\lambda d\lambda = C_1 \int_0^\infty \frac{d\lambda}{\lambda^5(e^{-c_2/\lambda T} - 1)}.$$

Using the standard technique of change of variables, this reduces to

$$B = \frac{c_1 T^4}{c_2^4} \int_0^\infty \frac{x^3 dx}{e^x - 1} = \frac{\pi^4}{15} \frac{c_1 T^4}{c_2^4},$$

and

$$F = \pi B = \frac{c_1 \pi^5 T^4}{15 c_2^4} \equiv \sigma T^4.$$

Thus the total energy radiated per unit area per unit time. Here

$$\sigma = \frac{2}{15} \frac{\pi^5 k^4}{c^2 h^3} = 5.67 \times 10^{-5} \text{ erg cm}^2 \text{ s}^{-1} \text{ K}^4.$$

Wien's displacement law

With B_λ given by (4.73), we can determine the wavelength at which it has its maximum. Denoting this wavelength by λ_m and $c_2/\lambda_m T = x$, we see that this maximum occurs at

$$5(1 - e^{-x}) = x.$$

It is found numerically that the solution is about $x = 4.965$. This gives

$$\lambda_m T = \frac{c_2}{x} = \frac{ch}{4.965k} = 0.2898 \text{ cm K}.$$

The experimental graphs of B_λ as well as the peak wavelength λ_m are often used to determine the temperature of a star or furnace.

Problems

4.1. In van der Waals equation of state, equation (4.3), a/v^2 is the internal pressure caused by van der Waals force between molecules (to be discussed in chapter V), and b is the reduction in volume cause partly (half) due to finite volume occupied by the molecules and partly (half) due to internal presure. If d is the diameter of the molecule, $b = 2d^3$. Then

(a) Show that (i) $a/v^2 p \approx b/v$, and (ii) at low temperature ($T < T_c$), p has a maximum, a minimum and a point of inflection, and that the three points merge at $v = 3b$ when $T_c = 8a/9bR$ and $p_c = a/3b^2$.

(b) For normal atmospheric air at $T = 0$ ^{0}C and $p = 1$ atm $= 1.033 \times 10^6$

dynes/cm^2, the value of a/v^2 is 0.0028 atm. Calculate the diameter of an average molecule (mass number $A = 28.97$ with 75% N_2 and 25% O_2 by weight). Take $N_A = 6.024 \times 10^{23}$ per gm-mole at $V = 22.414$ litres.

4.2. Draw isotherms and adiabats in the $p-v$ plane for a monatonic perfect gas at $T = 100$ K, 500 K and 1000 K.

4.3. Explain the difference between isothermal and adiabatic expansion in terms of work done, exchange of heat and change in entropy with the help of a diagram in the $p - v$ plane.

4.4. Calculate the change in entropy during melting of 1 gm of water and during boiling of 1 gm of water at atmospheric pressure. Take latent heat of melting = 80 cal/gm, latent heat of boiling = 540 cal/gm, specific density of ice = 0.9; 1 cal = $R/2 = 4.185 \times 10^7$ erg/mol.

4.5. Hawking has shown that a black hole of mass M has a temperature $T = \hbar K/2\pi c$, where $K = GM/R^2$ is surface gravity and $R = 2GM/c^2$ is the Scharzchild radius. Taking $dQ = c^2 dM$, show that the entropy of the black hole is $S = A/4L_p^2$, where A is the surface area of the black hole and $L_p = (G\hbar/c^3)^{1/2}$ is the Planck's length. Further, calculate values of Planck length, Planck time $T_p = L_p/c$ and Planck energy $E_p = \hbar/T_p$.

4.6. A body has a constant heat capacity C and is at an initial temperature T_1. It is placed in contact with a heat reservoir at a temperature $T_2 > T_1$ and comes to equilibrium with it at constant pressure. Calculate the entropy change of the universe.

4.7. Some scientists are marooned on an island where pressure and temperature are constant everywhere. They run out of power. They find that there is a cavity in which gas is leaking slowly, which is also at the same pressure and temperature. They have two membranes, one of which passes the gas but inhibits air, and the other one does the opposite. Create an engine for generating power.

4.8. Show that in vacuum the specific intensity of radiation remains constant along a ray.

4.9. Show that the flux of radiation varies as the inverse square of the distance in empty space.

4.10. Derive the equations of transfer for radiation passing through a layer of hot gas at temperature T, and solve it.

4.11. Black body radiation in a volume v and temperature T may be treated as a substance obeying the laws

$$pv = u/3, \quad u = 4\sigma v T^4/c.$$

If the temperature changes by dT and volume by dv, show that the heat entering the system is

$$\frac{16\sigma T^4}{3c} dv + \frac{16\sigma v T^3}{c} dT.$$

4.12. What would be the dominant radiation in the spectra of the following objects?

(i) Cosmic background at 2.7 K; (ii) hydrogen molecular cloud at 10 K; (iii) Jupiter at 125 K; (iv) Earth's surface at 300 K; (v) stars with surface temperatures of 4000 K, 6000 K, 10 000 K and 30 000 K; (vi) interstellar clouds at 10^6 K and 10^8 K.

5

Atomic Spectra and Quantum Mechanics

IT is highly creditable to the scientists of late nineteenth century that they made extremely careful observations of the wavelengths and intensities of emission and absorption lines of almost all chemical elements, under magnetic field and electric field, in laboratory conditions and in the light coming from the stars. The attempt to explain all these observations has lead to the Bohr's theory of hydrogen atom, Sommerfeld's modification to elliptical orbits and relativistic corrections, Uhlenbeck and Goudsmit's spin hypothesis, theories of coupling of angular momenta, Heisenberg's and Schrodinger's quantum mechanics, Pauli exclusion principle, Dirac's relativistic wave equation for an electron and electron spin, etc., all in a short span of a quarter century. We give a brief description of this development in this chapter. We also discuss simple quantum mechanical systems such as one-dimensional motion, harmonic oscillator, hydrogen and hydrogen-like atoms, and some other aspects.

5.1 Bohr's Theory of Hydrogen Atom

Bohr's model of the simplest atom, the hydrogen atom, was a step away from classical ideas of radiation. Rutherford had earlier showed that the atom contained a small massive part of positive charge whose linear size was about 10^{-5} times that of the atom. This section outlines the historical background, followed by an overview of the early developments leading to Bohr's proposal and its consequences.

5.1.1 Introduction

All our information regarding celestial bodies—stars, planets, galaxies—is derived mainly by studying the electromagnetic radiations coming from them, and in particular, by studying their spectra. Newton in 1664 obtained the first continuous spectrum of the sun, and Fraunhoffer in 1817 found many dark lines superposed on it. These features, viz., dark Fraunhoffer lines superposed on a continuous spectrum, are common to most stellar spectra. The chromosphere and corona of the sun and the gaseous nebulae show emission spectra.

On the other hand, in the laboratory also, there was tremendous activity to record spectra of practically all elements and compounds. Whether in laboratory conditions or in cosmic conditions, it is important to understand the origin of spectra in general. Kirchoff's laws formulated on the basis of the experiments he carried out with Bunsen between 1859–61 help us in our preliminary understanding of spectra. They state that (i) hot solids, liquids and dense gases produce a continuous spectrum; (ii) hot tenuous gases produce an emission spectrum; (iii) cool gases produce absorption spectrum when the continuous spectrum of a hotter source is passed through them. Accordingly, in the case of the sun and the stars, the continuous spectrum is produced by the hot dense gas of the photosphere which appears as the limb of the sun; see Fig. 5.1. The dark Fraunhoffer lines are produced when this continuous spectrum passes through the overlying cooler gas which forms what is called the reversing layer. But the reversing layer is itself quite hot as can be inferred from the emission spectrum displayed by it during a total solar eclipse. In this so-called flash spectrum, the wavelengths of emission lines are identical with those of dark Fraunhoffer lines.

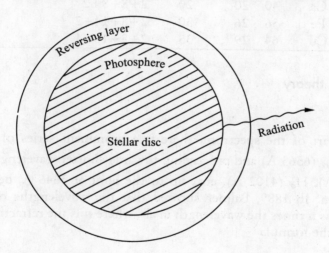

Fig. 5.1 *Structure of the stellar atmosphere*

Kirchoff and Bunsen also found that atomic gases produce line spectra while molecules produce band spectra. Further, each species has its own characteristic spectrum which can appear in absorption or emission depending on circumstances. These characteristic spectra can be used for identifying the constituents of a flame or star or interstellar space.

According to Rutherford's model (see Chapter 9) of the atom as modified by later discoveries, an atom consists of a small massive nucleus of atomic weight A, containing Z positively charged protons and $A - Z$ chargeless neutrons. Z, the *atomic number*, also gives the number of negatively charged electrons which revolve around the nucleus in a neutral atom. The electrons are arranged in shells called K, L, M, N, etc. The electrons in the outermost incomplete shell are called valence electrons; they determine the chemical properties of the atom as well as the nature of its spectrum. Table 5.1 gives the structure of some of the common elements. Hydrogen, which contains one proton and one electron, is the simplest of all atoms, and its spectrum is also the simplest and easiest to understand.

Table 5.1   *Structure of some atoms*

Element	Symbol	A	Z	$A - Z$	Electrons K, L, M, N, $\cdots$
Hydrogen	H	1	1	0	1
Helium	He	4	2	2	2
Lithium	Li	7	3	4	2+1
Carbon	C	12	6	6	2+4
Sodium	Na	23	11	12	2+8+1
Calcium	Ca	40	20	20	2+8+8+2
Iron	Fe	56	26	30	2+8+14+2
Copper	Cu	64	29	35	2+8+18+1

5.1.2   Bohr's theory

Background

In the visible part of the spectrum, there is a prominent series of hydrogen lines starting from H_α (6563 Å) and proceeding towards shorter wavelengths as H_β (4861 Å), H_γ (4343 Å), H_δ (4102 Å), etc., up to the limit at 3646 Å, beyond which we get a continuum. In 1889, Balmer showed that the wavelengths of these lines in vacuum (which is n times the wavelength in air, where n is the refractive index) can be represented by the formula

$$\lambda = \frac{n^2 G}{n^2 - 4}, \qquad n = 3, 4, 5, \cdots, \qquad G = 3647 \text{Å}$$

Since then it is called *Balmer series* of hydrogen.

In 1896, Rydberg put Balmer formula in the form

$$\sigma = \frac{1}{\lambda} = R_{\mathrm{H}} \left[\frac{1}{2^2} - \frac{1}{n^2} \right], \qquad n = 3, 4, 5, \cdots,$$

where $R_{\mathrm{H}} = 109677.581$ cm^{-1} is called the *Rydberg constant*; $\sigma = 1/\lambda$ is called the *wavenumber* and represents the number of waves that can be fitted in 1 cm. Ritz called R/n^2 as *terms* and showed that we can obtain any line or series by combining terms.

This is known as *Ritz combination principle*. Thus $1/\lambda = R[1/n_1^2 - 1/n_2^2]$ also gives all the other series of hydrogen. The terms suggest the existence of energy levels which were derived by Bohr in 1913.

Bohr's postulates

Bohr based his theory on three postulates which we state and discuss here.

1. The electron moves around the proton in discrete circular orbits without emission of radiation. This was a departure from classical electromagnetic theory according to which an accelerated charge like the electron in a circular orbit should emit radiation and thus lose energy. Consequently, it would ultimately spiral in and merge with the proton. This postulate prevents the electron from doing so.

2. The size of the orbit is determined by the condition that the angular momentum of the electron around the nucleus should be an integral multiple of $h/2\pi \equiv \hbar$, where h is the Planck's constant. If the angular momentum is $n\hbar$, where n is a positive integer, then n is called the *quantum number* of the orbit.

3. Radiation is emitted or absorbed when the electron jumps from one discrete orbit to another one, and the frequency of radiation is given by $h\nu = E_2 - E_1$ ($E_2 > E_1$).

5.1.3 Energy levels of hydrogen

In order to derive the energy levels, we write down the equation of motion of the electron by equating the product of mass and centripetal acceleration to the attractive force between the proton and the electron. Now, there are two kinds of force acting between these particles: electrostatic (e^2/r^2) and gravitational (Gm_pm_e/r^2). Here e is the magnitude of charge on electron or proton, m_e the mass of the electron, m_p the mass of the proton, r the distance between them and G the universal constant of gravitation. It can be seen that the electrostatic force is about 10^{39} times stronger than the gravitational force. Hence we can neglect the latter. Then looking at Fig. 5.2, where C is the centre of mass, we can write

$$m_e r_e \omega^2 = e^2/r^2, \qquad r = r_e + r_p,$$

where r_e and r_p are respectively the distances of the electron and the proton from the nucleus. But,

$$r_e = m_p r/(m_p + m_e), \tag{5.1}$$

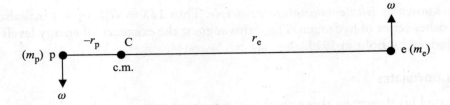

Fig. 5.2   *Proton and electron in a hydrogen atom*

so that we have

$$\mu r \omega^2 = e^2/r^2, \quad \text{with} \quad \mu = m_e m_p/(m_p + m_e). \tag{5.2}$$

The reduced mass μ of the electron–proton pair can be seen to be slightly smaller than the free electron mass. (Note that the electron mass was not known at that time. It was obtained by Millikan soon after.)

According to Bohr's second postulate, we have

$$(m_e r_e^2 + m_p r_p^2)\omega = n\hbar \Rightarrow \mu r^2 \omega = n\hbar. \tag{5.3}$$

Eliminating μ between Eqs. (5.2) and (5.3), we get

$$r_n = \frac{n^2 \hbar^2}{\mu e^2} \equiv a_0 n^2, \tag{5.4}$$

where $a_0 = 0.529\text{Å}$ and we have replaced r by r_n for the nth orbit. Thus the radii of the Bohr orbits are proportional to n^2. The energies in these orbits can now be obtained as under.

The total energy of the hydrogen atom is equal to the sum of kinetic energies of the electron and the proton and their mutual (electrostatic) potential energy. This gives

$$E = \frac{1}{2}(m_p r_p^2 + m_e r_e^2)\omega^2 - e^2/r.$$

A little algebra shows that the nth energy level is given by

$$E_n = -\frac{2\pi^2 \mu e^4}{n^2 h^2} \equiv -\frac{E_0}{n^2}, \tag{5.5}$$

which defines $E_0 = 2\pi^2 \mu e^4/h^2$. Substituting the now-known values of electronic charge, mass and proton mass, we see that $E_0 = 13.6$ eV. K, L, M, N, $\cdots$ shells of Table 5.1 correspond to $n = 1, 2, 3, 4, \cdots$. The total energy is negative for all n, i.e., the electron is bound to the proton. The energy is lowest for $n = 1$, the *ground state*; it increases with n and becomes zero as $n \to \infty$, when the electron becomes free. This process is called *ionisation* because we are left with a positively charged proton and a free electron. (Here zero of energy is the energy of a stationary proton and electron far away from each other.) Thereafter the total energy is equal to the sum of their kinetic energies which can take any value in a continuous manner.

5.1.4 Spectra of hydrogen

The energy levels of hydrogen are shown in Fig. 5.3. According to the third postulate of Bohr, the frequency of radiation emitted through a transition between levels n_1 and n_2 $(n_2 > n_1)$ is given by

$$h\nu = E_0 \left(\frac{1}{n_1^2} - \frac{1}{n_2^2} \right) \Rightarrow \frac{1}{\lambda} = R_{\mathrm{H}} \left(\frac{1}{n_1^2} - \frac{1}{n_2^2} \right) \Rightarrow \frac{1}{\lambda} = R \left(\frac{1}{n_1^2} - \frac{1}{n_2^2} \right). \qquad (5.6)$$

It can be seen that $E_n = -R_{\mathrm{H}} hc / n^2$. When a transition takes place from an upper level to a lower level, we get an emission line, and in the other case, we get an absorption line. The series of lines which have a definite n_1 and different $n_2 > n_1$ are given specific names after scientists who discovered or worked on them. The first few series are indicated below:

$n_1 = 1, n_2 \geq 2$; *Lyman series*, in ultraviolet, with limit at $\lambda = 1/R = 912$ Å;

$n_1 = 2, n_2 \geq 3$; *Balmer series*, in visible region, with limit at $\lambda = 4/R = 3647$ Å;

$n_1 = 3, n_2 \geq 4$; *Paschen series*, in infrared, with limit at $\lambda = 9/R = 8208$ Å;

$n_1 = 4, n_2 \geq 5$; *Bracket series*, in far infrared, with limit at $\lambda = 16/R = 14590$ Å.

Beyond each series (wavelength smaller than the series limit), we get a continuum due to transitions between n_1 and a free state as shown in Fig. 5.3.

5.1.5 Other hydrogen-like atoms

When an atom is given more and more energy, it loses more and more electrons, giving rise to singly-ionised atoms, doubly-ionised atoms, etc. Atoms or ions such as neutral hydrogen (H), singly-ionised helium (He^+), doubly-ionised lithium (L_i^+), etc., have only one electron moving around a central nucleus; hence they are called *hydrogen-like atoms/ions*. Their spectra are as simple as that of the hydrogen atom, and can be predicted from Bohr's theory with the following modifications:

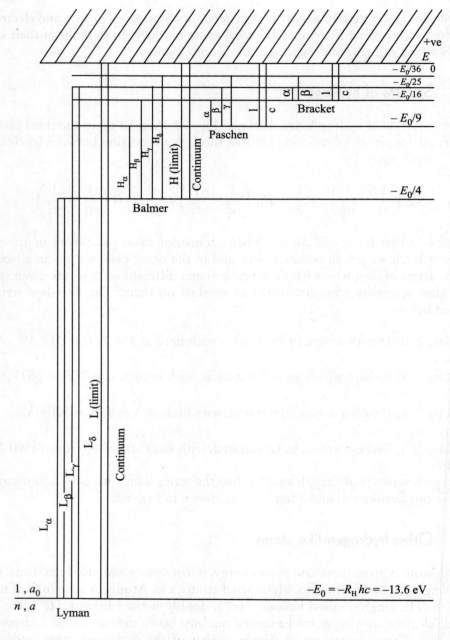

Fig. 5.3   *Energy levels of hydrogen, some lines and series*

(i) Since the charge on the nucleus is Ze, the electrostatic force of attraction on the electron will be Ze^2/r^2, so e^2 should be replaced by Ze^2 in all the equations of Bohr model.

(ii) The mass of nucleus is m_A which will change the reduced mass of the electron which will be $\mu_A = m_e m_A/(m_e + m_A)$. It is clear that even in the case of the lightest atom, hydrogen, this reduced mass is smaller than the electron mass only by 0.05%, and the reduced mass tends to the electron mass for heavier atoms.

Thus for a hydrogen-like atom, (5.4), (5.5) and (5.6) become

$$r_n = \frac{a_0 n^2}{Z}, \tag{5.7a}$$

$$E_n = -\frac{E_0 Z^2}{n^2}, \tag{5.7b}$$

$$\frac{1}{\lambda} = R_A Z^2 \left(\frac{1}{n_1^2} - \frac{1}{n_2^2} \right), \tag{5.7c}$$

where a_0, E_0 and R_A are now given by

$$a_0 = \frac{\hbar^2}{\mu_A e^2}, \quad E_0 = \frac{2\pi^2 \mu_A e^4}{h^2}, \quad R_A = \frac{2\pi^2 \mu_A e^4}{h^3 c}. \tag{5.8}$$

As $m_A \to \infty, \mu_A \to m_e, R_A \to R_\infty = 2\pi^2 m_e e^4/h^3 c$. This Rydberg constant for a heavy hydrogen-like atom has a limiting value 109737.312 cm^{-1}. With this, we can write $R_A = R_\infty/(1 + m_e/m_A)$ and $R_H = R_\infty/(1 + m_e/m_p)$. As an example, for He$^+$, we have $R_{\text{He}} = 109722.391$ cm^{-1}.

The case of the ion Ue$^+$, with $Z = 2$, is important. In this case, we would have

$$\frac{1}{\lambda} = R_{\text{He}} \left(\frac{4}{n_1^2} - \frac{4}{n_2^2} \right).$$

If we put $n_1 = 4$ and $n_2 = 5, 6, 7, \cdots$, we get what is known as the *Pickering series*. As $R_{\text{He}} \approx R_H$, alternate lines of the series will coincide with the lines of Balmer series, that is, $n_2 = 6$ will coincide with H$_\alpha$, $n_2 = 8$ will coincide with H$_\beta$, etc., of Balmer series. Of course the small difference between R_{He} and R_H will not make the coincidence exact. The wavelengths of the corresponding lines will differ by a fraction

$$\frac{\Delta\lambda}{\lambda_H} = \frac{\lambda_H - \lambda_{\text{He}}}{\lambda_H} = \frac{R_{\text{He}} - R_H}{R_{\text{He}}} \simeq 4.1 \times 10^{-4}.$$

This gives the wavelength difference between corresponding lines as

$$\lambda(H_\alpha) - \lambda \text{ (Pickering 6)} = 2.7 \text{ Å},$$

$$\lambda(H_\beta) - \lambda \text{ (Pickering 8)} = 2.0 \text{ Å},$$

$$\lambda(H_\gamma) - \lambda \text{ (pickering 10)} = 1.8 \text{ Å, etc.}$$

It may be mentioned that (5.7a) gives the radii of K, L, M, N, etc. shells by putting $n = 1, 2, 3, 4, \cdots$, etc, respectively, and can thus be used for estimating the size of various atoms. We list in Table 5.2 the wavelengths of some lines of H and He^+.

5.2   Sommerfeld's Modification of Bohr's Theory

While Bohr's theory agreed with the broad features of the spectral lines of hydrogen, discrepancies were noted between predictions and observations, before and after Bohr's theory. Bohr's theory had suggested that the successive energy levels with $n = 1, 2, 3$, etc., should have a degeneracy of 2, 8, 18, etc. A good deal of splitting was seen in these levels, which suggested the existence of sub-levels. Sommerfeld suggested a modification which required the electrons to move in elliptical orbits, and there were as many different elliptical orbits corresponding to each n as the degeneracy of that level. We shall now discuss these developments in this section.

Table 5.2   *Spectra of H and He^+ (λ in Å)*

Transition line	$1 \leftrightarrow n$		$2 \leftrightarrow n$		$3 \leftrightarrow n$		$4 \leftrightarrow n$	
	H	He^+	H	He^+	H	He^+	H	He^+
α	1215	303	6563	1640	18751	4686	40511	10123
β	1026	256	4861	1215	12815	4203	26252	6560
γ	972	243	4340	1084	10938	–	21656	5411
δ	949	237	4102	1025	10044	–	19445	4859
$\vdots$	$\vdots$	$\vdots$	$\vdots$	$\vdots$	$\vdots$	$\vdots$	$\vdots$	$\vdots$
Limit	912	220	3646	911	8204	2050	14584	3644
Series name	Lyman		Balmer		Paschan		Bracket	Picturing

5.2.1   Wilson–Sommerfeld quantization rule

A particle moving under a central force in a bound state can also move in elliptical orbits. In the plane of the orbit, the position of the electron can be specified by the polar coordinates r an ψ. Correspondingly we have two components of velocity, radial velocity $v_r = \dot{r}$ and angular velocity $v_\psi = r\dot{\psi}$, and two momenta, radial momentum

$p_r = \mu \dot{r}$ and angular momentum $p_\psi = \mu r^2 \dot{\psi}$. Now Bohr's condition $p_\psi = n\hbar$ can be recast as

$$\int_0^{2\pi} \mu r^2 \dot{\psi} d\psi = \int_0^{2\pi} p_\psi d\psi = kh,$$

where k is an integer. Generalising for the other coordinate, we can write

$$\int p_r dr = mh,$$

where m is also an integer. In general, if q is a coordinate and p the corresponding momentum, then we have

$$\int p dq = nh.$$

This is known as Wilson-Sommerfeld quantization rule. The numbers k and m are called *radial* and *azimuthal* quantum numbers.

5.2.2 Sommerfeld's elliptical orbits

We discuss in this section the shape, size and energy associated with an elliptical orbit.

Shape of the ellipse

For a central force, the transverse acceleration is

$$r\ddot{\psi} + 2\dot{r}\dot{\psi} = \frac{1}{r}\frac{d}{dt}(r^2\dot{\psi}) = 0.$$

Hence

$$p_\psi = \mu r^2 \dot{\psi} = \text{ constant } = k\hbar. \qquad (5.9)$$

An ellipse of eccentricity ϵ and semimajor axis a can be represented by the equation

$$r = \frac{a(1 - \epsilon^2)}{1 + \epsilon \cos \psi}.$$

Then,

$$dr = \frac{a(1 - \epsilon^2)\epsilon \sin \psi \, d\psi}{(1 + \epsilon \cos \psi)^2}$$

and

$$\dot{r} = \frac{a(1 - \epsilon^2)\epsilon \sin \psi}{(1 + \epsilon \cos \psi)^2} \quad \dot{\psi} = \frac{\epsilon \sin \psi}{a(1 - \epsilon^2)} \frac{p_\psi}{\mu}, \qquad (5.10)$$

where we have used the first part of (5.9). Therefore,

$$\oint p_r dr = \oint \mu \dot{r} dr = \int_0^{2\pi} \frac{\epsilon^2 \sin^2 \psi p_\psi d\psi}{(1 + \epsilon \cos \psi)^2} = mh.$$

Using (5.9) for p_ψ, we then get

$$m = \frac{k}{2\pi} \int_0^{2\pi} \frac{\epsilon^2 \sin^2 \psi \, d\psi}{(1 + \epsilon \cos \psi)^2}.$$

We now split the integrand into two factors, $\epsilon \sin \psi$ and $\epsilon \sin \psi / (1 + \epsilon \cos \psi)^2$ and integrate by parts to get

$$m = \frac{k}{2\pi} \left[\frac{\epsilon \sin \psi}{1 + \epsilon \cos \psi} \Big|_0^{2\pi} - \int_0^{2\pi} \frac{\epsilon \cos \psi}{1 + \epsilon \cos \psi} d\psi \right].$$

After some algebra, we can put this in the form

$$m = k \left[\frac{1}{\pi} \int_0^\pi \frac{\sec^2(\psi/2) \, d\psi}{(1 + \epsilon) + (1 - \epsilon) \tan^2(\psi/2)} - 1 \right].$$

Using the transformation $(1 - \epsilon)^{1/2} \tan(\psi/2) = (1 + \epsilon)^{1/2} x$, we finally get

$$\left(\frac{m + k}{k} \right)^2 = \frac{1}{1 - \epsilon^2} = \frac{a^2}{b^2},$$

where b is the semiminor axis of the ellipse. Putting $m + k = n$, where n is called the total or *principal quantum number*, we see that

$$k/n = b/a.$$

Thus for each value of n we get $k = 1, 2, 3, \cdots, n$, corresponding to decreasing eccentricities given by $\epsilon = (n^2 - k^2)^{1/2}/n$. They are called s, p, d, f, $\cdots$, orbits, respectively. In this model, $k = 0$ is not allowed because then the electron will have to pass through the nucleus.

Size of the orbit

The radial acceleration is $\ddot{r} - r\dot{\psi}^2$, which gives the equation of motion for the electron

$$\mu(\ddot{r} - r\dot{\psi}^2) = -Ze^2/r. \tag{5.11}$$

But from (5.10),

$$\mu \dot{r} = p_r = \frac{\epsilon \sin \psi}{a(1 - \epsilon^2)} p_\psi,$$

so that

$$\mu \ddot{r} = \epsilon \cos \psi \; p_\psi \; \frac{\dot{\psi}}{[a(1-\epsilon^2)]}.$$

At perinucleus (the point nearest to the nucleus),

$$\psi = 0, r = a(1-\epsilon), \dot{r} = 0, \dot{\psi} = \frac{p_\psi}{[\mu a^2 (1-\epsilon)^2]}. \tag{5.12}$$

Therefore (5.11), after some algebra becomes

$$\frac{p_\psi^2}{\mu a(1-\epsilon^2)} = Ze^2.$$

This, in turn, gives

$$a = \frac{k^2 \hbar^2}{\mu Z e^2 (1-\epsilon^2)}.$$

Since $1 - \epsilon^2 = k^2/n^2$, we get the semimajor axis for the nth level as

$$a_n = a_0 n^2 / Z, \tag{5.13}$$

which is the same as (5.4) for $Z = 1$.

Energy levels

The total energy of the atom in the orbit is given by

$$E = \frac{1}{2}\mu \dot{r}^2 + \frac{1}{2}\mu r^2 \dot{\psi}^2 - Ze^2/r,$$

which must be a constant. Using (5.10), we then get

$$E = \frac{\mu a e^2 Z (1-\epsilon^2)}{2\mu a^2 (1-\epsilon)^2} - \frac{Ze^2}{a(1-\epsilon)} = -\frac{Ze^2}{2a} = -\frac{E_0 Z^2}{n^2}, \tag{5.14}$$

which is similar to (5.5) of Bohr's model. Thus in the non-relativistic Sommerfeld's model, the energy depends only on the principal quantum number. All the sub-orbits (allowed elliptical orbits) with $k = 1, 2, \cdots, n$ are degenerate.

5.2.3   Removal of degeneracy

Sommerfeld used special theory of relativity to modify Bohr's model and obtained an expression for the energy level for the (n, k) orbit. The degeneracy for different k

(same n) no longer appears in his relativistic formula. Although we skip Sommerfeld's relativistic treatment, we discuss here some typical resulting features.

The angular momentum in the n-th orbit in the Bohr atom is given by

$$\mu r_n^2 \dot{\psi} = \mu r_n v_n = n\hbar.$$

Dividing this by c and using Eqs (5.7) and (5.8), we find that

$$v_n/c = Ze^2/n\hbar c.$$

This ratio for the first Bohr orbit in hydrogen atom is

$$\alpha = e^2/\hbar c = 1/137.29. \tag{5.15}$$

This universal constant is called Sommerfeld's *fine structure constant*. These equations show that the velocity of the electron in the first Bohr orbit is about 0.73% of the speed of light, and goes on decreasing for higher orbits.

As relativistic effects become significant, the reduced mass must be replaced by $\gamma\mu$, where γ is the relativistic enhancement factor. As E_0 is directly proportional to mass, all energy levels are slightly pulled down (become more negative), the lowering being maximum for $n = 1$.

In the case of elliptical orbits, the velocity varies from point to point, and it is maximum at the perinucleus. The increase in velocity, for the same n, also increases with eccentricity. This produces a rosette-shaped orbit as in Fig. 5.4. Hence s level is lowered mroe than p, and so on. Thus the degeneracy of sub-levels s, p, d, f, $\cdots$ is removed.

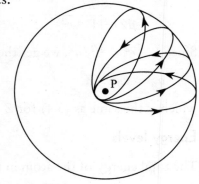

Fig. 5.4   *A rosette-shaped elliptical orbit*

Sommerfeld's relativistic calculation showed that the total energy in the (n, k) orbit is given by

$$E(n, k) = \mu c^2 \left\{ 1 - \left[1 + \frac{\alpha^2 Z^2}{n - k + (k^2 - \alpha^2 Z^2)^{1/2}} \right]^{-1/2} \right\}.$$

As α^2 is a small quantity with a value 5.3×10^{-5}, we can write

$$E(n, k) = \mu c^2 - \frac{RhcZ^2}{n^2} \left\{ 1 + \frac{\alpha^2 Z^2}{n^2} \left(\frac{n}{k} - \frac{3}{4} \right) \right\}. \tag{5.16a}$$

Leaving out the relativistic mass energy, we find that the energy of the hydrogen atom in the state (n, k) is

$$E(n, k) = -\frac{E_0}{n^2} - \frac{E_0 \alpha^2 Z^2}{n^4} \left(\frac{n}{k} - \frac{3}{4} \right).$$ (5.16b)

It is clear that the shifts of levels become smaller and smaller with increasing n. Thus for $n = 2$, the shifts in s and p levels are $5R\alpha^2/64 = 0.455$ cm^{-1} and $R\alpha^2/64 = 0.091$ cm^{-1}, respectively. For $n = 3$ level, we have three sub-levels s, p, d, with $k = 3, 2, 1$ respectively, and the lowering of the s level is $9R\alpha^2/(81 \times 4) = 0.162$ cm^{-1}. The differences between s, p, d levels are

$$\Delta\sigma_{\text{sp}} = \frac{R\alpha^2}{81} \left(\frac{3}{1} - \frac{3}{2} \right) = 0.108 \text{ cm}^{-1},$$

$$\Delta\sigma_{\text{pd}} = \frac{R\alpha^2}{81} \left(\frac{3}{2} - \frac{3}{3} \right) = 0.036 \text{ cm}^{-1}.$$

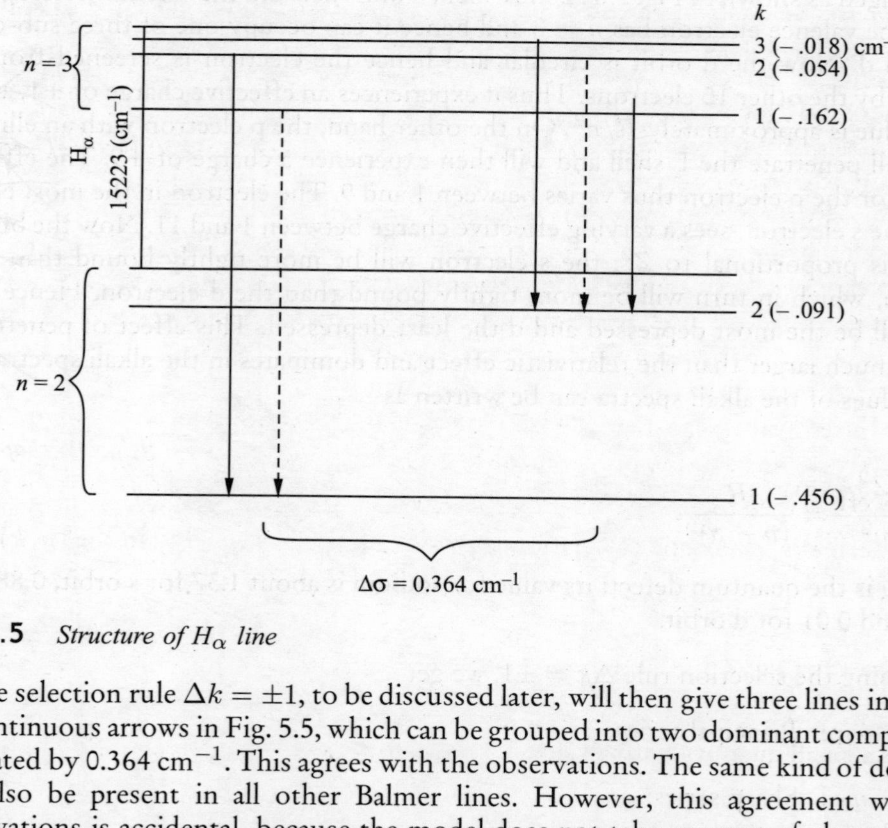

Fig. 5.5 *Structure of H_α line*

The selection rule $\Delta k = \pm 1$, to be discussed later, will then give three lines indicated by continuous arrows in Fig. 5.5, which can be grouped into two dominant components separated by 0.364 cm^{-1}. This agrees with the observations. The same kind of doubling will also be present in all other Balmer lines. However, this agreement with the observations is accidental, because the model does not take account of electron spin. When this is done, we get two more components indicated by the broken lines in Fig. 5.5. The levels which produce H_α line at 15223 cm^{-1} are Bohr levels. We shall discuss electron spin later.

5.2.4 Alkali spectra

Alkali atoms like Li, Na, K, have one valence electron; hence their spectra should also be simple and should resemble the spectrum of hydrogen-like atoms. But they show four different types of series — principal, sharp, diffuse and fundamental, which can be represented by formulae like

$$\frac{1}{\lambda} = R \left[\frac{1}{(n_1 - \mu_a)^2} - \frac{1}{(n_2 - \mu_b)^2} \right],$$

where n_1 and n_2 are the principal quantum numbers for lower and upper levels, and μ_a, μ_b stand for $\mu_s, \mu_p, \mu_d, \mu_f$, etc., which represent what are known as *quantum defects*. This equation can be understood on the basis of Sommerfeld's elliptic orbits. Let us take the example of sodium in which the nucleus has a charge of $+11$ and the electrons are arranged as shown in Fig. 5.6 : 2 in K shell, 8 in L shell and the valence electron in M shell. The valence electron has $n = 3$ and hence it can occupy one of three sub-orbits s, p and d. Now the d orbit is circular and hence the electron is screened from the nucleus by the other 10 electrons. Thus it experiences an effective charge of $+1$, and its term value is approximately R/n^2. On the other hand, the p electron with an elliptical orbit will penetrate the L shell and will then experience a charge of $+9$. The effective charge for the p electron thus varies between 1 and 9. The electron in the most elliptic orbit, the s electron, sees a varying effective charge between 1 and 11. Now the binding energy is proportional to Z^2; the s electron will be more tightly bound than the p electron, which in turn will be more tightly bound than the d electron. Hence the s level will be the most depressed and d the least depressed. This effect of penetration is very much larger than the relativistic effect and dominates in the alkali spectra. The term values of the alkali spectra can be written as

$$\frac{RZ^2_{\text{eff}}}{n^2} = \frac{R}{(n - \mu)^2},$$

where μ is the quantum defect; its value for sodium is about 1.37 for s orbit, 0.88 for p orbit, and 0.01 for d orbit.

Then using the selection rule $\Delta k = \pm 1$, we get

$n_1\text{s} - n_2\text{p}$: Principal series;

$n_1\text{p} - n_2\text{s}$: Sharp series;

$n_1\text{p} - n_2\text{d}$: Diffuse series;

$n_1\text{d} - n_2\text{f}$: Fundamental series.

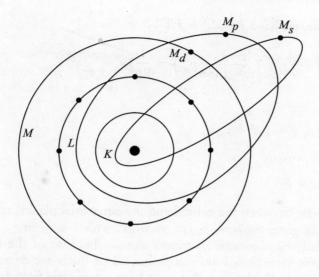

Fig. 5.6 *Schematic representation of electron orbits in sodium atom*

For Na, this gives the following lines:

3s −np: 5890.6 Å, 3302.3Å;

3p −ns: 11404Å, 6160Å;

3p −nd: 8194.8Å, 5688Å;

3d −nf: 18459Å, etc.

The last series corresponds to circular Bohr orbits and hence resembles the hydrogen spectrum most. However, all lines show a doublet structure which can be explained only when the electron spin is taken into account. We will discuss this next.

5.2.5 Additional degrees of freedom

Space quantization

In reality, the electron moves in three dimensions and has three degrees of freedom. Therefore using the spherical coordinates (r, θ, φ), with $p_r = \mu \dot{r}$, $p_\theta = \mu r^2 \dot{\theta}$ and $p_\varphi = \mu r^2 \sin^2 \theta \, \dot{\varphi}$, we should have the three conditions

$$\int p_r dr = nh, \quad \int p_\theta d\theta = th, \quad \int p_\varphi d\varphi = mh.$$

In the orbital plane, we have $\int p_\psi d\psi = kh$. Now

$$T = \frac{1}{2\mu}\left(p_r^2 + \frac{1}{r^2}p_\theta^2 + \frac{1}{r^2\sin^2\theta}p_\varphi^2\right) = \frac{1}{2\mu}(p_r^2 + \frac{1}{r^2}p_\psi^2).$$

This gives

$$p_r\dot r + p_\theta\dot\theta + p_\varphi\dot\varphi = p_r\dot r + p_\psi\dot\psi,$$

or $\quad p_\theta d\theta + p_\varphi d\varphi = p_\psi d\psi,$

which gives $t + m = k$.

Also if α is the angle between the orbital and the equatorial planes, then $p_\varphi = p_\psi\cos\alpha$ is a constant. This gives $\cos\alpha = p_\varphi/p_\psi = m/k$, with $-k \le m \le k$. This quantum number m is called the *magnetic quantum number* because of the following reason. Each k level is split into $2k+1$ substates, but all of them are degenerate so long as the z-axis is not physically defined. But if we have a magnetic field H along the z-axis, the levels are split due to the interaction between the magnetic field and the magnetic moment μ of the atom.

If A is the area of the electron orbit and T the periodic time for the electron to go around it, the magnetic moment associated with the current loop would be $\mu = Ae/cT$. Noting that A/T is the areal velocity, which is $r^2\dot\psi/2$, we have

$$\mu = ep_\psi/2\mu c = k(eh/4\pi\mu c) \equiv k\mu_B,$$

which defines μ_B; it is is called the *Bohr magneton*. It has a numerical value 9.18×10^{-21} erg/gauss and is the smallest unit of magnetic moment. As $m = k\cos\alpha$, space quantization implies discrete orientation of the orbit with respect to H. The energy associated with orientation m is

$$\Delta E = mH\frac{eh}{4\pi\mu c} \equiv mh\nu_L,$$

where $\nu_L = eH/4\pi\mu c$ is called the *Larmor frequency*, with which the orbit precesses around H. Thus the degeneracy with respect to m is removed. As there are $2k+1$ possible orientations of the orbit, we will get $2k+1$ magnetic sublevels, and transitions between them will produce several components giving rise to the so called Zeeman effect, which will be discussed in Section 5.7. Comparison with observed Zeeman components indicates that we get the correct number of levels only if we replace k by $l = k-1$, that is the total number of states is $2l+1 = 2k-1$ rather than $2k+1$.

Electron spin

We have noted above that each level of alkali atoms is found to have a doublet structure, which indicates an additional degree of freedom. Goudschamidt and Whlenbeck suggested that the electron may have a spin angular momentum and they suggested a spin quantum number $s = 1/2$. In the atomic magnetic field produced by the orbital motion of the electron, the spin axis orients in two directions, giving the projection $m_s = \pm 1/2$. On combining spin with $l = k - 1$, we get $j = l \pm 1/2$. Thus each level is split into two, and the selection rule for transitions $\Delta j = 0, \pm 1$ then gives the lines of various series as shown in Fig. 5.7. Electron spin is a natural outcome of relativistic quantum mechanics developed by Dirac, as we will see later in Section 5.6.

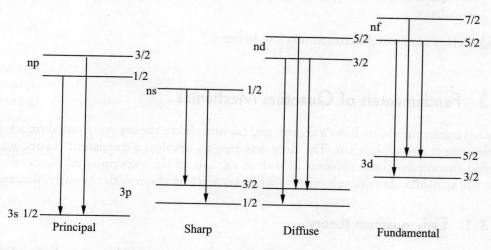

Fig. 5.7   *Schematic production of the four series in alkali atoms*

5.2.6   Evaluation of Bohr–Sommerfeld theory

Let us go back to the period near 1920s and evaluate Sommerfeld's modification of Bohr's model by incorporating velocity-dependent electron mass. Since v/c is of the order of 10^{-2} for an electron in hydrogen atom, which is also the order of the fine structure constant α, the observed shift in hydrogen energy levels agrees fairly well with that given by Sommerfeld's model, though there are also discrepancies in details. In the development of quantum mechanics, it was soon realised that the fine structure of hydrogen atom arises from spin–orbit interaction. A fully satisfactory theory was developed by Dirac in 1928. It was a relativistic theory which resulted in an understanding of electron spin and associated effects.

Even so, it is worth taking a look at the achievements and shortcomings of Sommerfeld's model which conjoins classical mechanics, Bohr's model and special relativity. We may summarise Sommerfeld's theory by saying that:

(i) It explains the spectra of hydrogen and hydrogen-like atoms including the gross features of fine structure.

(ii) It involves arbitrary selection rules.

(iii) For alkali spectra, the parameter k has to be replaced by $l = k - 1$.

(iv) On applying Wilson-Sommerfeld quantization rule to a rotator and a simple harmonic oscillator, and comparing with the observed rotational and vibration–rotation bands of diatomic molecules, Sommerfeld's quantization rule has to be modified to $\oint p\,dq = [n(n+1)]^{1/2}\hbar$ for a rotator and $\oint p\,dq = (n+1/2)\hbar$ for an oscillator.

(v) It fails for two-electron atoms like helium.

5.3   Fundamentals of Quantum Mechanics

We have seen that both Bohr's theory and Sommerfeld's theory were based on a lot of arbitrariness and ad-hocism. The time was ripe to develop a consistent theory which would encompass these theories as well as explain all the experimental observations. We will consider these developments while connecting them with the early history.

5.3.1   Early quantum theory

Before the advent of twentieth century, the wave nature of light was well established through (a) the phenomena of interference and diffraction, (b) the attribute of polarisation, and (c) Maxwell's theory of electromagnetic radiation. But its duality in the form of particle nature became apparent in the twentieth century on account of

(i) Planck's idea of energy quantum associated with light waves of frequency ν, viz, $E = h\nu = \hbar\omega$. This hypothesis was required for explaining the energy distribution in the continuous spectrum of a black body.

(ii) Einstein's explanation of photoelectric effect on the basis of light quanta which could account for the facts that photoelectrons are emitted only by radiation of frequency $\nu > \nu_c$, where ν_c is the critical frequency, irrespective of the intensity of incident radiation, and that these photoelectrons are stopped by applying a definite reverse potential V_c, called the cut-off potential. The cut-off potential can be written as $eV_c = h\nu - e\varphi$, where φ is the work function.

(iii) Bohr and Sommerfeld could explain most features of hydrogen, hydrogen-like and alkali atoms as well as some aspects of the Zeeman effect. The Bohr–Sommerfeld theory was to some extent a patchwork between classical mechanics and quantum hypothesis; it failed completely for molecules and two-electron systems like He. Hence there was a need for a fresh approach.

In 1924 de Broglie put forward the idea of matter waves. He argued that if light can exhibit both wave and particle nature, then matter should also do so. Now for photons,

$$E = h\nu, \quad p = h\nu/c.$$

Therefore,

$$\lambda = c/\nu = h/p,$$

where the symbols have their usual meaning. De Broglie suggested that wavelength of matter waves be also given by

$$\lambda = h/p = \hbar/mv$$

for non-relativistic particles. In effect, de Broglie suggested that matter particles of mass m travelling with a velocity v, which normally behave like particles having momentum p, should also behave like waves of wavelength $\lambda = h/p$. This prediction was confirmed soon in 1927 for electrons by the experiments of

(i) G. P. Thompson who obtained diffraction rings for thermoelectrons due to passage through a metal foil, and

(ii) Davisson and Germer who observed scattering of electrons by a single crystal of nickel which behaves like a diffraction grating.

Even before experimental verification of de Broglie's ideas, Heisenberg had formulated his *matrix mechanics* in 1925, and the ideas were actually used by Schrödinger in 1926 to formulate his *wave mechanics*. Later Schrödinger showed that both these approaches were identical in essence; both together are now referred to as *quantum mechanics*. We shall follow the methods of Schrödingers wave mechanics which is easier to handle.

5.3.2 Schrödinger wave equation

Starting from the wave equation for classical mechanical waves, Schrödinger developed his wave equation for matter waves. The phase velocity of matter waves is

$$u = \lambda\nu = (h/p)(E/h) = E/p.$$

Therefore the wave equation for matter waves becomes

$$\nabla^2 \Psi = \frac{1}{u^2}\frac{\partial^2 \Psi}{\partial t^2} = \frac{p^2}{E^2}\frac{\partial^2 \Psi}{\partial t^2},$$

where $\Psi(\mathbf{r}, t)$ is some function representing matter waves. Let us use the method of separation of variables and write

$$\Psi(\mathbf{r}, t) = \psi(\mathbf{r})f(t).$$

Then we have

$$f\nabla^2\psi(\mathbf{r}) = \frac{p^2}{E^2}\psi(\mathbf{r})\frac{d^2f}{dt^2}.$$

Therefore

$$\frac{E^2}{p^2}\frac{\nabla^2\psi}{\psi} = \frac{1}{f}\frac{d^2f}{dt^2} = \text{constant.}$$

Equations such as

$$\frac{1}{f}\frac{d^2f}{dt^2} = \text{constant}$$

are well-known and often occur, and it is well-known that physically acceptable solutions result only if the constant is less than zero. Hence let us take

$$\frac{1}{f}\frac{d^2f}{dt^2} = -\omega^2,$$

where ω is a real number. Then

$$f(t) = Ae^{i\omega t} + Be^{-i\omega t} \quad \text{with} \quad \omega = 2\pi\nu = E/\hbar.$$

Here A and B are arbitrary. One puts $A = 0$, $B = 1$; this was the historical choice, and it gives clockwise motion. Therefore

$$\Psi(\mathbf{r}, t) = \psi(\mathbf{r})e^{-iET/\hbar} \tag{5.17}$$

and

$$\nabla^2\psi(\mathbf{r}) = -(p^2/\hbar^2)\psi(\mathbf{r}).$$

But energy $E = p^2/2m + V(\mathbf{r})$, so that $p^2 = 2m(E - V)$. This gives

$$\nabla^2\psi = -\frac{2m(E - V)}{\hbar^2}\psi. \tag{5.18}$$

This is the Schrödinger equation for stationary state. It may be noted that it is not fundamental because (i) it depends on E, which is not known a priori, and (ii) it does not depend on time, so it is not an equation of motion.

The complete time-dependent Schrödinger equation is obtained as follows. From (5.17), we have

$$\nabla^2 \Psi = \nabla^2 \psi e^{-iEt/\hbar} = -\frac{2m(E-V)}{\hbar^2}\Psi.$$

Also,

$$\frac{\partial \Psi}{\partial t} = -\frac{iE}{\hbar}\Psi.$$

Eliminating E between these two equations, we get

$$-\frac{\hbar^2}{2m}\nabla^2\Psi + V(\mathbf{r})\Psi = i\hbar\frac{\partial \Psi}{\partial t}. \tag{5.19}$$

This is the fundamental equation of wave mechanics.

This Schrödingers time-dependent equation can be obtained from the equation of classical mechanics,

$$H\Psi = \left(\frac{p^2}{2m} + V(\mathbf{r})\right)\Psi,$$

using the correspondence

$$H \to i\hbar\frac{\partial}{\partial t}, \quad p_i = -i\hbar\frac{\partial}{\partial q_i}, \tag{5.20}$$

where p_i and q_i are canonically conjugate variables. If $\mathbf{p}$ is the momentum in three-dimensional space, then the operator corresponding to it becomes

$$\mathbf{p} \to -i\hbar\boldsymbol{\nabla}, \tag{5.21}$$

where $\boldsymbol{\nabla}$ is the gradient operator with respect to space coordinates. These operators are linear but not necessarily commutative. Also H becomes the operator

$$H = -\frac{\hbar^2}{2m}\nabla^2 + V(\mathbf{r}). \tag{5.22}$$

The potential energy depends on position $\mathbf{r} \equiv (x, y, z)$, and the coordinates represent their own operators.

5.3.3 Commutation rules for coordinates and momenta

We see that, with the canonical coordinates and momenta being replaced by the corresponding operators, we have

$$xp_x\psi = -i\hbar x\frac{\partial\psi}{\partial x},$$

but

$$p_x x\psi = -i\hbar\frac{\partial}{\partial x}(x\psi) = -i\hbar x\frac{\partial\psi}{\partial x} - i\hbar\psi.$$

Thus

$$(xp_x - p_x x)\psi = i\hbar\psi.$$

Since this holds good for any arbitrary function ψ, we may write the operator equation

$$xp_x - p_x x = i\hbar,$$

which is written, using the commulator symbol, as

$$[x, p_x] = i\hbar. \tag{5.23}$$

Thus, in general, any canonical coordinate q and its canonical momentum p do not commute. But if we have a set of coordinates, a coordinate q_i would commute with the canonical momentum p_j corresponding to another coordinate q_j. Thus

$$xp_y\psi = -i\hbar x\frac{\partial\psi}{\partial y},$$

and

$$p_y x\psi = -i\hbar\frac{\partial}{\partial y}(x\psi) = -i\hbar x\frac{\partial\psi}{\partial y}.$$

Thus $[x, p_y] = 0$.

If we have a set of generalised coordinates q_i and their canonical momenta p_i, we would have

$$[q_i, q_j] = 0, \quad [p_i, p_j] = 0, \quad [q_i, p_j] = i\hbar\delta_{ij}, \tag{5.24}$$

where δ_{ij} is the Kronecker delta function.

Thus essentially in wave mechanics, we deal with operators instead of physical variables.

5.3.4 Solution of the fundamental equation

Let us start from the Schrödinger equation, (5.19), and solve it by the separation of variables method. We let $\Psi(\mathbf{r}, t) = \psi(\mathbf{r})T(t)$, separate the variables $\mathbf{r}$ and t, and take the separation constant to be E. Then we get the two equations

$$\frac{i\hbar}{T}\frac{\partial T}{\partial t} = E, \tag{5.25a}$$

$$-\frac{\hbar^2}{2m}\nabla^2\psi + V(\mathbf{r})\psi = E\psi(\mathbf{r}). \tag{5.25b}$$

The first of these gives

$$T(t) = e^{-iEt/\hbar}.$$

With the identification of H as in (5.21), (5.25b) may be written as

$$H\psi = E\psi. \tag{5.26}$$

We see here that the separation constant E is the total energy, and it occurs in the solution for $T(t)$. (5.25b) is the same as (5.18), and is the equation for a stationary state.

Equation (5.25b) is a linear homogeneous equation, and has the trivial solution $\psi(\mathbf{r}) = 0$ for any value of E. We wish to find a non-zero solution of this equation which is physically acceptable. We note that it is a second order homogeneous differential equation containing one arbitrary constant E. If we wish $\psi(\mathbf{r})$ to represent a physical function, it should be an analytic function throughout the (infinite) volume, which means it should be continuous, finite and single-valued everywhere.

Being a second order equation, it will have an infinite number of solutions for any value of E. But when we impose the conditions of analyticity, it turns out that the equation has a solution only for certain values of E, say $E_n, n = 1, 2, 3, \cdots$. These are then called the *eigenvalues* and the corresponding solutions ψ_n are called the *eigenfunctions*. In terms of matrix mechanics, E_n are the eigenvalues of H and ψ_n the eigenvectors.

Suppose ψ_n is a solution of (5.25b) with the corresponding parameter E_n. Then $c\psi_n$, where c is a scalar, will also be a solution because it is a linear equation. But there would be no point in calling it a different solution because that way we would have an infinite number of solutions which would be just multiples of each other. Therefore when we talk of solutions of (5.25b) or (5.19), we would mean linearly independent solutions.

Now if ψ_n are (linearly independent) solutions of (5.25b), then since (5.19) is a linear equation, its most general solution would be

$$\Psi(\mathbf{r}, t) = \sum_n a_n\psi_n(\mathbf{r})e^{-itE_n/\hbar}. \tag{5.27}$$

This is known as the *principle of superposition*.

5.3.5   Formalisation of quantum mechanics

In order to make sense out of the solutions of Schrödingers equation, we have to make certain assumptions which form the basis of quantum mechanics. We describe these below.

(i)   Each physical observable q, like position, momentum, energy, etc, is associated with a hermitian operator Q, which has its eigenfunctions $\psi_n(\mathbf{r})$ and corresponding real eigenvalues λ_n.

(ii)   A measurement of the observable q can only result in one of the eigenvalues λ_n of the associated operator Q.

(iii)   The eigenvalues and eigenfunctions are solutions of the eigenvalue equation $Q\psi_n(\mathbf{r}) = \lambda_n\psi_n(\mathbf{r})$.

(iv)   The eigenfunctions ψ_n are well-behaved, which means they are continuous, single-valued, and finite. They can be normalised over the entire space, so that $\int_V \psi_n^*\psi_n d\tau = 1$. In that case, $|\psi(\mathbf{r})|^2$ represents the probability density of finding the particle near $\mathbf{r}$ in state n.

(v)   When the particle is in a particular state ψ_n, it has the specific value λ_n of the variable q.

(vi)   Any general state of the particle can be represented by a linear combination of ψ_n such as

$$\psi = \sum_n a_n\psi_n,$$

which is a consequence of the principle of superposition. On making $\int_V |\psi|^2 d\tau = 1$, $|a_n|^2$ will be the probability of finding the particle in state n.

(vii)   The expectation value of a variable q in a measurement is given by $\langle q \rangle = \int \psi^*Q\psi d\tau$ if $\int |\psi|^2 d\tau = 1$ and by

$$\langle q \rangle = \frac{\int \psi^*Q\psi d\tau}{\int \psi^*\psi d\tau}$$

if $\int |\psi|^2 d\tau \neq 1$. If $\int |\psi|^2 d\tau = 1$, the function $\psi(\mathbf{r})$ is said to be *normalised*.

(viii)   Since the result of any measurement of an observable must be real, the expectation value $\langle q \rangle$ must be real. This requires that Q must satisfy

$$\int u^*Qv d\tau = \int (Qu)^*v d\tau, \tag{5.28}$$

where u and v are arbitrary well-behaved functions. An operator satisfying (5.28) is called a *hermitian operator*. Its properties make it useful in quantum mechanics.

We shall now make use of these postulates to show that the eigenvalues of (5.25b) are real and that the solutions ψ_n of that equation for distinct eigenvalues are orthogonal to each other. Thus, let E_n be an eigenvalue corresponding to the solution ψ_n. Then we can write (5.28) with both u and v replaced by ψ_n and Q replaced by H as

$$\int \psi_n^* H \psi_n d\tau = \int (H\psi_n)^* \psi_n d\tau$$

$$\Rightarrow E_n \int \psi_n^* \psi_n d\tau = E_n^* \int \psi_n^* \psi_n d\tau$$

$$\Rightarrow (E_n - E_n^*) \int \psi_n^* \psi_n d\tau = 0.$$

Since we are looking for non-zero solutions ψ_n, the second factor in the above equation will be non-zero. Hence $E_n^* = E_n$, showing that all the eigenvalues of H are real.

Next, consider two distinct eigenvalues E_n and E_m, $E_n \neq E_m$, of (5.25b). then we can write (5.28) in the form

$$\int \psi_n^* H \psi_m d\tau = \int (H\psi_n)^* \psi_m d\tau$$

$$\Rightarrow E_m \int \psi_n^* \psi_m d\tau = E_n \int \psi_n^* \psi_m d\tau$$

$$\Rightarrow (E_m - E_n) \int \psi_n^* \psi_m d\tau = 0,$$

where we have used the fact that E_n is real. But since $E_m \neq E_n$, we must have $\int \psi_n^* \psi_m d\tau = 0$, showing that the eigenfunctions belonging to distinct eigenvalues are orthogonal to each other.

5.3.6 Expectation value and time evolution

Quantum mechanics is a probabilistic theory. Suppose q is a physical variable such as momentum, energy, angular momentum, etc., and we set up an experiment to measure the value of q at an instant of time. Then, quantum mechanics can at best give us probabilities for getting different possible values of q. If we make a large number of measurements on q, then each measurement may give us a different value of q, and quantum mechanics tells us that so many of them are likely to be q_1, so many likely to be q_2, etc. This probabilistic description is contained in the wavefunction $\psi(\mathbf{r})$ associated with the state of the system. We now discuss these aspects.

Expectation value

Consider a system described by a stationary state $\Psi(\mathbf{r})$. Let q be a dynamical observable and Q the hermitian operator associated with it. Then the *expectation value* of q in a measurement is given by

$$\langle q \rangle = \frac{\int \Psi^* Q \Psi d\tau}{\int \Psi^* \Psi d\tau} = \int \Psi^* Q \Psi d\tau \tag{5.29}$$

if $\Psi(\mathbf{r})$ is normalised to unity. Generally,

$$\langle q^2 \rangle \neq \langle q \rangle^2, \quad \text{or} \quad \langle q^n \rangle \neq \langle q \rangle^n, \quad \text{etc.}$$

for n, a positive integer. But if $\langle q^n \rangle = \langle q \rangle^n$ for all positive integers n, then q is an eigenvalue of Ψ, i.e., $Q\Psi = q\Psi$. For example, since $H\Psi_n = E_n\Psi_n$, E_n is an eigenvalue of energy.

Evolution of expectation value

Consider an observable q which does not depend explicitly on time. Then,

$$\frac{d}{dt}\langle q \rangle = \frac{\partial}{\partial t} \int \Psi^* Q \Psi d\tau. \tag{5.30}$$

Or,

$$\left\langle \frac{dq}{dt} \right\rangle = \int \left\{ \frac{\partial \Psi^*}{\partial t} Q\Psi + \Psi^* Q \frac{\partial \Psi}{\partial t} \right\} d\tau.$$

Putting

$$H\Psi = i\hbar \frac{\partial \Psi}{\partial t} \quad \text{and} \quad H\Psi^* = -i\hbar \frac{\partial \Psi^*}{\partial t},$$

we get

$$\left\langle \frac{dq}{dt} \right\rangle = \frac{i}{\hbar} \int (H\Psi^* Q\Psi - \Psi^* Q H\Psi) d\tau$$

$$= \frac{i}{\hbar} \int \{(H\Psi^* - \Psi^* H)Q\Psi + \Psi^*(HQ - QH)\Psi\} d\tau$$

$$\equiv I_1 + I_2, \quad \text{say}, \tag{5.31}$$

which defines the two terms I_1 and I_2.

Using (5.21) for H and noting that V is only a function of coordinates, we find that

$$I_1 = -\frac{\hbar^2}{2m} \int (\nabla^2 \Psi^* - \Psi^* \nabla^2)\Psi d\tau$$

$$= \frac{\hbar^2}{2m} \oint_S (\Psi^* \nabla(Q\Psi) - Q\Psi \nabla \Psi^*) \cdot \mathbf{dS},$$

where we have used the Green's theorem. Since Ψ and $\nabla \Psi$ vanish at infinity, I_1 approaches zero as the volume τ tends to infinity. Then we are left only with the second term in (5.31), giving

$$\left\langle \frac{dq}{dt} \right\rangle = \frac{i}{\hbar} \int \Psi^*[H, Q]\Psi d\tau. \tag{5.32}$$

This gives the equation of motion for the variable q which does not intrinsically depend upon time.

For example, if we take the coordinate x for q, we have

$$\left\langle \frac{dx}{dt} \right\rangle = \frac{i}{\hbar} \int \Psi^*[H, x]\Psi d\tau.$$

We note that H is given by (5.21) and x commutes with $V(\mathbf{r})$. Also

$$\nabla^2 \Psi(\mathbf{r}) = x\nabla^2 \Psi + 2\frac{\partial \Psi}{\partial x}.$$

This gives

$$\left\langle \frac{dx}{dt} \right\rangle = -\frac{i\hbar}{m} \int \Psi^* \frac{\partial \Psi}{\partial x} d\tau.$$

But $\quad \langle p_x \rangle = \int \Psi^* p_x \Psi d\tau = -i\hbar \int \Psi^* \frac{\partial \Psi}{\partial x} d\tau.$

Therefore,

$$\left\langle \frac{dx}{dt} \right\rangle = \frac{\langle p_x \rangle}{m}.$$

Similar equations for the other two components would lead to

$$\left\langle \frac{d\mathbf{r}}{dt} \right\rangle = \frac{\langle \mathbf{p} \rangle}{m}.$$ (5.33a)

Similarly, using p_x, p_y, p_z for q in (5.32), it can be shown that

$$\left\langle \frac{d\mathbf{p}}{dt} \right\rangle = -\langle \boldsymbol{\nabla} V \rangle = \langle \mathbf{F} \rangle,$$ (5.33b)

where $\mathbf{F}$ is the force acting on the system. Equation (5.33) together are called *Ehrenfest's* theorem. They tell us that Newton's equations of motion are valid for average values of variables $\mathbf{r}, \mathbf{p}$ and $\mathbf{F}$.

Constants of motion

If an observable q is such that $[H, Q] = 0$, then (5.32) shows that $\langle dq/dt \rangle = 0$, that is, q is a constant of motion. In this case, Ψ is an eigenfunction of Q. Also, the necessary and sufficient condition for any two operators to have the same set of eigenfunctions is that they should commute with each other. The observables corresponding to commuting operators can be determined simultaneously. As x and p_x, treated as operators, do not commute with each other, they cannot be determined simultaneously, as discussed in Section 5.3.3.

Putting $q = 1$ in (5.32), we get $\frac{d}{dt} \int \Psi^* \Psi d\tau = 0$. Thus the total probability is conserved over the entire space. But over a limited volume τ, we see that

$$\frac{d}{dt} \int_\tau \Psi^* \Psi d\tau = \int_\tau \left(\Psi^* \frac{\partial \Psi}{\partial t} + \frac{\partial \Psi^*}{\partial t} \Psi \right) d\tau.$$

Using (5.19) for $i\hbar \partial \Psi / \partial t$, this reduces to

$$\frac{d}{dt} \int_\tau \Psi^* \Psi d\tau = \int_\tau (\Psi^* \nabla^2 \Psi - \Psi \nabla^2 \Psi^*) d\tau$$

$$= \oint_S (\Psi^* \boldsymbol{\nabla} \Psi - \Psi \boldsymbol{\nabla} \Psi^*) \cdot d\mathbf{S},$$ (5.34)

where we have used Green's theorem, and S is the surface enclosing τ. Comparing (5.34) with the equation of charge conservation, (3.22), which reads

$$\int_V \frac{\partial \rho}{\partial t} d\tau = -\oint_S \mathbf{J} \cdot \mathbf{n} dS,$$

we see that $\Psi^* \Psi$ represents the probability density and

$$\mathbf{J}_p = \frac{i\hbar}{2m}(\Psi^* \boldsymbol{\nabla} \Psi - \Psi \boldsymbol{\nabla} \Psi^*) \tag{5.35}$$

represents the probability current density. The left-hand side of (5.34) represents the rate of increase of probability (of finding the system in state Ψ) in volume τ and $-\oint_S \mathbf{J}_p \cdot \mathbf{n} dS$ represents the probability entering into the closed surface S.

For example, consider a free particle of mass m travelling in the x-direction, whose matter wave is represented by e^{ikx} where $k = 2\pi\lambda$ is the wave vector. Then the magnitude of the probability current density is found from (5.35) to be

$$J_p = hk/2\pi m = h/\lambda m = p/m = v,$$

where v is the velocity of the particle.

5.3.7 Angular momentum and other operators

Angular momentum in Cartesian coordinates

Classically, angular momentum is $\mathbf{L} = \mathbf{r} \times \mathbf{p}$. Replacing $\mathbf{p}$ by $-i\hbar\boldsymbol{\nabla}$, we get the quantum mechanical angular momentum operators as $\mathbf{L} = -i\hbar\mathbf{r} \times \boldsymbol{\nabla}$. In Cartesian coordinates, we have

$$L^2 = L_x^2 + L_y^2 + L_z^2$$

$$= -\hbar^2 \left[\left(y\frac{\partial}{\partial z} - z\frac{\partial}{\partial y} \right)^2 + \left(z\frac{\partial}{\partial x} - x\frac{\partial}{\partial z} \right)^2 + \left(x\frac{\partial}{\partial y} - y\frac{\partial}{\partial x} \right)^2 \right].$$

The Cartesian components of $\mathbf{L}$ and L^2 are found to satisfy the commutation rules

$$[L_p, L_q] = i\hbar \sum_r \epsilon_{pqr} L_r, \ [L^2, L_p] = 0, \tag{5.36}$$

where p, q, r take values $(x, y, z) = (1, 2, 3)$ and ϵ_{pqr} is the fully antisymmetric tensor of rank 3. Thus ϵ_{pqr} is $+1$ if (p, q, r) is an even permutation of $(1, 2, 3)$ and -1 if it is an odd permutation of $(1, 2, 3)$.

These equations show that L^2 commutes with each component of $\mathbf{L}$ separately, but no two components of $\mathbf{L}$ commute with each other. This means that while L^2 and any one component of $\mathbf{L}$ can be determined simultaneously, but no two components of $\mathbf{L}$ can be determined simultaneously.

Angular momentum in spherical coordinates

In spherical coordinates, the Cartesian components of $\mathbf{L}$ are given by

$$L_x = -i\hbar \left\{ -\sin\varphi \frac{\partial}{\partial\theta} - \cos\varphi \cot\theta \frac{\partial}{\partial\varphi} \right\},$$

$$L_y = -\hbar \left\{ \cos\varphi \frac{\partial}{\partial\theta} - \sin\varphi \cot\theta \frac{\partial}{\partial\varphi} \right\},$$

$$L_z = -i\hbar \frac{\partial}{\partial\varphi}, \tag{5.37a}$$

and $L^2 = -\hbar^2 \left[\frac{1}{\sin\theta} \frac{\partial}{\partial\theta} \left(\sin\theta \frac{\partial}{\partial\theta} \right) + \frac{1}{\sin^2\theta} \frac{\partial^2}{\partial\varphi^2} \right].$ (5.37b)

It may be noted that when we choose the spherical system (r, θ, φ), we have already choose a special axis—the polar axis from where we measure the polar angle θ and around which we measure φ. When we relate this to Cartesian coordinates, we identify the polar axis with the z-axis. Therefore, L_z has a different standing than L_x and L_y.

For an axially symmetric system, its potential $V(\mathbf{r}) \equiv V(r, \theta)$ and is independent of φ. It is found that L_z commutes with the Hamiltonian of such a system. Further, if $V(\mathbf{r}) \equiv V(r)$, that is $V(\mathbf{r})$ is spherically symmetric, then L^2 commutes with the Hamiltonian

$$H = -\frac{\hbar^2}{2m}\nabla^2 + V(r).$$

This is true for Coulomb potential, as in a hydrogen-like atom. Hence L^2 and L_z are constants of motion for such a system while L_x and L_y are not. The expectation values of L_x^2 and L_y^2 would be equal and would be given by

$$\langle L_x^2 \rangle = \langle L_y^2 \rangle = (L^2 - L_z^2)/2.$$

Ladder operators

If we construct linear combinations

$$L_+ = L_x + iL_y, \quad L_- = L_x - iL_y,$$

it will found that they satisfy the commutation relations

$$[L_\pm, L_\mp] = \pm\hbar L_z,$$

$$[L_z, L_\pm] = \pm\hbar L_\pm.$$

We also have

$$L_\pm L_\mp = L^2 - L_z^2 \pm \hbar L_z,$$

$$[L^2, L_\pm] = 0.$$

Now, let ψ be a common eigenfunction of L^2 and L_z so that

$$L^2\psi = A^2\hbar^2\psi, L_z\psi = l\hbar\psi.$$

Then,

$$L^2(L_\pm\psi) = A^2\hbar^2(L_\pm\psi),$$

$$L_z(L_\pm\psi) = (l \pm 1)\hbar(L_\pm\psi). \tag{5.38}$$

Thus $L_\pm\psi$ are the other common eigenfunctions of L^2 and L_z, with no change in the eigenvalue of L^2 but a change in the eigenvalue of L_z by $\pm\hbar$. That is why $L_\pm$ are called *ladder operators*.

5.3.8 The inversion operator

We define the inversion operator as one which reverses all the coordinate axes, taking x to $-x$, y to $-y$, and z to $-z$, or taking $\mathbf{r}$ to $-\mathbf{r}$. If we denote it by R, then $R\psi(\mathbf{r}) = \psi(-\mathbf{r})$ for any function of space coordinates.

Let us try to work out the commutator of R with the Hamiltonian. H has two terms, one containing the laplacian ∇^2 and the other containing the potential $V(\mathbf{r})$. If $V(\mathbf{r})$ has inversion symmetry, which is mostly the case, then $V(-\mathbf{r}) = V(\mathbf{r})$. Therefore

$$RV(\mathbf{r})\psi(\mathbf{r}) = V(-\mathbf{r})\psi(-\mathbf{r}) = V(\mathbf{r})\psi(-\mathbf{r}) = V(\mathbf{r})R\psi(\mathbf{r}).$$

Since this holds for an arbitrary function $\psi(\mathbf{r})$, we have the operator relation

$$RV(\mathbf{r}) = V(\mathbf{r})R.$$

It is also easily seen that

$$\nabla^2 R\psi(\mathbf{r}) = R\nabla^2\psi(\mathbf{r}).$$

These two equations together lead us to conclude that

$$HR = RH.$$

Thus R is a constant of motion. Let it have eigenvalues R_e, so that we have

$$R\psi(\mathbf{r}) = R_e\psi(\mathbf{r}), \qquad R^2\psi(\mathbf{r}) = R_e^2\psi(\mathbf{r}).$$

But $R^2\psi(\mathbf{r}) = R\psi(-\mathbf{r}) = \psi(\mathbf{r})$. This shows that $R_e^2 = 1$ and $R_e = \pm1$. Thus if $\psi(\mathbf{r})$ is an eigenfunction of R, then we may have

$$R\psi(\mathbf{r}) = \psi(\mathbf{r}) \quad \text{or} \quad R\psi(\mathbf{r}) = -\psi(\mathbf{r}).$$

In the first case, we say that ψ has *even parity*, whereas in the second case, ψ has *odd parity*. Since R is a constant of motion, parity is maintained so long as no external forces are acting.

5.4   One-Dimensional Motion

To understand how the eigenfunctions behave, we first consider one-dimensional motion in a potential $V(x)$. As in classical mechanics, the total energy is the sum of potential energy and kinetic energy, $E = T + V(x)$. Thus kinetic energy at a point x is $T(x) = E - V(x) = p^2/2m$. In quantum mechanics, $p = h/\lambda = \hbar k$, where λ is the wavelength of matter waves associated with the particle and k the corresponding wave vector. As the particle moves in this potential with a constant energy E, when the potential rises, the kinetic energy decreases, and the wavelength of the particle increases, and vice versa. We now deal with some general considerations followed by some examples.

5.4.1   General considerations

The Schrödinger equation for the one-dimensional motion of a particle of mass m and total energy E is given by (5.18) as

$$\frac{d^2\psi}{dx^2} = \frac{2m}{\hbar^2}(V(x) - E)\psi(x).$$

Suppose $V(x)$ is a slowly-varying function of x. For the two cases, $V - E < 0$ and $V - E > 0$, we let

$$k = [2m(E - V(x))/\hbar^2]^{1/2}, \quad \text{and} \quad \alpha = [2m(V(x) - E)/\hbar^2]^{1/2},$$

respectively, so that both k and α are real. Then the general solution of the Schrödinger equation will be

$$\psi(x) = Ae^{ikx} + Be^{-ikx}, \quad E > V(x), \tag{5.39a}$$

$$\psi(x) = Ae^{\alpha x} + Be^{-\alpha x}, \quad E < V(x). \tag{5.39b}$$

The solution of (5.39a) is oscillatory with a (variable) wave vector $k(x)$ while (5.39b) is a combination of exponential functions. The first term of (5.39b) is well-behaved for $x < 0$ and the second term is well behaved for $x > 0$. On the other hand, the first term diverges as $x \to \infty$ and the second term diverges as $x \to -\infty$.

We now consider some typical simple examples.

Free particle

A particle is said to be free when it moves in a constant potential or under no force; $V(x)$ = constant. Therefore E must be greater than V, and the stationary eigenfunctions (solutions of (5.18)) are

$$\psi(x) = e^{ikx} \quad \text{or} \quad \cos kx \quad \text{or} \quad \sin kx,$$

with $k = (2m(E-v))^{1/2}/\hbar = 2\pi/\lambda = p/\hbar$.

The time-dependent eigenfunctions would be

$$\Psi(x,t) = Ae^{i(kx-\omega t)}.$$

For a given p, this represents a sinusoidal wave moving in the $+x$ direction, as shown in Fig. 5.8(a). The probability density at any point at any time is $|\psi|^2 = |A|^2$; thus in a measurement, the particle could be found anywhere with equal probability. It also implies that the probability of finding the particle in a small interval x to $x + dx$ is negligible. Thus for a fixed E, p is fixed but x is completely indeterminate.

This is a limiting, ideal case, for no particle would ever find a large (infinite) force-free region to move. In order to represent a Newtonian particle, we have to superpose many such waves, over a vast range of k or p, resulting in a wave packet.

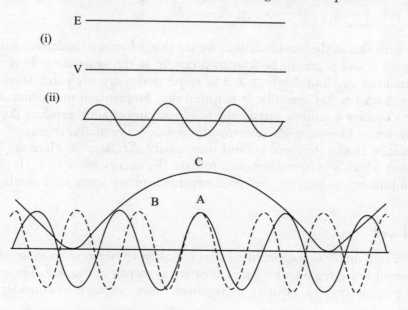

Fig. 5.8 *(a) (i) Potential V and (ii) eigenfunction for a free particle. (b) Superposition of two waves with slightly different wave vectors: A — wave 1, B — wave 2, C - superposed (group) wave.*

Let us consider two waves of nearly the same wave vectors k and k' and frequencies ω and ω' and the same amplitude. Taking the two waves as $\Psi_1 = \sin(kx - \omega t)$ and $\Psi_2 = \sin(k'x - \omega't)$, their sum will be

$$\Psi = \sin(kx - \omega t) + \sin(k'x - \omega't)$$

$$= 2\sin(kx - \omega t)\cos[(x\Delta k - t\Delta\omega)/2], \qquad (5.40)$$

where $\Delta k = k' - k$, $\Delta\omega = \omega' - \omega$, and we have assumed that $k + k' \simeq 2k$. This function is shown in Fig. 5.8(b). The first factor in (5.40) gives the wave motion with phase velocity $v_p = \omega/k$. The second factor gives the variation of the amplitude of the group which travels with group velocity

$$v_g = d\omega/dk = d(v_p k)/dk$$

$$= v_p - \lambda dv_p/d\lambda. \qquad (5.41)$$

It can be verified that both in classical and relativistic mechanics, $dE/dp = v_g$.

The group velocity represents the velocity of the wave packet, which is still unlocalised but has some finite spatial extension. At the other extreme, for complete localisation, k has to span the whole range from $-\infty$ to ∞. In such a case, we put

$$\Psi(x,t) = \int_{-\infty}^{\infty} a_k e^{-i(kx - \omega(k)t)} dk. \qquad (5.42)$$

But then, with this as the wavefunction, we see that p becomes indeterminate.

Thus both x and p cannot be known precisely at the same time. It is found that the uncertainties Δx and Δp in x and p, respectively, are such that their minimum product is $\Delta x \Delta p \simeq \hbar$. Generally, it is found that in quantum mechanics canonically conjugate variables q_j and p_j satisfy the minimum uncertainly product $\Delta q_j \Delta p_j > \hbar$. This is known as Heisenberg's *uncertainty principle*. Similarly, energy E and time t are canonically conjugate variables and they satisfy $\Delta E \Delta t \simeq \hbar$. Here Δt is the time interval over which a system continues to have the energy close to E. It also implies that the minimum volume in the six-dimensional phase space of a single particle is $d^3r d^3p \simeq \hbar^3$.

Potential well

Consider a one-dimensional potential that is (taken to be) zero in most of the range except a small finite region where it is negative as shown in Fig. 5.9. The total energy E of the particle may be positive or negative. Then we have to consider two cases separately.

(a) $E > 0$; the total energy E is more than the potential energy everywhere. We have seen that the wave vector k depends on $E - V(x)$. Thus in most of the region, the time-independent wavefunction is a sinusoidal function with $k = (2mE)^{1/2}/\hbar$, except in the region of the 'potential well'. Inside the well also it is sinusoidal with a variable $k(x) = [2m(E - V(x))]^{1/2}/\hbar$. In other words, the wavelength λ is constant away from the well but it decreases as the particle passes over the well.

(b) $E < 0$ but E is above the potential minimum; here $E < V$ in most of the region but $E > V$ in some region inside the potential well. Therefore the solution is oscillatory inside this region, with a variable k, but is exponentially damped outside this region. The three parts will join smoothly at both the boundaries of the potential well (where $E = V$) only for certain values of E which are called the eigenvalues of the potential well. The corresponding solutions are called the eigenfunctions of the system.

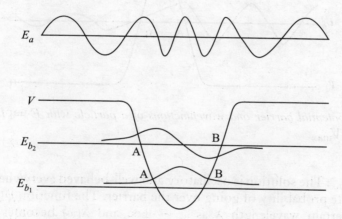

Fig. 5.9   *A potential well and wavefunctions for a particle with $E = E_a > 0$ and $E = E_{b1}$ or E_{b2}, both < 0*

For the lowest such energy level $E_{b1} \equiv E_1$, the eigenfunction $\psi_1(x)$ will have no node. The successive higher energy levels will have eigenfunctions with $1, 2, 3, \cdots$, nodes. In general, the eigenfunction corresponding to nth energy level will have $n - 1$ nodes. Thus we get a spectrum of discrete levels $E_n < 0$ and a continuum with $E > 0$. Even for discrete levels $\psi_n(x)$ has a nonzero value outside the potential well, which means there is a finite probability of finding the particle in the region where $E < V(x)$, which is not possible in classical mechanics.

A particle with energy $E < 0$ is confined more or less to the region of the potential well. A particle with $E > 0$ may lose energy by emitting radiation and become confined to the potential well (because it is a more stable state). For this reason, a potential well is also called a centre of attraction. We shall later consider in Section 3.5 a simple harmonic oscillator which has a potential well of this type.

Potential barrier

Here the potential and the eigenfunctions are shown in Fig. 5.10. Let us consider a particle with energy E moving from the left to right. Let $V = 0$ for most of the region and $V(x) > 0$ in a small finite interval. Such a potential is called a barrier. Let V_{max} be the $\max(V(x))$; V_{max} is then called the height of the potential barrier. Here also we have to consider two cases.

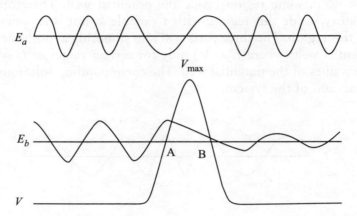

Fig. 5.10 *A potential barrier and wavefunctions of a particle with $E = E_a > V_{max}$ and $0 < E = E_b < V_{max}$*

(a) $E_a > V_{max}$: The solution is oscillatory and well-behaved everywhere. The particle has a finite probability of going over the barrier. The function $\psi(x)$ is sinusoidal with a certain wavelength λ as $x \to \pm\infty$ and $\lambda(x)$ becomes longer as one approaches the potential barrier, as shown in Fig. 5.10. There is also a finite probability of reflection of the particle at the barrier even if $E > V_{max}$.

(b) $E_b < V_{max}$: Classically, a particle with such an energy approaching this potential barrier from either side cannot cross the barrier and go to the other side. But quantum mechanically, $\psi(x)$ is oscillatory on either side of the barrier with the same constant wavelength as $x \to \pm\infty$, though with different amplitudes depending on which way the particle is traveling. In the barrier region, it will show an exponential decay. We can find a well-behaved solution for any energy $E_b > 0$. The transmission probability diminishes as E falls below V_{max} and approaches zero.

A potential barrier is also known as a centre of repulsion. We shall meet the phenomenon of barrier penetration in Chapter 9 in the context of nuclear physics.

Infinite square-well potential

This is an ideal but very useful case. It is one of the few analytically solvable problems in quantum mechanics. We take the one-dimensional potential to be $V(x) = 0$ for $0 < x < L$, and $V(x) \to \infty$ at $x \to 0$ and $x \to L$. Since a particle cannot cross an infinite potential barrier, the particle is confined to the region $0 \le x \le L$. The Schrödinger equation in this case reduces to

$$\frac{d^2\psi(x)}{dx^2} = -\frac{2mE}{\hbar^2}\psi(x), 0 < x < L. \tag{5.43}$$

Since the potential is zero, the total energy E is just the kinetic energy, hence $E > 0$. Equation (5.43) then has simple solutions

$$\psi(x) = Ae^{ikx} + Be^{-ikx}, \tag{5.44}$$

where A, B and k are constants to be determined. Since the potential reaches infinity at the boundaries of the region, we must have $\psi(0) = \psi(L) = 0$, meaning that the particle cannot be found at these points. Using this boundary condition in Eqs. (5.44), we see that the wavefunction takes the form

$$\psi(x) = A \sin kL, \quad k = n\pi/L, \tag{5.45}$$

where n is a positive integer. Substituting this in (5.43), we see that it has well-behaved non-zero solutions only for values of energy given by

$$E_n = \frac{\hbar^2 n^2 \pi^2}{2mL^2} = \frac{n^2 h^2}{8mL^2}. \tag{5.46}$$

Comparing this with the case of a free particle dealt with earlier in this section, we see that it corresponds to a free particle in a limited, finite region in one dimension.

5.4.2 Simple harmonic oscillator

We are familiar with a classical simple harmonic oscillator which is best illustrated by an oscillating mass–spring system or a pendulum. In the micro world, consider a diatomic molecule with a certain equilibrium distance between the atoms, which is the bond length. At any temperature the atoms vibrate along the line joining them. If the change in bond length is small as compared to the bond length, the oscillation is simple harmonic. We now discuss the quantum mechanical oscillator.

The equation of motion of simple harmonic oscillator of mass m is

$$m\ddot{x} = -kx,$$

which has a solution

$$x = A \sin[\omega_0(t - \delta)], \quad \text{with} \quad \omega_0 = (k/m)^{1/2}.$$

The force $F(x) = -kx$ implies a potential $V(x) = \frac{1}{2}kx^2$. Thus $V(x) = \frac{1}{2}m\omega_0^2 x^2$, and the wave mechanical equation of steady state, (5.18), becomes

$$\frac{d^2\psi}{dx^2} - \frac{m^2\omega_0^2 x^2}{\hbar^2}\psi + \frac{2mE}{\hbar^2}\psi = 0. \tag{5.47}$$

To solve this equation, we let

$$z = ax, \quad \text{with} \quad a^4 = \omega_0^2 m^2/\hbar^2$$

so that (5.47) becomes

$$\frac{d^2\psi}{dz^2} - z^2\psi + \frac{zE}{\hbar\omega_0}\psi(z) = 0.$$

Finally, let

$$2E/\hbar\omega_0 = \beta + 1 \tag{5.48}$$

so that the equation becomes

$$\frac{d^2\psi}{dz^2} - z^2\psi + (\beta + 1)\psi = 0. \tag{5.49}$$

We want solutions for $\psi(z)$ which are well-behaved for large z. So we must have

$$\psi(z) \to 0 \quad \text{as} \quad |z| \to \infty.$$

Now for large z, (5.49) reduces to

$$\frac{d^2\psi}{dz^2} = z^2\psi. \tag{5.50}$$

Let us try a solution $\psi(z) = e^{bz^2}$. Then the above equation becomes

$$e^{bz^2}(4b^2 - 1)z^2 = 0.$$

Since this should be identically true for all z, we have $b = \pm 1/2$. But with $b = 1/2$, the solution $e^{z^2/2}$ diverges as $|z| \to \infty$. So $e^{-z^2/2}$ is the only well-behaved solution of (5.50).

Having considered this limiting case, we now expect a solution of (5.49) of the form

$$\psi(z) = F(z)e^{-z^2/2}, \tag{5.51a}$$

where $F(z)$ is such that, as $|z| \to \infty, \psi(z) \to 0$. Substituting this in (5.49), we see that $F(z)$ must satisfy

$$F'' - 2zF' + \beta F = 0, \tag{5.51b}$$

which we recognise as the Hermite equation. This equation has solutions, which as $|z| \to \infty$, diverge faster than $e^{z^2/2}$ so that the product solution of (5.51a) would diverge for large $|z|$ except when $\beta = 2n$, with $n = 0, 1, 2, \cdots$. For $\beta = 2n$, one of the solutions of (5.51a) reduces to a polynomial of degree n. These are the Hermite polynomials given by

$$H_n(z) = (-1)^n e^{z^2} \frac{d^n}{dz^n}(e^{-z^2}). \tag{5.52}$$

The first few Hermite polynomials are seen to be

$$H_0(z) = 1, \quad H_1(z) = 2z,$$

$$H_2(z) = 4z^2 - 2, \quad H_3(z) = 8z^3 + 2z,$$

$$H_4(z) = 16z^4 - 48z^2 + 12, \quad \text{etc.} \tag{5.53}$$

It is seen that the polynomial $H_n(z)$ contains even/odd powers of z depending on whether n is even/odd. Also, $H_n(z)$ has n nodes on the entire real z line, which must also be the number of nodes of $\psi_n(z)$.

From (5.48), we see that the Schrödinger equation for a linear harmonic oscillator, (5.47), has the eigenvalues

$$E_n = \hbar\omega_0(n + 1/2). \tag{5.54}$$

The minimum energy is seen to be $E_0 = \hbar\omega_0/2$ which is nonzero. This is called the *zero-point energy* of the oscillator.

The normalised eigenfunctions are seen to be

$$\psi_n(x) = \mathcal{N}_n H_n(ax)e^{-a^2x^2/2},$$

$$\text{with} \quad \mathcal{N}_n = \left(\frac{a}{2^n n! \pi^{1/2}}\right)^{1/2}. \tag{5.55}$$

In fact, using the orthonormality of Hermite polynomials,

$$\int_{-\infty}^{\infty} H_n(x)H_m(x)e^{-x^2}\, dx = 2^n n! \pi^{1/2}\delta_{nm},$$

it can be seen that $\psi_n(x)$ of (5.55) are also orthonormal.

The lowest two eigenfunctions are

$$\psi_0(z) \propto e^{-z^2/2} \quad \text{and} \quad \psi_1(z) \propto ze^{-z^2/2},$$

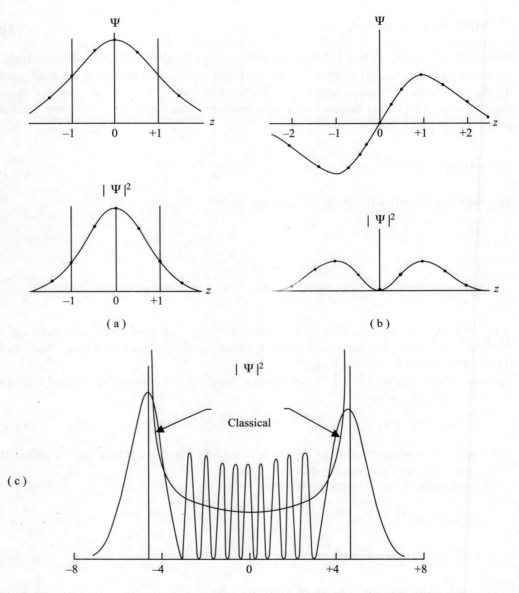

Fig. 5.11   *Schematic representation of eigenfunctions and probability distributions for three values of n; (a) n = 0, (b) n = 1, (c) n = 10.*

which have respectively zero and one node. Figure 5.11 shows these two wavefunctions and the corresponding probability distributions. This figure also shows the probability density for a large value of n such as $n = 10$. It is seen that for large n, $|\psi_n|^2$ approaches the probability distribution of a classical oscillator. For a classical oscillator following the equation $z = B \sin \omega t$, the probability of being found in the interval z to $z + dz$ is proportional to the time spent there, which is inversely proportional to the velocity, that is $1/v \propto (B^2 - z^2)^{-1/2}$. Here B is the amplitude so that $\pm B$ are the classical turning points. The classical amplitude in a quadratic potential well depends on the energy of the oscillator. Thus B will be the positive value of $z = ax$ when $E_n = \frac{1}{2}kx^2$, for the nth energy level, and this is seen to be

$$B = (2n + 1)^{1/2}.$$

Thus for $n = 10$, the classical turning points are at $z = \pm\sqrt{21}$.

5.4.3 Three-dimensional simple harmonic oscillator

In reality, an atom in a molecule of matter would perform oscillations in the three-dimensional space about its equilibrium position. For small displacements, these oscillations would be simple harmonic. The potential may be isotropic or anisotropic depending on the position of the other neighbouring atoms. We consider these cases here.

Isotropic case

Here $\mathbf{F} = -k\mathbf{r}, V = kr^2/2$, where $\mathbf{r}$ is the position vector from the equilibrium position of the atom and $r^2 = x^2 + y^2 + z^2$. Then the Schrödinger equation becomes

$$\nabla^2\psi - \frac{m^2\omega_0^2}{\hbar^2}(x^2 + y^2 + z^2)\psi(\mathbf{r}) + \frac{2mE}{\hbar^2}\psi = 0. \tag{5.56}$$

Letting $\psi(\mathbf{r}) = X(x)Y(y)Z(z)$, this equation is seen to be equivalent to three equations, each in one of the Cartesian coordinates, and each of the form of (5.47). Therefore, each of them will have eigenvalues of the form of (5.54) with the same value of ω_0. Thus the total energy is

$$E(n_x, n_y, n_z) = (n_x + n_y + n_z + 3/2)\hbar\omega_0. \tag{5.57}$$

It is clear that there is a large degeneracy for higher and higher states. The ground state is $3\hbar\omega_0/2$ and is non-degenerate with $n_x = n_y = n_z = 0$. The first excited state has $E = 5\hbar\omega_0/2$ and is three-fold degenerate with the quantum numbers being permutations of $(1, 0, 0)$. The next state has $E = 7\hbar\omega_0/2$ and is six-fold degenerate with the quantum numbers being permutations of $(2, 0, 0)$ or of $(1, 1, 0)$.

Anisotropic case

This case is closer to reality. But for small displacements, the potential can still be truncated at the second order terms and taken as a bilinear combination of the coordinates around the equilibrium point. We can then determine the principal axes and express the potential as a quadratic form and take the Cartesian coordinates along the principal axes. The curvature of potential along the three axes will be different, and the restoring force along them can be expressed as

$$F_x = -k_x x, \quad F_y = -k_y Y, \quad F_z = -k_z z.$$

Let the three frequencies of oscillations along these axes be $\omega_x, \omega_y, \omega_z$, respectively, given by $\omega_x = (k_x/m)^{1/2}$, etc. Then a similar treatment as above gives the energy levels as

$$E(n_x, n_y, n_z) = (n_x + \frac{1}{2})\hbar\omega_x + (n_y + \frac{1}{2})\hbar\omega_y + (n_z + \frac{1}{2})\hbar\omega_z. \tag{5.58}$$

We see that the degeneracy is now lifted.

5.5 Hydrogen-like Atom in Quantum Mechanics

There are very few systems in quantum mechanics which are exactly, analytically solvable. Most systems require the use of approximation methods and numerical methods. Such exactly solvable systems in the one-dimensional case are the infinite square well potential and the simple harmonic oscillator. Here we discuss the hydrogen-like atom which is a three-dimensional exactly solvable system. This atom consists of a nucleus of charge Ze and mass M_n at $\mathbf{R} \equiv (X, Y, Z)$ and an electron of charge $-e$ and mass m_e at $\mathbf{r} \equiv (x, y, z)$, in Cartesian coordinate system, as shown in Fig. 5.12.

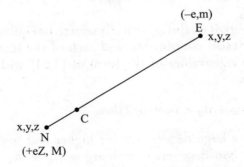

Fig. 5.12 *A hydrogen-like atom with nucleus at* $\mathbf{R}$ *and electron at* $\mathbf{r}$ *(with respect to some origin)*

5.5.1 Schrödinger equation for stationary state

The potential energy between the two charges is $V(r_0) = -Ze^2/r_0$, where $\mathbf{r}_0$ is the vector from the nucleus to the electron. Thus we have

$$r_0 = [(x - X)^2 + (y - Y)^2 + (z - Z)^2]^{1/2}.$$

The Hamiltonian of the system is

$$H = \frac{p_N^2}{2M_n} + \frac{p_e^2}{2m_e} + V(r_0)$$

$$= -\frac{\hbar^2}{2M_n}\nabla_R^2 - \frac{\hbar^2}{2m_e}\nabla_r^2 - \frac{Ze^2}{r_0},$$

where the subscripts R and r refer to differentiation with respect to **R** and **r**, respectively. So the steady-state Schrodinger equation becomes

$$-\frac{\hbar^2}{2M_n}\nabla_R^2\psi' - \frac{\hbar^2}{2m_e}\nabla_r^2\psi' - \frac{Ze^2}{r_0}\psi' = E'\psi'(\mathbf{R}, \mathbf{r}). \tag{5.59}$$

Since r_0 contains both **r** and **R**, it is not possible to separate the variables. So we consider the centre of mass C, which is at $\boldsymbol{\alpha} = (\xi, \eta, \eta)$ with respect to the chosen origin, which is given by

$$\boldsymbol{\alpha} = \frac{M_n\mathbf{R} + m_e\mathbf{r}}{M_n + m_e},$$

and the relative vector $\mathbf{r}_0 = \mathbf{r} - \mathbf{R}$ from the nucleus to the electron. Also let $M = M_n + m_e$ be the total mass and $\mu = M_n m_e/(M_n + m_e)$ the reduced mass. With this, we can rewrite (5.59) as

$$-\frac{\hbar^2}{2M}\nabla_\alpha^2\psi' - \frac{\hbar^2}{2\mu}\nabla_0^2\psi' + V(r_0)\psi' = E'\psi',$$

where ∇_0 refers to differentiation with respect to $\mathbf{r}_0$.

The wavefunction ψ' depends on the centre of mass coordinates and relative coordinates. It turns out that this equation can be separated into these two sets of coordinates. Thus, let us assume that $\psi'(\boldsymbol{\alpha}, \mathbf{r}_0) = \varphi(\boldsymbol{\alpha})\psi(\mathbf{r}_0)$. With this the above equation reduces to two equations, each a Schrödinger equation in one set of coordinates, of the form

$$-\frac{\hbar^2}{2M}\nabla_\alpha^2\varphi(\boldsymbol{\alpha}) = W\varphi(\boldsymbol{\alpha}), \tag{5.60a}$$

$$-\frac{\hbar^2}{2\mu}\nabla_0^2\psi(\mathbf{r}_0) + V(r_0) = E\psi(\mathbf{r}_0), \tag{5.60b}$$

where $W + E = E'$. Equation (5.60a) shows that the atom as a whole moves like a free particle of mass M having energy W. Equation (5.60b), with $V(r_0)$ as the Coulomb potential, is the Schrödinger equation for a hydrogen-like atom with a fixed nucleus and an electron of mass μ, with total energy E. Thus separation is also valid for any central potential.

Due to the spherical symmetry of the system, which is reflected in the Schrödinger equation (5.60b), it would be convenient to use spherical coordinates. Since we can now solve (5.60b) independently of (5.60a), we might as well drop the subscript zero and replace $\mathbf{r}_0$ by $\mathbf{r}$, with components x, y, z. Hence, we shall use $\mathbf{r}$ hereafter as the vector from the nucleus to the electron.

▶ **Guided Exercise 5.1**   Obtain the angular part of the eigenfunctions for a three-dimensional spherically symmetric potential.

Hints

(a) Using spherical coordinates $\mathbf{r} = (r, \theta, \varphi)$, write the Schrödinger equation, similar to (5.60b) with $\mathbf{r}_0$ replaced by $\mathbf{r}$ and $V(\mathbf{r}_0)$ replaced by $V(r)$, in the form

$$-\frac{\hbar^2}{2\mu}\left[\frac{1}{r^2}\frac{\partial}{\partial r}\left(r^2\frac{\partial\psi}{\partial r}\right) + \frac{1}{r^2\sin\theta}\frac{\partial}{\partial\theta}\left(\sin\theta\frac{\partial\psi}{\partial\theta}\right) + \frac{1}{r^2\sin^2\theta}\frac{\partial^2\psi}{\partial\varphi^2}\right] + V(r)\psi$$

$$= E\psi(\mathbf{r}). \tag{5.61}$$

(b) In accordance with the method of separation of variables, try a solution of the form

$$\psi(\mathbf{r}) = R(r)P(\theta)F(\varphi), \tag{5.62}$$

and rewrite (5.61) in the form

$$\frac{\sin^2\theta}{R}\frac{d}{dr}\left(r^2\frac{dR}{dr}\right) + \frac{\sin\theta}{P}\frac{d}{d\theta}\left(\sin\theta\frac{dP}{d\theta}\right) - \frac{2\mu}{\hbar^2}(V(r) - E)r^2\sin^2\theta$$

$$= -\frac{1}{F}\frac{d^2F}{d\varphi^2}. \tag{5.63}$$

(c) Note that this equation has variables r, θ on one side and φ on the other side. Hence both sides must be equal to a constant, independent of the coordinates. Let both sides be equal to some constant k. Then the φ-equation becomes

$$\frac{d^2F}{d\varphi^2} = -kF. \tag{5.64}$$

(d) While this equation has mathematical solutions for any value of k, positive, zero or negative, we must look for physically acceptable solutions. φ is a physically meaningful variable, the azimuthal angle, and $F(\varphi)$ is a factor in the wavefunction

$\psi(\mathbf{r})$. Hence $F(\varphi)$ must be a periodic function, $F(\varphi + 2\pi) = F(\varphi)$ for all φ. This is possible only if k takes positive values of the form of m^2 where m is an integer. Then (5.64) has well-behaved acceptable solutions of the form

$$F_m(\varphi) = e^{im\varphi}, \quad m = 0, \pm 1, \pm 2, \cdots.$$

The φ-part thus contains one parameter, m.

(e) With both sides of (5.63) equal to m^2, write its (r, θ)-part as

$$\frac{1}{R}\frac{d}{dr}\left(r^2\frac{dR}{dr}\right) - \frac{2\mu}{\hbar^2}(V(r) - E)r^2 = -\frac{1}{P\sin\theta}\frac{d}{d\theta}\left(\sin\theta\frac{dP}{d\theta}\right) + \frac{m^2}{\sin^2\theta}.$$

(f) Since r and θ are separated in this equation, both sides must be equal to a constant, say β. Write the θ-equation in the form

$$\frac{1}{\sin\theta}\frac{d}{d\theta}\left(\sin\theta\frac{dP}{d\theta}\right) + \left(\beta - \frac{m^2}{\sin^2\theta}\right)P = 0. \tag{5.65}$$

(g) Make a change of variables $x = \cos\theta$ and rewrite (5.65) in the form

$$(1 - x^2)\frac{d^2P}{dx^2} - 2x\frac{dP}{dx} + \left(\beta - \frac{m^2}{1 - x^2}\right)P(x) = 0.$$

(Note that x here has nothing to do with the Cartesian component of $\mathbf{r}$, but is related to the spherical coordinate θ. Also note that the equation is even in m.)

(h) Note that this is the associated Legendre equation. It has regular singular points at $x = \pm 1$. It has standard series solutions $P(x)$ and $Q(x)$ which are convergent for $|x| < 1$, but diverge at $x = \pm 1$ for arbitrary values of β. But θ is a physical variable, the polar angle, and $x = \pm 1$ correspond to $\theta = 0$ and π, respectively. $U(\theta)$ is a factor in the wavefunction $\psi(\mathbf{r})$ and we cannot allow it to tend to infinity along points on the polar axis. It can be seen that one of the series truncates to a finite polynomial for $\beta = l(l + 1)$, where l is an intger satisfying $l \geq |m|$. The series $P(x)$ or $Q(x)$ reduces to a polynomial of degree l depending on whether l is even or odd. We denote this polynomial as $P_l^m(x)$ or $P_l^m(\cos\theta)$, which is now well-behaved for $-1 \leq x \leq 1$ or $0 \leq \theta \leq \pi$.

The θ-part thus contains two related parameters, l, m.

(i) Write down the remaining r-equation, with $\beta = l(l + 1)$, in the form

$$\frac{1}{R}\frac{d}{dr}\left(r^2\frac{dR}{dr}\right) - \frac{2\mu}{\hbar^2}(V - E)r^2 - l(l + 1) = 0. \tag{5.66}$$

The solution of this equation depend on the form of the potential $V(r)$. Thus the angular part of the eigenfunctions of the Schrodinger equation is seen to be of the form

$$P(\theta)F(\varphi) = P_l^m(\cos\theta)e^{im\varphi}. \tag{5.67}$$

◀

5.5.2   Eigenfunctions of the hydrogen-like atom

We now consider the r-part, equation (5.66), with $V(r) = -Ze^2/r$. On replacing $R(r) = U(r)/r$, it reduces to the form

$$-\frac{\hbar^2}{2\mu}U'' - \frac{Ze^2}{r}U + \frac{\hbar^2}{2\mu}l(l+1)\frac{U}{r^2} = EU(r), \tag{5.68}$$

where primes denote differentiation with respect to the variable. For large r, this equation reduces to

$$\frac{\hbar^2}{2\mu}U''(r) = -EU(r).$$

For $E > 0$, this leads to oscillatory solutions for $U(r)$. These solutions correspond to the continuum of the H-like atom.

For $E < 0$, we get bound states (because the highest value of the potential is 0). For this case, we let

$$E = -\frac{\hbar^2}{2\mu}\alpha^2, \tag{5.69}$$

with α real. We now make a change of variables and define a few parameters for later convenience as

$$r = \rho/2\alpha, \quad \lambda = \mu Ze^2/\hbar^2\alpha. \tag{5.70}$$

With this, for $E < 0$, (5.68) becomes

$$\frac{d^2U}{d\rho^2} + \left[\frac{\lambda}{\rho} - \frac{l(l+1)}{\rho^2} - \frac{1}{4}\right]U(\rho) = 0. \tag{5.71}$$

It may be noted that α has dimensions of inverse length while λ and ρ are dimensionless. Thus (5.71) has been put in entirely dimensionless form. It is seen that as $\rho \to \infty$, the solutions of (5.71) go as $e^{-\rho/2}$. With a suitable change of the dependent variable, (5.68) can be put in the form of the associated Laguerre equation. It has two infinite series solutions for any value of λ which are, in general, divergent. It is seen that one of the two series reduces to a polynomial for particular values of λ. When λ is equal to a positive integer greater than l, we get a polynomial solution which is called the associated Laguerre polynomial. Thus finally (5.71) has the well-behaved solutions

$$U(\rho) = \rho^{l+1} L_{n+l}^{2l+1}(\rho) e^{-\rho/2}, \quad n > l. \tag{5.72}$$

The Laguerre polynomials and the corresponding associated Laguerre polynomials are given by[15]

$$L_n(\rho) = e^{\rho} \frac{d^n}{d\rho^n} (\rho^n e^{-\rho}), \quad n = 0, 1, 2, \cdots, \tag{5.73a}$$

$$L_n^k(\rho) = (-1)^k \frac{d^k}{d\rho^k} L_n(\rho), \quad 0 \le k \le n. \tag{5.73b}$$

The polynomial expressions which are consistent with Eqs. (5.73a) are

$$L_n(\rho) = \sum_{r=0}^{n} \frac{(-1)^r (n!)^2}{(r!)^2 (n-r)!} x^r, \tag{5.73c}$$

$$L_n^k(\rho) = \sum_{r=0}^{n-k} \frac{(-1)^r (n!)^2}{r!(k+r)!(n-k-r)!} x^r. \tag{5.73d}$$

Thus for $E < 0$, the eigenfunctions for the stationary states of hydrogen atom, in terms of the dimensionless parameter ρ, are of the form

$$\psi_{nlm}(\boldsymbol{\rho}) \propto \rho^l L_{n+l}^{2l+1}(\rho) e^{-\rho/2} \rho_l^m (\cos\theta) e^{im\varphi}, \tag{5.74a}$$

where $\boldsymbol{\rho}$ is related to $\mathbf{r} \equiv (r, \theta, \varphi)$ as in (5.70). The properly normalised eigenfunction, in terms of $\mathbf{r}$, can then be written as

$$\psi_{nlm}(\mathbf{r}) = N_{nlm} r^l L_{n+l}^{2l+1}(r) e^{-\alpha r} P_l^m (\cos\theta) e^{im\varphi},$$

$$N_{nlm} = \left[\frac{(n-l-1)!}{2n\{(n+l)!\}^3} \left(\frac{2Z}{na_0} \right)^3 \right]^{1/2} \left[\frac{2l+1}{4\pi} \frac{(l-|x|)!}{(l+|m|)!} \right]^{1/2}, \tag{5.74b}$$

where a_0 is the Bohr radius.

The condition on λ, that it must be equal to a positive integer, say n ($n = 1, 2, 3, \cdots$), leads to quantization of energy. With $\lambda = n$, (5.70) gives

$$\lambda = \frac{\mu Z e^2}{\hbar^2 \alpha} = n \quad \Rightarrow \quad \alpha_n = \frac{\mu Z e^2}{n\hbar^2},$$

---

[15]Various authors define these in various ways, with or without $n!$, and also denote associated Laguerre polynomials by L_n^k or L_{n+k}^k.

where α_n indicates the allowed values of α. The energy eigenvalues are then found from (5.69) to be

$$E_n = -\frac{\mu Z^2 e^4}{2n^2 \hbar^2} = -\frac{E_0 Z^2}{n^2}, \tag{5.75}$$

where E_0 is the ground state energy of the hydrogen atom. It is seen that this expression agrees with that obtained on the Bohr model for a hydrogen-like atom; see Eqs. (5.7).

The angular part of the eigenfunctions in (5.74a), with suitable normalisation, is seen to be

$$Y_{lm}(\theta, \varphi) = \left[\frac{(2l+1)}{4\pi}\frac{l-|m|!}{(l+|m|!)}\right]^{1/2} e^{im\varphi} P_l^m(\cos\theta)$$

$$= \left[\frac{(2l+1)}{4\pi}\frac{(l-|m|)!}{(l+|m|)!}\right]^{1/2} e^{im\varphi} \sin^m\theta \left(\frac{d}{d(\cos\theta)}\right)^m P_l(\cos\theta), \tag{5.76a}$$

where $P_l(\cos\theta)$ is the Legendre polynomial. These functions are called spherical harmonics. It is seen that they form a complete orthonormal set of functions over the unit sphere. Thus

$$\int_\Omega Y_{lm}^*(\theta, \varphi) Y_{l'm'}(\theta, \varphi) d\Omega = \delta_{ll'}\delta_{mm'};,$$

$$\sum_{l=0}^\infty \sum_{m=-l}^l Y_{lm}^*(\theta, \varphi) Y_{lm}(\theta', \varphi') = \delta(\varphi - \varphi')\delta(\cos\theta - \cos\theta'). \tag{5.76b}$$

5.5.3 Quantum numbers and angular momenta

We shall now consider the physical significance of the parameters n, l, m introduced above in the solution of the Schrodinger equation. We see from (5.75) that the energy in the different states depends only on the parameter n, which is therefore called the *principle quantum number*. All states ψ_{nlm} with common n and different values of l and m are degenerate. As seen earlier in another context, spin and relativity corrections remove the degeneracy of different l states, and magnetic field removes the degeneracy of m states.

Since H commutes with L^2, ψ_{nlm} is also an eigenfunction of L^2. Writing ψ_{nlm} from (5.62) in the form

$$\psi_{nlm}(\mathbf{r}) = N_{nlm} R_{nl}(r) P_l^m(\theta) F_m(\varphi),$$

we see that

$$L^2\psi_{nlm} = \mathcal{N}_{nlm}R_{nl}(r)F_m(\varphi)L^2P^m l(\theta)$$

$$= N_{nlm}R_{nl}(r)F_m(\varphi)\left[\frac{1}{\sin\theta}\frac{\partial}{\partial\theta}\left(\sin\theta\frac{\partial P}{\partial\theta}\right) - \frac{m^2}{\sin^2\theta}\right]\hbar^2$$

$$= N_{nlm}R_{ln}(r)F_m(\varphi)l(l+1)\hbar^2 P(\theta)$$

$$= l(l+1)\hbar^2\psi_{nlm}. \tag{5.77}$$

Therefore the eigenvalues of L^2 are $[l(l+1)]^{1/2}\hbar \equiv l^*\hbar$, where we have denoted $[l(l+1)]^{1/2}$ by l^*. Thus l is the *azimuthal quantum number*, but the angular momentum is $l^*\hbar$.

Comparing this with Sommerfeld's theory, we see that l^* here replaces k of Sommerfeld's theory. Then $l = 0, 1, 2, \cdots, n-1$ give s, p, d, f, $\cdots$, etc states, respectively. In Bohr's theory, the electron in the n-th orbit has angular momentum $L = n\hbar$. But that level splits into n sublevels with $l = 0, 1, 2, \cdots, n-1$, and the electrons in the sublevel have the angular momentum $l^*\hbar$. Further, each suborbit can orient itself in $2l+1$ directions with $L_z = l\hbar$.

As H and L_z also commute with each other, ψ_{nlm} is also an eigenfunction of L_z. We have

$$L_z\psi_{nlm} = N_{nlm}R_{nl}P_e^m(\theta)\left(-i\hbar\frac{\partial F}{\partial\varphi}\right) = m\hbar\psi_{nlm}. \tag{5.78}$$

Therefore L_z has eigenvalues $m\hbar$, where m is known as the *magnetic quantum number*, which has $2l+1$ values from $-l$ to l; m is related to the magnetic moment μ_m as we shall see below.

When the electron moves in a circular orbit, it produces a current $i = -e/\tau$ where $\tau = 2\pi r/v$ is the period. It gives rise to a magnetic moment μ_m given by

$$\mu_m = \frac{\pi r^2(-ev)}{2\pi rc} = -\frac{erv}{2c}. \tag{5.79}$$

It is related to the angular momentum $L = \mu rv$ as

$$\mu_m = -\frac{e}{2\mu c}L = -\frac{e\hbar}{2\mu c}l^*.$$

The z-component of μ_m will be

$$\mu_z = -\frac{e\hbar}{2\mu c}m.$$

We see from here that $e\hbar/2\mu c$ is the smallest unit of magnetic moment; it is called

Bohr magneton, and has a value of 9.18×10^{-21} erg/gauss. Thus the z-component of the magnetic moment is m Bohr magnetons.

As the spin angular momentum is $s = \hbar/2$, we see that in this case $\mu_s = es/\mu c$, that is, twice the orbital motion.

We show in Fig. 5.13 the vector representation of angular momentum for p and d states of the $n = 3$ level. The various quantum numbers and the angle α ($= \cos^{-1}(L_z/|L|)$) which the angular momentum makes with the z-axis are also shown in Table 5.3. This vector atom model was prevalent up to about 1925.

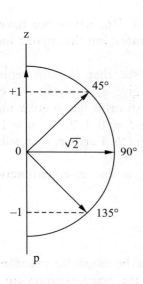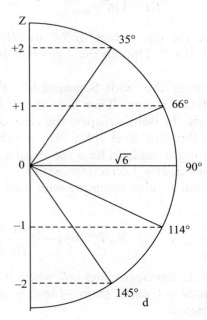

Fig. 5.13   *Vector representation of angular momentum for* $n = 3$ *(p and d states)*

Table 5.3   *Vector atom model for* $n = 3$ *level*

State	l	$|\mathbf{L}|$	m	α
s	0	0	0	0
p	1	$\sqrt{2}\hbar$	1	45°
			0	90°
			−1	135°
d	2	$\sqrt{6}\hbar$	2	35°
			1	66°
			0	90°
			−1	114°
			−2	145°

5.5.4   Relation to Bohr–Sommerfeld model

The maximum value of l in the above theory is $n - 1$ and, as we have seen, l is related to the angular momentum. Then the eigenfunction corresponding to $l = n - 1$, from (5.74a), becomes

$$\psi_{nlm}(\boldsymbol{\rho}) = N_{nlm} P_l^m(\cos\theta) e^{im\varphi} \rho^{n-1} L_{2n-1}^{2n-1}(\rho) e^{-\rho/2}.$$

We see from (5.73b) that

$$L_n^n(\rho) = \frac{d^n}{d\rho^n}\left[e^\rho \frac{d^n}{d\rho^n}(\rho^n e^{-\rho}) \right],$$

which is seen to be the pth derivative of a pth degree polynomial, and hence a constant. Hence the probability density $P(\rho) = |\psi_{nlm}(\boldsymbol{\rho})|^2$ of finding the electron at around a distance ρ is proportional to $\rho^{2n-2} e^{-\rho}$. Hence the probability of finding the electron between ρ and $\rho + d\rho$, which is related to $4\pi\rho^2 |\psi|^2 d\rho$, will be proportional to $\rho^{2n} e^{-\rho}$. This will have an extremum when

$$(2n - \rho)\rho^{2n-1} e^{-\rho} = 0.$$

Thus the probability, apart from being zero at $\rho = 0$ and $\rho \to \infty$, is a maximum at $\rho = 2n$, that is, at

$$r = n/\alpha = Zn^2 a_0,$$

which is the Bohr–Sommerfeld formula for the radius of the nth orbit. Thus these radii represent the distance at which the probability of finding the electron according to Schrödingers theory is maximum. But the probability at other values of r is not zero. Hence the quantum mechanical model represents an electron as a charge cloud rather than as a point charge. Because of this, chemists call it an *orbital*.

We shall now study the radial and angular distributions of these orbitals.

Radial distribution

For a general orbital, the radial probability distribution would be

$$P(\rho)d\rho \propto \rho^{2l+2}\left[L_{n+l}^{2l+1}(\rho) \right]^2 e^{-\rho}.$$

It has $n - l - 1$ minima at solutions of $L_{n+l}^{2l+1}(\rho) = 0$ and $n - l$ maxima when

$$(2l + 2 - \rho)L_{n+l}^{2l+1}(\rho) + 2\rho L_{n+l}^{2l+2}(\rho) = 0.$$

For example, when $n = 2, l = 0$,

$$L_2'(\rho) = 2\rho - 4, \quad \alpha = Z/2a_0.$$

Therefore there is one minimum at $\rho = 2, r = 2a_0/Z$, and maxima at $(2 - \rho)(2\rho - 4) + 4\rho = 0$, or

$$\rho = 3 \pm \sqrt{5}, \quad r = (3 \pm \sqrt{5})a_0/Z.$$

Some Laguerre and associated Laguerre polynomials are:

$L_0(\rho) = 1,$ $L_1(\rho) = -\rho + 1,$

$L_2(\rho) = \rho^2 - 4\rho + 2,$ $L_3(\rho) = -\rho^3 + 9\rho^2 - 18\rho + 6,$

$L_4(\rho) = \rho^4 - 16\rho^3 + 72\rho^2 - 96\rho + 24,$

$L_1^1(\rho) = -1,$

$L_2^1(\rho) = 2\rho - 4,$ $L_2^2(\rho) = 2,$

$L_3^1(\rho) = -3\rho^2 + 18\rho - 18,$ $L_3^2(\rho) = -6\rho + 18, \quad L_3^3(\rho) = -6.$

Then the radial distribution of a few lowest states and the resulting probability distribution are shown in Table 5.4. Figure 5.14 shows the normalised radial functions to scale.

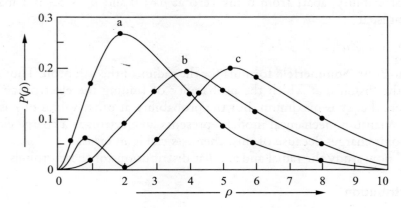

Fig. 5.14 *Radial probability functions for some low quantum number orbitals*

Table 5.4 *Associated Legendre polynomials and radial probability distribution for some orbitals*

n	l	$L_{n+l}^{2l+1}(\rho)$	$P(\rho)$ (unnormalised)
1	0	-1	$\rho^2 e^{-\rho}$
2	1	-6	$36\rho^4 e^{-\rho}$
2	0	$2(\rho - 2)$	$4\rho^2(\rho - 2)^2 e^{-\rho}$

Azimuthal dependence

Since the φ-dependence occurs through the factor $e^{im\varphi}$, $|\psi_{nlm}(\mathbf{r})|^2$ would be independent of φ. Thus the electron cloud has axial symmetry around the z-axis.

Polar distribution

The θ-dependence of the eigenfunction comes through the factor $P_l^m(\cos\theta)$. Normalisation of the θ-part shows that the probability distribution $P(\theta)$, of finding the electron at a point with polar angle near θ is

$$P(\theta) \propto \frac{(l-|m|)!}{(l+|m|)!}(P_l^m(\cos\theta))^2.$$

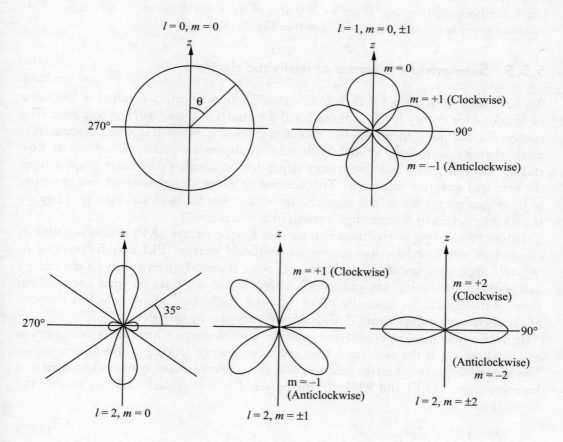

Fig. 5.15 *Some θ distributions of eigenfunctions for $l = 0, 1, 2$*

Some of the first few associated Legendre polynomials are:

$$l = 0: \quad P_0^0 = 1,$$

$$l = 1: \quad P_1^0 = \cos\theta, \quad P_1^1 = \sin\theta,$$

$$l = 2: \quad P_2^0 = \tfrac{1}{2}(3\cos^2\theta - 1),$$

$$P_2^1 = 3\sin\theta\cos\theta = 3(\sin 2\theta)/2,$$

$$P_2^2 = 3(1 - \cos^2\theta) = 3\sin^2\theta.$$

Thus for $l = 0, m = 0, P(\theta)$ is constant, and we have spherical symmetry. When $l = 1$, we have $P(\theta) \propto \cos^2\theta$ for $m = 0$ and $P(\theta) \propto \sin^2\theta$ for $m = \pm 1$. Some $P(\theta)$ distributions for $l = 0, 1$ and 2 are shown in Fig. 5.15.

5.5.5   Sommerfeld's treatment of relativistic electrons

We have seen in Section 5.2 that Sommerfeld's elliptical orbits resulted in the same expressions for energy levels as those for Bohr model in the non-relativistic case. The energy E_n did not depend on the eccentricity of the orbit. But when Sommerfeld used relativity with Bohr model, with velocity-dependent mass, he arrived at Eqs. (5.16a) according to which the energy depended on another parameter k apart from the principal quantum number n. This seemed to 'explain' the observed fine structure of hydrogen energy levels, although the agreement was far from satisfactory. Here we show some details of Sommerfeld's relativistic treatment.

Before proceeding to the theory, note that kinetic energy (KE) is non-negative in classical as well as relativistic mechanics. Potential energy (PE) may be positive or negative depending on where we choose its zero. It can be given a shift on the energy axis without producing any observable effect. When we refer to total energy E in classical mechanics, we generally mean the total mechanical energy which is the sum of the above two components. This energy E may also be negative.

In relativistic mechanics, there is another contribution, which at its minimum is $m_0 c^2$, where m_0 is the rest mass. This rest mass energy (RME) is also non-negative. The energy E given by relativistic formula $E^2 = p^2 c^2 + m_0 c^4$ consists of relativistic kinetic energy (RKE) and RME. In the presence of a potential $V(\mathbf{r})$, we modify the formula as

$$(E - V)^2 = p^2 c^2 + m_0^2 c^4. \tag{5.80}$$

The square root of each side of the above equation represents the sum of RKE and RME.

It can now be shown that by taking $V(\mathbf{r})$ to be the hydrogenic potential $-Ze^2/r$ and writing a Schrodinger equation with total energy E (RME + RKE) given by (5.80) we can arrive at an equation analogous to Sommerfeld's Eqs. (5.17). To this end, after a slight rearrangement of (5.80), and replacing p^2 by $-\hbar^2\nabla^2$, we write

$$-\frac{\hbar^2}{2m_0}\nabla^2\psi + \frac{1}{2m_0c^2}(E-V)^2\psi(\mathbf{r}) = \frac{1}{2}m_0c^2\psi(\mathbf{r}). \qquad (5.81)$$

Using the method of separation of variables and following a treatment similar to that used in arriving at (5.68), we get

$$\frac{d^2U}{dr^2} - \left[l(l+1) - \frac{Z^2e^4}{\hbar^2c^2}\right]\frac{U}{r^2} + \frac{2Ze^2E}{\hbar^2c^2}\frac{U}{r} = -\frac{m_0^2c^2}{\hbar^2}\left[\frac{E^2}{m_0^2c^4} - 1\right]U(r). \qquad (5.82)$$

If $E > m_0c^2$, this equation has free state solutions for any E, while for $E < m_0c^2$ it will have bound state solutions in potential $V(\mathbf{r})$ for quantized energies.

Therefore for $E < m_0c^2$, let us define α' and l' through

$$\frac{m_0^2c^2}{\hbar^2}\left[1 - \frac{E^2}{m_0^2c^4}\right] = \alpha'^2, \qquad (5.83a)$$

$$l(l+1) - \frac{Z^2e^4}{\hbar^2c^2} = l'(l'+1). \qquad (5.83b)$$

Using these in (5.81) and changing the independent variable to $\rho = 2\alpha'r$, we arrive at

$$\frac{d^2U}{d\rho^2} + \left[\frac{\lambda'}{\rho} - \frac{l'(l'+1)}{\rho^2} + \frac{1}{4}\right]U(\rho) = 0, \qquad (5.84a)$$

where

$$\lambda' = \frac{m_0Ze^2}{\alpha'\hbar^2}\left(1 - \frac{\alpha'^2\hbar^2}{m_0^2c^2}\right)^{1/2}. \qquad (5.84b)$$

These equations may be compared with (5.70) and (5.71). The first of these equations has well-behaved solutions when

$$\lambda' = n + (l' - l),$$

which gives

$$\alpha'^2 = \frac{m_0^2c^2}{\hbar^2\{1 + \hbar^2c^2[n + (l'-l)]^2/Z^2e^4\}}. \qquad (5.85)$$

This finally gives, after somewhat lengthy algebra,

$$E - m_0 c^2 = E_n - \frac{|E_0|\alpha^2 Z^4}{n^3} \left\{ \frac{1}{(l+1/2)} - \frac{3}{4n} \right\},$$

where $\alpha = e^2/\hbar c$ is the fine structure constant introduced in (5.16) (not to be confused with that introduced in (5.69) and (5.70)), and E_n is the hydrogenic energy level of (5.14). This can be seen to be similar to (5.16b) with the identification $l + 1/2 \equiv k$.

5.6 Electron Spin

In this section we shall discuss electron spin as proposed by Uhlenbeck and Goudsmit, Pauli spin matrices and their standard representation. Then we shall go back to energy levels of the hydrogen atom and splitting of levels due to orbital and spin angular momenta. We shall then discuss Dirac's relativistic theory and apply it to an electron.

5.6.1 Pauli's theory

According to Uhlenbeck and Goudsmit, the electron has a spin angular momentum s whose component in any direction can have only two values, $\pm\hbar/2$, and which has a spin magnetic moment $\boldsymbol{\mu} = -(e/mc)\mathbf{s}$. They put forward this hypothesis in order to explain the fine structure and Zeeman effect of spectral lines.

Very soon Heisenberg developed his matrix mechanics, with hermitian operators for observables. It was realised that each component of spin can have eigenvalues $\pm\hbar/2$ and $s^2 = s_x^2 + s_y^2 + s_c^2$ can have eigenvalues $3\hbar^2/4 = s(s+1)\hbar^2$. Pauli introduced dimensionless operators $\sigma_x, \sigma_y, \sigma_z$, which were related to spin angular momentum by

$$s_i = \frac{1}{2}\hbar\sigma_i, \quad i = x, y, z.$$

It was found that Pauli spin matrices must satisfy the relations

$$\sigma_x\sigma_y + \sigma_y\sigma_x = 0, \quad \sigma_y\sigma_z + \sigma_z\sigma_y = 0, \quad \sigma_z\sigma_x + \sigma_x\sigma_z = 0, \tag{5.86a}$$

and $\quad \sigma_x^2 = \sigma_y^2 = \sigma_z^2 = 1.$ $\tag{5.86b}$

The anticommutation relations of (5.86a) show that the σ's cannot be simple algebraic numbers, and we need to have at least 2×2 matrices to represent them. Also, at most one of them can be a diagonal matrix. Pauli choose σ_z to be the diagonal matrix, and the only possibility consistent with Eqs. (5.86) is that it must be represented by

$$\sigma_z = \begin{bmatrix} 1 & 0 \\ 0 & -1 \end{bmatrix}, \tag{5.87}$$

apart from the trivial interchange of 1 and -1 which results from a unitary transformation. Then trying the matrix $\begin{bmatrix} a & b \\ c & d \end{bmatrix}$ for σ_x and σ_y and using Eqs. (5.86), we get the following Pauli spin matrices, in a representation in which σ_z is diagonal, as

$$\sigma_x = \begin{bmatrix} 0 & 1 \\ 1 & 0 \end{bmatrix}, \quad \sigma_y = \begin{bmatrix} 0 & -i \\ i & 0 \end{bmatrix}, \quad \sigma_z = \begin{bmatrix} 1 & 0 \\ 0 & -1 \end{bmatrix}. \tag{5.88}$$

Owing to (5.86b), all the three matrices have eigenvalues ± 1. The commutators of these matrices turn out to be

$$[\sigma_x, \sigma_y] = 2i\sigma_z, \text{ and cyclic permutation of } x, y, z. \tag{5.89}$$

Spin wavefunctions

The eigenstate of these operators would be a column vector of order 2×1 and can be represented by $\psi = \begin{bmatrix} \psi_1 \\ \psi_2 \end{bmatrix}$. Thus we have

$$\sigma_z \psi = i \begin{bmatrix} 1 & 0 \\ 0 & -1 \end{bmatrix} \begin{bmatrix} \psi_1 \\ \psi_2 \end{bmatrix} = i \begin{bmatrix} \psi_1 \\ -\psi_2 \end{bmatrix}.$$

This gives $\sigma_z \psi_1 = i\psi_1, \quad \sigma_z \psi_2 = -i\psi_2$.

Thus ψ_1 and ψ_2 are eigenfunctions of σ_z with eigenvalues $\pm i$, respectively. The two eigenstates are denoted by

$$\alpha = \begin{bmatrix} 1 \\ 0 \end{bmatrix}, \quad \beta = \begin{bmatrix} 0 \\ 1 \end{bmatrix}.$$

These are two orthonormal spin wavefunctions.

Spin matrices and eigenvalues

In this representation, all operators are represented by 2×2 matrices. They are obtained by multiplying each operator by the unit matrix. Thus,

$$x = \begin{pmatrix} x & 0 \\ 0 & x \end{pmatrix}, \quad p_x = \begin{pmatrix} -i\hbar \partial/\partial x & 0 \\ 0 & -i\hbar \partial/\partial x \end{pmatrix}, \quad l_x = \begin{pmatrix} l_x & 0 \\ 0 & l_x \end{pmatrix}, \tag{5.90}$$

etc. Further, we now have an operator for the total angular momentum $\mathbf{j} = \mathbf{l} + \mathbf{s}$. So[16],

---

[16]It may be noted that we often denote the angular momentum and the corresponding quantum number by the same symbol. For example, electron spin could be denoted by $s = \hbar/2$ or $s = 1/2$.

$$j_x = \begin{pmatrix} l_x & \hbar/2 \\ \hbar/2 & l_x \end{pmatrix}, \quad j_y = \begin{pmatrix} l_y & -i\hbar/2 \\ i\hbar/2 & l_y \end{pmatrix}, \quad l_z = \begin{pmatrix} l_z + \hbar/2 & 0 \\ 0 & l_z - \hbar/2 \end{pmatrix}. \quad (5.91)$$

This gives

$$j^2 = \begin{pmatrix} l^2 + l_z\hbar + 3\hbar^2/4 & (l_x - il_y)\hbar \\ (l_x + il_y)\hbar & l^2 - l_z\hbar + 3\hbar^2/4 \end{pmatrix}. \quad (5.92)$$

The operators j_z, j^2, l_z and l^2 commute with each other and hence possess a common set of eigenfunctions. They have the eigenvalues

$$l^2 \; : \; l(l+1)\hbar^2 \text{ or } l : l^*\hbar, \; l = 0, 1, 2, \cdots;$$

$$l_z \; : \; m\hbar, \; m = 0, \pm 1, \pm 2, \cdots, \pm l;$$

$$s_z \; : \; \pm\hbar/2;$$

$$j^2 \; : \; j(j+1)\hbar^2 \text{ or } j : j^*\hbar, j = l \pm 1/2, j > 0.$$

Thus, each level for $l \neq 0$ (p, d, f, etc) splits into two, except for the s level for which we have only $j = 1/2$. The different angular momenta are

$$l = l^*\hbar = [l(l+1)]^{1/2}\hbar,$$

$$s = s^*\hbar = [s(s+1)]^{1/2} = \sqrt{3}\hbar/2,$$

$$j = j^*\hbar = [j(j+1)]^{1/2}\hbar,$$

which defines l^*, s^* and j^*.

The wavefunction of electrons now has two parts, the orbital part ψ_{nlm} and spin part α or β. They operate in two different domains.

Hydrogen energy levels

Consider atoms which have only one electron in the outermost shell. The orbital motion of the electron has a magnetic moment $\mu_l = (e\hbar/2mc)l^*$ while the spin contributes a magnetic moment $\mu_s = (e\hbar/mc)s^*$. The mutual effect of these gives rise to spin–orbit interaction which causes a shift in the energy levels. According to Pauli's theory, it is given by

$$\Delta E_{ls}(n, l, j = l + 1/2) = \frac{|E_0|\alpha^2 Z^4}{n^3(l+1)(2l+1)},$$

$$\Delta E_{ls}(n, l, j = l - 1/2) = -\frac{|E_0|\alpha^2 Z^4}{n^3 l(2l+1)}.$$

Thus each (n, l) level splits into two, except the s level. For example, the p level splits into $p_{1/2}$ and $p_{3/2}$, the d level into $d_{3/2}$ and $d_{5/2}$, etc. Adding to this Sommerfeld's relativistic correction

$$\Delta E_r = -\frac{|E_0|\alpha^2 Z^4}{n^4}\left(\frac{n}{l+1/2} - \frac{3}{4}\right),$$

we get

$$\Delta E(n, l, j = l \pm 1/2) = -\frac{|E_0|\alpha^2 Z^4}{n^4}\left(\frac{n}{j+1/2} - \frac{3}{4}\right). \tag{5.93}$$

Thus the energy for $s_{1/2}$ state ($l = 0, m_s = 1/2, j = 1/2$) and $p_{1/2}$ state ($l = 1, m_s = -1/2, j = 1/2$) or $p_{3/2}$ state ($l = 1, m_s = 1/2, j = 3/2$) and $d_{3/2}$ ($l = 2, m_s = -1/2, j = 3/2$) are identical. Hence the selection rules $\Delta l = \pm 1, \Delta j = 0, \pm 1$ give five components for H_α line which were indicated in Fig. 5.5 in another context. In the present context, the three $n = 3$ levels have term values $d_{5/2}$, ($p_{3/2}$ and $d_{3/2}$) and ($s_{1/2}$ and $p_{1/2}$), while the two $n = 2$ levels correspond to $p_{3/2}$ and ($s_{1/2}$ and $p_{1/2}$), respectively. These split levels form the fine structure of H_α line; that is why α is known as the fine structure constant. As we shall see, $k = l + 1/2$ forms a new quantum number in Dirac's theory which is the subject matter of the next sub-section. But the energy for $\pm k$ is the same, hence the relativistic energy obtained in Section 5.5.5 becomes

$$E = m_0 c^2\left[1 + \frac{Z^2\alpha^2}{\{n - |k| + \sqrt{k^2 + \alpha^2}\}^2}\right]^{-1/2}.$$

5.6.2 Dirac's theory of relativistic electron

In this section, we shall discuss Dirac's modification of Schrödinger equation for a relativistic particle. Then we discuss Dirac's introduction of α and Σ matrices, the emergence of spin angular momentum and the energy levels.

Dirac equation

This Schrödinger equation was developed for nonrelativistic particle in the Newtonian framework of space and time. While it could explain the potential steps, potential hills and the wells, the harmonic oscillator and the hydrogen atom, it could not explain electron spin as proposed by Uhlenbeck and Goudsmid and taken further by Klein and Gordon and Pauli. The theory of electron spin developed by the latter three was, in a sense, an adjunct to Schrödinger equation.

Dirac developed a completely relativistic theory of the electron, and spin angular momentum emerged as a consequence of this. Dirac noticed that Schrödinger equation contained second space derivative but first time derivative. He wanted to put space and time on the same footing in accordance with special relativity. According to it, we have

$$E - V = \frac{m_0 c^2}{(1 - \beta^2)^{1/2}}.$$

This can be recast in the form

$$E = V + c[p_x^2 + p_y^2 + p_z^2 + m_0^2 c^2]^{1/2}, \tag{5.94}$$

where we have retained only the positive square root for consistency. Dirac rationalised the square root on the right hand side and put it in the form

$$(p_x^2 + p_y^2 + p_z^2 + m_0^2 c^2)^{1/2} = \alpha_x p_x + \alpha_p \alpha_y + \alpha_z p_z + \alpha_t m_0 c, \tag{5.95}$$

where α_μ, $\mu = x, y, z, t$ are entities to be determined. On squaring both sides of (5.95), equating coefficients of respective terms and noting that p_x, p_y, p_z commute with each other, it can be seen that the αs satisfy

$$\alpha_\mu^2 = 1,$$

$$\{\alpha_\mu, \alpha_\nu\} = 0, \quad \mu, \nu = x, y, z, t. \tag{5.96}$$

Dirac matrices and relativistic Schrödinger equation

Since the αs anticommute with each other, they cannot be algebraic numbers. We next try matrices. We need four mutually anticommuting matrices. It can be seen that if we consider 2×2 matrices, there can be at most three mutually anticommuting matrices. In fact, three such matrices are taken as Pauli spin matrices, as we have seen in Section 5.6.1. Thus αs must be matrices of order greater than 2. It can be further seen[17] that 3×3 matrices cannot satisfy (5.96). So αs have to be at least 4×4 matrices. Because of anticommutation, at most one of them can be a diagonal matrix in any representation. Let α_t be the diagonal matrix. Then $\alpha_t^2 = 1$, where 1 is the 4×4 unit matrix I, shows that the diagonal elements must be ± 1, not all of the same sign. After a bit of matrix algebra and using our experience with Pauli spin matrices, we finally arrive at the following four α matrices in a representation which α_t is diagonal:

$$\alpha_t = \begin{bmatrix} I & 0 \\ 0 & -I \end{bmatrix} = \begin{bmatrix} 1 & 0 & 0 & 0 \\ 0 & 1 & 0 & 0 \\ 0 & 0 & -1 & 0 \\ 0 & 0 & 0 & -1 \end{bmatrix},$$

---

[17]See Joshi, A. W. (1995).

$$\alpha_x = \begin{bmatrix} 0 & \sigma_x \\ \sigma_x & 0 \end{bmatrix} = \begin{bmatrix} 0 & 0 & 0 & 1 \\ 0 & 0 & 1 & 0 \\ 0 & 1 & 0 & 0 \\ 1 & 0 & 0 & 0 \end{bmatrix},$$

$$\alpha_y = \begin{bmatrix} 0 & \sigma_y \\ \sigma_y & 0 \end{bmatrix} = \begin{bmatrix} 0 & 0 & 0 & -i \\ 0 & 0 & i & 0 \\ 0 & -i & 0 & 0 \\ i & 0 & 0 & 0 \end{bmatrix},$$

$$\alpha_z = \begin{bmatrix} 0 & \sigma_z \\ \sigma_z & 0 \end{bmatrix} = \begin{bmatrix} 0 & 0 & 1 & 0 \\ 0 & 0 & 0 & -1 \\ 1 & 0 & 0 & 0 \\ 0 & -1 & 0 & 0 \end{bmatrix}. \tag{5.97}$$

These four are known as *Dirac matrices*.

Then we can write the *Dirac Hamiltonian*, using (5.94) and (5.96), as

$$H = V + c(\alpha_x p_x + \alpha_y p_y + \alpha_z p_z + \alpha_t m_0 c), \tag{5.98}$$

and the corresponding relativistic Schrödinger equation for a relativistic particle as

$$[V + c(\alpha_x p_x + \alpha_y p_y + \alpha_z p_z + \alpha_t m_0 c)]\psi = E\psi. \tag{5.99}$$

Since αs are 4×4 matrices, the wavefunction ψ must be understood as a 4×1 column vector with components $\psi_{\dot\mu}, 1 \le \dot\mu \le 4$. Thus, we take

$$\psi = \begin{bmatrix} \psi_1 \\ \psi_2 \\ \psi_3 \\ \psi_4 \end{bmatrix}, \quad \psi^+ = [\psi_1^* \ \psi_2^* \ \psi_3^* \ \psi_4^*]; \tag{5.100}$$

ψ would be a normalised wavefunction if $\sum_\mu \psi_\mu^* \mu_\mu = 1$.

Equation (5.99) is then equivalent to the four coupled equations

$$(E - V - m_0 c^2)\psi_1 - c p_z \psi_3 - (p_x - i p_y)\psi_4 = 0,$$

$$(E - V - m_0 c^2)\psi_2 - c(p_x + i p_y)\psi_3 + c p_z \psi_4 = 0,$$

$$(E - V - m_0 c^2)\psi_3 - c p_z \psi_1 - c(p_x - i p_y)\psi_2 = 0,$$

$$(E - V - m_0 c^2)\psi_4 - c(p_x + i p_y)\psi_1 + c p_x \psi_3 = 0. \tag{5.101}$$

These are linear homogeneous equations and nontrivial solutions for ψ_i exist if and only if the determinant of coefficients vanishes. It gives $E = V \pm (c^2 \mathbf{p}^2 + m_0^2 c^4)^{1/2}$, which is the relativistic energy consisting of potential, kinetic and rest energies.

Existence of spin

Let us recast the Dirac Hamiltonian by defining new 4×4 matrices using 2×2 Pauli spin matrices, unit matrix and null matrix, as

$$\Sigma_x = \begin{bmatrix} \sigma_x & 0 \\ 0 & \sigma_x \end{bmatrix} = \begin{bmatrix} 0 & 1 & 0 & 0 \\ 1 & 0 & 0 & 0 \\ 0 & 0 & 0 & 1 \\ 0 & 0 & 1 & 0 \end{bmatrix},$$

$$\Sigma_y = \begin{bmatrix} \sigma_y & 0 \\ 0 & \sigma_y \end{bmatrix} = \begin{bmatrix} 0 & -i & 0 & 0 \\ i & 0 & 0 & -1 \\ 0 & 0 & 0 & -i \\ 0 & 0 & i & 0 \end{bmatrix},$$

$$\Sigma_z = \begin{bmatrix} \sigma_z & 0 \\ 0 & \sigma_z \end{bmatrix} = \begin{bmatrix} 1 & 0 & 0 & 0 \\ 0 & -1 & 0 & 0 \\ 0 & 0 & 1 & 0 \\ 0 & 0 & 0 & -1 \end{bmatrix}. \tag{5.102}$$

and

$$\rho = \begin{bmatrix} 0 & I \\ I & 0 \end{bmatrix} = \begin{bmatrix} 0 & 0 & 1 & 0 \\ 0 & 0 & 0 & 1 \\ 1 & 0 & 0 & 0 \\ 0 & 1 & 0 & 0 \end{bmatrix}. \tag{5.103}$$

The spin operators are related to the Σ matrices in the same manner as to the Pauli spin matrices, that is,

$$s_x = \frac{1}{2}\hbar\Sigma_x, \quad s_y = \frac{1}{2}\hbar\Sigma_y, \quad s_z = \frac{1}{2}\hbar\Sigma_z. \tag{5.104}$$

The α matrices can be written as

$$\alpha_x = \rho\Sigma_x, \quad \alpha_y = \rho\Sigma_y, \quad \alpha_z = \rho\Sigma_z. \tag{5.105}$$

It can be seen that the Σ matrices anticommute with each other but they commute with ρ. Also

$$\Sigma_i^2 = i, \quad i = x, y, z;$$

$$\Sigma_x\Sigma_y - \Sigma_y\Sigma_x = 2i\Sigma_z, \text{ and cyclic permutation of } x, y, z. \tag{5.106}$$

Now, in the context of Dirac equation, we can treat the spin matrices s_x, s_y, s_z as four-dimensional matrices defined by

$$s_x = \frac{1}{2}\hbar\Sigma_x, \quad s_y = \frac{1}{2}\hbar\Sigma_y, \quad s_z = \frac{1}{2}\hbar\Sigma_z. \tag{5.107}$$

In the non-relativistic case, we have seen that l^2 and l_z are constants of motion. This is not the case with the Dirac Hamiltonian, as we will see through a guided exercise.

▶ **Guided Exercise 5.2**   Show that l_z is not a constant of motion for the Dirac relativistic Hamiltonian, but $j_z = l_z + s_z$ is. Also show that $[j^2, H] = 0$.

Hints

(a) With $l_z = -\hbar(x\partial/\partial y - y\partial/\partial x)$, show that $[l_z, H] = -i\hbar c\rho(\Sigma_y p_x - \Sigma_x p_y) \neq 0$, where H is given by (5.98).

(b) Show that $[\Sigma_z, H] = 2ic\rho(\Sigma_y p_x - \Sigma_x p_y)$.

(c) Therefore show that

$$\left[\left(L_z + \frac{1}{2}\hbar\Sigma_z\right), H\right] = 0.$$

(d) The above equation also shows that $[j_z^2, H] = 0$. Now show that $[j_x^2 + j_y^2, H] = 0$. Thus,

$$[j_z, H] = 0, \quad [j^2, H] = 0. \tag{5.108}$$

◀

Thus $l_z + s_z$ is a constant of motion. Hence $\frac{1}{2}\hbar\Sigma_z = s_z$ can be considered to be the spin contribution to the total angular momentum of the particle in the z-direction. It arises naturally as the fourth degree of freedom in the four-vector space.

Hydrogen energy levels

It turns out that there is another operator which is also a constant of motion for the Dirac Hamiltonian. It is

$$K = \alpha_t\left(\frac{2}{\hbar}\hbar^2\, \mathbf{l}\cdot\mathbf{s} + 1\right). \tag{5.109}$$

On writing $\mathbf{l}\cdot\mathbf{s}$ as $l_x s_x + l_y s_y + l_z s_z$, and using $s_x = \frac{1}{2}\hbar\Sigma_x$, etc., we see that

$$K = \frac{1}{\hbar} \begin{bmatrix} l_z + \hbar & l_x - il_y & 0 & 0 \\ l_x + il_y & -l_z + \hbar & 0 & 0 \\ 0 & 0 & -l_z - \hbar & l_x - il_y \\ 0 & 0 & l_x + il_y & l_z - \hbar \end{bmatrix}. \tag{5.110}$$

The eigenvalues of K are found to be $k = \pm1, \pm2, \cdots$, etc. Then the energy can be found out after a somewhat lengthy derivation, which we skip here. It is seen that we get

$$E = m_0 c^2 \left[1 + \left(\frac{Z^2 \alpha^2}{n - |k| + \sqrt{k^2 - Z^2 \alpha^4}} \right)^2 \right]^{-1/2}$$

and

$$\Delta E_n = -\frac{|E_0| \alpha^2 Z^4}{n^4} \left(\frac{n}{|k|} - \frac{3}{4} \right). \tag{5.111}$$

5.6.3   Dirac's theory of free electron

We now write the time-dependent Dirac equation by replacing observables E and $\mathbf{p}$ in (5.99) by their corresponding operators in the form

$$i\hbar \frac{\partial \psi}{\partial t} = [-ic\hbar\rho(\mathbf{\Sigma} \cdot \mathbf{\nabla}) + \alpha_t m_0 c^2 + V]\psi, \tag{5.112}$$

where $\mathbf{\Sigma}$ stands for the triplet of Σ matrices. We now further define new matrices by

$$\gamma_x = -i\alpha_t \rho \Sigma_x = \begin{bmatrix} 0 & 0 & 0 & -i \\ 0 & 0 & -i & 0 \\ 0 & i & 0 & 0 \\ i & 0 & 0 & 0 \end{bmatrix},$$

$$\gamma_y = -i\alpha_t \rho \Sigma_y = \begin{bmatrix} 0 & 0 & 0 & -1 \\ 0 & 0 & 1 & 0 \\ 0 & 1 & 0 & 0 \\ -1 & 0 & 0 & 0 \end{bmatrix},$$

$$\gamma_z = -i\alpha_t \rho \Sigma_z = \begin{bmatrix} 0 & 0 & -i & 0 \\ 0 & 0 & 0 & i \\ i & 0 & 0 & 0 \\ 0 & -i & 0 & 0 \end{bmatrix},$$

$$\gamma_t = \alpha_t. \tag{5.113}$$

With this the time-dependent equation can be written as

$$\left[\gamma_x \frac{\partial}{\partial x} + \gamma_y \frac{\partial}{\partial y} + \gamma_z \frac{\partial}{\partial z} + \gamma_t \frac{\partial}{\partial(ict)}\right]\psi + \frac{m_0 c}{\hbar}\psi + V\psi = 0.$$

For a free particle, $V = 0$. So we finally have

$$(\gamma \cdot \square)\psi + \frac{m_0 c}{\hbar}\psi = 0, \tag{5.114}$$

where γ stands for the four components $(\gamma_x, \gamma_y, \gamma_z, \gamma_t)$ and $\square$ is the four-vector gradient operator. We now consider a few consequences of this equation.

Electron at rest

In this case, we have $p \to 0$ and $E \to \pm m_0 c^2$. The above equation reduces to

$$\gamma_t \frac{\partial \psi}{\partial(ict)} = -\frac{m_0 c}{\hbar}\psi, \tag{5.115}$$

which gives the solution

$$\psi = e^{-iEt/\hbar} = \exp(\mp i m_0 c^2 t/\hbar). \tag{5.116}$$

Substituting this back in (5.116), we see that

$$\gamma_t(\mp m_0 c/\hbar)\psi = -(m_0 c/\hbar)\psi. \tag{5.117}$$

Realising that ψ is a column vector having four components as in (5.100), we get different sets of solutions depending on whether $E > 0$ or $E < 0$. It can be seen that

for $E > 0$, $\psi_3 = \psi_4 = 0$;

for $E < 0$, $\psi_1 = \psi_2 = 0$. $\tag{5.118}$

For $E > 0$, we have the two independent solutions:

$$\psi_\alpha = \begin{bmatrix} 1 \\ 0 \\ 0 \\ 0 \end{bmatrix} e^{-iEt/\hbar},$$

$$\psi_\beta = \begin{bmatrix} 0 \\ 1 \\ 0 \\ 0 \end{bmatrix} e^{-iEt/\hbar}. \tag{5.119}$$

In other words, ψ_α is the spin-up wavefunction $\alpha = \begin{bmatrix} 1 \\ 0 \end{bmatrix}$ and ψ_β is the spin-down wavefunction $\beta = \begin{bmatrix} 0 \\ 1 \end{bmatrix}$. We see that for $E < 0$, the two independent solutions are $\psi_3 = \alpha$ and $\psi_4 = \beta$. The negative energy solutions represent antiparticles, as was realised after the discovery of positron in 1932.

Plane wave solutions

We have seen in Section 5.4 that a free particle travelling in a direction **r** with momentum **p** is represented by a wavefunction $\psi(\mathbf{r}) = \exp(i\mathbf{p} \cdot \mathbf{r}/\hbar)$. On combining time dependence, we can write

$$\psi(\mathbf{r}, t) = \exp[i(\mathbf{p} \cdot \mathbf{r} - Et)/\hbar].$$

So Dirac' equation for a free particle becomes

$$(i\gamma_x p_x + i\gamma_y p_y + i\gamma_z p_z - \gamma_t E/c)\psi = -m_0 c\psi. \tag{5.120}$$

On using the respective matrix representations, we obtain four equations similar to Eqs. (5.102) with $V = 0$. It gives rise to four independent solutions:

$$\psi_A \text{ with } \psi_1 = 1, \psi_2 = 0, \psi_3 = \frac{p_z c}{E + m_0 c^2}, \psi_4 = \frac{(p_x + ip_y)c}{E + m_0 c^2};$$

$$\psi_B \text{ with } \psi_1 = 0, \psi_2 = 1, \psi_3 = \frac{(p_x - ip_y)c}{E + m_0 c^2}, \psi_4 = \frac{-p_z c}{E + m_0 c^2};$$

$$\psi_C \text{ with } \psi_3 = 1, \psi_4 = 0, \psi_1 = \frac{-p_z c}{-E + m_0 c^2}, \psi_4 = \frac{-(p_x + ip_y)c}{-E + m_0 c^2};$$

$$\psi_D \text{ with } \psi_3 = 0, \psi_4 = 1, \psi_1 = \frac{-(p_x - ip_y)c}{-E + m_0 c^2}, \psi_2 = \frac{p_z c}{-E + m_0 c^2}.$$

All of them have the normalisation factor N such that

$$N = [(|E| + m_0 c^2)/2|E|]^{1/2}.$$

Thus the four independent solutions can be written in the explicit matrix form

$$
\psi_A = N \begin{bmatrix} 1 \\ 0 \\ p_z c/(E + m_0 c^2) \\ [(p_x + ip_y)c/(E + m_0 c^2)] \end{bmatrix} \exp[i(\mathbf{p} \cdot \mathbf{r} - Et)/\hbar];
$$

$$
\psi_B = N \begin{bmatrix} 0 \\ 1 \\ -(p_x - ip_y)c/(E + m_0 c^2) \\ -p_z c/(E + m_0 c^2) \end{bmatrix} \exp[i(\mathbf{p} \cdot \mathbf{r} - Et)/\hbar];
$$

$$
\psi_C = N \begin{bmatrix} -p_z c/(-E + m_0 c^2) \\ -(p_x + ip_y)c/(-E + m_0 c^2) \\ 1 \\ 0 \end{bmatrix} \exp[i(\mathbf{p} \cdot \mathbf{r} - Et)/\hbar];
$$

$$
\psi_D = N \begin{bmatrix} -(p_x - ip_y)c/(-E + m_0 c^2) \\ p_z c/(-E + m_0 c^2) \\ 0 \\ 1 \end{bmatrix} \exp[i(\mathbf{p} \cdot \mathbf{r} - Et)/\hbar].
$$

Solutions ψ_A and ψ_B hold good for $E > 0$. Since the minimum value of E for a free particle is $m_0 c^2$, we have solutions for $m_0 c^2 \leq E < \infty$. ψ_A represents a right-handed plane wave in which the spin angular momentum of the electron is parallel to the direction of propagation of the wave and ψ_B represents the left-handed plane wave in which the spin angular momentum is antiparallel to the direction of motion. This phenomenon is similar to the right circular and left circular polarizations of a photon beam.

Solutions ψ_C and ψ_D hold good for negative values of E, $-m_0 c^2 \geq E > -\infty$. They represent plane waves for antiparticles. Again we have right handed and left handed beams of antiparticles.

5.7 Spin Angular Momentum in Other Atoms

We have discussed above spin angular momentum in hydrogen-like atoms, its initial formulation by Pauli, its relativistic formulation by Dirac, and the similarity of the resulting energy levels with Sommerfeld's expressions. We shall now discuss electron spin in other atoms, beginning with alkali atoms and then go on to two-electron and many-electron atoms.

5.7.1 Alkali atoms

In the case of alkali atoms and other one-electron atoms, the outermost electron spends most of the time outside the filled shells, though it also penetrates the inner shells to some extent. Talking in terms of the wavefunction, the outermost electron will have a major fraction of probability outside the filled shell electrons (due to orthogonality of wavefunctions) and a small probability of being in the inner regions. Therefore we have to replace the factor of Z^4 by Z_{eff}^4, where Z_{eff} is an effective value of nuclear charge as seen by the outermost electron. If we neglect the penetration of the outermost electron in inner shells, then we have $Z_{\text{eff}} = 1$ because the nucleus will be shielded by $Z - 1$ filled-shell electrons. But in general, $Z_{\text{eff}} > 1$. Then the spin–orbit interaction and relativity corrections for $l > 0$, give

$$\Delta E(n, l, j = l + 1/2) = -\frac{|E_0|\alpha^2 Z_{\text{eff}}^4}{n^4}\left(\frac{1}{l+1} - \frac{3}{4}\right),$$

$$\Delta E(n, l, j = l - 1/2) = -\frac{|E_0|\alpha^2 (Z_{\text{eff}}^4}{n^4}\left(\frac{1}{l} - \frac{3}{4}\right). \tag{5.122}$$

Therefore the doublet separation, units of wavenumbers (for $l \neq 0$), would be

$$\Delta\sigma = \frac{|E_0|\alpha^2 Z_{\text{eff}}^4}{n^3 hcl(l+1)}. \tag{5.122}$$

Here Z_{eff} could have a value between 1 and Z.

The doublet separation increases with Z, that is, from H to Li to Na, and it decreases as we go from s to p to d, etc., because Z_{eff} becomes smaller and smaller. Hence $p_{3/2}$ does not coincide with $d_{3/2}$, and so on, as happens in hydrogen-like atoms. Thus the energy diagram takes a form as shown in Fig. 5.16.

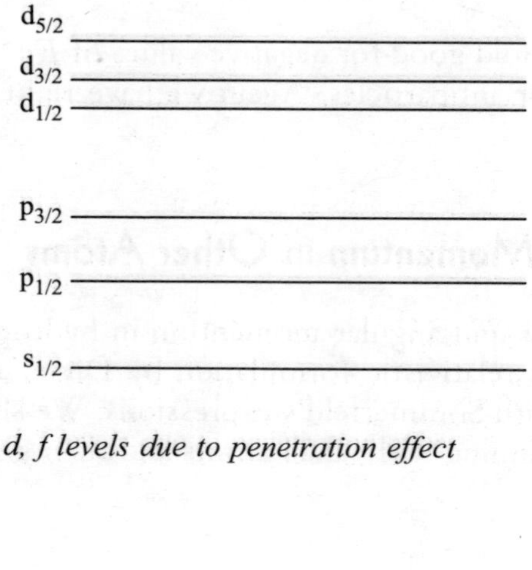

Fig. 5.16 *Splitting of p, d, f levels due to penetration effect*

5.7.2 Two- and many-electron systems

Pauli's exclusion principle

With the above developments in theory and experiments coming one after another in quick succession, the enigmatic electronic structure of atoms was explained by Pauli by making a hypothesis — every electron in an atom is characterised by a set of four quantum numbers n, l, j, m_j and *no two electrons can have the same set of quantum numbers*. Thus as we build up atoms of the periodic table by filling in electrons, each electron has to go to a state which differs from all the previous electrons by at least one quantum number, even if it has to go to a higher energy state — and this is the ground state of the atom, at absolute zero of temperature. Thus the Pauli exclusion principle, as it is called, explains the electronic structure of atoms and the periodic table at one stroke.

Now we can find the number of electrons that can be accommodated in an orbital with a given principal quantum number n. We have seen that n takes values $0, 1, 2, \cdots$; then for each n, l can take values $0, 1, 2, \cdots, n - 1$; for each l, j takes values $l - 1/2$ and $l + 1/2$, except for $l = 0$ when j takes the only value $1/2$; and for each j, m_j can take values from $-j$ to j in integral steps. The number of electrons that can be accommodated in each l state is shown in Table 5.5. Therefore the total number of states in a given level n is

$$2 + 6 + 10 + \cdots + (4n - 2) = 2n^2.$$

Table 5.5   *Number of electrons in various l states*

l	0		1		2	$\cdots$	$n - 1$	
j	1/2	1/2	3/2	3/2	5/2	$\cdots$	$n - 3/2$	$n - 1/2$
m_j	$\pm 1/2$	$\pm 1/2$	$\pm 3/2$	$\pm 3/2$	$\pm 1/2$	$\cdots$	$\pm 1/2$	$\pm 1/2$
					$\pm 3/2$		$\pm 3/2$	$\pm 3/2$
					$\pm 5/2$		$\vdots$	$\vdots$
							$\pm(n - 3/2)$	$\pm(n - 1/2)$
Number of states	2		2 + 4 = 6		4 + 6 = 10	$\cdots$	$2n - 2 + 2n = 4n - 2$	

In spectroscopic notation, $n = 1, 2, 3$, etc shells are denoted by K, L, M, etc. Thus we see that the K, L, M, $\cdots$ shells have at most 2, 8, 18, etc electrons, respectively.

However, as we have often remarked, not all sub-levels belonging to a certain n are degenerate in energy. For larger values of n, the upper sub-levels for a certain n may have higher energy than the lower sub-levels of the next higher n. This happens due to various effects such as spin–orbit splitting, the different nature (and resulting overlap) of the s, p, d, etc., wavefunctions, etc. For example, the $Z = 18$ (argon) atom has an electronic structure $1s^2\, 2s^2 2p^6 3s^2 3p^6$. One would expect the next electron for $Z = 19$ (potassium) to go in the 3d shell (with $n = 3, l = 2$). But it goes into the 4s shell (with $n = 4, l = 0$) because the $4s^1$ structure happens to have lower energy than $3d^1$ structure.

Quantum mechanical interpretation

Let us consider an interchange operator $\mathcal{J}$ which interchanges two particles such as electrons. Mathematically $\mathcal{J}\psi_{12} = \psi_{21}$; then $\mathcal{J}^2\psi_{12} = \mathcal{J}\psi_{21} = \psi_{12}$. Thus $\mathcal{J}^2$ has eigenvalue 1 and $\mathcal{J}$ has eigenvalues ± 1. In one case, $\mathcal{J}\psi_{12} = \psi_{21} = \psi_{12}$, that is the interchange keeps the same sign as the original ψ which is thus a symmetric function. In the other case, $\mathcal{J}\psi_{12} = \psi_{21} = -\psi_{12}$; thus an interchange of two particles changes the sign of ψ, which is therefore an antisymmetric function.

A particle which has symmetric eigenfunction is called a *boson*; such particles obey Bose–Einstein statistics. Thus photons, deuterons, etc., which have integral spin (angular momentum, in units of $\hbar$) are bosons. A particle which has antisymmetric wavefunction is called a *fermion*; such particles obey Fermi–Dirac statistics. Electrons, protons, neutrons, etc., which have half integral spin (angular momentum in units of $\hbar$) are fermions. Pauli's exclusion principle applies to fermions and not to bosons. Several bosons may occupy the same quantum state simultaneously.

One may appreciate the import of Pauli's path breaking hypothesis if one looks a little at the developments around 1924–27. Bose realised that photons do not follow the classical Maxwell–Boltzmann statistics. He developed a new statistics which was further crystallised by Einstein. It is now known as Bose–Einstein statistics. Then the ideas of electron spin were developed and it was realised that they are different from photons. Then came Heisenberg's and Schrödingers quantum and wave mechanics, respectively. Thus hydrogen atom was dealt with from several angles, by Bohr, Sommerfeld, Heisenberg an Schrödinger, to name a few. But there was no proper theory for the many-electron atoms. It is at this juncture that Pauli enumerated his exclusion principle.

LS and JJ coupling

Pauli's principle, in the form that the total wavefunction must be antisymmetric in the interchange of two electrons, must be applied to a many-electron system to obtain the possible quantum states. In doing so, it is assumed that the wavefunction can be factorised into a space (orbital) part and a spin part. In other words the orbital angular momenta $l^* = (l(l+1))^{1/2}\hbar$ of individual electrons can be combined to give a resultant orbital angular momentum $L^* = (L(L+1))^{1/2}\hbar$, where $L = \Sigma m_{l_i}$. Similarly, the spin angular momenta can be combined to give a resultant spin angular momentum of the system $S^* = (S(S+1))^{1/2}\hbar$, where $S = \Sigma m_{s_i}$. The total angular momentum of the system $\mathbf{J}$ can then be obtained by considering all allowed vector combinations of $\mathbf{L}$

and $\mathbf{S}$ giving $\mathbf{L} + \mathbf{S}$, as shown in Fig. 5.17. This is known as Russell–Saunders or LS coupling. Here $\mathbf{L}$ and $\mathbf{S}$ precess around $\mathbf{J}$.

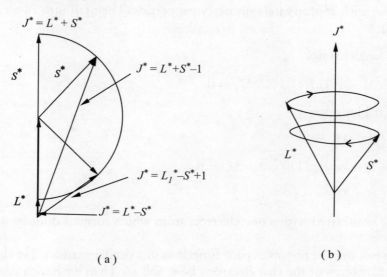

(a)

(b)

Fig. 5.17 *The LS coupling: (a) vector representation, (b) precession of* $\mathbf{L}$ *and* $\mathbf{S}$ *around* $\mathbf{J}$.

The wavefunction in this scheme can be written as $\psi = \psi_L \psi_S$, where ψ_L and ψ_S are the orbital and spin wavefunctions. Since it must be antisymmetric, we can have two situations – (a) space function ψ_L is symmetric and spin function ψ_S is antisymmetric, or (b) space function ψ_L is antisymmetric and spin function ψ_S is symmetric. Both functions cannot be simultaneously symmetric or antisymmetric. The total number of states is $2S + 1$, which is the multiplicity of a given L state.

JJ coupling

In the heavier atoms we have to first combine $\mathbf{l}_i$ and $\mathbf{s}_i$ of each electron to get $\mathbf{j}_i = \mathbf{l}_i + \mathbf{s}_i$ and then combine $\mathbf{j}_i$ to give $\mathbf{J} = \Sigma \mathbf{j}_i$. This is known as JJ coupling. Thus vectors $j_i^* = (j_i(j_i + 1))^{1/2}\hbar$ combine to give vector $J^* = (J(J + 1))^{1/2}\hbar$, and individual $\mathbf{j}_i$ precess around $\mathbf{J}$.

5.7.3 Eigenfunctions for a two-electron system

Here we will consider the possible spin and space functions of a two-electron atom. We shall label the two electrons as 1 and 2, and their spin functions as α and β, with $m_s = \pm 1/2$. Thus a function $\alpha(1)\alpha(2)$ would mean that both the electrons are in state α with $\Sigma m_s = 1$. The other possibilities are $\beta(1)\beta(2)$, with $\Sigma m_s = -1$, and $\alpha(1)\beta(2)$ and $\alpha(2)\beta(1)$, both with $\Sigma m_s = 0$. Among these $\alpha(1)\alpha(2)$ and $\beta(1)\beta(2)$ are symmetric because interchange of electrons does not change the eigenfunction. But $\alpha(1)\beta(2)$ and $\alpha(2)\beta(1)$ are neither symmetric nor antisymmetric. We can construct their linear combinations with appropriate symmetry properties. They fall into two categories as shown here.

$s = 0$, singlet state,

$$[\alpha(1)\beta(2) - \alpha(2)\beta(1)]/\sqrt{2}, M_s = 0; \tag{5.123a}$$

$s = 1$, triplet state,

$$\alpha(1)\alpha(2), \quad M_s = 1,$$

$$[\alpha(1)\beta(2) + \alpha(2)\beta(1)]/\sqrt{2}, \quad M_s = 0,$$

$$\beta(1)\beta(2), \quad M_s = -1. \tag{5.123b}$$

This may be contrasted with a one electron atom which forms a doublet with states α and β.

We can work out the possible space functions in a similar manner. Let the space part of the wavefunctions of the two electrons be ψ and ψ'. Then we have a symmetric and an antisymmetric linear combination which are

Symmetric wavefunction: $[\psi(1)\psi'(2) + \psi(2)\psi'(1)]/\sqrt{2}$;

Antisymmetric wavefunction: $[\psi(1)\psi'(2) - \psi(2)\psi'(1)]/\sqrt{2}$.

The symmetric space function combines with the antisymmetric spin function to produce a singlet, while the antisymmetric space function combines with the three symmetric spin functions and gives rise to a triplet state. It may happen that both the electrons have the same space wavefunction (same n, l, m_l), then the antisymmetric space function vanishes. The two electrons must differ in their spin projection m_s and thus we must have the antisymmetric spin function, giving rise to a singlet. The possible states for a two-electron system are enumerated in Table 5.6. All the space functions and all the spin functions are orthogonal to each other.

Table 5.6   *States of a two-electron system*

Space function	Spin function	M_s	Nature of state
$[\psi(1)\psi'(2) + \psi(2)\psi'(1)]/\sqrt{2}$	$[\alpha(1)\beta(2) - \alpha(2)\beta(1)]/\sqrt{2}$	0	Singlet
$[\psi(1)\psi'(2) - \psi(2)\psi'(1)]/\sqrt{2}$	$\alpha(1)\alpha(2)$	1	
	$[\alpha(1)\beta(2) + \alpha(2)\beta(1)]/\sqrt{2}$	0	Triplet
	$\beta(1)\beta(2)$	-1	

For example, in the ground state of He atom, both electrons have the space function with $n = 1, l = 0$, hence $m_l = 0$. Thus $\psi = \psi'$. So we have a singlet with $L = S = J = 0$. If one of the electrons goes to a higher state, giving rise to one of the several excited states of the atom, we get both singlet and triple states. Thus in the 1s 2s state of He, we get a singlet with $L = S = J = 0$ and a triplet with $L = 0, L = 1$ and $J = 0$ or 1.

5.7.4 General case of n electrons

In this section we shall discuss various aspects of many electron atoms, such as multiplicity, electron configuration, spectroscopic terminology, along with line intensities and selection rules for transitions.

Multiplicity

It would be convenient to represent the spin states of electrons by up and down arrows; thus ↑ for α state and ↓ for the β state. Then we list the possible combinations of spin states and the resulting multiplicities in Table 5.7.

Table 5.7 *Possible spin states of n electrons*

Number of electrons	Spin configuration	S	Multiplicity
1	↑	1/2	Doublet
2	↑ ↓	0	Singlet
	↑ ↑	1	Triplet
3	↑ ↓ ↑	1/2	Doublet
	↑ ↑ ↑	3/2	Qudruplet
4	↑ ↓ ↑ ↓	0	Singlet
	↑ ↓ ↑ ↑	1	Triplet
	↑ ↑ ↑ ↑	2	Quintet
n (odd)		$n/2$	Even
n (even)		$n/2$	Odd

Electron configuration

The electronic configuration of an atom is specified by the principal quantum number n, followed by the orbital quantum number l (through symbols s, p, d, etc) and a superscript indicating the number of electrons in that state. We give the electronic configuration of a few neutral atoms in Table 5.8.

Table 5.8 *Electronic structure of some neutral atoms*

Atom	Electronic configuration
H	$1s^1$
He	$1s^2$
B	$1s^2\,2s^2\,2p$
Na	$1s^2\,2s^2\,2p^63s^1$
Fe	$1s^2\,2s^22p^63s^23p^64s^23d^6$
Ni	$1s^2\,2s^2\,2p^6\,3s^23p^64s^23d^8$
Cu	$1s^2\,2s^2\,2p^6\,3s^23p^6\,3d^{10}4s^1$

It can be seen that all closed shells and sub-shells have $L = 0$ and $S = 0$. Hence the multiplicity is determined by the valence electrons in the outermost unfilled sub-shells only. An atom with a filled sub-shell has zero valence. In the first excited state of an atom, there is one less electron in the outermost sub-shell. The valence and the properties of such an ion charge accordingly.

Spectroscopic terminology

A spectroscopic terminology has been developed to indicate the various angular momenta and the multiplicity of an atom or ion. The L-value of a state is first indicated through a letter symbol. Thus $L = 0, 1, 2, \cdots$, etc are indicated by the symbol S, P, D, F, etc. Then the multiplicity $2S + 1$ of the state is indicated as a left numerical superscript. Finally the J-value, which goes from $|L - S|$ to $L + S$, is indicated as a right numerical subscript. The symbol thus takes the form $2S + 1$ L_J where L is replaced by the corresponding letter symbol and $2S + 1$ and J are given numerical values. Thus for $S = 1/2$ and $L = 1$, the symbol becomes $^2P_{1/2}$, $^2P_{3/2}$, for $S = 1, L = 2$, we may have atomic/ionic states $^3D_1$, $^3D_2$, and $^3D_3$, etc. This notation is known as the *spectroscopic term value* of a state.

We have earlier discussed the parity operator R which takes $\mathbf{r}$ to $-\mathbf{r}$. Thus $R\psi(\mathbf{r}) = \psi(-\mathbf{r})$. We have also seen that R has only two eigenvalues ± 1. The wavefunction is said to have even or odd parity according as $\psi(-\mathbf{r}) = \pm\psi(\mathbf{r})$. For hydrogenic wavefunctions, the parity depends on $(-1)^l$. Since the total wavefunction can be written as the product of individual electronic wavefunctions in the first approximation, the parity depends on whether Σl_i is even or odd. A P state with odd parity is denoted by a symbol such as P^0.

Selection rules for transitions

We shall discuss these is detail in Chapter 7. However, we may state here that the selection rules for transitions under electric dipole perturbation are $\Delta J = 0, \pm 1$, with $J = 0 \leftrightarrow J = 0$ forbidden, and $\Delta(\Sigma l_i) = \pm 1$, which means that only even $\leftrightarrow$ odd parity transitions are allowed. The latter is known as *Laporte rule*.

If electric dipole transitions are forbidden, higher order transitions may still occur, though with a weaker intensity. For these to occur, It is enough to have $\Delta L = \pm 1$ and $\Delta S = 0$. The last condition means that both states must have the same multiplicity. If an atomic system is held in a magnetic field, the selection rules for transitions depends on whether it is a weak or a strong magnetic field, that is, whether the magnetic energy is smaller or larger than other characteristic energies. For a weak magnetic field, the selection rule is $\Delta M_J = 0, \pm 1$, with $M_J = 0 \rightarrow M_J = 0$ forbidden, while for a strong magnetic field, it is $\Delta M_L = 0, \pm 1$ and $\Delta M_S = 0$.

The selection rules for electric dipole transitions described in the beginning are called *rigorous selection rules* because a transition between two states under the influence of electric dipole radiation cannot take place unless they are satisfied. Transitions between two states not satisfying these rules can still take place with weaker intensities. They are called *forbidden transitions*. They may take place under the influence of electric quadrupole, magnetic dipole and other higher order perturbations. The selection rules for such transitions are called *approximate* or *weaker selection rules*.

Line intensities

For electric dipole (allowed) transitions, these are governed by a sum rule which says: The sum of intensities of lines arising from a level with total angular momentum J is proportional to $2J + 1$. Similarly the sum of intensities of lines ending in a level with total angular momentum J is also proportional to $2J + 1$. Transitions from or to a level with quantum number J may have different L. For these specific cases, it turns out that for $(L - 1) \rightarrow L$ transitions, the intensity is proportional to

$$J - 1 \rightarrow J \quad : \quad I = \frac{(L+J+S+1)(L+J+S)(L+J-S)(L+J-S-1)}{J},$$

$$J \rightarrow J \quad : \quad I = \frac{(L+J+S+1)(L+J-S)(L-J+S)(L-J-S-1)(2J+1)}{J(J+1)},$$

$$J + 1 \rightarrow J \quad : \quad I = \frac{(L-J+S)(L-J+S-1)(L-J-S-1)(L-J-S-2)}{J+1}. \quad (5.124)$$

For $L \rightarrow L$ transitions the intensity is proportional to

$$J - 1 \rightarrow J \quad : \quad I = \frac{(L+J+S+1)(L+J-S)(L-J+S+1)(L-J-S)}{J},$$

$$J \rightarrow J \quad : \quad I = \frac{[L(L+1)+J(J+1)-S(S+1)]^2(2J+1)}{J(J+1)},$$

$$J + 1 \rightarrow J \quad : \quad I = \frac{(L+J+S+2)(L+J-S+1)(L-J+S)(L-J-S-1)}{J+1}. \quad (5.125)$$

Term splitting

The interaction between spin and orbital angular momenta causes splitting of energy levels resulting in the splitting of lines. The energy difference is proportional to the scalar product of the angular momenta. Thus for $J - J$ coupling of two electrons, we would have $\mathbf{J} = \mathbf{j}_1 + \mathbf{j}_2$ giving $\mathbf{j}_1 \cdot \mathbf{j}_2 = (J^2 - j_1^2 - j_2^2)/2$. Hence the energy splitting is

$$\Delta E_j \propto J^{*2} - j_1^{*2} - j_2^{*2}.$$

The corresponding splittings for orbital and spin angular momenta are

$$\Delta E_L \propto L^{*2} - l_1^{*2} - l_2^{*2},$$

$$\Delta E_S \propto S^{*2} - s_1^{*2} - s_2^{*2}.$$

In the case of $L - S$ coupling, we have

$$\Delta E_J \propto J^{*2} - l^{*2} - s^{*2}. \tag{5.126}$$

It is found that the spin–orbit splitting increases with L and hence S, P, D, F levels go up in that order. Secondly levels with higher multiplicity are below the lower multiplicity levels. Thus triplets lie below singlets, quartets lie below doublets, etc. Finally in a given multiplet, levels go up with increasing value of J.

5.7.5 Some typical spectra

We will now discuss some typical spectra such as those of one-electron systems, two-electron systems, etc.

One-electron systems

We have already discussed in Section 5.1 spectra of H and H-like atom such as He$^+$, Li^{2+} etc. The fine structure of H$_\alpha$ line was also discussed earlier. Neutral alkali atoms Na, K, etc, singly ionised Mg$^+$, Ca$^+$, etc, doubly ionised B^{2+} etc, triply ionised C^{3+} etc, are some species which have one valence electron in the s shell. They will have the principal, sharp, diffuse and fundamental series like Na as discussed in Section 5.2.

For such atoms, the ground state will have $L = 0, S = 1/2$ and the only term will be $^2S_{1/2}$. As the electron makes transitions to higher states with $L = 1, 2, 3, \cdots$, respectively, we will get doublet terms ($^2P^o_{1/2}$, $^2P^o_{3/2}$), ($^2D_{3/2}$, $^2D_{5/2}$), ($^2F_{5/2}$, $^2F_{7/2}$), etc. Transitions between them, with $\Delta L = 0, \Delta J = 0, \pm 1$, produce multiplets with two or three components as shown in Fig. 5.7. The two lines of the doublet differ very little in wavelength or energy and are called *resonance lines*. Resonance lines of some one-electron systems having the structure ns^1 are shown in Table 5.9. Their intensities are in the ration 1:2. The two lines of sodium are known as D_2 (5896Å) and D_1 (5890Å), respectively.

Table 5.9   *Resonance lines of alkali-like atoms (λ in Å)*

Transition	Na	K	Mg^+	Ca^+	Sr^+
$^2S_{1/2} - {^2P_{1/2}}$	5896	7699	2803	3968	4216
$^2S_{1/2} - {^2P_{3/2}}$	5890	7665	2796	3934	4018

Two-electron system of He

Neutral He and such other ions have a ground state with structure nS^2 and $L = S = 0$, giving a $^1S_0$ term. When one of the electrons makes a transition to a higher state, we get the following configurations and terms:

$$nsn's - L = 0, \quad S = 0, 1; \quad {^1S_0}, \; {^3S_1};$$

$$nsn'p - L = 1, \quad S = 0, 1; \quad {^1P_1^0}, \; {^3P_0^0}, \; {^3P_1^0}, \; {^3P_2^0};$$

$$nsn'd - L = 2, \quad S = 0, 1; \quad {^1D_2}, \; {^3D_1}, \; {^3D_2}, \; {^3D_3}; \; \text{etc.}$$

Selection rules allow transitions from ground state $ns^2({^1S_0})$ to $^1P_1$ state of $nsn'p$ excited state. In the case of He, they give rise to lines at 584 Å, 537 Å, 522 Å, etc, as shown in Fig. 5.18(a). The transition $^1S_0 \to {^3P_{0,1,2}}$ is forbidden, though it may occur with smaller probability due to higher order perturbations. Therefore the ^3P excited states with electronic structure $nsn'p$ are metastable states. Fig. 5.18(a) also shows the weaker 591Å line of the $^1S_0 \to {^3P}$ transition.

Some other transitions of neutral He are given here.

Singlet series:

Principal, $1s2s \to 1snp$; $^1S_0 \to {^1P_1^0}$: 20581, 5075, 3964, 3613 Å, etc;

Diffuse, $1s2p \to 1snd$; $^1P_0^0 \to {^1D_2}$: 6678, 4921, 4387, 4143, 3926 Å, etc;

Sharp, $1s2p \to 1sns$; $^1P_0^0 \to {^1S_0}$: 7281, 5047, 4437, 4168 Å, etc.

Triplet series :

Principal, $1s2s \rightarrow 1snp$; $^3S_1 \rightarrow {}^3P^0$: $10830, 3888, 3187$Å, etc;

Diffuse, $1s2p \rightarrow 1snd$; $^3P^0 \rightarrow {}^3D$: $5875, 4471, 4026, 3819$Å, etc;

Sharp, $1s2p \rightarrow 1sns$; $^3P \rightarrow {}^3S_1$: $7065, 4713, 4120, 3867$Å, etc.

These lines are shown in Fig. 5.18(b). The line 5875 Å belonging to the diffuse triplet occurs very close to the sodium D lines in the solar spectrum. It was discovered during a total solar eclipse and is known as the D_3 line. It was later found in the laboratory and characterised as belonging to the element helium.

Other two-electron systems

Neutral atoms Mg, Ca, Sr and ions Al^+, C^{2+} are some other species with two electrons in the s subshell in their ground state. Their spectra are similar to those of He with singlet and triplet lines. Some typical transitions and lines arising in Mg, Ca, Sr are listed in Table 5.10. The term level diagram and transition in Mg is shown in Fig. 5.19. It may be noted that the $^3P_0 \rightarrow {}^1S_0$ is a forbidden transition and hence weak. The last transition shown in Table 5.10 really gives rise to six lines with intensities shown in Table 5.11. We see that the sums of the intensities are in the ratio of $2J + 1$ for lines arising or ending in a given term. The wavelength of lines on the diagonal are almost identical; hence we get only three lines as shown in this table.

Table 5.10 *Some lines in the spectra of Mg, Ca and Sr* $(\lambda$ *in* Å$)$

Transition	Mg	Ca	Sr
$ns^2({}^1S_0) \rightarrow nsnp({}^3P^0_{1,2,3})^*$	4571	6572	6593
$ns^2({}^1S_0) \rightarrow nsnp({}^1P^0_1)$	2852	4227	4607
$nsn'p({}^3P_0) \rightarrow nsns({}^3S_1)$	5184	6162	7070
	5173	6122	6878
	5167	6102	6791
$nsn'p({}^3P_0) \rightarrow nsnd({}^3D)$	3838	4455	4962
	3832	4435	4872
	3829	4425	4832

* Forbidden, $n = 3$ for Mg, $n = 4$ for Ca, $n = 5$ for Sr.

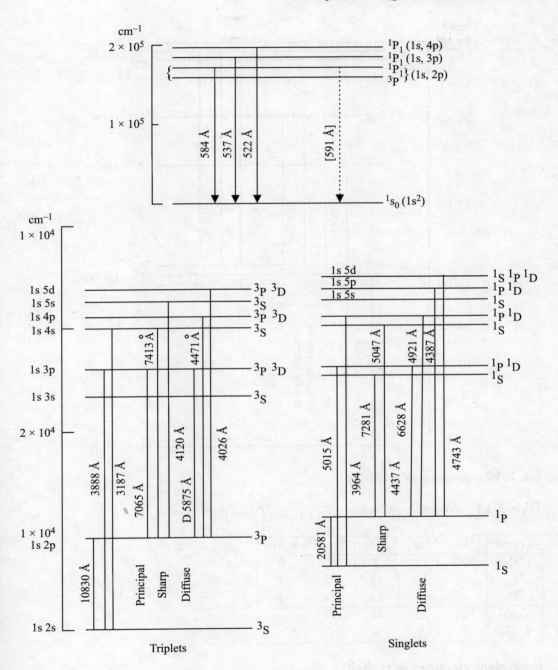

Fig. 5.18 *Spectrum of neutral He: (a) some transitions from the ground state; (b) the principal, sharp and diffuse series.*

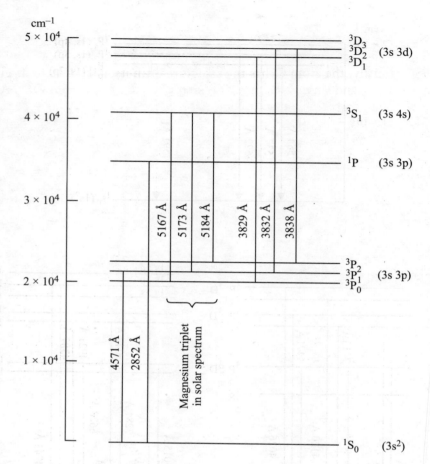

Fig. 5.19   *Spectrum of neutral Mg*

Table 5.11   *Intensities of lines of $^3P^0_{0,1,2} \rightarrow \, ^3D_{1,2,3}$ transitions in Mg*

Terms	$^3D_3$	$^3D_2$	$^3D_1$	Total	Ratio
$^3P_2$	100	18	2	120	5
$^3P_1$	×	54	18	72	3
$^3P_0$	×	×	24	24	1
Total	100	72	44		
Ratio	7	5	3		

Equivalent electrons in p shell

The ions N^+ and O^{2+} have two electrons in the p shell in their ground state. They give rise to triplet $^3P_{0,1,2}$ and singlets $^1S_0$ and $^1D_2$ terms. the atom/ion N and O^+

have three electrons in the p shell in their ground state. They give rise to doublets $^2P^0_{1/2}$, $^2P^0_{3/2}$ and $^2D^0_{3/2}$, $^2D^0_{5/2}$, as well as a quartlet $^4S_{3/2}$ terms, resulting in some weak spectral lines. Finally, the atom O has the structure p^4 in its ground state. It gives rise to the triplet $^3P_{0,1,2}$ and singlets $^1S_0$ and $^1D_2$ states, much like N^+ and O^{2+}. All these systems give rise to forbidden lines listed in Table 5.12.

Table 5.12   *Some forbidden lines of N, O and their ions* (λ *in* Å)

Transition	Intensity	System	
		N^+	O^{2+}
$^3P_{2,1} \rightarrow {}^1D_2$	5	6583	5007
	3	6548	4959
$^1D_2 \rightarrow {}^1S_0$		5755	4363
		N	O^+
$^4S_{3/2} \rightarrow {}^2D^0$	4	5200	3729
	6	5198	3726
$^4S_{3/2} \rightarrow {}^2P^0$		3966	2470
$^2D^0 \rightarrow {}^2P^0_{3/2}$	1	10396	7320
	2	10406	7330
			O
$^3P_{0,1,2} \rightarrow {}^1D_2$			6300
			6364
$^3P_{0,1,2} \rightarrow {}^1S_0$			5517
$^1D_2 \rightarrow {}^1S_0$			2972

5.7.6 Hyperfine structure

Just as an electron has a spin angular momentum $s = \hbar/2$, the nucleus too has a spin angular momentum of I in multiples of $\hbar/2$. This I combines with J of the electron to give F with total angular momentum $F^* = [F(F+1)]^{1/2}\hbar$. Here the vectors $I^* = [I(I+1)]^{1/2}\hbar$ and $J^* = [J(J+1)]^{1/2}\hbar$ precess around vectors F^*. The magnetic moment associated with a nucleus of spin angular momentum I is $\mu_N = g_I I^* e\hbar/2Mc$, where M is the mass of the nucleus, and g_I is the spectroscopic spilling factor. Thus the magnetic moment μ_N of a nucleus is about $1838A$ times smaller than that of an electron, where A is the nuclear mass number. Then the interaction energy becomes

$$dW \propto g_I(F^{*^2} - I^{*^2} - J^{*^2})/1838A.$$

For hydrogen H, $I = 1/2$, $g_I = 5$. As $J = S = 1/2$ in the ground state, we have $F = 0$ or 1. Then the selection rule $\Delta F = 0, \pm 1$ gives for the transition $F = 0$ to $F = 1$, the 21-cm line in radio region.

In general the number of levels is $2I + 1$ ranging from $J + I$ to $J - I$ if $I < J$ and $2J + 1$ ranging from $I + J$ to $I - J$ if $J < I$.

5.8 Zeeman and Stark Effects

A quantum mechanical system remains in a stationary state unless perturbed by some agency. A perturbation has the effect of changing the potential in which the system existed until the perturbation was applied. We consider in this section effects of perturbations caused by a static magnetic field, giving rise to Zeeman effect, and static electric field, giving rise to Stark effect. To begin with, we consider some basics of perturbation theory.

5.8.1 Perturbation theory

Suppose we have solved the quantum mechanical eigenvalue problem of a system which is represented by the unperturbed Hamiltonian H_0. This means that we know the eigenfunctions ψ_j^0 and the corresponding eigenvalues E_j^0. Then the system is perturbed and the perturbation is represented by a perturbing Hamiltonian H', so that the system is now represented by the Hamiltonian

$$H = H_0 + H'. \tag{5.127}$$

Let us denote the eigenfunctions of the new Hamiltonian by ψ_j. If $H' \ll H_0$, then the new eigenfunctions can be written as

$$\psi_j = \psi_j^0 + \psi_j'.$$

Since ψ_j' is also a function belonging to same Hilbert space of the eigenfunctions of H_0, it can be expanded as a linear combination in the form

$$\psi_j' = \sum_n a_{jn} \psi_n^0. \tag{5.128}$$

If ψ_j' is normalised to unity, then $\sum_n |a_{jn}|^2 = 1$. The new eigenfunction ψ_j corresponds to the eigenvalues $E_j = E_j^0 + E_j'$, where $E_j' \ll E_j^0$. Thus we have

$$(H_0 + H')(\psi_j^0 + \psi_j') = (E_j^0 + E_j')(\psi_j^0 + \psi_j'). \tag{5.129}$$

In this, H_0, ψ_j^0 and E_j^0 are zero-order quantities while H', ψ_j' and E_j' are first-order quantities. The zeroth approximation gives the unperturbed eigenvalue equation

$$H_0 \psi_j^0 = E_j^0 \psi_j^0. \tag{5.130}$$

Equating the first order terms in (5.128), along with (5.127), results in the equation

$$H'\psi_j^0 + \sum_n a_{jn}E_n\psi_n^0 = E_j'\psi_j^0 + E_j^0 \sum_n a_{jn}\psi_n^0. \tag{5.131}$$

Multiplying both sides of the equation by ψ_j^{0*} and integrating over the appropriate space, this results in

$$E_j' = \int \psi_j^{0*} H' \psi_j^0 d\tau \equiv H_{jj}', \tag{5.132}$$

which defines H_{jj}' as the (j,j) element of the perturbation. Thus the first order changes in the energy levels are equal to the corresponding diagonal elements of the perturbation matrix.

A perturbation, like a static magnetic field, may be time-independent and may be applied for a certain duration of time. On the other hand, the perturbation may itself be time-dependent, like radiation falling on the system. So the coefficients a_{jn} may depend on time and we would have to apply the time-dependent Schrödinger equation $H\psi = i\hbar\partial\psi/\partial t$. In that case, a_{jn} give the transition probabilities, which we shall discuss in Chapter 7.

In the case of relativistic systems, we must use the relativistic Hamiltonian including electromagnetic energies which, by (3.64), is

$$H = e\phi + m_0c^2 \left[1 + \frac{(\mathbf{p} - e\mathbf{A}/c)^2}{m_0^2c^2}\right]^{1/2} + V. \tag{5.133}$$

Since m_0c^2 is a constant, it can be left out for most considerations. In the absence of electric field, we will have $\varphi = 0$ and in the absence of magnetic field, we shall have $\mathbf{A} = 0$.

5.8.2 Zeeman effect

Let a steady magnetic field $|\mathbf{B}|$ be applied to a system. Let us take the direction of the magnetic field as the z-axis, and let it correspond to a vector potential $\mathbf{A}$. In the case when magnetic field energy is smaller than the rest mass energy, and in the absence of an electric field, the Hamiltonian can be written as

$$H = H_0 + H',$$

where

$$H_0 = m_0c^2 + p^2/2m_0 + V,$$

$$H' = -\frac{e}{m_0 c}\mathbf{p}\cdot\mathbf{A} + \frac{e^2}{2m_0 c^2}A^2. \tag{5.134a}$$

Here H_0 is the unperturbed Hamiltonian and H' the perturbation due to the magnetic field. In the case of atoms, we replace m_0 by the reduced mass μ and write

$$H' = -\frac{e}{\mu c}\mathbf{p}\cdot\mathbf{A} + \frac{e^2}{2\mu c^2}A^2. \tag{5.134b}$$

Perturbation in energy

The magnetic field is related to $\mathbf{A}$ by $|\mathbf{B}| = \nabla \times \mathbf{A}$. It would be convenient to choose the spherical coordinate system with the polar axis along the z-axis. Then in terms of unit vectors $\hat{\mathbf{r}}$, $\hat{\boldsymbol{\theta}}$, $\hat{\boldsymbol{\varphi}}$, we can write $\mathbf{p}$ and $\mathbf{B}$ as

$$\mathbf{p} = \mu\dot{r}\hat{\mathbf{r}} + \mu r\dot{\theta}\hat{\boldsymbol{\theta}} + \mu r\sin\theta\dot{\varphi}\hat{\boldsymbol{\varphi}},$$

$$\mathbf{B} = \frac{1}{r^2\sin\theta}\begin{vmatrix} \hat{\mathbf{r}} & r\hat{\boldsymbol{\theta}} & r\sin\theta\hat{\boldsymbol{\varphi}} \\ \partial/\partial r & \partial/\partial\theta & \partial/\partial\varphi \\ A_r & rA_\theta & r\sin\theta A_\varphi \end{vmatrix}. \tag{5.135}$$

Since $\mathbf{B}$ is along the z-axis, with magnitude $|\mathbf{B}|$, we may choose the vector potential of the form

$$A_r = A_\theta = 0, \quad A_\varphi = \frac{1}{2}|\mathbf{B}|r\sin\theta. \tag{5.136}$$

Using (5.134b) and (5.135), the perturbation becomes

$$H' = -\frac{e|\mathbf{B}|}{2c}r^2\sin^2\theta\dot{\varphi} = \frac{ie\hbar|\mathbf{B}|}{2\mu c}\frac{\partial}{\partial\varphi}. \tag{5.137}$$

For hydrogen atom, the eigenfunctions have the form given in (5.74b). Therefore

$$i\partial\psi_{nlm}/\partial\varphi = -m\psi_{nlm}.$$

Therefore the change in the energy of the level E_{nlm} due to the application of a magnetic field $\mathbf{B}$, according to (5.107), is found to be

$$E'_{nlm} = -e\hbar m|\mathbf{B}|/2\mu c. \tag{5.138}$$

We realise that $\mu_B = -e\hbar/2\mu c$ is the Bohr magneton and $-\mu_B m$ is the magnetic moment of the electron in the z-direction. If we represent the latter by μ^M, then the change in energy can be written as

$$E' = \mu^M|\mathbf{B}|. \tag{5.139}$$

(5.139) holds for all atoms.

Normal Zeeman effect

In singlet states, $S = 0$ and $J = L$, which effectively means that the effect of spin angular momentum is suppressed. Then the orbital moment and its z-component may be written as

$$|\mu_J^M| = [J(J+1)]^{1/2} e\hbar/2\mu c, \quad \mu_z^M = -m_J e\hbar/2\mu c.$$

In such a case the perturbation in energy level becomes

$$E' = \nu_L m_J h,$$

where $\nu_L = e|\mathbf{B}|/4\pi\mu c$

(5.140)

is the *Larsmor frequency* of precression of $\mathbf{J}^*$ around the magnetic field $\mathbf{B}$. Thus the energy level splits into $2J + 1$ states. Transitions between two levels give rise to lines with frequencies

$$\nu = dE/h = (\Delta E_0 + \Delta E')/h$$
$$= \nu_0 + \nu_L \Delta m_J.$$

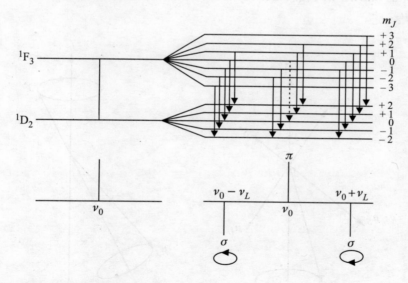

Fig. 5.20   *The normal Zeeman effect*

Let us consider the case of $^1D_2 - {}^1F_3$ transition. The two levels are split respectively in five ($-2 \leq m_J \leq 2$) and seven ($-3 \leq m_j \leq 3$) levels. The selection rule for transitions is $\Delta m_J = \pm 1, 0$ ($0 \rightarrow 0$ forbidden). So we get three sets of five or four transitions though each set has only one frequency. Thus we get three lines, at ν_0 corresponding to $\Delta m_J = 0$ and $\nu_0 \pm \nu_L$ corresponding to $\Delta m_J = \pm 1$. The central line at ν_0 is linearly polarized along the field, that is, the electron motion is along the magnetic field. It is visible when viewed at right angles to **B** but not in the longitudinal direction. It forms the π component of the Zeeman pattern. The lines at $\nu_0 \pm \nu_L$ are circularly polarized when viewed in the longitudinal direction and linearly polarised perpendicular to the field when viewed transversely. They are known as σ components of the Zeeman pattern. Here the electron has a circular motion around **B** with Larmor frequency, clockwise when $\Delta m_J = 1$ and anticlockwise when $\Delta m_J = -1$. The energy levels and transitions of this case are shown in Fig. 5.20.

Anomalous Zeeman effect

In non-singlet states, we have both orbital and spin angular momenta. They give rise to respective magnetic moments

$$|\mu_L| = [L(L+1)]^{1/2}(-e\hbar/2\mu c) = \mu_B L^*,$$

$$|\mu_S| = [S(S+1)]^{1/2}(-e\hbar/2\mu c) = \mu_B S^*.$$

As a result, μ_J is not parallel to J^* and it precesses around J^* while J^* itself precesses around **B**, as shown in Fig. 5.21. Hence we have to take its component along J^* to compute μ_z.

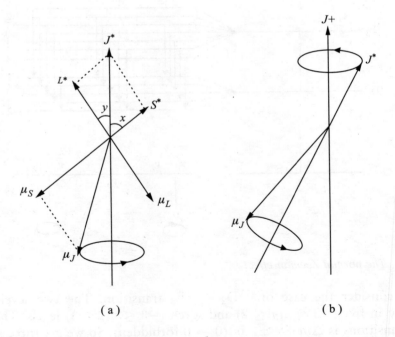

(a)   (b)

Fig. 5.21 *Angular and magnetic moment vectors. (a) μ_J precesses around J^*, (b) J^* precesses around* **B**.

Referring to Fig. 5.21, we have

$$\mu_J \cos(\mu_j, J^*) = \mu_S \cos x + \mu_L \cos y,$$

where the angles x, y that S^* and L^* respectively make with J^* are shown in Fig. 5.21(a). This works out to be

$$\mu_J \cos(\mu_J, J^*) = [2(S(S+1))^{1/2}\cos x + (L(L+1))^{1/2}\cos y]\mu_B$$

$$= [(J(J+1))^{1/2} + (S(S+1))^{1/2}\cos x]\mu_B$$

$$= (J(J+1))^{1/2}\left[1 + \frac{(S(S+1))^{1/2}}{(J(J+1))^{1/2}}\cos x\right]\mu_B.$$

Therefore we have to scale up the term splitting by the factor

$$g = 1 + \frac{(S(S+1))^{1/2}}{(J(J+1))^{1/2}}\cos x. \tag{5.141}$$

It is known as the *spectroscopic splitting factor* or *Lande g-factor*. Now from Fig. 5.21(a),

$$L(L+1) = S(S+1) + J(J+1) - 2[S(S+1)J(J+1)]^{1/2}\cos x.$$

Using this in (5.141), the Lange g-factor becomes

$$g = 1 + \frac{J(J+1) + S(S+1) - L(L+1)}{2J(J+1)}. \tag{5.142}$$

It may be noted that (i) when $S = 0$ and $J = L, g = 1$; that (ii) when $L = 0, J = S, g = 2$; and that (iii) when $L = S, g = 3/2$.

With this scaling factor for $S \neq 0$, the perturbation is a magnetic field and may be written as

$$E' = g\mu_B M_J|\mathbf{B}|.$$

It is seen to depend upon the quantum number M_J as well as L, S, J. Therefore different levels give rise to different splitting and we get more lines with $\nu = \nu_0 + \nu_L\Delta(gM_J)$.

To illustrate this case, we consider the transition $^3P_1 - {}^3S_1$ shown in Fig. 5.22. In this case, we have

$$^3P_1 : L = S = J = 1, \quad g = 3/2;$$

$$^3S_1 : L = 0, S = J = 1, \quad g = 2.$$

Thus we get two π components at $\nu_0 \pm \nu_L/2$ and four σ components at $\nu_0 \pm 3\nu_L/2$ and $\nu_0 \pm 2\nu_L$. The σ components on opposite sides of ν_0 are circularly polarised in opposite directions as shown. The relative intensities were calculated by Van Vleck and are found to be given by

$$J \to J \begin{cases} M_J \to M_j \pm 1: & I = A(J \pm M_J + 1)(J \mp M_J), \\ M_j \to M_J: & I = 4AM_j^2, \end{cases}$$

$$J \to J+1 \begin{cases} M_J \to M_J \pm 1: & I = B(J \pm M_J + 1)(J \pm M_J + 2), \\ M_J \to M_J: & I = 4B(J + M_J + 1)(J - M_J + 1), \end{cases} \tag{5.143}$$

where A and B are constants. In the case of π components the intensity is half that given by (5.118) in both transverse and longitudinal direction. Usually, the sum of the intensities of π components is equal to the sum of intensities of σ components. In the case under consideration, both π components have intensity $4A$ whereas each of the σ components has intensity $2A$.

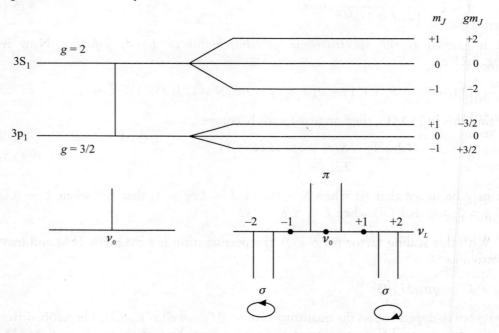

Fig. 5.22   *Anomalous Zeeman effect*

Paschen–Back effect

In strong magnetic fields, the coupling between L and S breaks down. Hence L^* and S^* precess independently around **B**. The total magnetic moment is then $M = M_L + 2M_S$, which goes from $L + 2S$ to $-L - 2S$. The energy is now

$$E = E_0 + M\nu_L h + aM_L M_S h.$$

The selection rules, $\Delta M_S = 0$, $\Delta M_L = 0, \pm 1$, then give three lines, one π component with $\Delta M_L = 0$ at ν_0 and two σ components for $\Delta M_L = \pm 1$ at $\nu_0 \pm \nu_L \Delta(gM)$. But the σ components have multiplicity $2S + 1$ due to residual spin–orbit interaction. This is known as Paschen–Back effect. The Lande g-factor is now given by

$$\text{Strong field;} \quad g = (M_L + 2M_S)/(M_L + M_S). \tag{5.144}$$

In a very strong field, all M_l and M_S get decoupled and we have $M = \Sigma M_l + 2\Sigma M_s$. Then we get *super Paschen–Back effect* in which we have again a triplet with π component at ν_0 and two σ components at $\nu_0 \pm \nu_L \Delta(gM)$, with

$$g = (\Sigma M_l + 2\Sigma M_s)/(\Sigma M_l + \Sigma M_s).$$

Figure 5.23 shows Zeeman, Paschen–Back and super Paschen–Back effects in sodium yellow lines.

Terms arising from equivalent electrons

Paschen–Back effect helps in determining the terms arising from equivalent electrons. Suppose we have two p electrons, with structure p^2. Then they can have $M_l = 1, 0, -1$, and $M_s = \pm 1/2$, which gives a total of six electronic states. Thus there would be 15 different ways of putting the two electrons in the six quantum states. Now two electrons in the p shell can give rise to $^1S_0$, $^1P_1$, $^1D_2$, $^3S_1$, $^3P_{0,1,2}$ and $^3D_{1,2,3}$ terms. It seems that all of them together can give rise to many more Paschen–Back states than 15. So some of the terms are not possible. But it can be seen that the only allowed terms giving rise to the desired 15 states are $^1S_0$ (1), $^3P_{0,1,2}$ $(1 + 3 + 5 = 9)$ and $^1D_2$ (5).

When we have three electrons in the p shell, with p^3 structure, we will have 20 possible atomic states. There are various possibilities such as $^2P_{1/2,3/2}$, $^2F_{5/2,7/2}$, $^2S_{1/2}$, etc., but it can be seen that the allowed terms from these 20 states are $^4S_{3/2}$ (4), $^2P_{1/2,3/2}$ $(2+4)$ and $^2D_{3/2,5/2}$ $(4+6)$.

Apart from structures p^2 and p^3, we have earlier described the terms arising from the electronic structure p^1. The structures p^4 and p^5 are mirror images of p^2 and p^1, respectively, having the same terms and similar states as described above. Finally, if the p shell is filled with six electrons, there is only one possible combination with $L = S = J = 0$. This is the case of inert atoms.

5.8.3 Stark effect

The permutation of energy levels when an atom is placed in an electric field is called Stark effect. We shall not solve the Schrödinger equation completely, but indicate the use of parabolic coordinates in obtaining the eigenfunctions and the perturbation. We shall also discuss Stark effect in hydrogen.

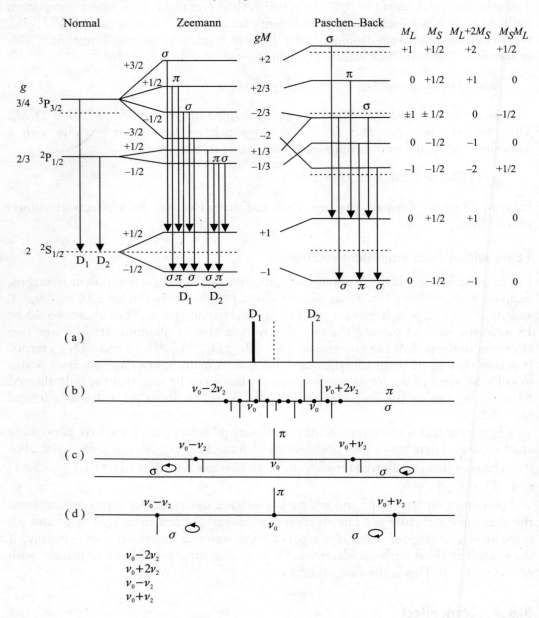

Fig. 5.23 *Various effects in sodium D lines: (a) normal spectrum, (b) anomalous Zeeman effect, (c) Paschen–Back effect in strong magnetic field, (d) super Paschen–Back effect in very strong magnetic field.*

Introduction

An electric field $\mathbf{E}_e$ applied to an atom polarises it and induces an electric dipole moment in it. The nucleus no longer coincides with the centre of the negative charge distribution. Let us take $\mathbf{E}_e$ along the z-axis. The actual separation of the nuclear charge and the centre of the negative charge depends upon the electric field, and let this separation be denoted by z. Then the atom has an additional electrostatic potential energy $-eE_e z$ and the perturbation on the atom can be written as

$$H' = -eE_e z.$$

The first order perturbation to energy is given by

$$E'_j = \int \psi_j^* H' \psi_j d\tau.$$

An electron in an atom already experiences a spherically symmetric potential $V(r)$. An electric field further creates an axially symmetric potential. Therefore the coordinate z along the field and the transverse coordinates x, y are not at the same footing. In such a situation it becomes convenient to use the two-dimensional parabolic coordinates ξ, η in order to deal with the Schrödinger equation and the perturbation. In relating these to Cartesian coordinates, we shall also use spherical coordinates r, θ, φ and 2-D polar coordinate ρ. Thus ξ, η are related to various other coordinates by

$$x = \sqrt{\xi\eta}\cos\varphi, \quad y = \sqrt{\xi\eta}\sin\varphi, \quad z = (\xi - \eta)/2. \tag{5.145}$$

We also have

$$z = r\cos\theta, \quad \rho = r\sin\theta = (x^2 + y^2)^{1/2}.$$

Further, we have

$$r = (x^2 + y^2 + z^2)^{1/2} = (\xi + \eta)/2;$$

$$\xi = r + z, \quad \eta = r - z, \quad \varphi = \tan^{-1}(y/x). \tag{5.146}$$

This gives

$$\rho = \sqrt{\xi\eta},$$

$$\xi^2 - \rho^2 = 2\xi z, \quad \eta^2 - \rho^2 = -2\eta z. \tag{5.147}$$

Since x, y are real and r and ρ are non-negative, the above relations imply that $\xi > 0, \eta > 0$.

Equatons (5.147) represent two sets of parabolas, one set with $\eta = $ constant, opening in the positive z-direction, and the other set with $\xi = $ constant, opening in the negative

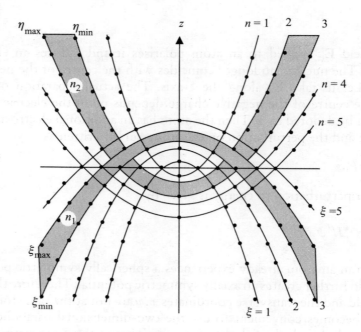

Fig. 5.24   *Parabolic coordinates (for the case $n_1 > n_2$)*

z-direction. These are shown in Fig. 5.24. The three-dimensional system is obtained by rotating the figure around the z-axis. Finally, the element of length ds, the element of volume $d\tau = d^3r$ and the Laplacian are given by

$$ds^2 = ((\xi + \eta)/4\xi)d\xi^2 + ((\xi + \eta)/4\eta)d\eta^2 + \xi\eta d\varphi^2, \tag{5.148a}$$

$$d\tau = ((\xi + \eta)/4)d\xi d\eta d\varphi, \tag{5.148b}$$

$$\nabla^2 = \frac{4}{\xi + \eta}\left[\frac{\partial}{\partial\xi}\left(\xi\frac{\partial}{\partial\xi}\right) + \frac{\partial}{\partial\eta}\left(\eta\frac{\partial}{\partial\eta}\right)\right] + \frac{1}{\xi\eta}\frac{\partial^2}{\partial\varphi^2}. \tag{5.148c}$$

Schrödinger equation in parabolic coordinates

For the hydrogen atom, on using (5.148c) and $V(r) = -e^2/r = -2e^2/(\xi + \eta)$, the Schrödinger equation becomes

$$\frac{4}{\xi + \eta}\left[\frac{\partial}{\partial\xi}\left(\xi\frac{\partial}{\partial\xi}\right) + \frac{\partial}{\partial\eta}\left(\eta\frac{\partial}{\partial\eta}\right)\right]\psi + \frac{1}{\xi\eta}\frac{\partial^2\psi}{\partial\varphi^2} + \frac{2\mu}{\hbar^2}\left(E + \frac{2e^2}{\xi + \eta}\right)\psi = 0. \tag{5.149}$$

As usual, the equation is to be solved by the method of separation of variables. Taking

$$\psi(\xi, \eta, \varphi) = f(\xi)g(\eta)F(\varphi), \tag{5.150}$$

we get the same solution $F(\varphi) = e^{im\varphi}$ as in spherical coordinates. In fact, the separation of the φ-part introduces an arbitrary constant which must be put equal to m^2, m being an integer, due to the constraint $F(\varphi + 2\pi) = F(\varphi)$. The remaining equation reduces to

$$\left[\frac{d}{d\xi}\left(\xi\frac{df}{d\xi}\right) - \frac{m^2}{4\xi}\right]g + \left[\frac{d}{d\eta}\left(\eta\frac{dg}{d\eta}\right) - \frac{m^2}{4\eta}\right]f + \frac{2\mu}{\hbar^2}\left[\frac{E(\xi+\eta)}{4} + \frac{e^2}{2}\right]fg = 0. \quad (5.151)$$

Dividing by the product fg, we get

$$\frac{1}{f}\left[\frac{d}{d\xi}\left(\xi\frac{df}{d\xi}\right) - \frac{m^2}{4\xi}\right] + \frac{1}{g}\left[\frac{d}{d\eta}\left(\eta\frac{dg}{d\eta}\right) - \frac{m^2}{4\eta}\right] + \frac{2\mu}{\hbar^2}\left[\frac{E(\xi+\eta)}{4} + \frac{e^2}{2}\right] = 0. \quad (5.152)$$

This can be separated into identical equations for $f(\xi)$ and $g(\eta)$ by taking $\frac{e^2}{2} = \frac{e^2}{2}(\beta_1 + \beta_2)$ with $\beta_1 + \beta_2 = 1$, β_1 being an arbitrary constant of separation. Thus we are lead to

$$\frac{d}{d\xi}\left(\xi\frac{df}{d\xi}\right) - \frac{m^2}{4\xi} + \left[\frac{\mu E\xi}{2\hbar^2} + \frac{\mu e^2\beta_1}{\hbar^2}\right]f = 0, \quad (5.153)$$

with a similar equation for g obtained on replacing f by g, ξ by η and β_1 by β_2. This can also be put in the form

$$\frac{d^2f}{d\xi^2} + \frac{1}{\xi}\frac{df}{d\xi} + \left[\frac{\mu E}{2\hbar^2} + \frac{\mu e^2\beta_1}{\hbar^2}\right]f - \frac{m^2}{4\xi} = 0. \quad (5.154)$$

We do not intend to solve this equation here. However, we shall point out that when appropriate boundary conditions are imposed on f and g as $\xi \to \infty$ and $\eta \to \infty$, it imposes certain constraints[18] on m and yields an acceptable polynomial solution only when $E \equiv E_n = -\mu e^4/2n^2\hbar^2$, with n being an integer. It also introduces two other integers n_1 and n_2, and leads to a condition relating n, β_1, β_2 and m, in the form

$$n\beta_1 = n_1 + (|m| + 1)/2, n\beta_2 = n_2 + (|m| + 1)/2$$

$$\Rightarrow n = n_1 + n_2 + |m| + 1. \quad (5.155)$$

It leads to an orthogonal set of eigenfunctions which can be labelled $\psi_{n_1n_2m}$, with $f(\xi) \equiv f_{n_1m}(\xi/n)$ and $g(\eta) \equiv g_{n_2m}(\eta/n)$.

Significance of parabolic quantum numbers

In the above discussion, n is the principal quantum number which determines the energy of the level, $|m|$ represents the z-component of angular momentum, but here

---

[18]See Sommerfeld, A. (1923).

only its magnitude is important and its positive and negative values are equivalent. The parameters n_1 and n_2 are determined by the extremum values of ξ and η, as shown in Fig. 5.24. It is found that $|\psi(z)|^2$ is greater than or less than $|\psi(-z)^2|^2$ depending on whether $n_1 > n_2$ or $n_1 < n_2$. Thus the centroid of the electric charge is shifted with respect to the proton in the positive or negative direction depending on whether n_1 is greater than or less than n_2, and $n_1 - n_2$ determines the dipole moment of the atom.

Stark effect in hydrogen

Since the perturbation is

$$H' = -eE_ez = -eE_e(\xi - \eta)/2,$$

the first order change in energy would be given by

$$E' = -\frac{eE_e}{8}\int_0^\infty d\xi \int_0^\infty d\eta \int_0^{2\pi} d\varphi(\xi^2 - \eta^2)|\psi_{n_1n_2m}|^2. \tag{5.156}$$

It is seen to be

$$E' = 3n(n_2 - n_1)a_0eE_e/2. \tag{5.157}$$

We see that the factor multiplying E_e in the above equation is the induced dipole moment. Dividing it by hc, we get the first order shift in energy in units of wavenumber as

$$\Delta(1/\lambda) = 3n(n_2 - n_1)a_0eE_e/2hc. \tag{5.158}$$

Thus it is seen that the first order shift is proportional to the applied electric field. This is known as linear Stark effect.

On taking higher order terms into account the shift in energy due to Stark effect can be put in the form

$$\Delta(1/\lambda) = AE_e + BE_e^2 + CE_e^3 + \cdots, \tag{5.159}$$

where[19]

$$A = \frac{3h}{8\pi^2 mec}n(n_2 - n_1),$$

$$B = \frac{h^5}{2^{10}\pi^6 m^3 e^6 c}n^4[17n^2 - 3(n_2 - n_1) - 9m^2 + 19],$$

$$C = \frac{3h^9}{2^{15}\pi^{10}m^5 e^{11}c}n^7[23n^2 - (n_2 - n_1)^2 + 11m^2 + 39].$$

---

[19]White, H.E. (1934).

Notice that the algebraic factors in B and C are much smaller than in A, though the large powers of n in them have some compensating effect. The linear term can be said to be the most dominant term unless the electric field is very high. The dependence of $\Delta(1/\lambda)$ on m shows that the number of levels is half that for Zeeman effect.

Discussion

The ground state of H atom has $n = 1$ and $n_1 = n_2 = 0$. So it has zero electric dipole moment and the coefficient $A = 0$. Thus there is no linear Stark effect, and the first non-vanishing contribution can only be quadratic. This is because the electron cloud is spherically symmetric, which is true for all s states. So the electric field first induces a separation of charges, that is, produces a dipole moment p proportional to E_e, which, in turn, causes a shift in the energy proportional to pE_e. Thus the shift is proportional to E_e^2.

The ground state of He also has no electric dipole moment because $n_1 = n_2$. So it also shows quadratic Stark effect. But higher states of H and He, which have electric dipole moment, exhibit linear Stark effect.

Figure 5.25 shows the Stark effect in H_α line, which has eleven Stark components. We see that the perturbation caused by the electric field lifts the degeneracy of $^2P_{1/2}$, $^2S_{1/2}$ and $^2P_{3/2}$, $^2D_{3/2}$ levels. Fig. 5.26 shows the Stark effect in sodium D_1 and D_2 lines. On comparing with Fig. 5.20, we now have three Stark components instead of six Zeeman components.

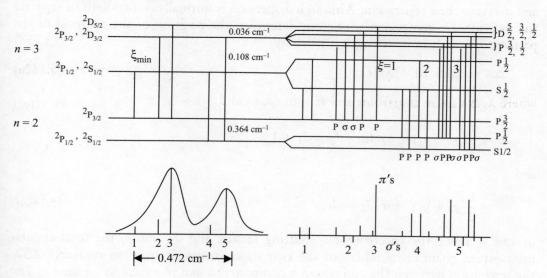

Fig. 5.25 *Stark effect in H_α line*

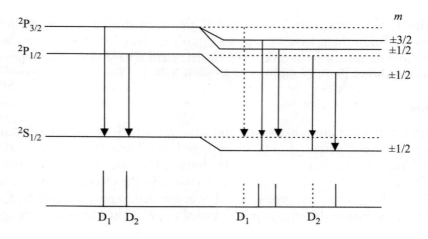

Fig. 5.26 *Stark effect in sodium D lines*

5.8.4 Zeeman and Stark effects in astrophysics

Zeeman effect

Zeeman effect can be used for determining the magnetic field in celestial objects like sunspots and magnetic stars. One can separate the two σ patterns by suitable devices and measure their separation. Although dispersion is normally not enough to separate the individual σ components, one can determine the position of the centroid of the pattern. The shift in wavelength is given by

$$\Delta\lambda = 4.6 \times 10^{-13}\lambda^2 Hz, \qquad (5.160a)$$

where λ, $\Delta\lambda$ are in angstroms and H is in gauss, and

$$z = \frac{1}{2}(g_2 - g_1) + g_2 \quad \text{for} \quad J_2 = J_1 + 1$$

and

$$z = \frac{1}{2}(g_2 + g_1) \quad \text{for} \quad J_2 = J_1, \qquad (5.160b)$$

g_1 and g_2 being the spectroscopic splitting factors and J_1 and J_2 the total angular momentum quantum numbers of the two states, respectively. One measures $2\Delta\lambda$, the separation between the composite σ components, and plots $\Delta\lambda/\lambda^2$ against z. The slope then gives the magnetic field H. It is in this way that astronomers have found magnetic fields of 3000 gauss in sunspots and thousands of gauss in magnetic stars.

Stark effect

Stark effect can be used for determining the electron pressure in the atmospheres of stars, which are plasmas consisting of electrons, protons and neutral atoms. The moving charged particles produce transient electric fields which give rise to Stark effect in spectral lines. As hydrogen is the most abundant element in the universe, most stars show Balmer lines in their spectra. In states higher than the first, hydrogen has an electric dipole moment $3a_0 en(n_2 - n_1)/2$, so the levels are broadened by Stark effect. In a stellar atmosphere, the radiating atoms experience a changing electric field produced by the passing electrons and ions. The electric field at the radiating atom due to a charge Ze at a distance r would be Ze/r^2, producing a line shift of $\Delta \nu = D_n Ze/r^2$, where D_n is different for each line. Since r is changing continuously, the line shift also changes accordingly, giving rise to what is called the *Stark broadening of the line*. As this is a statistical effect, it is also called *statistical broadening*.

Calculations show that in this case, the line profile shows $I \propto (\Delta \lambda)^{-5/2}$ dependence in the wings, where $\Delta \lambda = \lambda - \lambda_0$, λ_0 being the line centre. This is indeed found to be the case for Balmer lines. The constant of proportionality gives the electron density and the electron pressure $p_e = n_e kT$.

There is another way of determining the electron pressure. Stark effect splits the hydrogen line, with an energy spread of about

$$\Delta E = 3n(n_2 - n_1)a_0 e E_e/2 \simeq 3a_0 en^2/2.$$

But the difference between successive Bohr levels is

$$\Delta E = \Delta(-e^2/2n^2 a_0) = e^2/a_0 n^3.$$

The two become equal to each other when

$$n \simeq (2e/3a_0^2 E_e)^{1/5}.$$

Consequently all levels above this value of n merge with the continuum due to level splitting. So the last visible line of Balmer series gives n and hence E_e. As the mean electric field depends on the electron density, the last visible line gives n_e and the electron pressure.

Stark effect in other atoms

As we have noted earlier, He lines show quadratic Stark effect. But the most important effect is the appearance of the forbidden line at 4469.9 Å in the wing of the permitted line at 4471.5 Å. The strength of this line is also a measure of the electron pressure in a star.

Most of the other atoms have no permanent electric dipole moment and they are perturbed by other atoms with no dipole moment. On close approach, they induce dipole moments proportional to $1/r$ on each other. They produce electric field proportional to $1/r^4$ and cause Stark effect which would be proportional to $1/r^6$. This process gives rise to van der Waals forces and causes departure from the perfect gas law. It is also responsible for producing collisional damping of spectral lines.

Problems

5.1 Calculate the vacuum wavelengths of the first five lines and the limit for Lyman, Balmer, Paschen and Bracket series of hydrogen using (5.6), with the vacuum Rydberg constant 109677.58 cm^{-1}. What will be their wavelengths in air?

5.2 Estimate the approximate sizes of atoms in various groups of elements taking $a_0 = 0.53$ Å.

5.3 Calculate the radius of the K shall for He, C, Na, Fe, Ba and U atoms.

5.4 Calculate the shifts of the first four lines of Balmer series in heavy hydrogen (deuterium).

5.5 Calculate the excitation potentials of the second levels of H, He$^+$ and Li^{2+}. (Excitation potential of a level represents energy required to raise the electron from the ground state to that level measured in electron volts.)

5.6 Calculate the velocity of the electron in the first Bohr orbit and from it obtain its relativistic mass in terms of its rest mass.

5.7 Make a scale drawing of all the Sommerfeld orbits of hydrogen for $n = 1, 2$ and 3 with the proton at the origin.

5.8 Obtain an expression for the period of the electron in any Bohr–Sommerfeld orbit. What conclusion do you draw from it?

5.9 Calculate the energy differences between s, p, d, f orbits for $n = 4$ using Sommerfeld's relativity corrections.

5.10 Calculate the de Broglie wavelength of (i) a car weighing one metric ton and moving with a speed of 60 km/hr, (ii) an electron in the first Bohr orbit.

5.11 Show that the angular momentum operators L_x, L_y, L_z where $L_x = yp_z - zp_y$, etc, in spherical coordinates are

$$L_x = -i\hbar\{-\sin\varphi\frac{\partial}{\partial\theta} - \cos\varphi\cot\theta\frac{\partial}{\partial\varphi}\},$$

$$L_y = -i\hbar\{\cos\varphi\frac{\partial}{\partial\theta} - \sin\varphi\cot\theta\frac{\partial}{\partial\varphi}\},$$

$$L_z = -i\hbar\frac{\partial}{\partial\varphi}.$$

Also show that

$$L^2 = -\hbar^2\left[\frac{1}{\sin\theta}\frac{\partial}{\partial\theta}\left(\sin\theta\frac{\partial}{\partial\theta}\right) + \frac{1}{\sin^2\theta}\frac{\partial^2}{\partial\varphi^2}\right].$$

5.12 (a) Show that x and p_x do not have a common eigenfunction. What is the implication of this result? (b) Show that L_z and L^2 are constants of motion for H atom with a spherically symmetric potential.

5.13 Show that Newton's laws of motion hold good in quantum mechanics for average quantities.

5.14 Show that $dE/dp = v$ in both classical and relativistic cases.

5.15 Write the full expressions for the eigenfunctions $\psi_{100}, \psi_{200}, \psi_{211}, \psi_{210}$, and $\psi_{2,1,-1}$ of hydrogen in spherical coordinates.

5.16 Draw the r and (r, θ) probability distributions for the eigenfunction of Problem 5.15.

5.17 Give vector representation for all the states of level $n = 3$. What are the θ-distributions for different m values corresponding to level 2?

5.18 Calculate the energy difference in wavelengths for s and p states of level 2 from quantum mechanical relativistic formula of Sommerfeld.

5.19 Find all the spectroscopic terms for two electrons with the electronic structure d^1f^1, that is $l_1 = 2, s_1 = 1/2, l_2 = 3, s_2 = 1/2$.

5.20 Calculate the Zeeman pattern in Fe at 4442.343 Å arising from $^5P_2-^5D_2^0$ transition.

5.21 Show that energy E and time t are canonically conjugate variables.

5.22 Verify the commutation rules for components of angular momentum mentioned in Section 5.3.4. Also, discuss the commutation rules of Hamiltonia H with L_z and L^2.

6

Molecular Spectra

H A ving studied atomic spectra in the previous chapter, this chapter introduces molecular spectra, restricting the discussion to diatomic molecules. Any system containing more than one atoms can have electronic, vibrational and rotational excitations. For example, the electronic ground state and excited states of a molecule may have different sets of vibrational and/or rotational levels. Pure rotation, vibration–rotation and electronic bands are discussed in some detail. Multiplet structure of electronic states, selection rules, and isotope effect are then considered. Some idea of the intensities of various bands and lines is given. Finally, the bands of some diatomic molecules and radicals of common interest are discussed. These are often present in the earth's atomsphere, in all matter as impurities or otherwise, and in comets, stars and the cosmos.

6.1 Introduction

Molecular spectra have been an important tool in the development and understanding of quantum mechanics. In this section, we shall discuss the mutual potential energy of a diatomic molecule, types of diatomic molecules and their energy considerations.

6.1.1 Molecular potential energy

Let us start by looking at the mutual potential energy of two atoms as a function of distance r between them. When the atoms are very far away from each other, $r \rightarrow \infty$, they have no influence on each other and the potential $V(r)$ between them tends to a constant. As the atoms approach each other, various forces come into play at intermediate distances: (i) Attraction between the electrons of one atom and the nucleus of the other, which is known as the van der Waal interaction, (ii) Coulomb repulsion between the two nuclei and also between their electrons, and (iii) since electrons are fermious and are quantum particles, there is also the quantum mechanical exchange interaction. Essentially, as the electronic charge clouds of the two atoms start overlapping, all the electrons belong to one quantum mechanical system, the molecule, and have to occupy higher and higher energy states in accordance with Pauli exclusion principle. The shape of the potential curve depends upon the relative strengths of these interactions. The various posible resulting potentials are shown in Fig. 6.1. If the Coulomb repulsion is predominant everywhere, we get curves like 4 and 5, in which the minimum potential energy of the molecule is that at infinite separation between the atoms. If the attraction is stronger at intermediate distances, we get curves with a potential well like 1, 2 and 3.

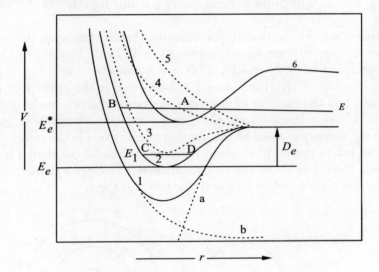

Fig.6.1 *Potential energy of a diatomic molecule*

Now suppose E is the total energy of the two atoms. Then if the potential is like curve 4, the two atoms will come as close as point A and then separate from each other due to a repulsion recoil. Hence no molecule will be formed. But if the potential is like curve 2, say, the two atoms can come as close as point B. If no other process takes place, the two atoms will again rebound as before. But if the two atoms give up some energy, either to a third particle in the neighbourhood or through emission of electromagnetic radiation, the atoms may be trapped in the potential and form a molecule with total energy E_1. The two atoms cannot now separate unless some energy is supplied to bring them above the potential well. This is the process of dissociation caused by absorption of energy either in the form of radiation or in the form of kinetic energy.

6.1.2 Types of diatomic molecule

A diatomic molecule with like atoms is called a *homonuclear* molecule; H_2, O_2, N_2

are obvious examples of this type. If, on the other hand, the two atoms are different, it is called a *heteronuclear* molecule; CO, OH, CN, HCl are examples of this type. The resultant electron distribution in the molecule leads to two kinds of bonds between the atoms. In molecules like NaCl, the electron in the outermost unfilled shell of Na is loaned to Cl for completing its outermost shell containing 7 electrons. Thus we get Na$^+$ and Cl$^-$ ions, both with filled shells, which are attracted to each other on account of their opposite polarity. The charge cloud of electrons shows a tendency to be concentrated near the two nuclear sites and very sparesely in the space between them; see Fig. 6.2 (a). The bond holding the two atoms together in this case is called a *bipolar bond.*

In the other case, H$_2$ molecule for example, both hydrogen atoms share their electron in an attempt to complete their own 1s shell. Both the electrons move around the space containing both the nuclei. This is a *homopolar bond*; Fig. 6.2 (b). In the first case the number of electrons loaned or borrowed is the valency of the atom. In the second case, half the number of electrons shared by the two nuclei is their valency. Thus Na, Cl, H have valency 1 each. Oxygen has valency 2 while Nitrogen has 3, and so on. Normally *valency* is equal to the number of electrons in the outermost shell if it is less than half full or the number of electrons needed to fill the shell if it is more than half full. When the valencies of the two atoms match, we get a stable molecule, while when they do not match, we get a radical like CN, OH, etc.

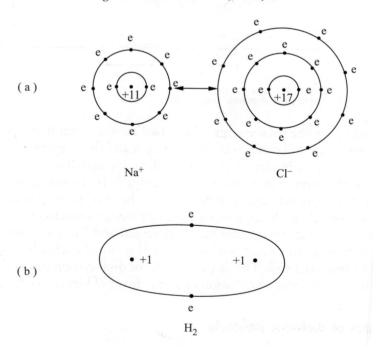

Fig. 6.2 *Molecular bonds: (a) bipolar, (b) homopolar*

6.1.3 Energy of a molecule

Let the curve 2 in Fig. 6.1 represent the mutual potential energy of two atoms when both are in their ground states. If one of the atoms is excited, we will get a different potential energy curve. Only a few of such curves have potential wells; for example, curve 6 of Fig. 6.1, which now represents the excited state of the molecule. The lowest point of the potential well represents the electronic energy of the molecule: E_e in the ground state and E_e^* in the excited state. The two nuclei of the diatomic molecule have six degrees of freedom. Three of them represent the kinetic motion of the centre of mass (CM) of the molecule. Molecular vibration takes up one degree of freedom, in which the distance between the two nuclei can oscillate around the average value, CD. The remaining two degrees of freedom are taken up by two rotations of the molecule around axes passing through its CM and perpendicular to the line of atoms, the two directions being themselves perpendicular to each other.

When all the degrees of freedom are active, we can denote the total energy of the molecule as

$$E = E_e + E_v + E_r, \tag{6.1}$$

where the terms stand respectively for electronic energy, vibrational energy and rotational energy when only these particular degrees of freedom are active. Generally, it is found that the magnitude of these terms decrease in that order, that is $|E_e| > E_v > E_r$. The term value, which is the reciprocal of the wavelength, is given by $T = 1/\lambda = E/hc$, and can also be written as

$$T = T_e + G + F, \tag{6.2}$$

where the three terms stand respectively for the three terms in (6.1). A transition may involve a combination of changes in electronic, vibrational and rotational states. Hence the wavenumber of an emission line can be written as

$$\sigma = \sigma_e + \sigma_v + \sigma_r$$
$$= (T_e{}' - T_e{}'') + (G' - G'') + (F' - F''). \tag{6.3}$$

Here $T_e{}', G'$ and F', etc., represent the term values for the upper state and $T_e{}'', G''$ and F'' for those of the lower state.

At low termperatures, only rotational transitions occur and we get pure rotational bands which are found in the far infrared, $\lambda > 20\ \mu$m, which corresponds to energy less than 60–70 meV. At intermediate temperatures, vibrational transitions take place, which are always accompanied by rotational transitions giving rise to vibration–rotation bands in the near infrared, $\lambda < 20\ \mu$m. At higher temperatures, the two atoms may detach and the molecule may dissociate. The energy required for this is the *dissocation energy* D_e, which is equal to the height of the flat portion of the $V(r)$ curve above E_e.

Finally, at very high temperatures, electronic transitions accompanied by vibrational and rotational transitions produce the very complex electronic bands in the visible part of the spectrum. Thus the molecular spectra, even for simple diatomic molcules, are more complex than the spectra of atoms and ions.

6.2 Pure Rotation Bands

In the first approximation, a diatomic molecule can be treated as two point masses separated by a fixed distance r. Since the masses are treated as point particles, there is no electronic excitation. Since the separation r is treated as constant, there is no vibrational excitation. The only possible excitation is rotational excitation. Again, due to the assumption of point masses, a rotation about the line of atoms has no meaning; it does not produce a new state. Thus we are left with two rotations, about axes normal to the line of atoms, and passing through the centre of mass. However, this is far from a trivial case. This system is called a rigid rotator. Quantum mechanically, it is a very important exactly solvable problem which leads to a good deal of understanding of various aspects.

If we allow the separation distance r to vary, we have the additional possibility of vibrations of the molecule. In this section, we deal with a rigid and a non-rigid rotator.

6.2.1 Rigid rotator

Consider a rigid rotator as explained above, with point masses m_1 and m_2, rotating about an axis perpendicular to line joining them with an angular velocity ω. Figure 6.3 shows this dumb-bell-like system. The distance between the masses is r, while the distance from the CM to the masses is r_1 and r_2, respectively.

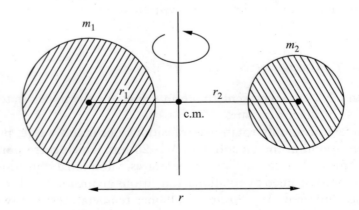

Fig. 6.3   *A rigid rotator*

Let I be the moment of inertia of the system about an axis through the CM and perpendicular to its axis. If the system rotates with an angular speed ω about such an axis, its energy, which would be entirely kinetic energy, would be

$$E = I\omega^2/2,$$

with $\quad I = m_1 r_1{}^2 + m_2 r_2{}^2.$

Relating r_1 and r_2 to r, we would see that

$$I = \mu r^2,$$

where $\mu = m_1 m_2/(m_1 + m_2)$ is the reduced mass of the molecule. Since the angular momentum is $L = I\omega$, we would have $E = L^2/2I$. The Schrodinger equation of the system then becomes

$$-\frac{\hbar^2}{2\mu}\nabla^2\psi(\mathbf{r}) + E\psi(\mathbf{r}) = 0. \tag{6.4}$$

We could expand the eigenfunction $\psi(\mathbf{r})$ in terms of spherical harmonies $Y_{lm}(\theta, \phi)$ in the form

$$\psi(\mathbf{r}) = \sum_{l,m} a_{lm}(r) Y_{lm}(\theta, \phi), \tag{6.5}$$

where the coefficients a_{lm} are given in terms of the angular integral by

$$a_{lm} = \int Y_{lm}^*(\theta, \phi)\psi(\mathbf{r})d\Omega.$$

Then, as in the case of hydrogen atom, the angular momentum operators are given by

$$L^2 = M^2 = -\hbar^2 \left[\frac{1}{\sin\theta}\frac{\partial}{\partial\theta}\left(\sin\theta\frac{\partial}{\partial\theta}\right) + \frac{1}{\sin^2\theta}\frac{\partial^2}{\partial\phi^2} \right],$$

and $\quad L_z = M_z = -i\hbar\partial/\partial\phi.$

They give

$$L = [J(J+1)]^{1/2}\hbar, J = 1, 2, 3, \ldots,$$

and $\quad L_z = m\hbar, m = J, J-1, \ldots, -J+1, -J.$

Then,

$$E_r = L^2/2I = J(J+1)\hbar^2/2I \tag{6.6}$$

It may be noted that L_x, L_y, L_z satisfy the same commutation rules as those for M_x, M_y, M_z. Thus,

$$[L_x, L_y] = i\hbar L_z, \ [L^2, L_x] = 0$$

and those obtained by a cyclic permutation of x, y, z, and

$$[L^2, L_\pm] = 0, L_z L_\pm \psi = (m \pm \hbar) L_\pm \psi.$$

The rotational terms are then given by

$$F = E_r/hc = (\hbar/4\pi cI) J(J+1),$$

so that the rotational spectrum contains lines with wave numbers[20]

$$\sigma_r = B[J'(J'+1) - J''(J''+1)]$$
$$= B(J' - J'')(J' + J'' + 1), \text{with} \quad B = \hbar/4\pi cI. \tag{6.7}$$

The selection rules are:

(i) There are no electric dipole transitions for homonuclear molecules because they do not have any permanent electric dipole moment.

(ii) Heterogeneous molecules have a permanent dipole moment and an electric dipole transitions can take place between states with $\Delta J = \pm 1$.

Then putting $J' = J'' + 1$ for a heterogeneous molecule, we get

$$\sigma_r = 2B(J'' + 1) \equiv 2Bm, \ m = 1, 2, 3, \ldots \tag{6.8}$$

Thus the spacing between lines is $2B$ as shown in Fig. 6.4. We shall discuss such selection rules in Chapter 7.

6.2.2 Effect of non-rigidity

The bond between the atoms may not be rigid, as is the real case in nature, and the potential may be anharmonic. The bond would stretch with increasing rotational speed. Hence r increase with J, resulting in an increase in I and decrease in B. Hence we may replace B by B_J and write

$$B_J = B[1 - uJ(J+1)],$$

---

[20]In accordance with the convention in spectroscopy, all parameters such as σ, B, D, ω, etc, have been expressed as wavenumbers (cm^{-1}).

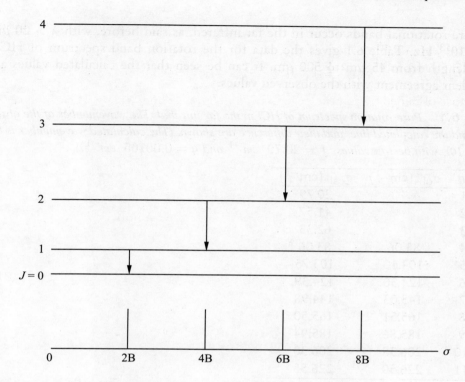

Fig. 6.4 *Energy levels and spectrum of a rigid rotator*

where u is a constant. Then the term becomes

$$F = B\left[1 - uJ(J+1)\right]J(J+1)$$
$$= BJ(J+1) - DJ^2(J+1)^2, \tag{6.9}$$

where D is another constant. It will be shown in Section 6.3.3 that

$$D = 4B^3/\omega_e^2 \text{ with } \omega_e = \nu_e/c.$$

Therefore

$$\sigma_r = B\left[J'(J'+1) - J''(J''+1)\right] - D\left[J'^2(J'+1)^2 - J''^2(J''+1)^2\right].$$

With $J' = J''+1$, this can be written as

$$\sigma_r = fm - gm^3, \quad m = 1, 2, 3\ldots, \tag{6.10}$$

with $m = J'' + 1$.

Thus the spacing between successive lines decreases with increasing m.

Pure rotational bands occur in the far infrared, as said before, with $\lambda > 20$ μm or $\nu \leq 10^{13}$ Hz. Table 6.1 gives the data for the rotation band spectrum of HCl for wavelength from 45 μm to 500 μm. It can be seen that the calculated values are in excellent agreement with the observed values.

Table 6.1 *Pure rotation spectrum of HCl in the far infrared. The wavenumber of the observed line and the calculated line, and their difference are shown. (The calculated wavenumber is based on (6.10), with best-fit values, $f = 20.79$ cm^{-1} and $g = 0.00165$ cm^{-1}.)*

m	$\sigma_{\text{obs}}(\text{cm}^{-1})$	$\sigma_{\text{calc}}(\text{cm}^{-1})$
1		20.79
2		41.57
3		62.33
4	83.06	83.06
5	104.10	103.75
6	124.30	124.39
7	145.03	144.98
8	165.51	165.50
9	185.86	185.94
10	206.38	206.30
11	226.50	226.55

6.3 Vibration–Rotation Bands

We have mentioned that vibrational energies are generally larger than rotational energies, by one or two orders of magnitude. Vibrational transitions are almost always accompanied by rotational transitions, with a change in the angular momentum. This results in vibration–rotation bands in the near infrared in which each vibrational line appears together with a series of rotational lines. We now consider a simple model for these bands.

6.3.1 Vibrational motion

The potential can be approximated by a parabola near the potential minimum, so that we can treat the molecule as a simple harmonic oscillation. Let r_e be the equilibrium separation (in a certain electronic state of the molecule) and r the actual separation between the two atoms. Then letting $x = r - r_e$, the equation of motion in the first approximation is

$$\mu d^2 x/dt^2 = -kx,$$

where k is a constant. The frequency of oscillations, again in that electronic state, is $\omega_0 = \sqrt{k/\mu}$. Then the Schrodinger equation corresponding to vibrational motion can be written as

$$-\frac{\hbar^2}{2\mu}\frac{d^2\psi}{dx^2} + \frac{1}{2}kx^2\psi = E_v\psi(x).$$ (6.11)

It has energy levels

$$E_v = \hbar\omega_e(v + 1/2), v = 0, 1, 2, \ldots,$$

where ω_e is the natural classical frequency of the oscillator in the electronic level. This can be written in units of cm^{-1} as

$$G_v = E_v/hc = \omega_e(v + 1/2).$$ (6.12)

The terms G_v are shown schematically in Fig. 6.5.

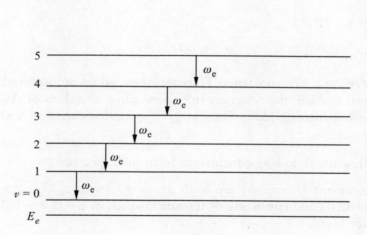

Fig. 6.5 *Energy levels of a simple harmonic oscillator with reference to the electronic level* E_e

The eigenfunctions are given by (5.41). For $v = 0$, it has a maximum at the centre of the potential well. For $v \neq 0$, they have extrema at the boundaries of the potential well, at r_{min} and r_{max} for the vibrational state in question. This property allows us to estimate the intensities of vibrational bands, which will be discussed in Section 6.7.

In a real system, the parabolic potential does not rise to infinity, but tapers off to a constant value, generally on one side of the equilibrium point. Therefore, the vibrational quantum number v has a maximum value d :ermined by the dissociation energy D_e. The transition lines will appear at energies (in cm^{-1}) given by $\sigma_v = \omega_e \Delta v$. The selection rule for a simple harmonic oscillator is $\Delta v = \pm 1$; see Section 7.6. Therefore, σ_v has only one value, $\sigma_v = \omega_e$. Thus we have only one line.

However, this does not agree with observations, which show overtones. This, in fact, is an indication of the fact that the potential well departs from a parabola, as shown in Fig. 6.1. The potential can be expanded in powers of $r - r_e$ starting from the quadratic term, taking the minimum of the potential as zero, in the form

$$V = f(r - r_e)^2 - g(r - r_e)^3 + \ldots, \tag{6.13}$$

where f and g are positive constants. Then we find that the energy level can be written as

$$G(v) = \omega_e(v + 1/2) - \omega_e x_e(v + 1/2)^2 + \omega_e y_e(v + 1/2)^3$$

$$+ \omega_e z_e(v + 1/2)^4 + \ldots, \tag{6.14}$$

where x_e, etc., are constants such that $1 > x_e > y_e > z_e$, etc. Then the transition line is given by

$$\sigma = G(v') - G(v'')$$

$$= (av' - bv'^2 + \ldots) - (a'v'' - b'v''^2 + \ldots), \tag{6.15}$$

where a, b, a', b', etc., are constants. Thus the term values are lowered, the lowering increasing with v. Also, the selection rules now allow all values of Δv, the intensity decreasing with increasing $|\Delta v|$. Thus we get the fundamental and overtone bands as shown in Fig. 6.6.

We may draw the following conclusions from such spectra:

(i) The dominant transitions are with $\delta v = \pm 1$, with $\sigma \simeq \omega_e$, which give the fundamental band consisting of transitions such as $1 - 0, 2 - 1, 3 - 2$, etc, with intensity 1, say.

(ii) Less intense transitions with $\Delta v = \pm 2$, with $\sigma \simeq 2\omega_e$, which give the first overtone band consisting of transitions such as $2 - 0, 3 - 1, 4 - 2$, etc., with intensity $\sim 10^{-2}$.

(iii) Still weaker transitions with $\Delta v = \pm 3\omega_e$, with $\sigma \simeq 3\omega_e$, which give the second overtone band consisting of transitions such as $3 - 0, 4 - 1, 5 - 2$, etc., with intensity $\sim 10^{-3}$, etc.

(iv) If the transition is to the ground state, $v'' = 0$, the line is given by

$$\sigma = av' - bv'^2, \tag{6.16a}$$

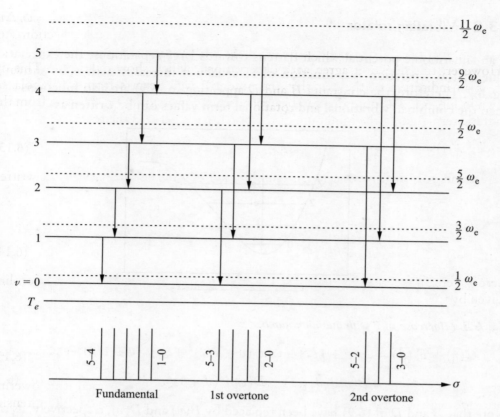

Fig. 6.6 *Energy levels and vibration spectrum of an anharmonic oscillator*

whereas if the transition is from the ground state, $v' = 0$, the line is given by

$$\sigma = -a'v'' + b'v''^2.$$ (6.16b).

Table 6.2 lists the vibrational bands of HCl in near-IR region. The calculated values are based on (6.16a) with best-fit values of parameters, $a = 2937$ cm^{-1} and $b = 51.60$ cm^{-1}. The excellent agreement may be noted.

Table 6.2 *Vibrational bands of HCl in the near-infrared*

v'	$\lambda(\mu m)$	$\sigma_{obs}(cm^{-1})$	$\sigma_{calc}(cm^{-1})$
1	3.46	2885.9	2885.7
2	1.76	5668.0	5668.2
3	1.20	8346.9	8347.5
4	0.92	10923.1	10923.6
5	0.75	13396.5	13396.5

6.3.2   Vibrating rotator

In an anharmonic potential which flattens towards large separations, the expectation value of the separation is a function of v; usually it increases with v, as shown in Fig. 6.7. Therefore, the constants B and D appearing in (6.9) are also functions of v. Then the combined vibrational and rotational term values can be written as

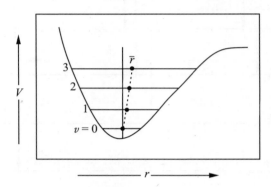

Fig. 6.7   *Increase of $\bar{r}$ with quantum number v*

$$G(v) + F_v(J) = \omega_e(v + 1/2) - \omega_e x_e(v + 1/2)^2 + \omega_e y_e(v + 1/2)^3 + \dots$$

$$+ B(v)J(J + 1) - D(v)J^2(J + 1)^2. \qquad (6.17)$$

Note that B and D of (6.9) have been replaced by $B(v)$ and $D(v)$, respectively, as they would depend on the vibrational quantum number. Therefore, the wavenumber of the vibration–rotation line is given by

$$\sigma = \sigma_0 + F_{v'}(J') - F_{v''}(J'')$$

with $\sigma_0 = G(v') - G(v'').$ \qquad (6.18)

σ_0 is called the *null line*; it corresponds to $J' = J'' = 0$, that is, a vibrational transition when both initial and final states are non-rotational. It may be noted that it is a non-existent line because the transition $\Delta J = 0$ is forbidden.

The branch of lines corresponding to $J' = J'' - 1$ is called the P branch, while that corresponding to $J' = J'' + 1$ is called the R branch. The wavenumber for the two branches can be written together as

$$\sigma = \sigma_0 + (B(v') + B(v''))m + (B(v') - B(v''))m^2, \qquad (6.19)$$

where $m = J'' + 1$ for the R branch and $m = -J''$ for the P branch. The production of these bands is shown in Fig. 6.8.

Equation (6.19) can be represented as a parabola between wavenumber and the parameter m which contains J'', the final rotational state. This is shown in Fig. 6.9. Depending on whether $B(v')$ is greater or smaller than $B(v'')$, we have the following two cases.

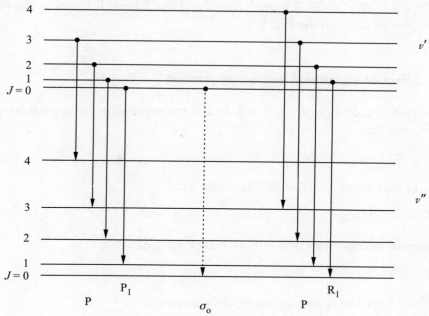

Fig. 6.8 *Production of P and R branches in VR bands*

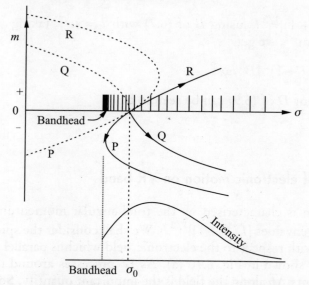

Fig. 6.9 *Fine structure of a V-R band. Case (i) dashed curve; Case (ii) continuous curve. Inset shows intensity pattern of rotational lines; see Section 6.7 (c).*

(i) It $B(v') < B(v'')$, the R branch closes to a head and the P branch opens in the other direction as shown by the dashed parabola in Fig. 6.9.

(ii) It $B(v') > B(v'')$, the R branch opens and the P branch closes as shown in the continuous curve in the Fig. 6.9.

6.3.3   Effect of vibrational motion on rotation

From the equation of motion $\mu r \omega^2 = k \Delta r$ and the expression for angular momentum $N = \mu r^2 \omega$, we get

$$\Delta r = N^2 / \mu k r^3.$$

Since the kinetic energy is $K = N^2 / 2\mu r^2$, we have

$$\Delta K = -N^4 / \mu^2 k r^6.$$

With potential energy expressed as $V = k(\Delta r)^2 / 2$, we have

$$\Delta V = N^4 / 2\mu^2 k r^6.$$

We can then write the change in the rotational term as

$$\Delta F = -N^4 / 2\mu^2 k r^6 hc.$$

Using $N = [J(J+1)]^{1/2} \hbar$, using B of (6.7) with $I = \mu r^2$, taking $\omega_e = (k/\mu)^{1/2}$ and convertng ω_e to cm^{-1}, we get

$$\Delta F = -J^2(J+1)^2 4B^3 / \omega_e{}^2.$$

Thus the coefficient D of (6.9) can be seen to be

$$D = 4B^3 / \omega_e{}^2. \qquad (6.20)$$

6.3.4   Effect of electronic motion on VR band

Electronic motion is characterised by the total angular momentum of the electrons $L^* \hbar$ which has eigenvalues $[L(L+1)]^{1/2} \hbar$. We shall consider the spin later. The vector $\mathbf{L}^*$ orients itself with respect to the electronic field which is parallel to the line joining the two atoms as shown in Fig. 6.10 (a). As $\mathbf{L}^*$ precesses around the field direction, only its components $\Lambda \hbar$ along the field is the important quantity. So we have different states with $\Lambda = 0, 1, 2, 3, ...$, etc., which are respectively designated as $\Sigma, \Pi, \Delta, \Phi$, etc. The rotational energy of the electrons associated with these states is given by

$E_\Lambda = \hbar^2 \Lambda^2 / 2I_A$ where I_A is the moment of inertia around the atomic axis due to electrons. The corresponding term values are $\hbar \Lambda^2 / 4\pi c I_A$.

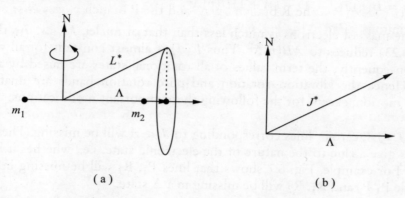

Fig. 6.10   *Combination of electronic and molecular angular moments*

The rotational motion of the molecule takes place around an axis perpendicular to the line joining the two atoms. In the absence of electrons, this nuclear angular momentum is $N = [J(J+1)]^{1/2} \hbar$ as discussed earlier. But now we have to combine **N** with **Λ** to get $\mathbf{N} + \mathbf{\Lambda} = \mathbf{J}^*$. So,

$$J^{*2} = J(J+1) + \Lambda^2 \Rightarrow J^* = \left[J(J+1) + \Lambda^2\right]^{1/2}. \qquad (6.21)$$

Occasionally, it so happens that I_B, the moment of inertia of the nuclei around the axis of rotation perpendicular to the atomic axis, is independent of the orientation of the axis of rotation. In that case we have the *symmetric top model* which is found to hold good for most diatomic and many polyatomic molecules. In that case the rotational energy of the molecule is

$$E_N = \left[J(J+1) + \Lambda^2\right] \hbar^2 / 2I_B. \qquad (6.22)$$

It has to be combined with the rotational energy of the electrons given by

$$E_e = \hbar^2 \Lambda^2 / 2I_A.$$

The total energy is then given by

$$E = \left[J(J+1) + \Lambda^2\right] \frac{\hbar^2}{2I_B} + \frac{\hbar^2 \Lambda^2}{2I_A}.$$

Defining $B(v) = \dfrac{\hbar}{4\pi c I_B}$ and $I_B / I_A = A$,

348 *An Overview of Basic Theoretical Physics*

the rotational terms can be written as

$$F(J) = B(v)J(J+1) + (A+1)B(v)\Lambda^2. \tag{6.23}$$

Since the mass of electrons is much less than that of nuclei, $I_A << I_B$, the second term in (6.23) reduces to $AB(v)\Lambda^2$. Thus $F(J)$ is almost constant for all vibrational states. Consequently, the term values of all rotational states are raised by a constant amount. Hence the vibration–rotation and pure rotation bands are unaffected by electronic motion, except for the following these important modifications:

(i) As $J \geq A$, certain lines corresponding to $J < A$ will be missing. These missing lines give a clue to the nature of the electronic state, i.e., whether it is Σ, Π, Δ, etc. For example, Fig. 6.7 shows that lines P_1, R_1 will be missing in a Π state, while $P_1 P_2$ and R_1, R_2 will be missing in a Δ state.

(ii) The term values of (6.24) are identical for $\pm\Lambda$, i.e. for each $\Lambda \neq 0$, we have two degenerate states corresponding to opposite rotations of electrons around the atomic axis. The degeneracy is removed when the interaction between Λ and N is taken into account. Thus we get what is called the Λ-type doubling of the rotational lines.

(iii) The selection rule is modified to $\Delta J = 0, \pm 1$ ($J = 0 \leftrightarrow J = 0$ forbidden) if $\Lambda \neq 0$, and $\Delta J = \pm 1$ if $\Lambda = 0$. So the VR bands of the Σ state are unaffected. But for other states a third branch corresponding to $\Delta J = 0$ is included, which is known as the Q-branch, shown in Fig. 6.8. Since the spacing of the rotational levels is roughly the same in both vibrational states, the Q branch is made up of very closely packed lines near σ_0. The wavenumbers of the three branches are given by

$$\sigma_P = \sigma_{P,\,ori} - (B(v') - B(v''))\Lambda^2,$$

$$\sigma_R = \sigma_{R,ori} - (B(v') - B(v''))\Lambda^2,$$

where $\sigma_{P,ori}$ and $\sigma_{R,ori}$ are the original term values given by (6.19), and
$$\sigma_R = \sigma_0 - (B(v') - B(v''))\Lambda^2 + dJ + eJ^2,$$
where d and e are some other constants. Thus the lines of the Q branch lie on another parabola opening in the same direction as that of the P and R branches, as shown in Fig. 6.8.

6.3.5 Study of levels by Raman effect

When a strong photon beam of frequency ν_0 is scattered by a molecular gas, the scattered light shows additional lines of frequency $\nu_0 \pm \Delta\nu_i$. This is known as the

Raman effect, discovered by C. V. Raman in 1928. The lines at $\nu_0 - \Delta\nu_i$ are called Stokes lines and those at $\nu_0 + \Delta\nu_i$ are called anti-Stokes lines.

This is a quantum effect arising from the inelastic collision between the incident photon of frequency ν_0 and a molecule of the gas. If E_1 and E_2 are two energy levels of the molecule before and after collision, the scattered photon will have a frequency ν' given by

$$\Delta\nu_i = \nu' - \nu_0 = (E_1 - E_2)/h.$$

Thus $\Delta\nu_i$ are determined by the energy levels of the molecule. So we can study rotational and vibration–rotational structure of the molecule by measuring the values of $\Delta\nu_i$ in the scattered light.

For example, when Hg line at 2536.5 Å ($\sigma = 39424 \text{ cm}^{-1}$) is scattered by HCl molecules, one finds a line at 2736.9 Å ($\sigma = 36538 \text{ cm}^{-1}$) with $d\sigma = 2886 \text{ cm}^{-1}$, which corresponds to the 0–0 vibrational band of HCl. Both 2536.5 Å and 2736.9 Å lines show satellite lines at short intervals on either side. They correspond to the rotational levels of v' and v'' vibrational states.

The selection rules for Raman effect are $\Delta v = \pm 1, \pm 2$, etc., though usually only the strongest band $\Delta v = \pm 1$ is seen. In the case of rotational levels the selection rules are $\Delta J = 0, \pm 1, \pm 2$, for $\Lambda \neq 0$ states, and $\Delta J = 0, \pm 2$, for $\Lambda = 0$ states. So in the first case ($\Lambda \neq 0$), we have Q ($\Delta J = 0$), P($\Delta J = 1$), R ($\Delta J = -1$), S ($\Delta J = 2$) and O ($\Delta J = -2$) branches. When $\Lambda = 0$, P and R branches are missing and only R, S and O branches are seen. Further, since $J \geq \Lambda$, lines corresponding to $J < \Lambda$ will be missing.

Now for $\Lambda = 0$ when $\Delta J = \pm 2$, we have for S and O branches, respectively,

$$\sigma \simeq B\left[(J'' + 2)(J'' + 3) - J''(J'' + 1)\right] = 4B(J'' + 3/2), \text{ for S,}$$

$$\sigma \simeq B\left[(J'' - 2)(J'' - 1) - J''(J'' + 1)\right] = -4B(J'' + 1/2), \text{ for O.}$$

On comparing with (6.7) and (6.8), we see that the spacing of lines is doubled with respect to that for pure rotation bands. This is indeed found to be the case with rotational Raman lines associated with 0-0 band of HCl, that is, $\Delta\sigma = 41.64 \text{ cm}^{-1}$ instead of $\Delta\sigma = 20.79 \text{ cm}^{-1}$ as shown in Table 6.1.

6.4 Electronic Bands

Here we will consider some features of electronic bands coupled with vibrational and rotational transitions.

6.4.1 Electronic transitions

We have noted earlier that electronic bands occur in the visible part of the spectrum due to transitions of electrons from one state to another. The selection rules for electronic transitions are

$$\Delta\Lambda = 0, \pm 1.$$

Thus we can have transitions like $\Sigma \rightarrow \Sigma, \Sigma \leftrightarrow \Pi, \Pi \rightarrow \Pi, \Pi \leftrightarrow \Delta$, etc. The equilibrium distance r_e differs from one electronic state to another; similarly the dissociation energy D_e will be different. Therefore, $\omega_e, \omega_e x_e, \omega_e y_e, B(v), D_v$ will also depend upon the electronic state. It is found that $B(v)/\omega_e$ is approximately constant. Then, as $B(v) \propto 1/r_e^2$ and $\omega_e \propto D_e$, we see that a small r_e is associated with a large D_e and vice versa. Thus more tightly bound atoms are closer at equilibrium distance, which should be more or less obvious. That is why D_e is called the *bond strength*.

The wave numbers of all transitions are given by

$$\sigma = T_e' - T_e'' + G(v') - G(v'') + F(J') - F(J'').$$

The first part $\sigma_e = T_e' - T_e''$ gives the general region of the spectrum where the band is located. The differences $\sigma_v = G(v') - G(v'')$ and $\sigma_r = F(J') - F(J'')$ determine, respectively, the vibrational and rotational structure of the band.

6.4.2 Vibrational structure

We may write the electronic plus vibrational contributions to the term value σ as

$$\sigma = \sigma_e + G(v') - G(v'') = \sigma_0 + \omega_e'(v' + 1/2) - \omega_e''(v'' + 1/2), \qquad (6.24)$$

where σ_0 is the band origin for the various VR bands and ω_e' and ω_e'' are the vibrational frequencies in cm^{-1} in the two vibrational states, respectively. As there is no selection rule for Δv, all values of Δv are allowed. We can arrange σ_0's in a tabular form as shown in Table 6.3, known as *Deslandre scheme*.

In this scheme, the bands are designated by $v'-v''$. Then, using (6.16a), we see that the columns are v'-progressions with $\sigma_0 = \sigma_{00} + av' - bv'^2$, where σ_{00} and σ_{00}' are given by (6.16a) and (6.16b). In this case, σ increases with v'. The rows, on the other hand, are v''-progressions, with $\sigma_0 = \sigma_{00}' - a'v'' + b'v''^2$. In this case, σ decreases with incrasing v'. The diagonals, with Δv=constant, are sequences. Transitions (0–0), (1–1), (2–2), etc., form the main sequence. In this case σ increases or decreases according to whether $\omega_e' > \omega_e''$ or $\omega_e' < \omega_e''$, that is, according as $D_e' > D_e''$ ($r_e' < r_e''$) or $D_e' < D_e''$ ($r_e' > r_e''$). The differences between the successive v-terms are denoted by $\Delta G(1/2)$, $\Delta G(3/2)$, etc., as shown in Table 6.3.

Table 6.3 *Deslandre scheme*

	v''	$\Delta G''(1/2)$	$\Delta G''(3/2)$	$\Delta G''(5/2)$	$\Delta G''(7/2)$
v'	0	1	2	3	4
	0 (0-0)	(0-1)	(0-2)	(0-3)	(0-4)→red
$\Delta G'(1/2)$					
	1 (1-0)	(1-1)	(1-2)	(1-3)	(1-4)→red
$\Delta G'(3/2)$					
	2 (2-0)	(2-1)	(2-2)	(2-3)	(2-4)→red
$\Delta G'(5/2)$					
	3 (3-0)	(3-1)	(3-2)	(3-3)	(3-4)→red
$\Delta G'(7/2)$					
	4 (4-0)	(4-1)	(4-2)	(4-3)	(4-4)→red
	↓	↓	↓	↓	↓
	violet	.	.	.	.

If $r_e' > r_e''$, as shown in Fig. 6.11 , then $D_e' < D_e''$ so that $\omega_e' < \omega_e''$. Also, $B_e' < B_e''$, B_e/ω_e=constant, and $\Delta G' < \Delta G''$, that is, the sequences run towards longer wavelengths. If $r_e < r_e''$, we have the opposite case. If $r_e' \simeq r_e''$ so that $\Delta G' \simeq \Delta G''$, the bands in a sequence are practically coincident; for example, *Swan bands*. The spectrum of C_2 (see Section 6.8) shows the appearance of such bands.

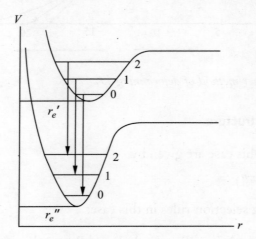

Fig. 6.11 *Potential wells with different r_e in different electronic states of a molecule*

How do we go about determining the binding energy D_e? As shown in Fig. 6.5, we note that it is the sum of energy differences from ground level state to $v_{\max}$. Thus,

$$D_e = G(0) + \sum_{v=0}^{\infty} \Delta G(v + 1/2) = G(0) + D(0),$$

where $D(0)$ is the area under the curve of Fig. 6.12 beyond $v=0$. Now

$$\Delta G(v + 1/2) = G(v + 1) - G(v)$$

$$= \omega_e - 2\omega_e x_e - 2\omega_e x_e v + \ldots.$$

Therefore, there is a linear relation between $\Delta G(v + 1/2)$ and v; see Fig. 6.12. Further, $\Delta G(v + 1/2)$ will be zero for a particular value of $v = v_{\max}$. Thus there are a finite number of vibrational levels. This $v_{\max}$ can be determined by extrapolating $\Delta G(v + 1/2)$ beyond the last observed band. Then the area under the line gives $D(0)$. This is *Birge–Sponer extrapolation method* for determining D_e.

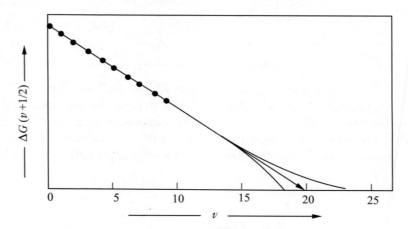

Fig. 6.12   *Birge–Sponer method of determining D_e*

6.4.3 Rotational structure

The wavenumbers in this case are given by

$$\sigma = \sigma_0 + (F' - F'').$$

We have the following selection rules in this case:

(i) $\Sigma \rightarrow \Sigma$ transition: $\Delta J = \pm 1$, so only P and R branches are present.

(ii) Other cases: $\Delta J = 0$ or ± 1, so P, R and Q branches will be present, but some lines will be missing. Some examples of such cases are shown in Fig. 6.13.

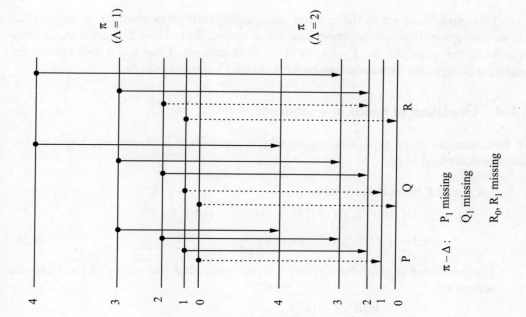

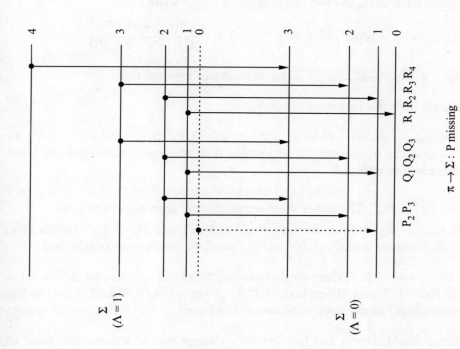

Fig. 6.13 *Missing lines in the rotational spectra of electronic bands*

If all the molecules are in the ground state, as happens to be the case in interstellar gas, we may get only one absoprtion line, for example, R_0 in $\Pi \to \Sigma$ transition, or only three lines, for example, R_2, P_2, Q_2 in $\Pi \to \Delta$ transition. Thus the whole electronic band may degenerate into only one or a few lines.

6.4.4 Gradation of bands in a sequence

We have various cases depending upon which parameters is dominant. Some of the cases are described here.

(i) For R and P branches, we have

$$\sigma = \sigma_0 + (B(v') + B(v''))\, m + (B(v') - B(v''))\, m^2$$
$$+ (A+1)(B(v')\Lambda'^2 - B(v'')\Lambda''^2). \tag{6.25}$$

The bandhead is given by $d\sigma/dm = 0$, showing that the vertex of the parabola occurs at

$$m_{\text{vertex}} = -\frac{B(v') + B(v'')}{2(B(v') - B(v''))}.$$

Putting this back in (6.25) and rearranging, we may write

$$\sigma_{\text{vertex}} - \sigma_0 - (B(v')\Lambda^2 - B(v'')^2)\Lambda''^2) = -\frac{(B(v') + B(v''))^2}{4(B(v') - B(v''))}.$$

Therefore, $\sigma_{\text{vertex}} > \sigma_0$ and the band degenerates towards red.

We again have the following sub-cases :

A. If $\sigma_{\text{vertex}} > \sigma_0$, we will have $B(v') < B(v'')$, $r_e' > r_e''$ and $D_e' < D_e''$. In this case the upper state is less stable than the lower state, and the band degrades towards red.

B. If $\sigma_{\text{vertex}} < \sigma_0$, that is, the band degrades towards violet, and $B(v') < B(v'')$ and $D_e' > D_e''$. The upper state is more stable than the lower state.

C. If $\sigma_{\text{vertex}} - \sigma_0 \to \infty$ as, we have $B(v') \simeq B(v'')$ and $D_e' \simeq D_e''$. In this case, both states are equally stable and the parabola tends to a straight line.

(ii) When $D_e' > (\text{or} <) D_e''$, then we also have $\omega_e' > (\text{or} <) \omega_e''$, so that $\Delta G'(1/2) > (\text{or } <) \Delta G''(1/2)$, etc. Therefore, $\sigma_0(1,1) > (\text{or} <) \sigma_0(0,0)$ and so on. So the sequence tails off in the same direction as the bands.

(iii) Sometimes $B(v')-B(v'')$ and $G(v')-G(v'')$ change sign at a particular value of v. Then the sequence returns from that point and the gradation of bands also changes direction. The returning sequence may go beyond the first band of the sequence giving rise to the tail bands; see Secction 6.8.

6.5 Multiplet Structure of Electronic States

We will discuss in this section the multiplet structure of bands arising from the electron spin angular momentum and its combination with other angular momenta. We will consider some related points here.

Spin multiplets

Let S be the spin quantum number representing the total spin of all the electrons. We have to combine it with the total orbital angular momentum L and the nuclear angular momentum N. There are five possible schemes due to Hund.

Case (a)   When both L and S are strongly coupled to the nuclear axis, we first take the component of $\mathbf{S}$ along the atomic axis, say Σ. Then $\Omega = \mathbf{L} + \Sigma$ gives the total angular momentum of the electron. Combining it with $\mathbf{N}$, we get

$$\mathbf{J} = \Omega + \mathbf{N} \text{ with } J \geq \Omega.$$

This gives

$$F_v(J) = B(v)[J(J+1) - \Omega^2].$$

The number of possible values of Ω are $2\Sigma + 1$ if $L > \Sigma$ and $2L + 1$ if $L < \Sigma$. Thus the combination of $L=1$, $\Sigma=1/2$ gives rise to a doublet state ${}^2\Pi_{1/2}, {}^2\Pi_{3/2}$, while the combination $L=2$, $\Sigma=1$ gives rise to the triplet ${}^3\Delta_1$, ${}^3\Delta_2$, ${}^3\Delta_3$. In such cases, the separation between the multiplet levels is proportional to the produe $L\Sigma$, as shown in Fig. 6.14.

Case (b)   In this case only Λ is strongly coupled to the nuclear axis. So only its component Λ along that axis is important. Then N and Λ first give $\mathbf{K} = \mathbf{N} + \mathbf{\Lambda}$, where K has values $\Lambda, \Lambda + 1, \Lambda + 2$, etc. Then we get $\mathbf{J} = \mathbf{K} + \mathbf{S}$, where $\mathbf{K}$ and $\mathbf{S}$ precess around $\mathbf{J}$. Since J takes values from $|K - S|$ to $K + S$, the multiplicity is $2S + 1$ or $2K + 1$ depending on whether $S < K$ or $S > K$. In this case, the separation is small and proportional to K, as shown in Fig. 6.15.

Case (c)   In this case, we have strong L–S coupling with J coupled to the nuclear axis. We first combine $\mathbf{L}$ and $\mathbf{S}$ to give $\mathbf{J}_a$, which processes around the atomic axis with component Ω. Then we have $\mathbf{J} = \mathbf{N} + \mathbf{\Omega}$ as in Case (a). Again, $2S + 1$ is the multiplicity of $\mathbf{J}_a$.

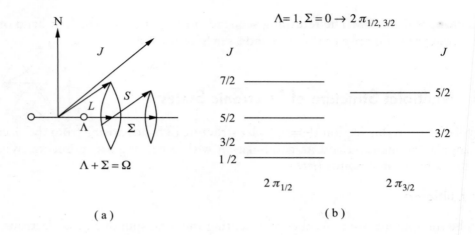

(a)

(b)

Fig. 6.14   *Hund's coupling, Case (a): vector representation (a); rotational splitting (b).*

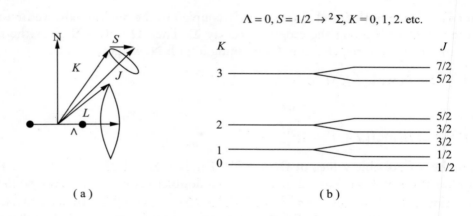

(a)

(b)

Fig. 6.15   *Hund's coupling, Case (b): vector representation (a), rotational splitting (b).*

Case (d)   Here **L** is strongly coupled to the atomic axis. So we combine **N** and **L** to give $\mathbf{K} = \mathbf{N} + \mathbf{L}$ and then form $\mathbf{J} = \mathbf{K} + \mathbf{S}$. Here K has $2L + 1$ values from $N-L$ to $N + L$.

Case (e)   In this case, L–S are coupled to get $\mathbf{J}_a$ and $\mathbf{J}_a$ is strongly coupled to the molecular axis. So we combine L–S to give $\mathbf{J_a} = \mathbf{L} + \mathbf{S}$ and couple $\mathbf{J}_a$ and N to give $\mathbf{J} = \mathbf{N} + \mathbf{J_a}$.

Cases (c), (d) and (e) are shown in Fig. 6.16. It is found that Cases (a) and (b) are the most important ones. Case (c) is found in heavy molecules. Case (d) is very rarely found while Case (e) has not been met with so far.

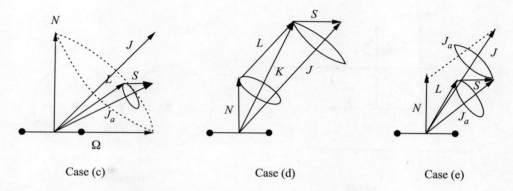

Fig. 6.16 *Hund's coupling: Cases (c), (d) and (e)*

Symmetries of a molecule

For nondegenerate states like Σ states, we may have symmetry or antisymmetry with respect to a plane passing through the nuclear axis. If z-axis coincides with the rotation axis, the eigenfunction may satisfy

$$\psi(x, y, z) = \pm\psi(x, y, -z).$$

The states are then respectively denoted by Σ^+ and Σ^-. For degenerate states like Π, Δ, etc., with eigenfunctions χ and $\overline{\chi}$ corresponding to Λ and $-\Lambda$, respectively, we would have

$$\chi(x, y, z) = \overline{\chi}(x, y, -z).$$

In this case we can construct symmetric and antisymmetric functions by writing

$$\psi_e^+ = \chi e^{iL\phi} + \overline{\chi}e^{-iL\phi}, \Pi^+ \text{ state,}$$

$$\psi_e^- = \chi e^{iL\phi} - \overline{\chi}e^{-iL\phi}, \Pi^- \text{ state.}$$

For homonuclear molecules we can consider symmetry with respect to the centre of mass. Thus we may have inversion leading to

$$\psi(-\mathbf{r}) = \pm\psi(\mathbf{r}),$$

and the states may be denoted by Σ_g, Σ_u or Π_g, Π_u, etc.[21] Thus we would have four combinations, $\Sigma_g^+, \Sigma_g^-, \Sigma_u^+, \Sigma_u^-$, etc. Figure 6.17 shows the $\psi(\mathbf{r})$ function for E_u^+ term.

---

[21]The subscripts g and u stand for 'gerade' and 'ungerade' which are German words for even and odd, respectively.

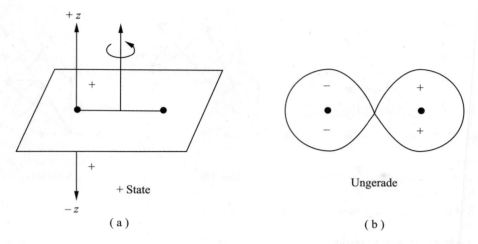

Fig. 6.17   *Symmetries of a molecule:* (a) $\psi(x, y, -z) = \pm\psi(x, y, z)$;   (b) $\psi(-\mathbf{r}) = \pm\psi(\mathbf{r})$.

Let us finally summarise the selection rules for these cases.

(1) The general rule is that we must have $\Delta J = 0$ or ± 1, when Λ_1 or Λ_2 is nonzero, with $J = 0 \to J = 0$ transition being disallowed.

(2) In addition, for homonuclear molecules, the transition between two gerade or two ungerade states is disallowed. So we must have $g \to u$ or $u \to g$.

(3) Various cases arise for Hund's (a) and (b) coupling schemes, as under:

    (I) $\Delta\Lambda = 0, \pm 1$, that is, $\Sigma \to \Sigma, \Sigma \longleftrightarrow \Pi, \Sigma$ to Δ not allowed

    (II) $\Sigma^+ \to \Sigma^-$, but Σ^- to Σ^+ not allowed

    (III) Both $\Sigma^+, \Sigma^- \to \Pi$ due to degeneracy of Π

    (IV) $\Delta S = 0$, multiplicity, must be the same

(4) For Hund's (a) coupling, we may have

    (I) $\Delta\Sigma = 0$

    (II) If $\Omega = 0$ for both states, $\Delta J \neq 0$, that is Q branch forbidden.

(5) For Hund's (b) coupling, $\Delta K = 0, \pm 1$; $\Delta K = 0$ gives Q branch and $\Delta K = \pm 1$ give P and R branches, respectively.

6.6 Isotope Effects

We shall discuss in this section isotope effects on the vibrational and rotational structure.

6.6.1 Vibrational structure

In the case of say C_2 molecule, carbon atom has three isotopes, C^{12}, C^{13}, C^{14}. Taking the mass for C^{12} atom as 12 amu, the reduced masses of $C^{12}C^{12}$, $C^{12}C^{13}$ and $C^{13}C^{13}$ molecules are seen to be $\mu = 6.00$, 6.24 and 6.50, respectively. Then the wavenumber ω_e of $C^{12}C^{13}$ and $C^{13}C^{13}$ molecules is seen to be related to that of $C^{12}C^{12}$ as

$$(\omega_e)_{12,13}/(\omega_e)_{12,12} = 0.98,$$

$$(\omega_e)_{13,13}/(\omega_e)_{12,12} = 0.96. \tag{6.26}$$

Then with $\omega_i = \rho_i\omega_e$, where ω_e is the wavenumber of $C^{12}C^{12}$ isotope and ρ_i is the quotient appearing in (6.26), equation (6.14) becomes, for the ith isotopic molecule,

$$G_i(v) = \rho_i G(v).$$

Then the corresponding wave numbers for vibrational lines, in accordance with (6.24) become

$$\sigma = \sigma_e + \sigma_v,$$

$$\sigma_i = \sigma_e + \rho_i\sigma_v.$$

This gives a shift of

$$\sigma - \sigma_i = (1 - \rho_i)\sigma_v$$

$$= (1 - \rho_i)(\omega'_e - \omega''_e)(v + 1/2) - (1 - \rho_i)\omega''_e\Delta v. \tag{6.27}$$

For fixed v'' we have v'-progression, with $\sigma > \sigma_i$ and $\lambda_i > \lambda$, that is, towards 0–0 band. On the other hand, for fixed v', we have v''-progression with $\sigma < \sigma_i$ and $\lambda_i < \lambda$. For fixed Δv sequence, $|\sigma - \sigma_i|$ increases with Δv. In particular, for $\Delta v = 0$ we have $\sigma >$ or $< \sigma_i$ for $\omega'_e > \omega''_e$. Therefore $\sigma - \sigma_i$ can be written as

$$\sigma - \sigma_i = (1 - \rho_i)\sigma_v.$$

Note that $\rho_i \propto \omega_i \propto 1/\sqrt{\mu_i}$, where μ_i refers to the reduced mass of the ith species.

In general the bands of the heavier isotopes are shifted towards the 0–0 band, the shift being proportional to Δv, as shown in Fig. 6.18.

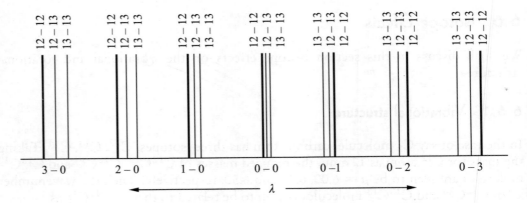

Fig. 6.18   *Isotope effect in the vibrational bands of $C^{12}C^{12}$, $C^{12}C^{13}$ and $C^{13}C^{13}$ molecules*

6.6.2 Rotational structure

From (6.7) we notice that the parameter B, which we have later denoted by $B(v)$, is inversely proportional to the reduced mass of the diatomic molecule. Therefore, the corresponding parameter for the isotopic molecule will have the ratios

$$B_{12,13}/B_{12,12} = 0.962, \quad B_{13,13}/B_{12,12} = 0.924. \tag{6.28}$$

Since B is proportional to $1/\mu_i$, we will have, for the ith isotope, $B_i \propto \rho_i^2$. Then (6.9) for the rotational lines becomes

$$F_J^i = [BJ(J+1) - DJ^2(J+1)^2]\rho_i^2 = \rho^2 F_J,$$

where ρ are the ratios as in (6.28). Therefore,

$$\sigma = \sigma_e + \sigma_v + F(J') - F(J'')$$

$$= \sigma_0 + (B(v') \pm B(v''))m + (B(v') - B(v''))m^2.$$

This finally gives

$$\sigma - \sigma^i = \sigma_0 - \sigma_0^i + (1 - \rho_i^2)[(B(v') \pm B(v''))m + (B(v') - B(v''))m^2]. \tag{6.29}$$

Thus, for a molecule with heavier isotopes, the parabola is more opened up. Combining with the shift of σ_0, we obtain parabolas shown in Fig. 6.19. The bandheads are shifted in the same direction as the band origins.

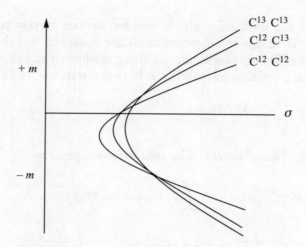

Fig. 6.19 *Isotope effect in rotational bands of the molecules $C^{12}C^{12}$, $C^{12}C^{13}$, $C^{13}C^{13}$*

Isotope effect is a very important tool in astrophysics for determining the relative abundances of various isotopes in the atmospheres of stars and planets.

6.7 Strengths of Bands and Lines

So far we have discussed the various types of molecular bands, electronic, vibrational and rotational, as well as their occurence as a mixture. We have also discussed selection rules for various cases. The intensity of all these bands and lines is a very important parameter which helps to characterise the spectrum and identify it with a particular molecule. Starting from the Frank–Condon principle, we shall now discuss certain aspects of emission and absorption causing these lines and bands and their relative intensities.

6.7.1 Frank–Condon principle

Transition times between molecular electronic transitions are usually small compared to the periods of vibrational motion. For example, the former could be of the order of femtoseconds to picoseconds, while the latter of the order of picoseconds to nanoseconds, with a ratio of two to three orders of magnitude. Hence in computing the transition probability, we can assume that the internuclear distance and the relative velocity do not change during the electronic transition. This is known as Frank–Condon principle. It can be formulated quantum mechanically as follows.

Consider two states of a diatomic molecule ψ' and ψ'', each being a product of electronic and vibrational functions of the form $\psi = \psi_e \psi_v$. Let **p** be the net electric dipole moment of the molecule, which may be written as $\mathbf{p} = \mathbf{p_e} + \mathbf{p_n}$, which are respectively the electric dipole moments of all the electrons and the nuclei. Since the dipole moments in a diatomic molecule are along the line of atoms, we drop the vector character. Then the transition probability R between states ψ' and ψ'' will be given by

$$R = \int \psi' p \psi'' d\tau,$$

where $d\tau = d\tau_e d\tau_n$. Therefore R can be split into two parts as

$$R = \int \psi'_e \psi'_v p_e \psi''_e \psi''_v d\tau_e d\tau_n + \int \psi'_e \psi'_v p_n \psi''_e \psi''_v d\tau_e d\tau_n \qquad (6.30)$$

$$= \int \psi'_e p_e \psi''_e d\tau_e \int \psi'_v \psi''_v d\tau_n + \int \psi'_v p_n \psi''_v d\tau_n \int \psi'_e \psi''_e d\tau_e.$$

The second term vanishes due to orthogonality of electronic wavefunctions. However, since the vibrational wavefunctions are restricted to the region between r_{min} and r_{max}, they may not be orthogonal to each other. Therefore we may write

$$R = \int \psi'_e p_e \psi''_e d\tau_e \int \psi'_v \psi''_v d\tau_n \equiv R_e R_\nu, \qquad (6.31)$$

where

$$R_e = \int \psi'_e p_e \psi''_e d\tau_e, \quad R_v = \int \psi'_v \psi''_v d\tau_n.$$

Note that R_e is the matrix element of the electronic dipole moment between two elecronic states of the molecule and R_v is the overlap integral of two vibrational functions. Following the Frank–Condon principle, we may assume a constant mean separation between the two nuclei. Accordingly, we replace R_e by the average matrix element $\overline{R_e}$, and write

$$R = \overline{R_e} \int \psi'_v \psi''_v d^3r. \qquad (6.32)$$

Thus the transition probability depends on the overlap of the vibrational wave functions.

Now ψ'_v and ψ''_v have maxima at the separation r_e for $v' = v'' = 0$, while they have maxima near the walls of the potential well for higher vibrational states; see Fig. 6.11. Therefore we can obtain the following qualitative results by applying Frank–Condon principle.

Absorption from $v'' = 0$ or emission from $v' = 0$

In Fig. 6.20(a), crosses denote the points of maximum probability of transition. So a vertical jump (in the V versus r diagram) corresponds to no variation in internuclear distance. This gives rise to cases A, B and C, when r'_e in the upper state is nearly equal to, greater than, and much greater than r''_e in the lower state, respectively. The variation of intensity of the line is shown in Fig. 6.20 (b).

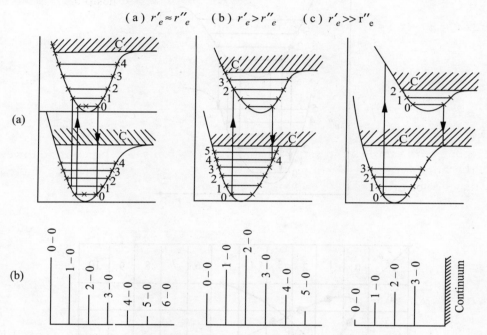

Fig. 6.20 *(a) Potential in upper and lower electronic states and the likely vibrational transitions; (b) relative intensities.*

We notice that in Case A, $r'_e \approx r''_e$, the 0–0 line has maximum intensity which goes on decreasing for 1–0, 2–0, etc transitions. The continuum may not exist in this case. In Case B, $r'_e > r''_e$, the 0–0 transition has some intensity, which increases for 1–0, etc, until it reaches a maximum for some $v' - 0$ and then again decreases. The continuum may or may not exist for this case. In Case C, $r'_e >> r''_e$, the band starts with weak 0–0 line, the intensity increases for 1–0, 2–0, etc., transitions, and the band may end in a possibly strong continuum.

Emission and absorption from higher vibrational states

In these cases the initial states have two probability maxima at $r_{\min}$ and $r_{\max}$, where $r_{\min}$ and $r_{\max}$ are the minimum and maximum values of r in a particular vibrational state. Hence two transitions will be strongest for each v' or v''. In this way we get the *Condon parabola* in the Deslandre Table 6.3. Due to the subsidiary maxima in the probability curve, we may get some satellite parabolas.

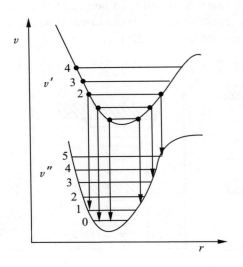

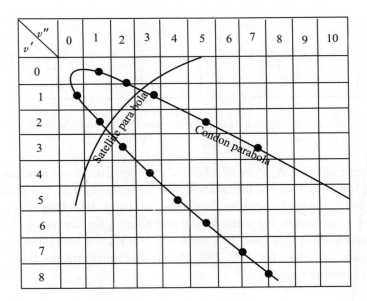

Fig. 6.21   *Condon parabola and a satellite parabola showing stronger vibrational transitions*

6.7.2 Relative intensities of bands

Let N_v' and N_v'' be the number densities of molecules in the vibrational state v' and v'', respectively, and let $R_{v'v''}$ be the overlap integral of the function ψ_v' and ψ_v''. Then the intensities of the $v' - v''$ transition line will be proportional to

$$I_{v'v''}^{\mathrm{em}} \propto N_{v'} A_{v'v''} h\nu,$$

$$I_{v'v''}^{\mathrm{abs}} \propto N_{v''} B_{v'v''} \hbar\nu I_0,$$

where $A_{v'v''}$ and $B_{v'v''}$ are the emission and absorption coefficients, respectively, to be discussed later in Section 7.4. These can be further seen to reduce to

$$I_{v'v''}^{\mathrm{em}} \propto N_{v'} \nu^4 |R_{v'v''}|^2,$$

$$I_{v'v''}^{\mathrm{abs}} \propto N_{v''} \, \nu |R_{v'v''}|^2. \tag{6.33}$$

If I_0 is the incident intensity, then we see that

$$\sum_{v''} I_{v'v''}^{\mathrm{em}}/\nu^4 \propto N_{v'},$$

$$\sum_{v'} I_{v'v''}^{\mathrm{abs}}/I_0\nu \propto N_{v''}. \tag{6.34}$$

The Boltzmann probability function gives the ratio of molecules in an excited vibrational state to that in the ground vibrational state as

$$N_v/N_0 = e^{-G(v)hc/kT},$$

where $G(v)$ is the wavenumber of the $v \to 0$ transition line. Since every molecule must be in some vibrational state, we must have

$$N = \sum_{v=0}^{\infty} N_v = N_0 \sum_{v=0}^{\infty} e^{-G(v)hc/k\tau} \equiv N_0 Z_v, \tag{6.35}$$

which defines Z_v, known as the *vibrational partition function*. Therefore,

$$N_v = \frac{Ne^{-G(v)hc/kT}}{Z_v}.$$

We could now write (6.33) as

$$\ln\left(\sum_{v''} I_{v'v''}^{\mathrm{em}}/\nu^4\right) = c_1 - G(v')hc/kT,$$

$$\ln \left(\sum_{v'} I^{\text{abs}}_{v'v''}/I_0 \nu \right) = c_2 - G(v'')hc/kT. \tag{6.36}$$

where c_1 and c_2 are constants. Thus a plot of the left-hand side of (6.36) against $G(v')$ and $G(v'')$, respectively, is a straight line with slope $-hc/k\tau$. The temperature obtained in this manner is called the *effective vibrational temperature*.

6.7.3   Relative intensities of rotational lines

In this case, we can write the number density of molecules in the Jth rotational state from among the molecules in the vth vibrational state as

$$N_J = \frac{N_v}{Z_r}(2J+1)e^{-F(J)hc/kT},$$

where $Z_r = \sum_{J=0}^{\infty}(2J+1)e^{-F(J)hc/kT}$

$$\simeq \sum_{J=0}^{\infty}(2J+1)\exp(-BJ(J+1)hc/kT) \tag{6.37}$$

is the *rotational partition function*. Note that B and $F(J)$ depend on the vibrational quantum number v.

First we note that N_J is maximum at J_{max} given by $dN_J/dJ = 0$, which gives $J_{\text{max}} = (-1 + (2kT/Bhc)^{1/2})/2$. This shows that J_{max} is approximately proportional to $T^{1/2}$ for $kT >> 2Bhc$. Thus we can determine the rotational temperature T by noting J_{max} of the initial state. See Fig. 6.9.

However, the standard method of obtaining the rotational temperature is as follows. For every value of J, there are $2J+1$ components. So we have to first calculate $\sum_{J'}|R_{J'J''}|^2 \equiv S$, and obtain $A_{J'J''}$ for one component. Since $A_{J'J''} \propto S/(2J+1)$, then it will be seen in Chapter 8 that

$$B_{J'J''} \propto \frac{c^3}{2h\nu^3}\frac{S}{2J''+1}. \tag{6.38}$$

Therefore, we have

$$\frac{I^{\text{em}}_{J'J''}}{S\nu^4} \propto \frac{N'_v}{Z'_r}\exp(-F(J')hc/kT),$$

$$\frac{I^{\text{abs}}_{J''J'}}{I_0 S\nu} \propto \frac{N''_v}{Z''_r}\exp(-F(J'')hc/kT). \tag{6.39}$$

So a logarithmic plot of the left-hand side and $F(J')$ or $F(J'')$ gives a straight line with a slope $-hc/k\tau$, which gives the rotational temperature.

For VR bands in Σ–Σ transition, $S = (J' + J'' + 1)/2$. So R branch with $J' = J'' + 1$ has $S = J'' + 1$, and P branch with $J' = J'' - 1$ has $S = J''$. Hence R branch is stronger than P branch. For other transitions we find that in Π–Σ transitions, R branch is stronger than P branch, in $\Sigma - \Pi$ transition, P branch is stronger, and in both cases, the Q branch is found to be the strongest. In Π–Π transition, R branch is stronger in emission, P branch is stronger in absorption and Q branch is the weakest.

6.7.4 Molecular continua

As we have seen in various cases above, in the vibrational and rotational spectra of diatomic molecules, the spectrum ends in a continuum. We will discuss some such cases here.

Absorption continuum

This occurs when the upper electronic state is a repulsive state as shown in Fig. 6.22. In this case, there will be a superposition of continua produced by transitions from various v-states of the lower electronic state. For each v-state ($v \neq 0$), there will be two maxima.

Emission continuum

This occurs when the lower electronic state is a repulsive state, as shown in Fig. 6.23. The lower state of H_2 is split into two branches, u and g, and u branch is repulsive. The H_2 molecule can be formed by the coming together of $H + H$ or $H^- + p$.

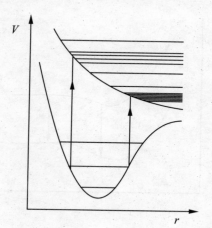

Fig. 6.22 *Origin of absorption continuum*

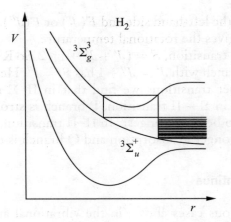

Fig. 6.23   *Emission continuum in H_2*

Incipient molecule

Many times both electronic states have shallow dips in their potential curves in which no vibrational levels can be accommodated. Thus two atoms may come close to each other, but on finding that there is not enough binding energy for them to settle down, keep going round each other with very little binding energy and no vibrations. In this case the electronic energy levels are displaced causing a displacement of the emission or absorption line; see Fig. 6.24. The collective effect is a broadening of the line around its normal position. The close approach of atoms is referred to as a collision, and collisions increase with pressure. This is, therefore, known as *pressure broadening* or *collisional broadening*. It is an important cause of line broadening in stellar atmosphere.

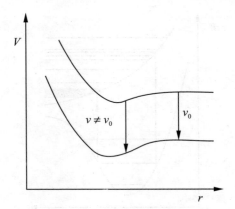

Fig. 6.24   *Collisional broadening*

6.8   Some Typical Examples of Molecular Spectra

Here we shall describe a few typical cases.

Aluminium oxide

Aluminium oxide (AlO) is a radical because the valancies of Al and O do not match. As the ground state of Al is $3p^1$ ($^2P_{1/2}$) and that of O is $2p^4$ ($^3P_2$), they combine to give

$$L_1 = 1, L_2 = 1, \Lambda = 0, 1, 2, (\Sigma, \Pi, \Delta);$$

$$S_1 = 1/2, S_2 = 1, \text{ giving } \Sigma = 1/2, 3/2 \text{ (doublet, quartet)}.$$

The ground state of AlO is $^2\Sigma$ and the next excited state is $^2\Sigma$. Transitions between them give bands in the wavelength region 4500-4000 Å as shown in Fig. 6.25 (a). We see that the bands are degraded towards red. Therefore, as discussed in Section 6.4.2 and Section 6.4.4, $r_e' > r_e''$ and $D_e' < D_e''$. Thus the upper state is less bound than the lower state, with $B(v') < B(v'')$ and $\omega_e' < \omega_e''$.

Nitrogen

Nitrogen (N_2) is a homonuclear molecule where the $2p^3$ electrons of each atom complete the p shell in the molecule. As the ground state of N is $^4S_{3/2}$, we have $L_1 = L_2 = 0$, giving $\Lambda = 0$, and $S_1 = S_2 = 3/2$, giving $\Sigma = 0, 1, 2, 3$. The ground state of N_2 is $^1\Sigma_g^+$. As one or more electrons are excited, we get the next excited states $^3\Sigma_u^+$ and $^3\Pi_g$, and further up the $^3\Pi_g$ state. Then the following transitions give rise to three bands which are as described here.

(i) $^3\Sigma_u^+ \rightarrow ^1\Sigma_g^+$ transition gives rise to *Vagard–Kaplan band*, in the UV region (2300–3400 Å),

(ii) $^3\Pi_g \rightarrow ^3\Sigma_u^+$ transitions give the first positive system in the red region at 5000–10000 Å, and

(iii) $^3\Pi_u \rightarrow ^3\Pi_g$ transitions give the second positive system in the violet region (2800–5000 Å).

Figure 6.25 (b) shows the second positive band system in which the bands are degraded toward violet. Here $r_e' < r_e''$ and $D_e' > D_e''$. So the upper state is more bound with $\omega_e' > \omega_e''$ and $B(v') > B(v'')$.

.

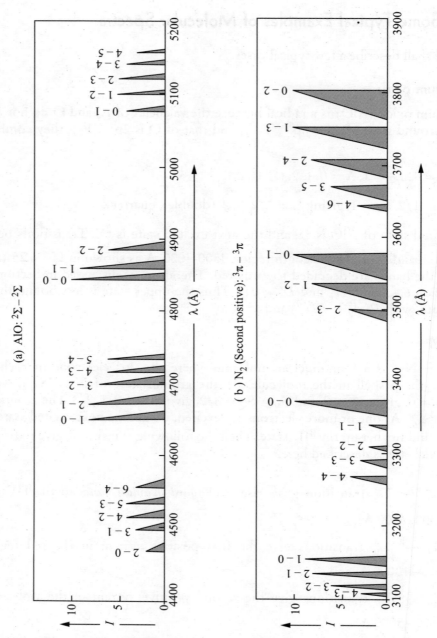

Fig. 6.25   *(a) Spectrum of AlO degraded towards red; (b) spectrum of N_2 degreaded to violet.*

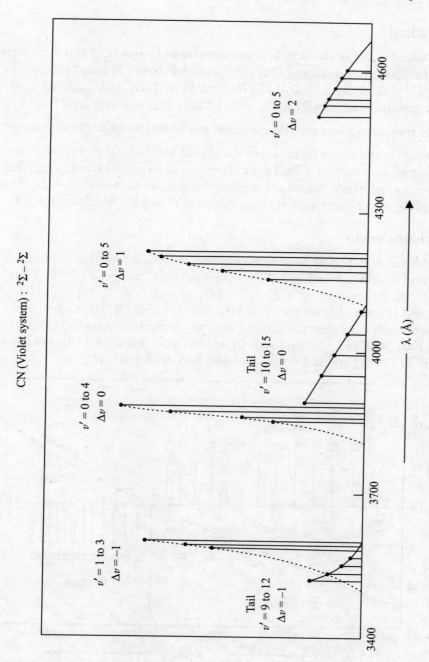

Fig. 6.26 *CN spectrum in violet showing tail bands*

Nitrile radical

Nitrile radical (CN) is also a radical as valencies of C and N, 4 and 3 respectively, do not match. The ground state of C is $2p^2$ ($^3P_0$) and that of N is $2p^3$ ($^4S_{3/2}$). So we have $L_1 = 1$, $L_2 = 0$, giving $\Lambda = 0, 1$ (Σ, Π), and $S_1 = 1$, $S_2 = 3/2$, giving $\Sigma = 1/2, 3/2$, 5/2. The ground state of CN is $^2\Sigma$ and the next higher states are $^2\Pi$ and $^2\Sigma$. So the $^2\Pi \rightarrow^2 \Sigma$ transition gives the red spectrum in the region $4400-9400$ Å, and $^2\Sigma \rightarrow^2 \Sigma$ transition gives the violet spectrum in the region $3600-4600$ Å. Figure 6.26 shows the violet system of bands of CN. In addition to the principal bands degraded towards violet, we see tail bands, which are degraded towards red. So $B(v') - B(v'')$ as well as $\Delta G(v') - \Delta G(v'')$ changes sign at some values of v' and v''. See Section 6.4.4.

Hydrochloric acid

Hydrochloric acid (HCl) is a heteronuclear molecule. As the ground states of H and Cl are $1s^1$ ($^2S_{1/2}$) and $3p^5$ ($^2P_{1/2}$), respectively, we get $L_1 = 0$, $L_2 = 1$, giving $\Lambda = 0, 1$ (Σ, Π), and $S_1 = 1/2$, $S_2 = 1/2$, giving $\Sigma = 0, 1$ (singlet, triplet). The ground and the next higher states of HCl are $^1\Sigma^+$ and $^2\Pi$, $^1\Pi$, respectively. We have noted in Section 6.3.2 that the ground state gives rise to vibrational bands, (1–0) 2886 cm^{-1}, (2–0) 5668 cm^{-1}, ... up to (5–0) 13396 cm^{-1}. Figure 6.27 shows the rotational structure of (1–0) band at 3.64 μm; see also Section 6.4.4.

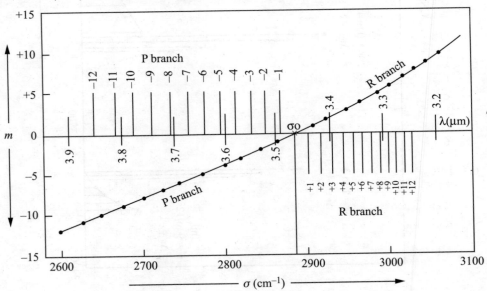

Fig. 6.27   *Rotational structure of $^2\Sigma \rightarrow^2 \Pi$ transition band of HCl around 3.46 μm. The curve corresponds to $\Sigma = 2885.90 + 20.577m - 0.3034m^2$ in cm^{-1}. The corresponding line spectrum is shown in the middle.*

Hydroxyl radical

Hydroxyl radical (OH) has $2p^4$ ($^3P_2$) and $1s$ ($^2S_{1/2}$) as the electronic structure for its atoms, respectively. Thus we have $L_1 = 1$, $L_2 = 0$, resulting in $\Lambda = 0, 1$ (Σ, Π) and $S_1 = 1$, $S_2 = 1/2$, resulting in $\Sigma = 1/2, 3/2$. But OH follows Hund's coupling (b) scheme. So we have to form $K = \Lambda, \Lambda + 1$, etc., and each K then gives a doublet with $J = K \pm 1/2$. Further, as $\pm\Lambda$ states are also split, we get four components in rotational lines. Characteristic signs of OH radical have been seen in the spectra of comets, night sky and late type M stars. VR bands of OH have been observed in 7000Å–9000Å region. They show (8–3), (4–0), (5–1), (6–2) and (7–3) vibrational bands. It has been inferred from those that $B(v') < B(v'')$ so that R branch closes at the head and P branch opens towards shorter wavelength. High resolution spectra of the supergiant star Betelguese M have been obtained with Fourier-transform spectrometer, which clearly show the four components of rotational lines in the OH spectrum between 2640 Å–3420 Å.

Problems

6.1. From the values of f and g given in the caption of Table 6.1, obtain (i) moment of inertia about an axis of rotation, (ii) intermolecular distance, (iii) natural frequency of vibrations, (iv) force constant k (as in $V(x) = kx^2/2$), for the $v = 0$ state of HCl molecule.

6.2. Taking the values of ω_e from Problem 6.1 and constants a and b from Section 6.3.1 for the HCl molecule, obtain the parameters x_e and y_e for it, using only quadratic terms in $G(v)$.

6.3. Taking B_0 from Problem 6.2, obtain B_1 from the best fit curve for σ given in Fig. 6.27, for the HCl molecule.

6.4. Enumerate the symmetry properties of the rotational energy levels of a diatomic molecule. Hence explain intensity alteration in the rotational Raman spectrum of a molecule with identical nuclei.

6.5 From the following list of bandheads of CN, construct a Deslandre scheme for the molecule and obtain its dissociation energy[22].

---

[22]Pearse and Gaydon (1950).

v'	v''	$\lambda(\text{Å})$	v'	v''	$\lambda(\text{Å})$
0	2	4604.1	11	10	3658.1
1	3	4578.0	10	9	3628.9
2	4	4553.1	9	8	3603.0
3	5	4531.9	1	0	3590.4
4	6	4514.8	2	1	3585.9
5	7	4502.2	3	2	3583.9
0	1	4216.0	13	11	3501.4
1	2	4197.2	12	10	3465.3
2	3	4181.0	11	9	3432.9
3	4	4167.8	10	8	3404.8
4	5	4158.1	9	7	3380.3
5	6	4152.4	8	6	3359.1
15	15	4078.7	7	5	3340.6
14	14	4029.3	6	4	3322.3
13	13	3984.7	5	3	3296.3
12	12	3944.6	10	7	3203.5
11	11	3909.5	9	6	3180.2
0	0	3883.4	8	5	3159.9
1	1	3871.4	7	4	3142.6
2	2	3861.9	6	3	3127.6
3	3	3851.7	5	2	3114.3
4	4	3850.9			

7

Quantum Theory of Radiation

WE studied the classical theory of radiation in Chapter 3. Having dealt with the rudiments of quantum mechanics and atomic and molecular spectra, it is appropriate to turn our attention to the quantum theory of radiation. There were a few decades in the second quarter of the twentieth century when the quantum mechanics of particles was fairly well understood (or so it was thought), but that of the radiation still remained at the heuristic level. We begin this chapter with quantization of electromagnetic radiation, followed by radiation together with matter. Then we deal with transition probabilities and amplitudes, in various orders, which are the most important aspects in making contact between theory and experiment. Then we introduce Einstein coefficients and oscillator strengths, ending the chapter with the Weisskopf–Wigner picture of broadening of energy levels.

7.1 Quantization of Pure Radiation

In this Chapter we shall discuss electromagnetic fields and their representation through various modes of vibration. Then we will discuss the harmonic oscillator model for radiation, the vector potential and Hamiltonian, creation and annihilation operators of the radiation field and some other aspects.

7.1.1 Fields and modes of vibration

We have seen in Chapter 3 that the energy density of the radiation field in a medium with permeability ϵ and permittivity μ is $u_{\mathrm{em}} = (\mathbf{E}\cdot\mathbf{D}+\mathbf{B}\cdot\mathbf{H})/8\pi = (\epsilon E^2 + \mu H^2)/8\pi$. So the total energy of radiation in a volume τ will be

$$H_{\mathrm{r}} = \frac{1}{8\pi}\int_{\tau}(\epsilon E^2 + \mu H^2)d\tau, \tag{7.1}$$

where $d\tau = d^3r$. The **E** and **B** fields in vacuum are defined through

$$\mathbf{B} = \boldsymbol{\nabla} \times \mathbf{A}, \quad \mathbf{E} = -\boldsymbol{\nabla}\varphi - \frac{1}{c}\frac{\partial \mathbf{A}}{\partial t}. \tag{7.2}$$

The potentials satisfy the wave equation (3.40) and are subject to Lorentz gauge. In the case of pure radiation, source-free vacuum, we have $\rho = 0, \mathbf{J} = \mathbf{0}$, and we can put $\varphi = 0$. Then these equations reduce to

$$\nabla^2 \mathbf{A} - \frac{1}{c^2}\frac{\partial^2 \mathbf{A}}{\partial t^2} = 0, \quad \boldsymbol{\nabla} \cdot \mathbf{A} = 0. \tag{7.3}$$

The radiation energy then becomes

$$H_{\mathrm{r}} = \frac{1}{8\pi} \int (\mathbf{E}^2 + \mathbf{B}^2) d\tau. \tag{7.4}$$

Let us obtain $\mathbf{A}$ for radiation inside a cubical box of size L^3. We assume periodic boundary conditions in each Cartesian direction to obtain the allowed wave vectors. Then we have to consider a multiplicity of modes with different wave vectors $\mathbf{k}$ and with two polarizations denoted by α. Also, let us separate the time and space variables by putting

$$\mathbf{A}(\mathbf{r},t) = \sum_{\mathbf{k}\alpha} \mathbf{A}_{\mathbf{k}\alpha}(\mathbf{r},t) = \sum_{\mathbf{k}\alpha} q_{\mathbf{k}\alpha}(t)\mathbf{A}_{\mathbf{k}\alpha}(\mathbf{r}). \tag{7.5}$$

Then for each mode, we have

$$q_{\mathbf{k}\alpha}\nabla^2\mathbf{A}_{\mathbf{k}\alpha}(\mathbf{r}) - \frac{1}{c^2}\ddot{q}_{\mathbf{k}\alpha}\mathbf{A}_{\mathbf{k}\alpha}(\mathbf{r}) = 0, \quad \boldsymbol{\nabla} \cdot \mathbf{A}_{\mathbf{k}\alpha}(\mathbf{r}) = 0. \tag{7.6}$$

We note from the above that $\nabla^2\mathbf{A}_{\mathbf{k}\alpha}(\mathbf{r})$ and $\mathbf{A}_{\mathbf{k}\alpha}(\mathbf{r})$ are in the same direction. The variables $\mathbf{r}$ and t will be separated by putting

$$\frac{c^2|\nabla^2\mathbf{A}_{\mathbf{k}\alpha}|}{|\mathbf{A}_{\mathbf{k}\alpha}|} = \frac{\ddot{q}_{\mathbf{k}\alpha}}{q_{\mathbf{k}\alpha}},$$

which should be a constant independent of $\mathbf{r}$ and t.
    For periodic variation, we take

$$q_{\mathbf{k}\alpha} = a_{\mathbf{k}\alpha}e^{i\omega t}, \tag{7.7}$$

where $a_{\mathbf{k}\alpha}$ are constants. Then we have

$$\ddot{q}_{\mathbf{k}\alpha} = -\omega^2 q_{\mathbf{k}\alpha}, \tag{7.8}$$

giving $\nabla^2\mathbf{A}_{\mathbf{k}\alpha}(\mathbf{r}) + \dfrac{\omega^2}{c^2}\mathbf{A}_{\mathbf{k}\alpha}(\mathbf{r}) = 0.$ \hfill (7.9)

The solution for $\mathbf{A}_{\mathbf{k}\alpha}(\mathbf{r})$ will be in the form of stationary transverse waves given by

$$\mathbf{A}_{\mathbf{k}\alpha}(\mathbf{r}) = b_{\mathbf{k}\alpha}\boldsymbol{\epsilon}_{\alpha}\exp(i\mathbf{k}\cdot\mathbf{r}), \tag{7.10}$$

where $\boldsymbol{\epsilon}_{\alpha}$ is a unit vector in the direction of polarization, satisfying the relation $\boldsymbol{\epsilon}_{\alpha}\cdot\mathbf{k} = 0$, which follows from the second of (7.6), and $b_{\mathbf{k}\alpha}$ are constants giving the amplitude of the $(\mathbf{k}\alpha)$-component. Equation (7.10) implies

$$\nabla^2\mathbf{A}_{\mathbf{k}\alpha}(\mathbf{r}) + k^2\mathbf{A}_{\mathbf{k}\alpha}(\mathbf{r}) = \mathbf{0}. \tag{7.11}$$

Equation (7.10) will be a solution of (7.9) if

$$k = \omega/c = 2\pi/\lambda. \tag{7.12}$$

For a stationary wave within a cube of side L, we must have

$$k_x = 2\pi n_x/L, \ \text{ etc.,}$$

where n_x etc are integers.

Using (7.5), (7.7) and (7.10) and letting

$$a_{\mathbf{k}\alpha}b_{\mathbf{k}\alpha} = c_{\mathbf{k}\alpha}, \tag{7.13}$$

the vector potential finally becomes

$$\mathbf{A}(\mathbf{r}, t) = \sum_{\mathbf{k}\alpha} c_{\mathbf{k}\alpha}\boldsymbol{\epsilon}_{\alpha}\exp[i(\mathbf{k}\cdot\mathbf{r} - \omega t)]. \tag{7.14}$$

7.1.2 Hamiltonian for the radiation field

The total energy of the radiation field in the cubic box is given by

$$H_{\mathrm{r}} = \frac{L^2}{8\pi}(\mathbf{E}^2 + \mathbf{B}^2).$$

The fields $\mathbf{E}$ and $\mathbf{B}$ are determined from the vector potential of (7.14) by using (7.2). Each state $\mathbf{k}\alpha$ of the electromagnetic radiation has fields which may be decomposed into components $\mathbf{E}_{\mathbf{k}\alpha}$ and $\mathbf{B}_{\mathbf{k}\alpha}$. Then the energy density can be written as

$$H_{\mathrm{r}} = \frac{1}{8\pi}\sum_{\mathbf{k}\alpha}(|\mathbf{E}_{\mathbf{k}\alpha}|^2 + |\mathbf{B}_{\mathbf{k}\alpha}|^2). \tag{7.15}$$

Since $\varphi = 0$, the potential energy is zero, and the Lagrangian (density) is same as the Hamiltonian (density). Now,

$$\mathbf{E}_{\mathbf{k}\alpha} = -\frac{1}{c}\frac{\partial \mathbf{A}_{\mathbf{k}\alpha}(\mathbf{r}, t)}{\partial t} = \frac{i\omega}{c}\mathbf{A}_{\mathbf{k}\alpha}(\mathbf{r}, t).$$

Also,

$$\mathbf{B}_{\mathbf{k}\alpha} = \nabla \times \mathbf{A}_{\mathbf{k}\alpha}(\mathbf{r}, t) = ib_{\mathbf{k}\alpha}q_{\mathbf{k}\alpha}(t)e^{i\mathbf{k}\cdot\mathbf{r}}(\mathbf{k} \times \epsilon_\alpha). \qquad (7.16)$$

Let $\mathbf{n}$ be a unit vector in the direction of propagation and ϵ_B a unit vector in the direction of the magnetic field, so that ϵ_α, ϵ_B and $\mathbf{n}$ make a right handed triad. Then $\mathbf{k} \times \epsilon_\alpha = k\epsilon_B$, and the magnetic field of (7.16) becomes

$$\mathbf{B}_{\mathbf{k}\alpha} = \frac{i\omega}{c}|\mathbf{A}_{\mathbf{k}\alpha}|\epsilon_B. \qquad (7.17)$$

Since $\dot{\mathbf{A}}(\mathbf{r}, t) = -i\omega\mathbf{A}_{\mathbf{k}\alpha}(\mathbf{r}, t)$, we now write the Hamiltonian density of (7.15) as

$$H_{\mathrm{r}} = \frac{1}{8\pi c^2}\sum_{\mathbf{k}\alpha}(|\dot{\mathbf{A}}_{\mathbf{k}\alpha}|^2 + \omega^2|\mathbf{A}_{\mathbf{k}\alpha}|^2). \qquad (7.18)$$

Let us define

$$\mathbf{Q}_{\mathbf{k}\alpha} = \mathbf{A}_{k\alpha}(\mathbf{r}, t)/c\sqrt{4\pi}. \qquad (7.19)$$

as the generalised coordinates of the radiation field. This allows us to write (7.18) in the form

$$H_{\mathrm{r}} = \frac{1}{2}\sum_{\mathbf{k}\alpha}(|\dot{\mathbf{Q}}_{k\alpha}|^2 + \omega^2|\mathbf{Q}_{k\alpha}|^2). \qquad (7.20)$$

Noting that the Lagrangian density $L_{\mathbf{r}}$ is same as the Hamiltonian density, the generalised momenta are seen to be

$$\mathbf{P}_{k\alpha} = \partial L_{\mathbf{r}}/\partial \dot{\mathbf{Q}}_{k\alpha} = \dot{\mathbf{Q}}_{k\alpha}. \qquad (7.21)$$

Then

$$L_{\mathrm{r}} = H_{\mathbf{r}} = \frac{1}{2}\sum_{\mathbf{k}\alpha}(|\mathbf{P}_{k\alpha}|^2 + \omega^2|\mathbf{Q}_{k\alpha}|^2). \qquad (7.22)$$

Equation (7.22) satisfies the Hamiltonian equations.

$$\frac{\partial H_{\mathrm{r}}}{\partial \mathbf{Q}_{\mathbf{k}\alpha}} = \omega^2\mathbf{Q}_{k\alpha} = -\dot{\mathbf{P}}_{k\alpha}, \quad \frac{\partial H_{\mathrm{r}}}{\partial \mathbf{P}_{\mathbf{k}\alpha}} = \mathbf{P}_{k\alpha} = \dot{\mathbf{Q}}_{k\alpha}. \qquad (7.23)$$

Thus H_{r} is the Hamiltonian of the radiation field.

7.1.3 The photon hypothesis

The Hamiltonian of (7.20) suggests that a pure radiation field can be considered as a system of infinite number of independent oscillators. However, they are not oscillators of matter particles. Each packet of radiation is a *photon* of energy $\hbar\omega = hc/\lambda$. We can have transitions of a matter particle between two energy levels by absorption or emission of energy $n\hbar\omega$. Thus, in a process involving radiation, the energy that can be transferred is

$$E = \sum_{\mathbf{k}\alpha} n_{\mathbf{k}\alpha} \hbar\omega_{\mathbf{k}\alpha}. \tag{7.24}$$

The radiation field gains or loses energy by addition or loss of one or more photons of energy $\hbar\omega$. The photon hypothesis is supported by experiments such as photoelectric effect and Compton scattering. It is also supported by the derivation of black body radiation spectrum based on it. A photon also carries a momentum equal to $\hbar\mathbf{k}$ where $\mathbf{k}$ is the wave vector in the direction of propagation of photon and magnitude equal to $2\pi/\lambda$. Thus the magnitude of the photon momentum is also equal to $h\nu/c = \hbar\omega/c$.

7.1.4 Some operators of the radiation field

The variables $Q_{\mathbf{k}\alpha}$ and $P_{\mathbf{k}\alpha}$ can be considered as quantum mechanical operators, so that $P_{\mathbf{k}\alpha} = -\hbar\partial/\partial Q_{\mathbf{k}\alpha}$. Then the Schrödinger equation for the radiation field can be written as

$$H_{\mathrm{r}}(Q_{\mathbf{k}\alpha}, P_{\mathbf{k}\alpha})\psi = E_n\psi.$$

It has eigenvalues $E_{\mathbf{k}\alpha} = n_{\mathbf{k}\alpha}\hbar\omega_{\mathbf{k}\alpha}$. The variables $Q_{\mathbf{k}\alpha}, P_{\mathbf{k}\alpha}$ satisfy the commutation relations

$$[Q_{\mathbf{k}\alpha}, Q_{\mathbf{k}'\alpha'}] = 0, \quad [P_{\mathbf{k}\alpha}, P_{\mathbf{k}'\alpha'}] = 0,$$

$$[Q_{\mathbf{k}\alpha}, P_{\mathbf{k}'\alpha'}] = i\hbar\delta_{\mathbf{k}\mathbf{k}'}\delta_{\alpha\alpha'}.$$

We now define operators

$$a_{\mathbf{k}\alpha} = (1/2\hbar\omega_{\mathbf{k}})^{1/2}(\omega_{\mathbf{k}}Q_{\mathbf{k}\alpha} + iP_{\mathbf{k}\alpha}),$$

$$a_{\mathbf{k}\alpha}^{\dagger} = (1/2\hbar\omega_{\mathbf{k}})^{1/2}(\omega_{\mathbf{k}}Q_{\mathbf{k}\alpha} - iP_{\mathbf{k}\alpha}), \tag{7.25a}$$

$$N_{\mathbf{k}\alpha} = a_{\mathbf{k}\alpha}^{\dagger}a_{\mathbf{k}\alpha}. \tag{7.25b}$$

They are seen to satisfy the relations

$$[a_{\mathbf{k}\alpha}, a_{\mathbf{k}'\alpha'}^{\dagger}] = \delta_{\mathbf{k}\mathbf{k}'}\delta_{\alpha\alpha'},$$

$$[a_{\mathbf{k}\alpha}, a_{\mathbf{k}'\alpha'}] = [a^\dagger_{\mathbf{k}\alpha}, a^\dagger_{\mathbf{k}'\alpha'}] = 0, \qquad\qquad (7.26a)$$

$$[a_{\mathbf{k}\alpha}, N_{\mathbf{k}'\alpha'}] = a_{\mathbf{k}'\alpha'}\delta_{\mathbf{k}\mathbf{k}'}\delta_{\alpha\alpha'}. \qquad\qquad (7.26b)$$

These commutation relations lead us to the following matrix representation for the operators a and $a^\dagger$ (dropping the subscripts when they are same on all operators). The nm-element of the three operators is seen to be

$$a_{nm} = \sqrt{n}\delta_{m,n+1}, a^\dagger_{nm} = \sqrt{n-1}\delta_{m,n-1}. \qquad\qquad (7.27a)$$

$$N_{nm} = (n-1)\delta_{nn}. \qquad\qquad (7.27b)$$

Written out fully, the matrices appear to be

$$a = \begin{bmatrix} 0 & 1 & 0 & 0 & 0 & \cdot & \cdot & \cdot \\ 0 & 0 & \sqrt{2} & 0 & 0 & \cdot & \cdot & \cdot \\ 0 & 0 & 0 & \sqrt{3} & 0 & 0 & \cdot & \cdot \\ \vdots & & & & & & & \end{bmatrix},$$

$$a^\dagger = \begin{bmatrix} 0 & 0 & 0 & 0 & 0 & \cdot & \cdot & \cdot \\ 1 & 0 & 0 & 0 & 0 & \cdot & \cdot & \cdot \\ 0 & \sqrt{2} & 0 & 0 & 0 & \cdot & \cdot & \cdot \\ 0 & 0 & \sqrt{3} & 0 & 0 & 0 & \cdot & \cdot \\ \vdots & & & & & & & \end{bmatrix},$$

$$N = \begin{bmatrix} 0 & 0 & 0 & 0 & 0 & 0 & 0 & \cdot & \cdot & \cdot \\ 0 & 1 & 0 & 0 & 0 & 0 & 0 & \cdot & \cdot & \cdot \\ 0 & 0 & 2 & 0 & 0 & 0 & 0 & \cdot & \cdot & \cdot \\ 0 & 0 & 0 & 3 & 0 & 0 & 0 & \cdot & \cdot & \cdot \\ \vdots & & & & & & & & & \end{bmatrix}. \qquad\qquad (7.28)$$

Let us now consider a column vector $|n\rangle$ given by

$$|n\rangle = \begin{bmatrix} 0 \\ 0 \\ \vdots \\ 1_{n+1} \\ 0 \\ 0 \\ \vdots \end{bmatrix}, \qquad\qquad (7.29)$$

which means it contains zeros in all rows except $(n + 1)$-row which has 1 in it. That is

$$|n\rangle_i = \delta_{i,n+1} \tag{7.30}$$

Thus we will have column vectors such as

$$|0\rangle = \begin{bmatrix} 1 \\ 0 \\ 0 \\ \vdots \end{bmatrix}, \quad |1\rangle = \begin{bmatrix} 0 \\ 1 \\ 0 \\ 0 \\ 0 \\ \vdots \end{bmatrix}, \quad |2\rangle = \begin{bmatrix} 0 \\ 0 \\ 1 \\ 0 \\ 0 \\ 0 \\ \vdots \end{bmatrix}, \text{ etc.} \tag{7.31}$$

With this representation, we have

$$a|n\rangle = \sqrt{n}|n-1\rangle, a^\dagger|n\rangle = \sqrt{n+1}|n+1\rangle, N|n\rangle = a^\dagger a|n\rangle = n|n\rangle. \tag{7.32}$$

7.1.5   Significance of em field operators

The pure radiation field can be considered as an assembly of independent oscillators with frequencies $\omega_{\mathbf{k}}$ and polarization vectors ϵ_α. The eigenstate $|n_{\mathbf{k}\alpha}\rangle$ represents a state where there are $n_{\mathbf{k}\alpha}$ photons of energy $\hbar\omega_{\mathbf{k}}$. Starting from the state

$$|0\rangle = \begin{bmatrix} 1 \\ 0 \\ 0 \\ \vdots \end{bmatrix},$$

we operate on it successively by $a^\dagger$ to get

$$a^\dagger|0\rangle = 1|1\rangle,$$

$$(a^\dagger)^2|0\rangle = a^\dagger|1\rangle = \sqrt{2}|2\rangle,$$

$$(a^\dagger)^3|0\rangle = \sqrt{2}a^\dagger|2\rangle = \sqrt{3!}|3\rangle, \text{ etc.,}$$

so that

$$(a^\dagger)^n|0\rangle = \sqrt{n!}|n\rangle.$$

This shows that the eigenstate $|n\rangle$ can be generated from the vacuum state $|0\rangle$ by

$$|n\rangle = (a^\dagger)^n |0\rangle / \sqrt{n!}. \tag{7.33}$$

Thus $a^\dagger$ represents the creation operator. Similarly the operator a (or $a_{\mathbf{k}\alpha}$) represents the annihilation operator. Finally the operator $N = a^\dagger a$ acting on a state $|n\rangle$ results in $n|n\rangle$. It preserves the state and shows how many photons ($n_{\mathbf{k}\alpha}$) of the wave vector and polarization $\mathbf{k}\alpha$ the radiation field has. The three operators $a^\dagger$, N and a may be likened to the trinity, Brahma, Vishnu, and Mahesh, of the Indian mythology.

The operators come into play when there is interaction of radiation with matter, leading to absorption and emission of photons. The uncertainty principle also allows for creation and destruction of virtual photons[23]. A microscopic piece of matter may emit a photon without satisfying energy conservation. If ΔE is the energy discrepancy in the interaction, such a photon must be re-absorbed within a time interval Δt satisfying $\Delta E \Delta t \simeq \hbar$.

7.2 Radiation and Matter

In this section we shall discuss the combined Hamiltonian of a system containing matter and radiation, and transition probabilities. The matter is represented by particles having mass m_l, charge q_l, momentum p_l, and the radiation is represented by the potentials $(\varphi_l, \mathbf{A}_l)$.

7.2.1 The Hamiltonian

The non-relativistic Hamiltonian for a system of particles in the presence of electro-static field, with potential φ, is

$$H = \sum_l (q_l \varphi_l + p_l^2 / 2m_l). \tag{7.34}$$

In the presence of electromagnetic radiation, we replace $\mathbf{p}$ by $\mathbf{p} - q\mathbf{A}/c$ (see Section 3.3.7), and add the Hamiltonian of the radiation field to get

$$H = \sum_l \left\{ q_l \varphi_l + \frac{1}{2m_l} \left(\mathbf{p}_l - \frac{q_l \mathbf{A}_l}{c} \right)^2 \right\} + \sum_{\mathbf{k}\alpha} \frac{1}{2} (\mathbf{P}_{\mathbf{k}\alpha}^2 + \omega^2 \mathbf{Q}_{\mathbf{k}\alpha}^2). \tag{7.35}$$

Note that q_l, $\mathbf{p}_l$ here are the charge and momentum of the particle and $\mathbf{Q}_{\mathbf{k}\alpha}, \mathbf{P}_{\mathbf{k}\alpha}$ the generalised coordinates of the radiation field. Also, $\mathbf{A}_l$ is the vector potential $\mathbf{A}(\mathbf{r}, t)$ at the position of the lth particle. This Hamiltonian can now be written as

$$H = H_0 + H', \tag{7.36}$$

---

[23]Philosophically speaking, a virtual photon is one which has not been detected.

where H_0 is the unperturbed Hamiltonian including matter and radiation and H' is the perturbation caused by the interaction between them. Thus,

$$H_0 = \sum_l \{q_l\varphi_l + p_l^2/2m_l\} + \sum_{\mathbf{k}\alpha} \frac{1}{2}(P_{\mathbf{k}\alpha}^2 + \omega^2 Q_{\mathbf{k}\alpha}^2)$$

$$H' = -\frac{1}{c}\sum_l \frac{q_l}{m_l}\mathbf{p}_l \cdot \mathbf{A}_l + \frac{1}{2c^2}\sum_l \frac{q_l^2}{m_l}|\mathbf{A}_l|^2. \tag{7.37}$$

The first term in (7.37) gives rise to emission and absorption, and the second term gives rise to electron scattering or Compton effect. Usually, the latter is much weaker than the former due to the factor of $1/c^2$. Therefore, we can drop that term and write

$$H' = -\frac{1}{c}\sum_l \frac{q_l}{m_l}\mathbf{p}_l \cdot \mathbf{A}_l.$$

Using (7.14), this becomes

$$H' = -\frac{1}{c}\sum_l \sum_{\mathbf{k}\alpha} \frac{q_l}{m_l}c_{\mathbf{k}\alpha}\boldsymbol{\epsilon}_\alpha \cdot \mathbf{p}_l \exp[i(\mathbf{k}_\alpha \cdot \mathbf{r}_l - \omega_k t)]. \tag{7.38}$$

We shall consider dipole approximation which is characterised by the neglect of phase change over the size of the system. Thus, we assume $|\mathbf{r}_l| \ll \lambda$, which reduces the perturbation to

$$H' = -\frac{1}{c}\sum_l \sum_{\mathbf{k}\alpha} \frac{q_l}{m_l}c_{\mathbf{k}\alpha}\boldsymbol{\epsilon}_\alpha \cdot \mathbf{p}_l \, e^{i\omega_k t}. \tag{7.39}$$

Let us write this as

$$H' = \mathbf{X}_{\mathrm{m}} \cdot \mathbf{X}_{\mathrm{r}}, \tag{7.40a}$$

where

$$\mathbf{X}_{\mathrm{m}} = -\frac{1}{c}\sum_l q_l\mathbf{p}_l/m_l, \quad \mathbf{X}_{\mathrm{r}} = \sum_{\mathbf{k}\alpha} \boldsymbol{\epsilon}_\alpha c_{\mathbf{k}\alpha}e^{-i\omega_\alpha t}. \tag{7.40b}$$

Here the subscripts m and r refer to matter and radiation, respectively.

We would now consider the effect of the perturbing Hamiltonian. As electrons are mainly responsible for the process of radiation, we shall put $q_l = -e$ and $m_l = m$, the electron charge and mass. Then we may write

$$\mathbf{X}_{\mathrm{m}} = \frac{e}{mc} \sum_l \mathbf{p}_l = \frac{e}{c} \sum_l \dot{\mathbf{r}} = \frac{\dot{\mathbf{P}}}{c}, \tag{7.41}$$

where $\mathbf{P}$ is the polarization.

7.2.2 Transition probabilities

Let the eigenvalues of the unperturbed Hamiltonian be E_k^0 and the eigenfunctions be

$$\Psi^0(\mathbf{r}, t) = \psi_k^0 \exp(-iE_k^0 t/\hbar). \tag{7.42}$$

Then for $H = H_0 + H'$, where $H' \ll H_0$, we get new eigenvalues $E_k = E_k^0 + E_k'$, and also new eigenfunctions corresponding to E_k. In the perturbation theory discussed in Chapter 5, we have seen how we get the perturbation shifts E_k' as the matrix elements of the perturbation in the form

$$E_k' = H_{kk}' = \int \Psi_k^{0*} H' \Psi_k^0 d\tau. \tag{7.43}$$

Now, let the new composite eigenfunctions be

$$\Psi(\mathbf{r}, t) = \sum_k a_k(t) \psi_k^0 \exp(-iE_k^0 t/\hbar), \tag{7.44}$$

that is, we have a sum of the unperturbed eigenfunctions with time-dependent coefficients. But $|a_k(t)|^2$ is the probability of finding the system in state k at time t. As we see, this probability changes with time, which means that transitions from one state to another take place due to absorption or emission of photons. In order to find the transition probabilities, we have to consider the fundamental time-dependent wave equation

$$H\Psi = i\hbar \partial \Psi / \partial t. \tag{7.45}$$

Using the operator equivalent of H, that is, $i\hbar \partial / \partial t$, $H = H_0 + H'$ and (7.44), we get the time dependence of the coefficients as

$$i\hbar \dot{a}_j(t) = \sum_k a_k H_{jk}' \exp[-i(E_k^0 - E_j^0)t/\hbar], \tag{7.46}$$

where

$$H_{jk}' = \int \psi_j^{0*} H' \psi_k^0 d\tau. \tag{7.47}$$

Assume that the unperturbed system is in state k and the perturbation is applied at time $t = 0$. Then we have to solve (7.46) with the initial condition

$$a_{kj} = \delta_{jk}. \tag{7.48}$$

It will be seen that H'_{jk} are responsible for transitions between various levels, giving rise to emission and absorption of photons. In other words, they activate the creation and annihilation operators.

7.2.3 Computation of matrix elements

Due to (7.40a), we can write

$$H'_{jk} = \int \psi^{0*}_{mj} \psi^{0*}_{rk} \mathbf{X}_m \cdot \mathbf{X}_r \psi^0_{mk} \psi^0_{rk} d\tau_m d\tau_r = \mathbf{X}_{m,jk} \cdot \mathbf{X}_{r,jk}, \tag{7.49}$$

where

$$\mathbf{X}_{m,jk} = \int \psi^{0*}_{mj} \mathbf{X}_m \psi^0_{mk} d\tau_m = \dot{\mathbf{P}}_{jk}/c, \tag{7.50}$$

with

$$\dot{\mathbf{P}}_{jk} = \frac{d}{dt} \int \psi^0_{mj} e^{iE^0_j t/\hbar} \; \mathbf{P} \; \psi^0_{mk} e^{-iE^0_k t/\hbar} d\tau_m$$

$$= \frac{i}{\hbar}(E^0_j - e^0_k)\mathbf{P}_{jk} = \pm i\omega_{jk}\mathbf{P}_{jk},$$

$$= \pm ie\omega_{jk}(\sum \mathbf{r}_l)_{jk}, \tag{7.51}$$

which defines ω_{jk} as the transition frequency. Here the sign depends on which state, E^0_j or E^0_k, is higher. The positive sign indicates absorption while the negative sign indicates emission.

It is convenient to define $Q_{kj,m}$ as

$$Q_{kj,m} = ie\omega\boldsymbol{\epsilon}_\alpha \cdot [\Sigma \mathbf{r_l}]_{jk}. \tag{7.52}$$

As the particle traverses the field of an atom, the change in its momentum can be related to the force it experiences, that is,

$$\mathbf{dp} = -\boldsymbol{\nabla}V dt.$$

Integrating this over a period T of the transition frequency, we have

$$\mathbf{p} = -\int_0^T \boldsymbol{\nabla}V dt = -\frac{\boldsymbol{\nabla}V}{\omega_{jk}}. \tag{7.53}$$

The various methods to determine the matrix element generally give different results because the wavefunctions are not known correctly. The method based on $\mathbf{p}$ favours large $\nabla\psi$ portions, the method based on $\mathbf{r}$ favours large r regions, and the method based on ∇V favours smaller r parts.

In the next section we evaluate the matrix element of the radiation field $\mathbf{X_r}$.

7.3 First Order Approximation for Transitions

7.3.1 Selection rules for hydrogen atom

We can evaluate $\mathbf{X_m}$ from (7.50). But we shall proceed more simply and directly. Let radiation of frequency ω_{kj} be incident along the x-axis; see Fig. 7.1. The force acting on the electron is given by $\mathbf{F} = e(\mathbf{E} + \mathbf{v} \times \mathbf{H}/c)$. As $|E| \sim |H|$, the effect of the magnetic field is much less due to the factor v/c. So we shall consider only the electric field associated with the radiation, that is, we shall neglect magnetic dipole transitions.

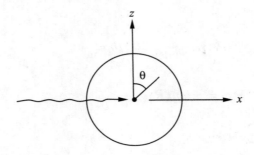

Fig. 7.1 *Geometry of radiation falling on a hydrogen atom*

Let the transverse components of $\mathbf{E}$ be given by

$$E_y = E_y^0 \sin(\omega t - kx), \quad E_z = E_z^0 \sin(\omega t - kx - \delta).$$

Since x varies over a range of $1\overset{\circ}{\mathrm{A}}$ and $\lambda \sim 5000\overset{\circ}{\mathrm{A}}$ for visible radiation, we have $kx \sim 10^{-3}$. Therefore we can neglect phase change within the atom. This is the electric dipole approximation. Then the perturbing Hamiltonian $\mathbf{F} \cdot \mathbf{r}$ can be taken as

$$\begin{aligned} H' &= -eE_z^0 z \sin(\omega t - \delta) - eE_y^0 y \sin\omega t \\ &= -eE_z^0 r \sin\theta \sin(\omega t - \delta) - eE_y^0 r \sin\theta \sin\varphi \sin\omega t \\ &= H_{\mathrm{I}}' + H_{\mathrm{II}}', \end{aligned} \tag{7.54}$$

which defines the two parts of the perturbation.

The hydrogen atom wavefunctions are of the form

$$\psi_{nlm} \propto R_{nl}(r) P_l^{|m|}(\cos\theta) e^{im\varphi}.$$

We shall now evaluate the matrix elements of H_{I} and H_{II} separately. First, we have

$$H'_{\mathrm{I}n'l'm',nlm} = \int \psi^*_{n'l'm'} H'_{\mathrm{I}} \psi_{nlm} d^3r. \tag{7.55}$$

This involves factors like

$$\int_0^\pi P_{l'}^{|m'|}(\cos\theta) P_l^{|m|}(\cos\theta) \cos\theta \sin\theta d\theta$$

and

$$\int_0^{2\pi} e^{i(m-m')\varphi} d\varphi. \tag{7.56}$$

On using well-known formulas involving Legendre polynomials and others, such as

$$\int_0^{2\pi} e^{i(m-m')\varphi} d\varphi = 2\pi\delta_{mm'},$$

$$\int_{-1}^1 P_l^{|m|}(x) P_{l'}^{|m|}(x) dx = \delta_{ll'},$$

and the recursion relation

$$x P_l^{|m|}(x) = \frac{l-|m|+1}{2l+1} P_{l+1}^{|m|}(x) + \frac{l+|m|}{2l+1} P_{l-1}^{|m|}(x), \tag{7.57}$$

we get the selection rules for electric dipole transitions,

$$\Delta m = 0, \quad \Delta l = \pm 1. \tag{7.58}$$

These are the selection rules for radiation polarized in the z direction.

Similarly, the matrix element of H'_{II} between states (nlm) and $(n'l'm')$ will involve factors like

$$\int_0^\pi P_{l'}^{|m'|}(\cos\theta) P_l^{|m|}(\cos\theta) \sin^2\theta d\theta$$

and

$$\int_0^{2\pi} e^{i(m-m')\varphi} \sin\varphi d\varphi.$$

This leads to the selection rules for radiation polarized in the y direction,

$$\Delta m = \pm 1, \ \ \Delta l = \pm 1. \tag{7.59}$$

The two together give the selection rules for radiation with any polarization,

$$\Delta m = 0, \ \ \pm 1, \ \ \Delta l = \pm 1. \tag{7.60}$$

In both cases the r-integral is the same, and it involves the overlap of $R_{n'l'}$ and R_{nl}. The overlap is maximum for $\Delta n = 0$ or ± 1, and drops off rapidly for higher values of $|\Delta n|$. Thus H_α radiation is stronger than H_β, which in turn is stronger than H_γ, etc.

If one of the above selection rules is not obeyed, it is called a forbidden transition. Such transitions can still take place under higher order interactions. The intensity of the emitted line goes on decreasing with increasing order. We may note the following in this case.

(i)   We have neglected the phase change of the incident wave within the region of the atom, by dropping the term kx in writing (7.54). In reality the various portions of the atomic electron cloud may move in different phases. This effectively gives rise to an electric quadrupole moment, and other higher moments, resulting in electric quadrupole and other higher order transitions.

(ii) The effect of the magnetic field gives rise to magnetic dipole and higher order transitions.

(iii) The intensity falls with increasing order. Also, the magnetic multipole transition of a certain order is weaker than the electric multipole transition of the same order.

7.3.2   Matrix elements of the radiation field

In several cases, it is possible to treat the material system as a collection of simple harmonic oscillators. Examples of such systems are a gas of diatomic molecules, lattice vibrations (phonons), electrons in an oscillating electric field, etc. We can obtain the matrix elements of the radiation Hamiltonian, and hence the selection rules. We treat q as the displacement variable in a one-dimensional parabolic potential, with a as the amplitude parameter of oscillations. Then, if we define a dimensionless variable $\eta = q/a$, we have

$$X_{r,jk} = H'_{jk} = \int_{-\infty}^{\infty} \psi_j^*(\eta)\eta\psi_k(\eta)d\eta = a^2 \int_{-\infty}^{\infty} \frac{H_j(\eta)H_k(\eta)e^{-\eta^2/2}}{(2^j j! 2^k k! \pi a^2)^{1/2}}, \tag{7.61}$$

where $H_j(\eta)$ are Hermite polynomials. Now if

$$u_n(\eta) = \frac{H_n(\eta)e^{-\eta^2/2}}{(2^n n! \pi^{1/2})^{1/2}},$$

then we have

$$\eta u_n(\eta) = \sqrt{n/2}u_{n-1} + \sqrt{(n+1)/2}u_{n+1}. \tag{7.62}$$

Further, the functions u_n are orthonormal in the interval $-\infty < \eta < \infty$. Then (7.61) gives us the selection rules $\Delta n = \pm 1$, with

$$X_r(n_{k\alpha} \rightarrow n_{k\alpha} + 1) = c(h(n_{k\alpha}+1)/2\pi\nu L^3)^{1/2}, \text{ for emission}$$

$$X_r(n_{k\alpha} \rightarrow n_{k\alpha} - 1) = c(hn_{k\alpha}/2\pi\nu L^3)^{1/2}, \text{ for absorption.} \tag{7.63}$$

The selection rule $\Delta n = \pm 1$ becomes $\Delta v = \pm 1$ for vibrational states of a diatomic molecule, where v is the vibrational quantum number in the parabolic potential. However, the actual potential curve for a diatomic molecule differs from a parabola. This produces overlap between eigenfunctions of various vibrational quantum states, and the selection rule is modified to $\Delta v = \pm 1, \pm 2, \pm 3$, etc. Here the transition probabilities become smaller and smaller for large values of $|\Delta v|$.

7.3.3   Summary of selection rules

We summarise the selection rules for electric dipole transitions for atoms and molecules.

Atoms

An atom could have one or more electrons in the outermost unfilled shell. The selection rules, in accordance with Section 7.3.1 become:

(a) In a one-electron atom, we have $s = 1/2$ both in initial and final states, hence $\Delta s = 0$. Also, we must have $\Delta l = \pm 1, 0$ and $\Delta m_s = \pm 1, 0$.

(b) In the case of other atoms (see Section 5.7.4 (iv)), l and s are replaced respectively by L and S. We then have $\Delta L = \pm 1, \Delta M_L = \pm 1, 0, \Delta S = 0, \Delta M_S = \pm 1, 0$. Further, since $J = L + S$, we get $\Delta J = \pm 1, 0$, except that $J = 0$ to $J = 0$ transition is forbidden. Also, $\Delta M_J = \pm 1, 0$ and $\Delta\Sigma|l_i| = \pm 1$, so that parity changes by 1.

Molecules

In the case of molecules, we have the following selection rules:

(a) For rotation (see Section 6.2.1, J replaces l so that we must have $\Delta J = \pm 1$.

(b) For Raman effect (see Section 6.3.5), there are two transitions so $\Delta J = 0, \pm 2$ when $\Lambda = 0$, but $\Delta J = 0, \pm 1, \pm 2$ when $\Lambda \neq 0$.

(c) For vibrational transitions of an unharmonic oscillator, $\Delta v = \pm 1, \pm 2$, etc.

(d) For electronic bands (see Section 6.3.4 and 6.4.1), Λ, Σ, Ω correspond to L, S, J, respectively. So $\Delta \Lambda = \pm 1, 0$. In the case of rotational structure, $\Delta J = \pm 1$ when $\Lambda = 0$ and $\Delta J = \pm 1, 0$ when $\Lambda \neq 0$.

7.4   Computation of Transition Probabilities

The time-rate of change of the coefficient a_j is given in (7.46). We obtain expressions for it by first assuming that the system is in a certain state k at time $t = 0$ so that $a_j = \delta_{jk}$ for all j. With time a_k decreases and the other coefficients rise (in magnitude). So we can write (7.46) as

$$i\hbar \dot{a}_k(t) = \sum_j a_j(t) H'_{kj} \exp[-i(E_j^0 - E_k^0)t/\hbar],$$

$$i\hbar \dot{a}_j(t) = a_k(t) H'_{jk} \exp[-i(E_j^0 - E_k^0)t/\hbar]. \tag{7.64}$$

In writing the second of the above equations, we have neglected second-order transitions. That is, in considering the population of the state j, we have taken account only of the $k \to j$ transition and disregarded the transitions via a third state.

7.4.1   The transition amplitude

In the beginning the population of the state k would fall exponentially. So let

$$a_k = e^{-\gamma t/2},$$

where $1/\gamma$ is a measure of the life time of the level. Hence

$$i\hbar \dot{a}_j = H'_{jk} \exp[\{i(E_j^0 - E_k^0)/\hbar - \gamma/2\}t]. \tag{7.65}$$

On integrating with respect to t, this gives

$$a_j(t) = H'_{jk} \frac{1 - \exp[\{i(E_j^0 - E_k^0)/\hbar - \gamma/2\}t]}{(E_j^0 - E_k^0) + i\gamma\hbar/2}. \tag{7.66}$$

As $t \to \infty$, we can write

$$|a_j(\infty)|^2 = \frac{|H'_{jk}|^2}{(E_j^0 - E_k^0)^2 + (\gamma\hbar/2)^2}.$$
(7.67)

It gives the final population of the state j under first-order transitions alone.

This transition probability is maximum when $E_j^0 = E_k^0$. When $E_j^0 \neq E_k^0$, the change in energy of matter is compensated by that in the radiation field. Equation (7.67) indicates that the probability of transition to other states falls off as $(\Delta E^0)^2$, where $\Delta E = E_j^0 - E_k^0$.

The perturbation also broadens the energy levels as shown in Fig. 7.2. So the transitions between these sublevels cause emission or absorption of photons with $\Delta E \neq \Delta E^0$. This results in broadening of spectral lines. The concerned frequency would then give $\nu \neq \nu_{jk}(= E_j^0 - E_k^0)$. The change in frequency is caused by the recoil of the atom during emission or absorption.

In Fig. 7.2, the initial energy E_k is

$$E_k = E_{k,\mathrm{m}}^0$$

and the final energy is

$$E_j = E_{j,\mathrm{m}}^0 + E'_{j,\mathrm{m}} + h\nu.$$

As $E_k = E_j$ near the peak in transition probability, we get

$$E'_{j,\mathrm{m}} = E_{k,\mathrm{m}}^0 - E_{j,\mathrm{m}}^0 - h\nu = h(\nu_0 - \nu),$$
(7.68)

where $h\nu_0 = E_{k,\mathrm{m}}^0 - E_{j,\mathrm{m}}^0$.

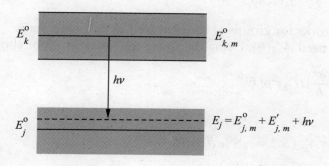

Fig. 7.2 *Energy balance in the transition $k \rightarrow j$*

It is now convenient to group together all transitions ending in energy $E_{j,\mathrm{m}}$ close to $E_{j,\mathrm{m}}^0$. If $\rho(E)$ is the density of states around $E_{j,\mathrm{m}}^0$, then

$$\sum' |a_j|^2 = \sum_{E \approx E_{j,m}} |a_j|^2 \rho(E) dE, \tag{7.69}$$

where prime on the left-hand side indicates sum over states $E \approx E_{j,m}$. Taking into account two states of polarization, the density of states $\rho(E)$ is twice the number of cells in phase space, that is,

$$\rho(E) dE = 2p^2 dp/h^3.$$

Since $p = E/c$ for electromagnetic radiation, we get

$$\rho(E) dE = 8\pi^2 E^2 dE/c^3 h^3. \tag{7.70}$$

Since $\rho(E)$ is a slowly varying function of E, we can sum (7.67) over states with $E_j = E_{j,k}^0$ and obtain

$$\sum' |a_j(\infty)|^2 = |H'_{jk}|^2 \rho(E_j) \int_0^\infty \frac{dE}{(E - E_k^0)^2 + (\gamma\hbar/2)^2}$$

$$= \frac{2\pi}{\gamma\hbar} |H'_{jk}|^2 \rho(E_j^0). \tag{7.71}$$

It is convenient to define

$$A'_{kj} = \sum' |a_j(t)|^2 \tag{7.72}$$

so that

$$A'_{kj}(\infty) = \sum' |a_j(\infty)|^2. \tag{7.73}$$

This equation works for groups of states as well as individual states.

Actually, we need $A_{kj}(0)$. Using the above treatment it can be shown that

$$A'_{kj}(0) = \frac{2\pi}{\hbar} |H'_{jk}|^2 \rho(E_j^0). \tag{7.74}$$

and

$$A'_{kj}(0) = \gamma A'_{kj}(\infty) = \gamma \sum' |a_j(\infty)|^2. \tag{7.75}$$

Then from (7.67) we get

$$A'_{kj}(0) = \frac{\gamma |H'_{jk}|^2}{(E_{j,m} - E_k^0)^2 + (\gamma\hbar/2)^2}. \tag{7.76}$$

Now,

$$E_{j,\mathrm{m}} - E_k^0 = \Delta E_\mathrm{m} + \Delta E_\mathrm{r} = \pm(\nu_{jk} - \nu),\tag{7.77}$$

where the upper sign stands for absorption and the lower one for emission. In either case

$$(E_j - E_k^0)^2 = h^2(\nu - \nu_{jk})^2.\tag{7.78}$$

Therefore (7.77) finally reduces to

$$A'_{kj} = \frac{\gamma|H'_{jk}|^2}{h[(\nu - \nu_{jk})^2 + (\gamma/4\pi)^2]}.\tag{7.79}$$

7.4.2 Meaning of the decay constant

Substituting for a_j and a_k in (7.64), we get

$$-\frac{i\hbar\gamma}{2} = \sum_j |H'_{jk}|^2 \frac{it}{\hbar} \frac{e^{-ix} - 1}{x},\tag{7.80}$$

where x is the dimensionless variable

$$x = [(E_j^0 - E_k^0) + i\gamma\hbar/2]t/\hbar.\tag{7.81}$$

The summation is equivalent to a summation over gross levels j around $E \sim E_j^0$, that is,

$$\sum{}''_{\mathrm{gross}\,j} \int_{E \sim E_j^0} \rho(E)dE.$$

Therefore

$$-\frac{i\gamma\hbar}{2} = \sum{}'' |H'_{jk}|^2 \rho(E_j^0) \int_{-\infty}^{\infty} \frac{e^{-ix} - 1}{x} dx.$$

Using

$$\int_{-\infty}^{\infty} \frac{\cos x - 1}{x} dx = 0, \qquad \int_{-\infty}^{\infty} \frac{\sin x}{x} = \pi,$$

we get

$$\gamma = \frac{2\pi}{\hbar} \sum{}'' |H'_{jk}|^2 \rho(E_j^0) = \sum{}'' A'_{kj}(0) \equiv A_{kj},\tag{7.82}$$

which defines A_{kj} as the sum of transition probabilities from level k. Thus $\tau = 1/\gamma$ is the mean life of state k.

7.4.3   The transition probability amplitudes

We want to relate transition probabilities to the intensities of radiation field. Using (7.40) and (7.63), we get, for emission and absorption:

$$H'_{jk} = -\left(\frac{\hbar(n_{\mathbf{k}\alpha} + 1)}{\nu_{jk}L^3}\right)^{1/2} Q_{kj,m} \text{ for emission,}$$

$$H'_{jk} = -\left(\frac{\hbar(n_{\mathbf{k}\alpha})}{\nu_{jk}L^3}\right)^{1/2} Q_{kj,\mathrm{m}} \text{ for absorption.} \tag{7.83}$$

Putting these in (7.80), we have

$$A_{kj,\mathbf{k}} = \frac{\gamma}{h^2[(\nu - \nu_{jk})^2 + (\gamma/4\pi)^2]} \frac{\hbar(n_{\mathbf{k}} + 1)}{\nu_{\mathbf{k}}L^3}|Q_{kj,\mathrm{m}}|^2 \text{ for emission,}$$

$$A_{kj,\mathbf{k}} = \frac{\gamma}{h^2[(\nu - \nu_{jk})^2 + (\gamma/4\pi)^2]} \frac{\hbar n_{\mathbf{k}}}{\nu_{\mathbf{k}}L^3}|Q_{kj,\mathrm{m}}|^2 \text{ for absorption.} \tag{7.84}$$

The density of states $\rho(E)dE$, on using $p = h\nu/c$, is given by

$$\rho(E)dE = 2p^2dpL3/h_3 = 2\nu^2 L^3 d\nu/c^3. \tag{7.85}$$

On multiplying (7.84) by $\rho(E)dE$ and integrating over the whole line, we find

$$A_{kj} = \frac{2\nu_{\mathbf{k}}}{\hbar c^3}(n_{\mathbf{k}} + 1)|Q_{kjm}|^2 \text{ for emission,}$$

$$A_{jk} = \frac{2\nu_{\mathbf{k}}}{\hbar c^3}n_{\mathbf{k}}|Q_{kjm}|^2 \text{ for absorption.} \tag{7.86}$$

7.5   Absorption, Emission and Einstein Coefficients

7.5.1   Intensity of radiation

Now $n_{\mathbf{k}}$ is related to intensity of radiation I_ν as

$$\frac{I_\nu d\Omega d\nu}{c} = 2n_{\mathbf{k}}h\nu\frac{\nu^2 L^3}{c^3}d\Omega\frac{d\nu}{L^3} \Rightarrow n_{\mathbf{k}} = \frac{c^2}{2h\nu^3}I_\nu. \tag{7.87}$$

Therefore, on suppressing the subscript, (7.86) become

$$A_{kj} = \frac{2\nu}{\hbar c^3}|Q_{jk,\mathrm{m}}|^2[1 + c^2 I_\nu/2h\nu^3] \text{ for emission}$$

$$A_{jk} = \frac{2\nu}{\hbar c^3}|Q_{jk,\mathrm{m}}|^2 c^2 I_\nu / 2h\nu^3 \text{ for absorption.} \tag{7.88}$$

The first of the above equations suggests that a part of emission depends on the intensity of ambient radiation of appropriate frequency present in the system while the other part is independent of radiation present. These two parts of emission are respectively called *induced emission* and *spontaneous emission*. The induced emission is proportional to I_ν. Absorption is only of one type, induced absorption, and it is proportional to I_ν. It is customary to consider the induced emission as negative (reverse) of absorption and include it in the so-called absorption coefficient. We can now define the following coefficients:

The spontaneous emission transition probability:

$$\mathcal{A}_{kj} = \frac{2\nu}{\hbar c^3}|Q_{kj,\mathrm{m}}|^2 \tag{7.89a}$$

The induced emission transition probability:

$$\mathcal{B}_{kj} = \frac{c^2}{2h\nu^3}\mathcal{A}_{kj} \tag{7.89b}$$

The absorption transition probability:

$$\mathcal{B}_{jk} = \mathcal{B}_{kj}. \tag{7.89c}$$

Now, from (7.52),

$$|Q_{kj,\mathrm{m}}|^2 = e^2\omega^2\{\epsilon_{\mathbf{k}\alpha} \cdot \sum_l \mathbf{r}_l\}^2.$$

This gives

$$A_{kj} = \frac{8\pi^2\nu^3 e^2}{\hbar c^3}\{\epsilon_{\mathbf{k}\alpha} \cdot (\sum \mathbf{r}_l)_{jk}\}^2.$$

Integrating over all directions, we see that

$$\int\{\epsilon_{\mathbf{k}\alpha} \cdot (\sum \mathbf{r}_l)_{jk}\}^2 d\Omega = \frac{4\pi}{3}\sum_l [x_l]_{jk}^2\}^2. \tag{7.90}$$

Using this integral, the coefficients of (7.89) reduce to

$$\mathcal{A}_{kj} = \frac{32\pi^3\nu^3 e^2}{3\hbar c^3}\{\sum_l [x_l]_{jk}\}^2,$$

$$\mathcal{B}_{kj} = \frac{c^2}{2h\nu^3} \mathcal{A}_{kj},$$

$$\mathcal{B}_{jk} = \mathcal{B}_{kj}. \tag{7.91}$$

7.5.2   Einstein coefficients

A level with total angular momentum J splits into $2J + 1$ levels in the presence of a magnetic field. Hence $g = 2J + 1$ is known as the *weight of the level*. Since magnetic field is not always present, or weak if present, it is customary to group all magnetic levels into one state of weight g.

Let a be the upper state and b the lower state with weights g_a and g_b, respectively. Also let k denote a microlevel in state a and j that in state b. Then the mean transition rate from these levels are

$$A_{ab} = \frac{1}{g_a} \sum \mathcal{A}_{kj},$$

$$B_{ab} = \frac{1}{g_a} \sum \mathcal{B}_{kj} = \frac{1}{g_a} \frac{c^2}{2h\nu^3} \sum \mathcal{A}_{kj},$$

$$B_{ba} = \frac{1}{g_a} \sum \mathcal{B}_{jk} = \frac{1}{g_b} \frac{c^2}{2h\nu^3} \sum \mathcal{A}_{kj}, \tag{7.92}$$

where all the summations are double summations over k in a and j in b. Therefore,

$$\frac{A_{ab}}{g_b 2h\nu^3/c^2} = \frac{B_{ab}}{g_b} = \frac{B_{ba}}{g_a}. \tag{7.93}$$

These are called Einstein transition probabilities or *Einstein coefficeints*. A_{ab} is the Einstein coefficient for spontaneous emission, B_{ab} the Einstein coefficient for stimulated emission and B_{ba} the coefficient for absorption. We see that all the coefficients are proportional to the weight of the final level. Also, note that A_{ab} is transition probability per unit time while B_{ab} and B_{ba} are transition probability per unit time per unit intensity of radiation. In magnitude, A_{ab} is about $2h\nu^3/c^2$ times B_{ab}. We shall see in Chapter 8 that they give Planck's formula for intensity of radiation in thermodynamic equilibrium.

7.5.3   Oscillator strengths

In classical electromagnetic theory, the total line scattering coefficient is

$$\int \sigma_\nu d\nu = \pi e^2/mc.$$

On the other hand, quantum theory gives

$$\int \sigma_\nu d\nu = h\nu_0 B_{\mathrm{ba}}.$$

So we put

$$h\nu_0 B_{\mathrm{ba}} = (\pi e^2/mc) f_{\mathrm{ba}}, \tag{7.94}$$

where f_{ba} is called the absorption oscillator strength. Thus,

$$f_{\mathrm{ba}} = \frac{h\nu_0 mc}{\pi e^2} B_{\mathrm{ba}}. \tag{7.95}$$

We may also define the stimulated and spontaneous oscillator strengths as

$$f_{\mathrm{ab}}^{\mathrm{stim}} = \frac{g_{\mathrm{b}}}{g_{\mathrm{a}}} f_{\mathrm{ba}},$$

$$f_{\mathrm{ab}}^{\mathrm{spont}} = \frac{2h\nu_0^3}{c^2} f_{\mathrm{ab}}^{\mathrm{stim}}. \tag{7.96}$$

They have the same relation with each other as the A and B coefficients.

7.6 Weisskopf–Wigner Picture

The phenomenon of spontaneous emission indicates that all levels except the ground state have a finite life time $\tau = 1/\gamma$. Hence from the uncertainty principle we conclude that these higher energy levels are broadened with a width $\Delta E \sim \hbar/\tau \sim \gamma\hbar$. Weisskopf and Wigner took this into account, and it may be seen as an approximation.

This means that on a statistical basis the atoms have an energy distribution of this width. So Weisskopf and Wigner attached a weight to each energy interval, E to $E + dE$, which is given by

$$W(E)dE = \frac{\gamma\hbar}{2\pi} \frac{dE}{(E - E_0)^2 + (\gamma\hbar/2)^2}, \tag{7.97}$$

where E_0 is the mean energy of the level. It can be seen that in the small width limit, $E_0 \gg \gamma\hbar$,

$$\int_0^\infty W(E)dE = 1.$$

Let us consider two levels centred at E_{a} and E_{b} and characterised by widths $\gamma_{\mathrm{a}}\hbar/2$ and $\gamma_{\mathrm{b}}\hbar/2$, respectively. See Fig. 7.3. If N_{a} is the total population of the level E_{a}, then that of a microlevel having energy between E' and $E' + dE'$ would be given by

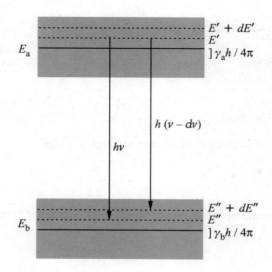

Fig. 7.3   *Transitions between naturally browdened energy levels*

$$\frac{N_{\mathrm{a}}(\gamma_{\mathrm{a}}\hbar/2\pi)dE'}{(E'-E_{\mathrm{a}})^2+(\gamma_{\mathrm{a}}\hbar/2)^2}.$$

In writing this, we have neglected the effect of temperature. It could be taken into account by the Boltzmann factor, $\exp[(E'-E_{\mathrm{a}})/kT]$, which is nearly unity if $E'-E_{\mathrm{a}} \ll kT$.

If A_{ab} is the Einstein coefficient for spontaneous emission for all sublevels of E_{a}, that for transition to a microlevel between E'' and $E''+dE''$ of level E_{b} would be

$$\frac{A_{\mathrm{ab}}(\gamma_{\mathrm{b}}\hbar/2\pi)dE''}{(E''-E_{\mathrm{b}})^2+(\gamma_{\mathrm{b}}\hbar/2)^2}.$$

Then the transition rate of emission from any level of E_{a} to any level of E_{b} would be seen to be

$$\mathcal{N}=\frac{N_{\mathrm{a}}A_{\mathrm{ab}}\gamma_{\mathrm{a}}\gamma_{\mathrm{b}}\hbar^2 dE'dE''}{4\pi^2\{(E'-E_{\mathrm{a}})^2+(\gamma_{\mathrm{a}}\hbar/2)^2\}\{(E''-E_{\mathrm{b}})^2+(\gamma_{\mathrm{b}}\hbar/2)^2\}}.$$

A transition from microlevel E' in E_{a} to E'' in E_{b} gives rise to a photon of energy $h\nu$. For a fixed E', the energy width of this line would be $dE''=-hd\nu$. So the contribution of these transitions to emission coefficient is given by

$$dj_{\nu}=\frac{N_{\mathrm{a}}A_{\mathrm{ab}}h^2\nu\gamma_{\mathrm{a}}\gamma_{\mathrm{b}}\hbar^2 dE'}{16\pi^4\{(E'-E_{\mathrm{a}})^2+(\gamma_{\mathrm{a}}\hbar/2)^2\}\{(E'-h\nu-E_{\mathrm{b}})^2+(\gamma_{\mathrm{b}}\hbar/2)^2\}}. \tag{7.98}$$

This can be integrated over all transitions giving rise to photons of frequency ν to get

$$j_\nu = N_a A_{ab} \nu_0 \frac{(\gamma_a + \gamma_a)}{4\pi^2[(\nu - \nu_0)^2 + (\gamma_a + \gamma_b)^2/16\pi^2]}, \tag{7.99}$$

where $\nu_0 = (E_a - E_b)/h$.

Similarly, the absorption coefficient will be

$$k_\nu = \frac{\pi e^2}{mc} \frac{N_b f_{ba}}{\pi} \frac{\gamma_{rad}}{4\pi[(\nu - \nu_0)^2 + \gamma_{rad}^2/16\pi^2]},$$

where $\gamma_{rad} = \gamma_a + \gamma_b$. Also,

$$\frac{j_\nu}{k_\nu} = \frac{N_a A_{ab}}{N_b B_{ba} - N_a B_{ab}}. \tag{7.100}$$

We shall see in Chapter 8 that $j_\nu/k_\nu = B_\nu$ in thermodynamic equilibrium.

Thus we find that the line has a half-width $\gamma_{rad}\hbar/2$, that is the sum of the widths of the two levels; γ_{rad} is called the *natural damping constant* of the line.

We may summarise the essential aspects as follows:

(i) Each level has a width, say $\gamma_a\hbar/2$, where $\tau_a = 1/\gamma_a$ is the lifetime of the state.

(ii) For transition between two levels, the line has a width $\gamma_{rad}\hbar$, where $\gamma_{rad} = \gamma_a + \gamma_b = 1/\tau_a + 1/\tau_b$.

(iii) The ground state is sharp in the absence of radiation and has a very long life time.

(iv) The damping constants γ_a are modified by collisions with other particles which reduce the life time and broaden the level. This is known as collisional broadening.

(v) If τ_{coll} is the average interval between successive collisions, the line will have a width equal to $\gamma\hbar$, where $\gamma = 1/\tau = 1/\tau_{rad} + 1/\tau_{coll}$.

Problems

7.1. Show that the number operator $N_k = a_k^\dagger a_k$ commutes with the Hamiltonian $H = \sum_k E_k a_k^\dagger a_k$ and also with any function of position such as potential $V(\mathbf{r})$.

7.2. For the Hamiltonian of (7.54), obtain the selection rules (7.58) for transitions of an electron in a hydrogen atom.

Statistical Mechanics

$\mathbb{S}$Tatistical mechanics and classical mechanics are two pillars of Newtonian mechanics which reached their pinnacle during the period 1660–1900, though some of the developments continued in the twentieth century. Newtonian mechanics deals with point particles and rigid bodies. But soon it was realised that any real system consisted of several particles, with different kinds of interactions among them. In that case, although possible in principle, the time and efforts required to obtain complete information about every particle in the system is not commensurate with the benefits. Ingenious methods have therefore been developed to obtain average properties and to relate them to measurable parameters. Beginning with kinetic theory of a perfect gas in this chapter, we go over to the Boltzmann probability distribution, and discuss the three ensembles of systems. We also discuss the statistics of photons, which is a discovery of the twentieth century.

8.1 Kinetic Theory of Gases

In Chapter 4 we discussed the bulk properties of gases from the thermodynamic point of view. Actually, these properties are the result of interactions between innumerable particles (atoms and molecules) of the gas. As it is not possible to consider such interactions in all the details, we have to take recourse to statistics. The first such attempt was made by Maxwell through his kinetic theory of gases.

Temperature is a measure of energy in the gas. For non-interactive particles, it can only be the kinetic energy of individual particles. In the simplest model of a gas, we neglect all interactions among atoms. We further disregard any structure of the atom (and even molecule) and treat them as a collection of point particles of mass m. This model is known as a *perfect gas* or an *ideal gas*.

8.1.1 Maxwell's hypothesis

In kinetic theory it is assumed that the particles of a gas are constantly in motion. According to Maxwell's hypothesis, their velocities have a normal or Gaussian distribution. Thus for each component of velocity v_i $(i = x, y, z)$, let $N(v_i)dv_i$ be the number of particles per unit volume having the i-component of their velocity between v_i and $v_i + dv_i$. Maxwell's hypothesis states that

$$N(v_i)dv_i = N(m/2\pi kT)^{3/2}\exp(-mv_i^2/2kT)dv_i, \qquad (8.1)$$

where N is the total number of particles per unit volume in the gas, T the temperature, and k the Boltzmann constant. Then the number of molecules per unit volume having their velocity in the neighbourhood of **v** in a velocity element d^3v will be

$$N(\mathbf{v})d^3v = N(m2\pi kT)^{1/2}\exp(-mv^2/2kT)d^3v. \qquad (8.2)$$

For most purposes, we are interested only in the magnitude of velocity of a particle, not its direction of travel. In order to obtain $N(v)dv$, the number of particle (per unit volume) having their magnitude of velocity between v and $v + dv$, we choose a spherical shell in velocity space with radius v and thickness dv, as shown in Fig. 8.1. The volume of the shell is $4\pi v^2 dv$ where $v^2 = v_x^2 + v_y^2 + v_z^2$. Therefore

$$N(v)dv = N(m/2\pi kT)^{3/2}4\pi v^2\exp(-mv^2/2kT)dv. \qquad (8.3a)$$

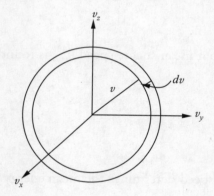

Fig. 8.1 *Spherical shell as a velocity element dv*

This is Maxwell's law of velocity distribution in an ideal gas in thermodynamic equilibrium, which is shown in Fig. 8.2. It is clear that $p(v)dv = N(v)dv/N$ gives the probability of finding particles having their speed between v and $v + dv$. It is given by

$$p(v)dv = (m/2\pi kT)^{3/2}4\pi v^2\exp(-mv^2/2kT)dv. \qquad (8.3b)$$

It can be verified that $p(v)$ is a probability distribution function, that is, $p(v)$ is non-negative in the entire range and $\int_0^\infty p(v)dv = 1$.

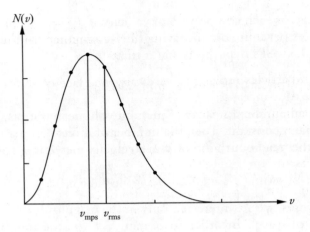

Fig. 8.2   *Maxwell's law of distribution of velocities in an ideal gas*

We see that for small v, $p(v) \propto v^2$, and for large v, it falls exponentially as $\exp(-mv^2/2kT)$. It has one maximum at son v_{mps}, which is the maximum probable speed; this can be found by differentiation to be

$$v_{mps} = (2kT/m)^{1/2}. \tag{8.4}$$

On the other hand, the root mean square speed v_{rms} is found from

$$v_{rms}^2 = \int_0^\infty v^2 p(v)dv,$$

which gives

$$v_{rms} = (3kT/m)^{3/2}. \tag{8.5}$$

The root mean square speed determines the mean energy of a particle through

$$\overline{E} = \frac{1}{2}mv_{rms}^2 = 3kT/2. \tag{8.6}$$

Thus we see that each component of velocity contributes $kT/2$ to the mean energy of a particle. Equations (8.1) and (8.3a) can be written in terms of v_{rms} as

$$N(v_i)dv_i = \frac{N}{v_{rms}\sqrt{\pi}} \exp(-v_i^2/v_{rms}^2)dv_i,$$

$$N(v)dv = \frac{4\pi N}{v_{rms}^3 \pi^{3/2}} v^2 \exp(-v^2/v_{rms}^2)dv. \tag{8.7}$$

We shall now see how this consideration leads us to the ideal gas law. Consider a cube whose sides have a unit length, containing an ideal gas with N particles. The average number of particles moving towards any one face of the cube would be $N/6$. Each of them has a root-mean-squared speed of v_{rms}, so the number of particles striking the face per unit time is $Nv_{rms}/6$. In an ideal gas, the assumption is that a particle is reflected without loss of energy and follows the laws of reflection. Thus on reflection, there is a reversal of speed which produces a change in momentum equal to $2mv_{rms}$ per particle. So the pressure on the wall, which is the force per unit area or rate of change of momentum per unit time at each surface of the cube is

$$p = (Nv_{rms}/6)(2mv_{rms}) = NkT,$$

which is the equation of state of perfect gas.

The perfect (or ideal) gas approximation becomes better and better when the number of particles per unit volume is small. The volume tends to infinity and is much larger than the particle (atom, molecule) size. This will be the case at low pressures. But when pressure is increased, N/V increases, and experiments show deviation from the ideal gas law. At high pressures, the particle size becomes significant. Due to lower inter-particle distance, they exert mutually attractive forces on each other. Van der Waals introduced two more parameters, one of which is the volume of the particle. We wrote an equation which prevents the volume per particle of the gas from going below the particle volume, which discussed this in Section 4.1.3, and gave the van der Waal equation in (4.3).

Microstates and macrostates

Consider a throw of four identical coins, each having two possible outcomes denoted by H (for head) and T (for tail). There are $2^4 = 16$ possible outcomes. If there are n coins in the 'system', there would be 2^n possible outcomes, of the form HHTHT$\cdots$ or TTHHT$\cdots$, etc. A specification such as this, where the state of each coin is given, is a *microstate* of the system. It is clear that such a specification would become cumbersome for large n. In such a case, we may not be interested in knowing whether a particular coin shows H or T. We may be satisfied to know how many coins show H and how many show T. The number r of Heads can take values from 0 to n. These are *macrostates* of the system. For n coins we could have $n+1$ macrostates, with $0 \leq r \leq n$. Each macrostate corresponds to one or more microstates of the system. For example, the 1H macrostate of four coins corresponds to the four microstates HTTT, THTT, TTHT and TTTH. A system of n spin-1/2 particles, each with two possible orientations, is a system to which the same mathematics as that of n coins applies. Also, the case of a one-dimensional random walk is similar.

Whether a system has a small and finite number of microstates like the system of four coins or a large and almost infinite number of microstates, there is no reason why a system should prefer one microstate over the other. Thus we logically expect every microstate of a system to be equally probable. This is known as the *hypothesis of equal a priori probability*. The probability of occurrence of a macrostate would of course depend on the number of microstates it contains. Thus in the case of four fair coins, each microstate would have a probability of 1/16, there being 16 possible microstates, while the 1H macrostate would have a probability of 1/4.

If our system is a gas of n particles, the complete specification of $3n$ position and $3n$ momentum coordinates gives a microstate of the system. Because of the large number of particles and continuous nature of the variables, even after dividing each coordinate in a finite number of cells, the total number of microstates would be enormous. However, to obtain the gross features like pressure, temperature, average energy and total energy of the system and their probabilistic variation with time, it is enough to know the statistical distribution of particles in different microstates. Such average and bulk properties can then be related to experimentally observed parameters of the system.

8.1.2   Boltzmann's law for excited states

Maxwell's law deals with the velocity distribution of free particles in a gas. Can it throw light on the energy distribution of bound electrons in various atomic levels?

For working out the number distribution of electrons in various energy levels in an atom, we have to borrow some concepts from quantum mechanics. The dynamical condition of a free electron is specified by three coordinates x, y, z and three momenta p_x, p_y, p_z with $p_x = mv_x$, etc., where m is the electron mass. As all of them change with time, it was seen in Chapter 1 that the representative point of the electron moves in the phase space of six dimensions. A volume element in phase space is denoted by $dxdydzdp_xdp_ydp_z$. The minimum volume of a cell, according to the uncertainly principle, can be h^3, where h is an arbitrary small number in classical mechanics and is related to the Planck's constant in quantum mechanics. So the volume element would contain d^3rd^3p/h^3 cells. Further, according to Pauli principle, each cell can accommodate at most two electrons of opposite spin. So the total number of states in the volume element will be $2d^3rd^3p/h^3$, which is taken as the statistical weight of that volume element. So a unit volume of electron gas has a weight of $2d^3p/h^3$. Hence the weight of a spherical shell of radius p and thickness dp will be $8\pi p^2dp/h^3$. Thus, in terms of velocity we get

$$g(v)dv = 8\pi m^3 v^2 dv/h^3.\tag{8.8}$$

Note that each cell of phase space represents a microstate.

Then (8.7) can be written as

$$N(v)dv = N[h^3/2(2\pi mkT)^{3/2}]g(v)e^{-E_v/kT}dv.\tag{8.9}$$

Therefore for two velocities v and v', the ratio of population densities would be

$$\frac{N(v)}{(Nv')} = \frac{g(v)}{g(v')}\exp(-(E_v - E_{v'})/kT).\tag{8.10}$$

Equation (8.10) can be generalised for the discrete bound states of an atom. Let N_A and N_B be the volume densities of atoms in states A and B, respectively, and let g_A and g_B the statistical weights of these levels given by $2J + 1$ Zeeman states; then

$$\frac{N_A}{N_B} = \frac{g_A}{g_B}e^{-(E_A - E_B)/kT}.$$

The energy of a level measured from the ground state is called the *excitation potential* χ. Then,

$$\frac{N_A}{N_B} = \frac{g_A}{g_B}e^{-(\chi_A - \chi_B)/kT}.\tag{8.11}$$

This is known as *Boltzmann law*. Putting $\chi_A - \chi_B = \chi_{AB}$, this can be put in the form

$$\ln(N_A/N_B) = -\theta\chi_{AB} + \ln(g_A/g_B),$$

where

$$\theta = 5040.4 \ (eV)/T \ (K).\tag{8.12}$$

Equation (8.11) can be used for finding the ratio of the member of atoms N_k in the state k in terms of the number of atoms N_1 in the ground state. If g_k is the statistical weight of the state k, then we have

$$N_k = N_1(g_k/g_1)e^{-\chi_k/kT}.$$

The total number of atoms in the excited states is then given by

$$N_{ex} = N_1 B(T)/g_1,\tag{8.13}$$

where

$$B(T) = \sum_k g_k e^{-\chi_k/kT}\tag{8.14}$$

is known as the *partition function*. It represents the total statistical weight of the atom. It gives

$$\frac{N_k}{N_{\text{ex}}} = \frac{g_k e^{-\chi_k/kT}}{B(T)}. \tag{8.15}$$

8.1.3   Saha's equation of ionization

We shall show how Saha's equation of ionization is an extension of Maxwell's and Boltzmann's laws. Let N be the number of neutral atoms per unit volume and N_{ion} and N_e the numbers of ionized atoms and electrons per unit volume, respectively. Then in the thermodynamic equilibrium, we have a balance between ionization and recombination, that is, the rates of the two reactions

$$N \rightleftharpoons N_{\text{ion}} + N_e$$

must be equal. Saha argued that this was exactly like a chemical reaction where two chemical elements A and B combine to give a compound AB, which again dissociates into A and B according to

$$A + B \rightleftharpoons AB.$$

Therefore the law of mass action must apply when the species are in equilibrium. Therefore we must have

$$N_{\text{ion}} N_e / N = \text{ constant},$$

which is known as the ionization constant.

We can determine this constant by using Boltzmann's equation for the ionized and neutral states. If χ_{ion} is the ionization potential, we can write

$$N_{\text{ion}}/N = (g_{\text{ion}}/g)e^{-\chi_{\text{ion}}/kT}, \tag{8.16}$$

where $g = B(T)$, the partition function of the neutral atom, and g_{ion} is the combined statistical weight of the ionized atoms and the resultant free electrons. Then $g_{\text{ion}} = B_{\text{ion}}(T)g_e$, where $B_{\text{ion}}(T)$ is the partition function of the ions and g_e is the average weight of a free electron. From Maxwell's law, we get

$$g_e = (1/N_e) \int_0^\infty g(v)e^{-E_v/kT}dv, \tag{8.17}$$

where

$$N_e = \int_0^\infty N_e(v)dv = \frac{N_e h^3}{2(2\pi mkT)^{3/2}} \int_0^\infty g(v)e^{-E_v/kT}dv. \tag{8.18}$$

Equations (8.17) and (8.18) together lead to

$$\frac{N_{\text{ion}}}{N} = \frac{2(2\pi mkT)^{3/2}}{N_e h^3} \frac{B_{\text{ion}}(T)}{B(T)} e^{-\chi_{\text{ion}}/kT}. \tag{8.19a}$$

In terms of electron pressure $p_e = N_e kT$, we have

$$\frac{N_{\text{ion}} p_e}{N} = \frac{2(2\pi m)^{3/2}(kT)^{5/2}}{h^3} \frac{B_{\text{ion}}(T)}{B(T)} e^{-\chi_{\text{ion}}/kT}. \tag{8.19b}$$

Substituting numerical values of constants, we get the Saha's equation of ionization

$$\ln \frac{N_{\text{ion}}}{N} = -\frac{5040.4}{T}\chi_{\text{ion}} + 2.5\ln T - \ln p_e + \ln \frac{2B_{\text{ion}}(T)}{B(T)} - 0.48, \tag{8.20}$$

where χ_{ion} is in electron volts and T in kelvin.

8.2 Fundamentals of Statistical Mechanics

Let us consider a system of n particles, which could be a block of ice or some gas in a container. The course of the system in phase space (the p–q space or Γ space) can be known as a function of time if we know the initial conditions. But since n is large, it will be very difficult to know all the $2n$ initial conditions. Also according to quantum mechanics the uncertainty principle applies, and the measurement of variables must satisfy $\Delta p \Delta q \simeq \hbar$ for each component of each particle. So it is not even possible to know the exact initial conditions; hence we cannot find out the actual course of events as a function of time. However, for most purposes it is not necessary to know the detailed motion of all the particles. Only a statistical knowledge of energy and space distribution among the particles is sufficient to obtain the gross properties of the system. Statistical mechanics aims at deriving these properties from a consideration of probability. We find the probability density W in phase space which gives the probability $W d^{3n}p d^{3n}q$ of finding the system in the volume element $d^{3n}p d^{3n}q$ of the phase space. The average of quantity X is then given by

$$\overline{X} = \int X W d^{3n}p d^{3n}q. \tag{8.21}$$

It will be our aim to calculate these averages.

Now, in order to know the average properties, we should take a time average. But it is more convenient and appropriate to consider many similar systems or replicas with different initial conditions, and take an average over all such systems, that is, an ensemble average. The two averages become equivalent under what is called the *ergodic hypothesis*.

Systems and ensembles

A system may contain n number of particles, where n is any positive integer. A simple pendulum of a definite length ($n = 1$) or a gas of hydrogen atoms in a flask ($n \approx 10^{20}$) is a system amenable to statistical mechanics. A system is defined by its external parameters, which are its fixed parameters, such as volume, temperature, number of particles, energy, etc. The state of a system at any instant is described by internal variables which vary with time. For a simple pendulum, its length and energy (hence amplitude) are fixed parameters. Its only internal variable is the phase ϕ ($0 \leq \phi < 2\pi$), which gives the instantaneous state of the pendulum. In the case of the gas, its volume is the fixed parameter, whereas the $3n$ position coordinates and $3n$ momentum coordinates are internal time-dependent variables which decide the instantaneous state of the system.

An ensemble is a collection of systems, possibly infinite number of them, all having the same external parameters but no two of them having the same values of internal variables. If the system is a simple pendulum, the ensemble would consist of a large number of identical pendulums with the same length and amplitude, but each one differing from the others in phase. As the actual pendulum evolves in time and goes through each phase, at every instant, it resembles some system of the ensemble. An instantaneous snapshot of the ensemble would show all the states of the pendulum. If a gas is the system, its ensemble would consist of a large number of identical replicas having the same external parameters but no two of them having the same values of all the position and momentum coordinates. As the real gas evolves in time and goes from one microstate to another, at every instant, it resembles some system of the ensemble.

If the number of particles in a system and its total energy are fixed, the resulting ensemble is called a *microcanonical ensemble*. The ensemble of a simple pendulum would be of this kind. Truly speaking, the total energy E lies in a short interval ΔE around E; this ΔE can be made arbitrarily small for a classical system, but is governed by the uncertainty principle for a quantum system. In the case of a gas in a flask, we may have different arrangements. The walls of the flask may be completely insulating, not allowing any exchange of energy between the particles inside with the external surroundings. The ensemble of such a system would again be a microcanonical ensemble. The walls may allow exchange of energy between the system and the surroundings, and thus may be in statistical equilibrium with the large surroundings. A system like a gas in a thin glass container kept in a large room has a fixed number of particles, but its instantaneous energy may vary over a wide range. In this case the temperature becomes an external parameter in addition to volume.

An ensemble of such a system is called a *canonical ensemble*. A sealed bottle of mercury or oxygen molecules kept in a large room would be a good example of this case. Finally, the walls may also allow exchange of particles with the surroundings so that neither the number n of particles nor the system energy E is fixed. An ensemble of such a system is called a *grand canonical ensemble*. A small empty container loosely covered with a light lid in a large room would serve as an example of this case. Both the container and the room would contain the same type of atoms or molecules but the number of particles in the container and their energy would vary with time.

In the case of a canonical ensemble and a grand canonical ensemble, the surroundings (also known as *bath*) are supposed to be much larger than the system, that is, the number of particles in the bath is much larger than that in the system and the energy of the bath is much larger than that of the system. The bath is thus capable of supplying as well as taking up particles and energy to and from the system.

A microstate of a system is represented by a point in the phase space, which is called the *representative point* of the system. As the system evolves in time, this representative point moves along a trajectory in the phase space. Due to the various constraints on the system put by the external parameters, the representative point may lie only in a certain region or volume of the phase space. This is then called the *volume of the phase space accessible to the system.*

The totality of similar systems is called an *ensemble*. Three kinds of ensembles are recognized.

8.2.1 Microcanonical ensemble

This is characterised by (a) constant total energy E, and (b) constant numbers of particles of each type, and hence also the total number of particles. If n_μ is the number of particles of species μ and $n = \Sigma n_\mu$, then n_μ as well as n are constants in time. It represents an idealised isolated system. Let $H(q, p, t)$ be the Hamiltonian of the system, where q and p is the shorthand notation for $3n$ canonical coordinates each. Then for each system of this ensemble, $H(q, p, t) = E$ is a constant. The representative points of all systems in the microcanonical ensemble will lie on a thin shell near a surface Σ in Γ space; see Fig. 8.3 (a). Also, any system with energy E will lie somewhere on Σ at time t, and it will occupy all points on Σ at different times.

It was realised by Gibbs around 1895, well before the advent of quantum mechanics, that the total energy E cannot have a sharply defined value, but must be allowed to have a range E to $E + \Delta E$ where, in classical mechanics ΔE can be made arbitrarily small. In quantum mechanics, it was given a physical meaning by the uncertainty principle.

Henceforth we shall use dq to mean $d^{3n}q$ and dp to mean $d^{3n}p$. Let $dpdq$ be an element of volume in the $6n$-dimensional phase space and $W\,dpdq$ be the probability of finding a system in the ensemble whose representative point lies in the volume element. Then $W\,dpdq$ will be constant for systems in the shell near Σ and zero elsewhere.

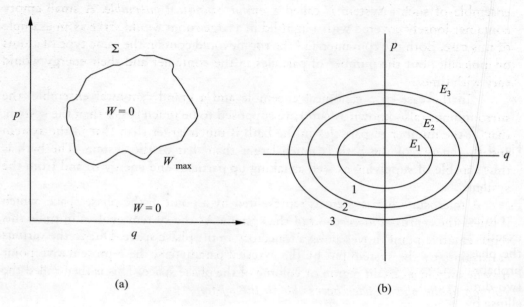

(a)   (b)

Fig. 8.3   *Probability in microcanonical ensemble: (a) the shell near surface Σ accessible to the ensemble; (b) phase space of an oscillator with different energies.*

As an example, consider a system of a single one-dimensional simple harmonic oscillator. It has an energy $E = (p^2 + \omega^2 q^2)/2$ per unit mass. So all oscillators of a given energy started at any initial phase will lie in a thin shell around an ellipse with semimajor axis $a = (2E/\omega^2)^{1/2}$ and semiminor axis $b = \sqrt{2E}$, as shown in Fig. 8.3 (b). Then the probability of finding an oscillator of energy E will be constant and maximum near ellipse 1 and will drop off to zero rapidly inside and outside the thin shell.

8.2.2 Canonical ensemble

Actually, there will be exchange of energy and particles between the system A and the surroundings A', as shown in Fig. 8.4. If we allow only exchange of energy between the system and the bath, we will still have n_μ and n constants in time, but the system energy E will fluctuate. In this case, the collection of similar systems is called a canonical ensemble. The systems in this ensemble are spread all over the phase space. In this case, W exists everywhere in the $6n$ dimensional phase space, and we must have

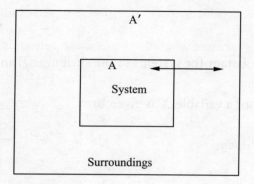

Fig. 8.4   *Exchange of energy between a system A and the surrounding bath A'*

$$\int_\Gamma W\, dp\, dq = 1.$$

On account of the classical or quantum uncertainty, we cannot specify each p_i, q_i exactly. So we have to consider an elementary cell of volume $dp\, dq = h^n$. Consequently the phase space has to be divided into cells of this size and we have to specify the probability density W for each cell. Thus for the case of a simple oscillator, we have a two-dimensional phase space and $E = (n + 1/2)h\nu$, which is related to the area of the ellipse by

$$\pi ab = 2\pi E/\omega = E/\nu = (n + 1/2)h.$$

Thus the elementary cells are elliptical shells of area h between successive energy ellipses.

8.2.3   Grand canonical ensemble

Finally, in a real situation, there will be both exchange of energy and particles between the system and the surrounding bath. So both E and n_μ, as well as n, will fluctuate. Such a collection of systems is called a grand canonical ensemble. In this case the members of the ensemble cannot be represented in a single phase space. Further, in several systems, and particularly in quantum mechanics, the numbers n_μ also fluctuate. So we have to consider an infinite number of phase spaces of several dimensions. Here

$$\int_\Gamma W\, dp\, dq \neq 1,$$

but

$$\sum_{\Gamma=1}^{\infty} \int_{\Gamma} W \, dp \, dq = 1.$$

The ensemble average of a variable X is given by

$$\overline{X} = \sum_{\Gamma=1}^{\infty} \int_{\Gamma} X W \, dp \, dq.$$

Our example of the simple harmonic oscillator in one dimension contains only one particle, and an ideal undamped oscillator cannot take or give energy to the surroundings. Hence it cannot be a part of canonical or grand canonical ensemble. But consider a diatomic molecule such as N_2 or O_2 freely floating in the atmosphere in a hall. It is a one-dimensional oscillator which can exchange energy with the surroundings.

8.2.4   Constancy of number of elementary cells

Let us consider a finite number ν of systems with different initial conditions occupying a volume $d\tau$ in the phase space at a certain instant t, as shown in Fig. 8.5(a). As the systems evolve, they traverse different non-intersecting paths in the phase space, and occupy a volume $\overline{d\tau}$ at time $t + \Delta t$. Then,

$$d\tau = dq_1 dq_2 \cdots dq_\nu dp_1 dp_2 \cdots dp_\nu$$

and

$$\overline{d\tau} = \overline{dq_1} \, \overline{dq_2} \cdots \overline{dq_\nu} \, \overline{dp_1} \, \overline{dp_2} \cdots \overline{dp_\nu}$$

are related by the Jacobian

$$\int d\tau = \int J(q, p/\overline{q}, \overline{p}) \overline{d\tau}.$$

Now, $\overline{q}_k = q_k + \dot{q}_k \Delta t + \mathrm{O}(\Delta t^2),$

or $\overline{q}_k = q_k + \dfrac{\partial H}{\partial p_k} \Delta t + \mathrm{O}(\Delta t^2).$ (8.22a)

Similarly,

$$\overline{p}_k = p_k - \frac{\partial H}{\partial q_k} \Delta t + \mathrm{O}(\Delta t^2).$$ (8.22b)

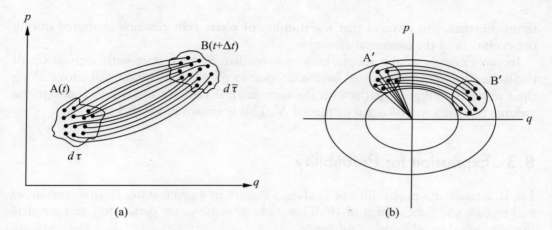

Fig. 8.5   *Constancy of volume in phase space: (a) general case; (b) phase space of a simple oscillator*

Then,

$$J^{-1} = \begin{vmatrix} d\bar{q}_i/dq_j & \vdots & d\bar{q}_i/dp_j \\ \cdots\cdots & \cdots & \cdots\cdots \\ d\bar{p}_i/dq_j & \vdots & d\bar{p}_i/dp_j \end{vmatrix},$$

where the determinant in each block has dimensions $\nu \times \nu$. On account of (8.22), we have, to first order in Δt,

$$J^{-1} = \begin{vmatrix} \delta_{ij} + (\partial^2 H/\partial p_i \partial q_j)\Delta t & \vdots & (\partial^2 H/\partial p_i \partial p_j)\Delta t \\ \cdots\cdots & \cdots & \cdots\cdots \\ -(\partial^2 H/\partial q_i \partial q_j)\Delta t & \vdots & \delta_{ij} - (\partial^2 H/\partial q_i \partial p_j)\Delta t \end{vmatrix}.$$

Working out this determinant, one sees that the first order terms vanish and

$$J^{-1} = 1 + O(\Delta t^2),$$

so that

$$\frac{d}{dt}\left(\frac{1}{J}\right) = 0,$$

giving $J = $ constant. But $J = 1$ at time $t = 0$, hence $J = 1$ for all t, to first order in Δt. Thus $d\tau = \overline{d\tau}$ at time $t + \Delta t$.

As a transformation to a state after a finite time can be considered as a resultant of many transformations over small intervals of time, we will have $d\tau = \overline{d\tau}$ at all times. Further, this ensures that the number of phase cells remains unaltered during the evolution of the canonical ensemble.

In our example of the simple harmonic oscillator, if we start with a given set of oscillators in some volume A$'$ in the accessible phase space, they will move along their respective elliptical paths with the same angular velocity and will be found in the volume B$'$ which will be equal to that of A$'$. This is shown in Fig. 8.5 (b).

8.3   Expression for Probability

Let W denote the probability of finding a system in a given state. In this section we will discuss the Gibbs form of W. This leads to some other parameters and we shall discuss here their physical significance.

8.3.1   Gibbs form of probability

We have seen that in microcanonical ensemble, $W\,dpdq$ is constant over the surface Σ. Then the constancy of the volume element $dpdq$ implies that W is independent of time. So we have

$$\frac{dW}{dt} = \frac{\partial W}{\partial t} + \sum_k \left(\frac{\partial W}{\partial q_k} \dot{q}_k + \frac{\partial W}{\partial p_k} \dot{p}_k \right) = 0.$$

In other words, W is a constant of motion. Hence it can be expressed as a function of other constants of motion. W is most likely to be a function of energy E, that is, $W = W(E)$, even when E fluctuates in the canonical ensemble.

Let n_1 and n_2 be the number of systems with energies E_1 and E_2, respectively. If n is the total number of systems in the ensemble, then $W(E_1) = n_1/n$ and $W(E_2) = n_2/n$. Then the probability of finding systems with energies E_1 and E_2 will be $W(E_1)W(E_2) = n_1 n_2/n^2$. But now the total energy of the system is $E_1 + E_2$. Therefore we have

$$W(E_1 + E_2) = W(E_1)W(E_2).$$

Then,

$$\frac{dW}{dE} = \lim_{\epsilon \to 0} \frac{W(E + \epsilon) - W(E)}{\epsilon}$$

$$= \lim_{\epsilon \to 0} \frac{W(E)[W(\epsilon) - 1]}{\epsilon}$$

$$= cW(E),$$

where

$$c = \lim_{\epsilon \to 0}[(W(\epsilon) - 1)/\epsilon].$$

Integrating with respect to E, we get

$$\ln W = cE + d \Rightarrow W(E) = e^{d+cE}.$$

Now, the probability of finding the system with large E is small, so c must be negative. Then putting $c = -\beta$ where $\beta > 0$, we get

$$W = e^{\beta(\Omega - E)}, \tag{8.23}$$

where we have put $d = \beta\Omega$. Equation (8.23) is the Gibbs form of $W(E)$. Here Ω has to be determined by the condition $\Sigma W(E) = 1$.

In the grand canonical ensemble, the numbers of particles n_j of each type j also fluctuate. They will also enter into the probability function W. By similar arguments as above, we get for the probability the expression

$$W = \exp\{\beta(\Omega - E) + \sum_j \gamma_j n_j\}, \tag{8.24}$$

where γ_j are dimensionless and may be positive or negative. It may be noted that the number of microstates increases with increasing n_j.

8.3.2 Identification of parameters with thermodynamic quantities

We have introduced a few parameters such as β, Ω and γ_i. We shall now relate these with thermodynamic quantities such as temperature, entropy, Gibbs free energy, etc. Consider the following aspects:

(i) Large values of energy E are favoured by large values of temperature T and small values of the parameter β, and vice versa. It is therefore reasonable to put $\beta = 1/k'T$, where k' is a constant. Now β has dimensions of E^{-1} and k' has dimensions of E/T, that is, those of Boltzmann constant introduced in Chapter 4.

(ii) Consider the expression $S' = k\overline{\ln W}$ where $\overline{\ln W}$ is the ensemble average of $\ln W$. Thus,

$$S' = -k \sum_j W_j \ln W_j,$$

where we have used the expression for average of an observable quantity A as

$$\overline{A} = \sum_j W_j A_j,$$

and W_j is the probability of the occurrence of macrostate j. (We take all microstates having energy around E to form a macrostate. Hence we shall often suppress the summation index j). Then using (8.24), we get

$$dS' = -k\delta\Sigma W\{\beta(\Omega - E) + \sum_k \gamma_k n_k\}$$

$$= -k\delta[\beta(\Omega - \overline{E}) + \sum_k \gamma_k \overline{n}_k]$$

$$= -k[\delta\beta(\Omega - \overline{E}) + \beta\delta(\Omega - \overline{E}) + \sum_k (\overline{n}_k\delta\gamma_k + \gamma_k\delta\overline{n}_k)]. \tag{8.25}$$

Since $\sum W(E) = 1$, we have

$$\sum_E \delta W(E) = 0 \quad \text{and} \quad \sum_E W(E)\delta(\ln W(E)) = 0.$$

Then

$$\sum W\delta\{\beta(\Omega - E) + \sum \gamma_k n_k\} = 0.$$

This results in

$$(\Omega - \overline{E})\delta\beta + \beta(\delta\Omega - \delta\overline{E}) + \overline{n}_k\delta\gamma_k + \gamma_k\delta\overline{n}_k = 0. \tag{8.26}$$

Now consider volume changes only, that is, that T and P are constant. Then $\delta\overline{E} = -P\delta V$ and $\delta\overline{n}_k = 0$. So (8.26) gives

$$(\Omega - \overline{E})\delta\beta + \beta(\delta\Omega + P\delta V) + \sum \overline{n}_k\delta\gamma_k = 0. \tag{8.27}$$

Then substituting (8.27) in (8.25), we get

$$dS' = k[\beta P\delta V + \beta\delta\overline{E} - \sum \gamma_k\overline{n}_k]. \tag{8.28}$$

Then replacing U by E in (4.37), we have

$$E = G - PV + TS. \tag{8.29}$$

The Gibbs free energy G can be written as

$$G = \sum_k g_k m_k = \sum_k \mu_k n_k,$$

where g_k is the Gibb's free energy per unit mass of the kth species, m_k its mass and μ_k the Gibb's free energy per particle of the kth species. Then keeping P and T constant, as before, we get

$$\delta E = T\delta S - P\delta V + \sum_k \mu_k \delta n_k, \tag{8.30}$$

since $\delta\mu_k = 0$. Thus,

$$\delta S = [P\delta V + \delta E - \sum \mu_k \delta n_k]/T. \tag{8.31}$$

Comparing (8.31) and (8.28), we can identify S' with $S = -k\overline{\ln W}$ if we put

$$k\beta = 1/T,$$

that is, $k/k'T = 1/T \Rightarrow k' = k,$

and $\gamma_k = \mu_k/kT.$ \hfill (8.32)

(iii)   Finally we have

$$S = -k \sum W \ln W$$

$$= -k \sum W\{[\Omega - E + \sum \mu_k n_k]/kT\} = (-\Omega + E - G)/T$$

$$\Rightarrow TS = -\Omega + E - G. \tag{8.33}$$

Comparing (8.33) and (8.30), we find

$$\Omega = -PV. \tag{8.34}$$

Hence

$$W = \exp\{(PV - E + \sum \mu_k n_k)/kT\}. \tag{8.35}$$

8.3.3   Application to non-interacting particles

We consider in this section some aspects of a system consisting of several species of non-interacting particles.

Law of partial pressures

Consider a system as described above where the species may be molecules, electrons, protons, excited states of atoms, photons, etc. We assume that they behave like a perfect gas. Then we can write the total energy E as $\sum E_k$, where E_k is the energy of the k-th species. So

$$W = \exp\{[\Omega - E + \sum \mu_k n_k]/kT\}.$$

Then putting $\Omega = \sum \Omega_k$, we have

$$W = \Pi_k W_k = \Pi_k \exp\{\Omega_k - E_k + \mu_k n_k)/kT\}, \tag{8.36}$$

which defines W_k. Then $\Omega = -PV$ gives $\Omega_k = -p_k V$. Thus $P = \sum p_k$, where p_k is the pressure of the kth species in volume V when no other species is present. This is the *law of partial pressures*.

Statistical weight of a species

A volume of phase space $dpdq$ contains $4\pi p^2 dpdV/h^3$ elementary cells. So if g_i is the statistical weight of a given quantum level, the statistical weight per unit volume will be $g = 4\pi p^2 dpg_i/h^3$. For example, $g_i = 2$ for electrons due to two spin states, $g_i = 2J + 1$ for an atomic state with total spin quantum number J, and $g_i = 2n^2$ for the nth energy level of a hydrogen-like atom. Thus for photons, we have $p = h\nu/c$ and $g(\nu) = 8\pi\nu^2 d\nu/c^3$ per unit volume, and for free electrons, we have $p = mv$ and $g(v)$ is given by (8.8). The statistical weight is thus the number of microstates associated with a macrostate.

Statistical weight in grand canonical ensemble

Let us consider a grand canonical ensemble of free particles of only one type. A state of the whole system will be specified by the systems in each single-particle state. We can combine these states into μ groups[24] containing systems with nearly the same energy E_μ and having nearly the same value X_μ of a parameter X. If g_μ is the weight or number of single-particle states in the group μ and n_μ is the number of particles in that cell, we can distribute n_μ particles in g_μ cells, subject to appropriate conditions depending on the nature of particles (fermions, bosons, photons, or classical particles) in a certain number of ways. (Note that a cell in phase space represents a microstate of the system.) Let us denote this number by $\left\{ \begin{array}{c} g_\mu \\ n_\mu \end{array} \right\}$. Then the total number of ways in which the total number of particles $n = \Sigma n_\mu$ can be distributed in μ groups such that $E = \Sigma n_\mu E_\mu$ will be

$$G = \Pi_\mu \left\{ \begin{array}{c} g_\mu \\ n_\mu \end{array} \right\}.$$

Here G represents the total weight of the groups. Then the grand canonical ensemble average of the parameter X is

---

[24]There should be no confusion between μ used as energy per particle and as a subscript or summation index.

$$\overline{X} = \sum XWG, \tag{8.37}$$

where the summation is over groups. It may be noted that $\Sigma WG = 1$.

Average number of particles in an energy level

Using (8.36), we now see that

$$\sum XWG = \sum X \exp\{(\Omega - E)/kT + \gamma n\} \Pi_\mu \left\{ \begin{array}{c} g_\mu \\ n_\mu \end{array} \right\}. \tag{8.38}$$

With $X = 1$, and $\Sigma WG = 1$, we have

$$e^{-\Omega/kT} = \sum \Pi_\mu \left\{ \begin{array}{c} g_\mu \\ n_\mu \end{array} \right\} \exp\{-n_\mu E_\mu/kT + \gamma n_\mu\}. \tag{8.39}$$

Also, with $X = n_\nu$, we find that the average number of particles is given by

$$\overline{n}_\nu = \frac{\Sigma n_\nu \Pi_\mu \left\{ \begin{array}{c} g_\mu \\ n_\mu \end{array} \right\} \exp\{-n_\mu E_\mu/kT + \gamma n_\mu\}}{\Sigma \Pi_\mu \left\{ \begin{array}{c} g_\mu \\ n_\mu \end{array} \right\} \exp\{-n_\mu E_\mu/kT + \gamma n_\mu\}}, \tag{8.40}$$

where, as before, the summation is over groups of systems. Differentiating (8.39) with respect to E_ν and using (8.40), we get

$$\frac{\partial \Omega}{\partial E_\nu} = \overline{n}_\nu. \tag{8.41}$$

We may say that Ω is the generating function for $\overline{n}_\nu$.

8.4 Population Functions

In this section we shall derive the average number of particles occupying an energy level or a quantum state in various cases when the particles are bosons, fermions and classical particles. We shall apply this to the case of massless photons, which are bosons. We shall also study their behaviour with respect to energy and temperature.

8.4.1 Distribution of particles in states

We must distinguish at the outset between fermions and bosons. We have seen earlier that particles such as electrons, protons, neutrons, obey Pauli exclusion principle and have antisymmetric wavefunction; they belong to the class of fermions. On the other hand, photons and deuterons which have integral spin, have symmetric wave function and do not obey Pauli exclusion principle; they are bosons.

In the following we shall determine the number $\left\{ \begin{array}{c} g \\ n \end{array} \right\}$ introduced earlier for the cases of fermions, bosons (with mass), photons and classical particles.

Fermions

In this case, not more than one particle can occupy one quantum state. If there are g cells with a maximum occupancy of one each and n particles to fill in, then the number of ways in which this can be done is

$$\left\{ \begin{array}{c} g \\ n \end{array} \right\} = {}^g C_n = \frac{g!}{n!(g-n)!}. \tag{8.42}$$

Naturally, g must be greater than or equal to n, otherwise n particles cannot be placed in a smaller number of states subject to the Pauli exclusion principle.

Bosons

Here any number of particles can occupy the same state, so we are concerned with combinations with repetition. We shall do this through a Guided Exercise.

▶ Guided Exercise 8.1

Obtain the number of ways in which n bosons with nonzero rest mass can be placed in g cells. Note that these particles are subject to the conservation of particles, that is, n is a constant.

Hints

(a) Put $n - p$ particles in the first cell and the remaining p particles in the remaining $g - 1$ cells. This can be done in ${}^{g-1}C_p$ ways.

(b) Since we can put any number of particles in the first cell, p can run through 0 to n. So

$$\left\{ \begin{array}{c} g \\ n \end{array} \right\} = \sum_{p=0}^{n} \left\{ \begin{array}{c} g-1 \\ p \end{array} \right\}.$$

(c) Write the term $p = n$ separately and rewrite the above equation as

$$\left\{ \begin{array}{c} g \\ n \end{array} \right\} = \sum_{p=0}^{n-1} \left\{ \begin{array}{c} g-1 \\ p \end{array} \right\} + \left\{ \begin{array}{c} g-1 \\ n \end{array} \right\} = \left\{ \begin{array}{c} g \\ n-1 \end{array} \right\} + \left\{ \begin{array}{c} g-1 \\ n \end{array} \right\}. \tag{8.43}$$

(d) Note that (8.43) looks like the familiar identity

$$^nC_r = {}^{n-1}C_r + {}^{n-1}C_{r-1}.$$ (8.44)

Hence we can identify $\left\{\begin{array}{c} g \\ n \end{array}\right\}$ with ^NC_r if we take

$$N = g + n + a, \quad r = g + b,$$

where a and b are constants to be determined. Thus we take

$$\left\{\begin{array}{c} g \\ n \end{array}\right\} = {}^{g+n+a}C_{g+b},$$ (8.45)

so that

$$\left\{\begin{array}{c} g \\ n-1 \end{array}\right\} = {}^{g+n+a-1}C_{g+b}, \quad \left\{\begin{array}{c} g-1 \\ n \end{array}\right\} = {}^{g+n+a-1}C_{g+b-1}.$$

(e) Verify that these are consistent with (8.43) and (8.44), suggesting that our suggestion, (8.45), is right.

(f) Now determine a and b in the following manner. If there is only one cell, $g = 1$, and n particles ($n \geq 0$), we can put these n particles in the only cell available in only one way. Thus,

$$\left\{\begin{array}{c} 1 \\ n \end{array}\right\} = 1 \Rightarrow {}^{n+a+1}C_{b+1} = 1.$$

(g) Since this is true for all n, write it for two consecutive values of n such as s and $s+1, s \geq 0$, and equate the two expressions

$$1 = \frac{(s+a+1)!}{(b+1)!(s+a-b)!} = \frac{(s+a+2)!}{(b+1)(s+a-b+1)!}$$

to get

$$b = -1.$$ (8.46)

Thus (8.45) reduces to

$$\left\{\begin{array}{c} g \\ n \end{array}\right\} = {}^{g+n+a}C_{g-1}.$$

(h) Now take $n = 1$, and note that one particle can be put in $g > 0$ cells in g ways, giving

$$\left\{\begin{array}{c} g \\ 1 \end{array}\right\} = g \Rightarrow {}^{g+a+1}C_{g-1} = g.$$

(i) Put $g = 2$ in the above to get

$$a = -1.$$ (8.47)

(j) Thus, we finally arrive at the distribution function for bosons,

$$\left\{ \begin{array}{c} g \\ n \end{array} \right\} = {}^{g+n-1}C_{g-1} = \frac{(g+n-1)!}{(g-1)!n!}. \tag{8.48}$$

This leads to Bose–Einstein statistics.   ◀

Classical case

Both boson and fermion statistics reduce to classical statistics when $g \gg n$, that is, when the number of cells in the phase space is much larger than the number of particles. The phase space is thus thinly populated. It can be seen that both (8.42) and (8.48) then reduce to

$$\left\{ \begin{array}{c} g \\ n \end{array} \right\} = \frac{g^n}{n!}. \tag{8.49}$$

The difference between quantum and classical statistics becomes important when $n \sim g$, that is when there is, a crowding of particles in phase cells.

8.4.2   Average number of particles

Let us put

$$x_\mu = \exp(-E_\mu/kT + \gamma). \tag{8.50}$$

Then (8.38) gives

$$e^{-\Omega/kT} = \sum \Pi_\mu \left\{ \begin{array}{c} g_\mu \\ n_\mu \end{array} \right\} x_\mu^{n_\mu}$$

$$= \sum \left\{ \begin{array}{c} g_1 \\ n_1 \end{array} \right\} x_1^{n_1} \left\{ \begin{array}{c} g_2 \\ n_2 \end{array} \right\} x_2^{n_2} \cdots,$$

where the summation is over groups. Summation over groups implies summation over n_1, n_2, etc. Hence we can interchange the summation and product in the above expression and write it as

$$e^{-\Omega/kT} = \left[\sum_{n_1} \left\{ \begin{array}{c} g_1 \\ n_1 \end{array} \right\} x_1^{n_1} \right] \left[\sum_{n_2} \left\{ \begin{array}{c} g_2 \\ n_2 \end{array} \right\} x_2^{n_2} \right] \cdots. \tag{8.51}$$

Now we are in a position to obtain population functions for various cases.

Fermi–Dirac statistics

In this case $\left\{ \begin{array}{c} g \\ n \end{array} \right\}$ is given by (8.42). Therefore

$$\sum_{n=0}^{\infty} \left\{ \begin{array}{c} g \\ n \end{array} \right\} x^n = \sum_{n=0}^{g} {}^g C_n x^n = (1+x)^g, \tag{8.52}$$

where we have used the fact that $\left\{ \begin{array}{c} g \\ n \end{array} \right\} = 0$ for $n > g$. Then (8.51) becomes

$$e^{-\Omega/kT} = \Pi_\mu (1 + e^{\gamma - E_\mu/kT})^{g_\mu}$$

$$\Rightarrow \Omega = -kT \sum_\mu g_\mu \ln(1 + e^{\gamma - E_\mu/kT}). \tag{8.53}$$

Then

$$\overline{n}_\nu = \frac{d\Omega}{dE_\nu} = -g_\nu \frac{e^{\gamma - E_\nu/kT}}{1 + e^{\gamma - E_\nu/kT}},$$

or, dropping the subscript ν, this can also be written as

$$\frac{\overline{n}}{g} = \frac{1}{e^{(E-\mu)/kT} + 1}. \tag{8.54}$$

Bose–Einstein statistics

Here

$$\sum_{n=0}^{\infty} \left\{ \begin{array}{c} g \\ n \end{array} \right\} x^n = \sum_{n=0}^{\infty} {}^{g+n-1} C_{g-1} x^n = (1-x)^{-g}. \tag{8.55}$$

Therefore (8.51) gives

$$\Omega = kT \sum g_\mu \ln \left(1 - e^{-\gamma - E_\mu/kT} \right).$$

Equation (8.41) then gives, again dropping the subscript ν,

$$\frac{\overline{n}}{g} = \frac{1}{e^{E/kT - \gamma} - 1} = \frac{1}{e^{(E-\mu)/kT} - 1}. \tag{8.56}$$

Maxwell–Boltzmann statistics

When $g \gg n$, we have

$$\sum \left\{ \begin{array}{c} g_\mu \\ n_\mu \end{array} \right\} x^{n_\mu} = \sum_{n_\mu} \frac{g_\mu^{n_\mu}}{n_\mu!} x^{n_\mu} = \exp(g_\mu x_\mu).$$

Then,

$$e^{-\Omega/kT} = \Pi_\mu \exp(g_\mu x_\mu), \tag{8.57}$$

or $$-\Omega/kT = \sum_\mu g_\mu e^{-E_\mu/kT+\gamma} = \sum_\mu \overline{n}_\mu = n. \tag{8.58}$$

Equation (8.58) also gives

$$-\frac{1}{kT}\frac{\partial\Omega}{\partial E_v} = -\frac{g_v}{kT}e^{-E_v/kT+\gamma}.$$

This further gives, dropping ν,

$$\frac{\overline{n}}{g} = \frac{1}{g}\frac{\partial\Omega}{\partial E} = e^{-E/kT+\gamma}. \tag{8.59}$$

Equation (8.59) combines Maxwell's law and Boltzmann's equation. Here γ is related to the number density of particles (see Section 8.7 below). In terms of the chemical potential μ_v introduced in (8.32), the population function becomes

$$\frac{\overline{n}}{g} = e^{-(E-\mu)/kT}. \tag{8.60}$$

Since $\Omega = -PV$, (8.58) also results in

$$PV = nkT,$$

which is the perfect gas law. If the volume V contains more than one types of particles, we have $n = \sum_\mu \overline{n}_\mu$, so that $PV = kT\sum_\mu \overline{n}_\mu$. This gives $p_\mu V = \overline{n}_\mu kT$ and $P = \sum p_\mu$, which is the law of partial pressures.

Each phase space cell represents a microstate of the system. The parameter $\overline{n}/g$ obtained in (8.54), (8.56) and (8.59) for the three statistics can then be seen as the probability of occupation of a microstate at a certain energy and temperature. The chemical potential μ is determined by the particle density (number of particles per unit volume of real space) and temperature.

8.4.3   Range of chemical potential

The chemical potential or the Gibb's potential μ, introduced earlier, represents the increase in energy of the system when the particle number of a particular species is increased by one, subject to keeping the entropy, volume and other particle numbers constant. The particles may follow any of the three statistics given by equation (8.54) for FD, equation (8.56) for BE, or equation (8.59) for MB. We shall consider a system consisting of non-interacting free particles of a single species. Thus the only energy of each particle is its kinetic energy. We take the zero of energy as the energy when all the particles are at rest. Then the energy of each particle will have a spectrum going from 0 to ∞.

As is clear from equation (8.54) for fermions, the average particle number per cell lies between 0 and 1 for any values of E, μ, T. From the point of view of chemical potential, μ may have any value, positive or negative, though with our choice of zero for energy, μ would be positive. Equation (8.56) for bosons shows that μ must be negative. For, if $\mu > 0$, the particle number will tend to infinity as $E \to \mu$. Therefore $\mu \le 0$ for bosons.

Both FD and BE statistics go over to classical MB statistics as the particle density becomes small. This means that the probability of occupation of a cell is lower in a classical gas as compared to a Bose gas. This shows that the chemical potential for a classical gas must be even lower (more negative) than a Bose gas.

When the particle number N in a single-phase system increases to $N + 1$, with N much larger than 1 for statistical mechanics to be applicable, the number of microstates increases much faster. Hence the entropy substantially increases. This number, W, is a very strong function of N and energy E. In order to obtain the chemical potential, we have to keep the entropy constant.

In the case of fermions, when we add a particle to the system, it has to go to a higher unoccupied state due to the exclusion principle. The fact that one has to keep the entropy constant has a weaker effect. There is little possibility of placing particles in lower energy states as most of them are occupied.

In the case of bosons and classical particles, when we place a new particle in a certain energy state, the entropy increases very fast. In order to bring it down to its original value, we have to push all the particles down the energy ladder, which is possible because there is no limit on the occupancy of a state. The result, in effect, is that the total energy of the system decreases, making μ negative.

8.4.4   Graphical representation of population functions

Let us look at the population function for all the three cases. They are plotted as functions of E or γkT in Fig. 8.6. We see that for large values of E, all of them give $\overline{n}/g = e^{-E/kT}$. The graph is also extended to negative values of E, although the parts below $E = 0$ must be taken only in a mathematical sense. At $E = \mu$, fermions have $\overline{n}/g = 1/2$ (point P in Fig. 8.6), that is, half the available cells are occupied by particles. For classical particles, $\overline{n}/g = 1$ (point Q in Fig. 8.6) while for bosons, $\overline{n}/g \to \infty$ as $E \to \mu$, though as said, this is an unphysical region. In general, $\overline{n}/g$ goes on increasing as E decreases, and approaches unity for fermions. The difference between fermions, bosons and classical particles becomes evident at small energies.

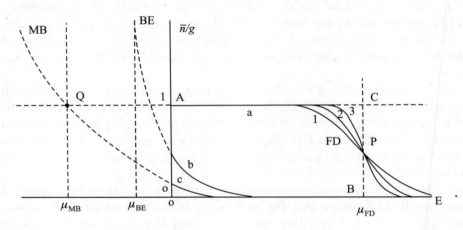

Fig. 8.6 *Population functions; (a) fermions (1, 2, 3 for $T_1 > T_2 > T_3$), (b) bosons, (c) classical particles.*

For photons, $\mu = 0$ but for particles with mass, μ is a function of particle density and temperature, that is, $\mu \equiv \mu(N/V, T)$. The slope of the population function at $E = \mu$ is found to be

$$\delta_P = -1/4kT \text{ for fermions, and}$$

$$\delta_Q = -1/kT \text{ for classical particles.}$$

In both cases the slope at points P and Q becomes more and more negative as T decreases, and tends to $-\infty$ as $T \to 0$, as shown in Fig. 8.6 for three temperatures for fermions. In the case of fermions, the population function is reduced to a rectangle OACB. This means all the states with $E < \mu$ are completely filled and those with $E > \mu$ are empty. This is known as a state of complete degeneracy, which plays an important part in the equation of state at low temperatures.

8.5 Equation of State for Fermions

The case of a fermion gas is interesting. Pauli exclusion principle and Fermi–Dirac statistics were major departures from classical statistics. Electrons in a metal provide one of the best examples of an electron gas. The ideal Fermi gas also provides the first approximation for electrons in conduction band and holes in valence band of a semiconductor, with different effective masses. Neutron stars in the universe can be treated as a relativistic Fermi gas of neutrons. The γ-factor, that is μ/kT, for electrons in a metal is of the order of 10^2, while for a neutron star, it could be of the order of 10^6. In this section, we will obtain the equation of state of a relativistic Fermi gas from which follows the case of a non-relativistic Fermi gas, and consider various cases.

8.5.1 General relativistic case

Equation (8.53) along with $\Omega = -PV$ gives

$$PV = kT \sum_{\mu} g_{\mu} \ln(1 + e^{\gamma - E_{\mu}/kT}). \tag{8.61}$$

Putting $g_{\mu} = 8\pi p^2 dp V/h^3$, and replacing summation over μ by integration over p, we get

$$P = \frac{8\pi kT}{h^3} \int_0^{\infty} p^2 \ln(1 + e^{\gamma - E/kT}) dp. \tag{8.62}$$

To evaluate this integral, we note that the kinetic energy of a relativistic particle can be written as

$$E = (m_0^2 c^4 + p^2 c^2)^{1/2} - m_0 c^2. \tag{8.63}$$

So $E \to \infty$ as $p \to \infty$ and $\ln(1 + e^{\gamma - E/kT}) \to 0$. Equation (8.62) can now be integrated by parts to give

$$P = \frac{8\pi}{3h^3} \int_0^{\infty} \frac{p^3}{e^{(E-\mu)/kT} + 1} \frac{dE}{dp}. \tag{8.64}$$

The total number of particles per unit volume (of real space) can be seen to be

$$n = \sum \overline{n}_{\mu} = \frac{8\pi}{h^3} \int_0^{\infty} \frac{p^2 dp}{e^{(E-\mu)/kT} + 1}. \tag{8.65}$$

Similarly, the internal energy density $U = \sum E_{\mu} \overline{n}_{\mu}$ can be seen to be

$$U = \frac{8\pi}{h^3} \int_0^{\infty} \frac{E p^2 dp}{e^{(E-\mu)/kT} + 1}. \tag{8.66}$$

Rewriting (8.63) in a suitable form, we get

$$dp = \frac{E + m_0 c^2}{c(E^2 + 2Em_0 c^2)^{1/2}} dE.$$

With these relations, we then have

$$P = \frac{8\pi}{3h^3 c^3} \int_0^\infty \frac{(E^2 + 2m_0 c^2 E)^{3/2}}{e^{(E-\mu)/kT} + 1} dE, \tag{8.67a}$$

$$n = \frac{8\pi}{h^3 c^3} \int_0^\infty \frac{(E^2 + 2m_0 c^2 E)^{1/2}(E + m_0 c^2)}{e^{(E-\mu)/kT} + 1} dE, \tag{8.67b}$$

$$U = \frac{8\pi}{h^3 c^3} \int_0^\infty \frac{(E^2 + 2m_0 c^2 E)^{1/2}(E^2 + m_0 c^2 E)}{e^{(E-\mu)/kT} + 1} dE. \tag{8.67c}$$

Equation (8.67) give the equation of state for relativistic fermions.

8.5.2 Complete degeneracy

We have seen that when T approaches zero, the population function for fermions tends to 1 and is close to 1 almost upto $E = \mu$. Then it drops from 1 to 0 in a short range of a few kT around μ, after which $\bar{n}/g \to 0$. Therefore (8.67) become

$$P = \frac{8\pi}{3h^3 c^3} \int_0^\mu (E^2 + 2m_0 c^2 E)^{3/2} dE,$$

$$n = \frac{8\pi}{h^3 c^3} \int_0^\mu (E^2 + 2m_0 c^2 E)^{1/2}(E + m_0 c^2) dE, \tag{8.68}$$

$$U = \frac{8\pi}{h^3 c^3} \int_0^\mu (E^2 + 2m_0 c^2 E)^{1/2}(E^2 + m_0 c^2 E) dE.$$

To evaluate these and simplify them, we take

$$m_0^2 c^4 y^2 = E^2 + 2m_0 c^2 E, \tag{8.69}$$

where we have introduced the parameter $y(E)$. Then

$$m_0^2 c^4 y \, dy = (E + m_0 E^2) dE.$$

From (8.69), we see that

$$y = [(E + m_0 c^2)^2 - m_0^2 c^4]^{1/2}/m_0 c^2$$

$$\Rightarrow E = m_0 c^2 [(1 + y^2)^{1/2} - 1].$$

It can be seen that y is the ratio of relativistic kinetic energy to rest mass energy. Then (8.68) becomes

$$P = \frac{\pi m_0^4 c^5}{3h^3} \int_0^x \frac{8y^4 dy}{(1+y^2)^{1/2}} \equiv \frac{\pi m_0^4 c^5}{3h^3} f(x), \qquad (8.70a)$$

$$N = \frac{8\pi m_0^3 c^3}{h^3} \int_0^x y^2 dy = \frac{8\pi m_0^3 c^3}{3h^3} x^3, \qquad (8.70b)$$

$$U = \frac{8\pi m_0^4 c^5}{h^3} \int_0^x [(1+y^2)^{1/2} - 1]y^2 dy \equiv \frac{\pi m_0^4 c^5}{3h^3} g(x), \qquad (8.70c)$$

where x is the value of y when $E = \mu$, and the functions $f(x)$ and $g(x)$ are defined through the integrals in (8.70a) and (8.70c). From (8.63) and (8.69), we have

$$y = \left[\frac{(E + m_0 c^2)^2 - m_0^2 c^4}{m_0^2 c^4} \right]^{1/2} = p/m_0 c.$$

So $x = p_F$ is p at $E = \mu$, that is, the Fermi momentum. Thus x is the ratio of the Fermi momentum to the momentum of a classical particle of mass m_0 if it were travelling at a speed c. Equation (8.70) give the equation of state of a completely degenerate Fermi gas.

We must now distinguish between non-relativistic and relativistic degeneracy.

Non-relativistic degeneracy

In this case $v \ll c$, so $p \leq m_0 c$, and x and y are small as compared to 1. Then,

$$f(x) = \int_0^x \frac{8y^4 dy}{(1+y^2)^{1/2}} \simeq \frac{8x^5}{5},$$

$$g(x) = \int_0^x 24[(1+y^2)^{1/2} - 1]y^2 dy \simeq 12x^5/5.$$

The number density, pressure and energy density in this case reduce to

$$n = 8\pi m_0^3 c^3 x^3 / 3h^3, \quad P = \frac{8\pi m_0^4 c^5 x^5}{15h^3}, \quad U = \frac{12\pi m_0^4 c^5 x^5}{15h^3}, \qquad (8.71)$$

showing that $P = 2U/3$, like a perfect Fermi gas.

Relativistic degeneracy

In this case $v \simeq c, p \geq m_0 c$, that is, x and y are large compared to 1. Then

$$f(x) \simeq \int_0^x 8y^4 dy \simeq 2x^4,$$

$$g(x) \simeq \int_0^x 24y^3 dy \simeq 6x^4.$$

Then the number density, pressure and energy density are given by

$$n = 8\pi m_0^3 c^3 x^3 / 3h^3,$$

$$P = 2\pi m_0^4 c^5 x^4 / 3h^3, \quad U = 2\pi m_0^4 c^4 x^4 / 3h^3, \tag{8.72}$$

showing that $P = U/3$, like radiation.

8.5.3 Departure from complete degeneracy

A Fermi gas will depart from complete degeneracy when $\exp[(E - \mu)/kT] > 1$, that is $E > \mu$. When the Fermi–Dirac distribution function of (8.54) has a small value, we can expand it in a series as follows:

$$\frac{1}{e^{(E-\mu)/kT} + 1} = \frac{1}{e^{(E-\mu)/kT}(1 + e^{(\mu-E)/kT})}$$

$$= \sum_{j=0}^{\infty} (-1)^j e^{(j+1)j(\mu-E)/kT}.$$

Then (8.65) gives

$$n = \frac{8\pi}{h^3} \sum_{j=0}^{\infty} (-1)^j \int_0^{\infty} p^2 e^{(j+1)(\mu-E)/kT} dp. \tag{8.73}$$

In the non-relativistic case, $E = p^2/2m$, so (8.73) becomes

$$n = \frac{8\pi\sqrt{2}m^{3/2}}{h^3} \sum_{j=0}^{\infty} (-1)^j \int_0^{\infty} E^{1/2} {}^{(j+1)(\mu-E)/kT} dE. \tag{8.74}$$

Now,

$$\int_0^{\infty} e^{-qE} E^{1/2} dE = \frac{\sqrt{\pi}}{2q^{3/2}}, \quad q > 0. \tag{8.75}$$

Using this in (8.74) with $q = j/kT$, we get

$$n = \frac{2(2\pi mkT)^{3/2}}{h^3} \sum_{j=0}^{\infty} (-1)^j \frac{e^{(j\mu/kT}}{j^{3/2}}.$$ (8.76)

The series in (8.76) will converge fast if $e^{\mu/kT} \ll 1$. Then taking only the first term, we see that

$$n \ll 2(2\pi mkT)^{3/2}/h^3.$$ (8.77)

8.6 Some Aspects of Bose Gas

In this section we briefly discuss photons in a black body enclosure, leading to Planck's law, and Bose–Einstein condensation.

8.6.1 Planck's law

Photons in an enclosure form a special case of bosons. They are massless particles with integral spin; hence they are not constrained by Pauli exclusion principle. Their total number (in a volume V at a temperature T) also does not remain a constant. They can be created or destroyed, subject to conservation of energy in a time-average sense. The only parameter that governs their energy distribution is temperature of the enclosure. In 1924, S. N. Bose applied his new statistics to photons and derived the Planck's law.

The chemical potential of photons is zero, $\mu = 0$, because a change in the total photon number by one does not change the total energy (two photons can be destroyed giving rise to one and vice versa). In the case of photons, the number of phase cells with a frequency ν and range $d\nu$ (or ω and $d\omega$) per unit solid angle per unit volume is given by

$$g_\nu d\nu d\Omega = 2\nu^2 d\nu d\Omega/c^3.$$

Then, from (8.56), the number of photons having their frequency in the interval $d\nu$ around ν will be

$$d\bar{n} = \frac{2\nu^2 d\nu d\Omega}{c^3(e^{h\nu/kT} - 1)}.$$

If we consider a cylindrical volume of area $d\sigma$ and length cdt, the distance travelled by photons in time dt, then these photons will have an energy of

$$dE = d\bar{n} \ h\nu \ cdt \ d\sigma d\Omega.$$

Then using (4.53) and setting $\cos\theta = 1$, we have

$$I_\nu = \frac{2h\nu^3}{c^2} \frac{1}{e^{h\nu/kT} - 1},$$

(8.78)

which is Planck's law.

8.6.2   Bose–Einstein condensation

In 1925, Einstein applied Bose's new statistics to indistinguishable Bose particles with mass. In the case of mass particles, their total number N (or number density per unit volume) is preserved, and we have seen that this results in a negative chemical potential, $\mu < 0$. Einstein realised that under suitable conditions of low temperature and high number density, the chemical potential would tend to zero. In such a situation, as $E \to 0$ from above and $\mu \to 0$ from below, the occupancy of the zero-energy state, $\overline{n}_0$, may diverge.

Let us write (8.56) in the form

$$\overline{n} = \frac{1}{z^{-1}e^{\beta E} - 1}, \quad z = e^{\beta\mu},$$

(8.79)

where $\overline{n}$ is now the average number of particles in a state with energy E at a temperature T in volume V, and the parameter z defined above is called *fugacity*. In general, the total number of particles N in volume V can be written as $N = \sum_{\mathbf{p}} \overline{n}_{\mathbf{p}}$, where $\mathbf{p}$ indicates the energy state in terms of momentum and $\overline{n}_{\mathbf{p}}$ is the average number of particles with momentum around $\mathbf{p}$. For a classical Maxwell–Boltzmann gas, $\mu < 0$ and $z < 1$. In this case $\overline{n}_{\mathbf{p}}$ varies slowly and we can replace the summation by integration over the phae space. It is given by

$$\overline{n}_{\mathbf{p}} = z \exp(-\beta E_{\mathbf{p}}).$$

Hence

$$N = \sum_{\mathbf{p}} \overline{n}_{\mathbf{p}} \simeq \frac{zV}{h^3} \int_0^\infty 4\pi p^2 \exp(-\beta p^2/2m)dp = zV/\lambda^3,$$

(8.80)

where $\lambda = h/(2mkT)^{1/2}$ is the average de Broglie wavelength of a free particle at temperature T. Equation (8.80) shows that

$$z = \lambda^3 N/V \equiv \lambda^3/v,$$

(8.81)

where $v = V/N$ is the average volume per particle. With the hindsight of de Broglie hypothesis, we may now say that fugacity is the ratio of de Broglie volume of a particle to its average volume.

But for a BE gas, (8.79) makes it clear that the number of particles diverges for the ground state $E = 0$ (and only for this state) as $\mu \to 0$. Let us denote the occupation number of the lowest energy state by $\overline{n}_0$. Now since $\overline{n}_0 \to \infty$ under a certain situation, we must take it out of the sum and change the remaining summation into integration. This then leads to

$$N = \sum_{\mathbf{p}} \overline{n}_{\mathbf{p}} = \overline{n}_0 + \sum_{\mathbf{p} \neq 0} \overline{n}_{\mathbf{p}}$$

$$= \overline{n}_0 + \frac{4\pi V}{h^3} \int_0^\infty \frac{p^2 dp}{z^{-1}e^x - 1},$$

where $x = \beta p^2/2m$.

Since $0 < z \leq 1$, we can expand the above in powers of z, in the form

$$N = \overline{n}_0 + \frac{2\pi V (2mkT)^{3/2}}{h^3} \sum_{n=1}^\infty I_n, \tag{8.82}$$

where

$$I_n = z^n \int_0^\infty x^{1/2} e^{-nx} dx.$$

This integral is related to the gamma function, and we finally get

$$N = \overline{n}_0 + \frac{V(2\pi mkT)^{3/2}}{h^3} g_{3/2}(z), \tag{8.83}$$

where

$$g_{3/2}(z) = \sum_{n=1}^\infty \frac{z^n}{n^{3/2}}, \quad 0 \leq z \leq 1 \tag{8.84}$$

is the Riemann zeta function. It may be noted that $g_{3/2}(z) \approx z$ for small z and $g_{3/2}(1) = 2.612$. It diverges for $z > 1$.

Putting $E = 0$, (8.79) shows that

$$\overline{n}_0 = z/(1-z). \tag{8.85}$$

Equation (8.83) can be written as

$$N \simeq \overline{n}_0 + \frac{V}{\lambda^3} g_{3/2}(z). \tag{8.86}$$

From these two equations, (8.85) and (8.86), we can understand the variation of the occupation numbers of the ground state and the rest of the states. We can see how all the particles tend to occupy the ground state as $\mu \to 0$. Note again that μ is a function of the particle density N/V and T; μ increases (approaches zero for BE gas) as N/V increases and as T decreases.

Equation (8.86) can be written as

$$\frac{\lambda^3 \overline{n}_0}{V} = \frac{\lambda^3}{v} - g_{3/2}(z).\tag{8.87}$$

The variation of these three terms with z is shown in Fig. 8.7. Note that generally λ^3/V is a small factor of the order to $1/N$, though λ^3/v will depend on N/V and T.

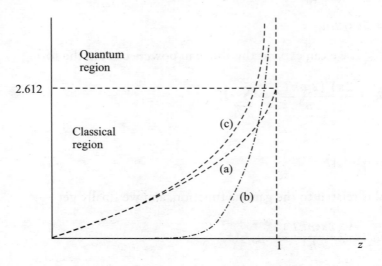

Fig. 8.7   *Variation of the three terms of (8.87) with z: (a) $g_{3/2}(z)$, (b) $\lambda^3 \overline{n}_0/V$, (c) λ^3/v (schematic).*

How do we determine z, μ and $\overline{n}_0$ in a particular case? For a given gas (BE or MB), λ^3/v is a known parameter, depending on V, N, m, T. The curve (c) in Fig. 8.7 gives the value of z. The curve (b) then gives the value of $\overline{n}_0$ and curve (a) gives the excited state occupation.

The phenomenon when more and more particles cascade down to the ground state under suitable conditions is called *Bose–Einstein condensation* (BEC). Note that it is a condensation in the phase space, not in real space.

Noting that $g_{3/2}(1) = 2.612$, we can define a critical temperature T_c, critical volume per particle v_c and critical de Broglie wavelength λ_c. Thus we define λ_c by

$$\lambda_c^3/v = g_{3/2}(1) = 2.612.\tag{8.88}$$

This means that when the ratio of de Broglie volume to the volume per particle is above 2.612, we are in the quantum region (BE gas) while when it is smaller than 2.612, we have a classical gas. The critical temperature can be seen to be

$$T_c = \frac{h^2}{2\pi m k}\left(\frac{1}{2.612v}\right)^{2/3}.$$

(8.89)

This means that for a given v, BEC starts when $T < T_c$. Finally, the critical (average) volume is given by

$$v_c = \frac{\lambda^3}{g_{3/2}(1)} = \frac{h^3}{2.612(2\pi m k T)^{3/2}}.$$

(8.90)

This shows that for a given temperature, BEC starts when the average volume per particle goes below v_c.

It is found that in the case of helium, $T_c = 2.18$ K when there is a phase change. Liquid helium becomes superfluid and is in equilibrium with helium vapour. So the atoms get piled up in the ground state, producing BE condensation. Hydrogen solidifies before liquid hydrogen shows BEC. A good popular account of BEC can be found in the article by N. Kumar[25].

8.7 Classical Non-Degenerate State

A system of particles which is far away from the state of degeneracy follows classical Maxwell–Blotzmann (M–B) statistics. Such a situation arises when N/V is small and the temperature is large. Particles like molecules have internal degrees of freedom. In such a case the total energy must be expressed as the sum of the energy of the external state of the particle (such as motion of the molecule as a whole) and the internal energy due to vibrations or rotation. In this section we shall discuss a classical gas of atoms or molecules. We shall discuss the partition function in these cases. Then we shall discuss ionization and dissociation, finally leading to Saha's equation of ionization.

8.7.1 Equilibrium state

Consider a system consisting of a single type of complex particle like a molecule. Let there be N particles of mass m in a volume V. Each particle has a translational motion as a whole and also has internal degrees of freedom. The total energy E_μ consists of the translational energy $p^2/2m$ and the internal energy E_j, that is

$$E_\mu = E_j + p^2/2m.$$

Let g_j be the statistical weight of the internal state j. The statistical weight of the translational state is $d^3p\,dV/h^3$. If $N_j(p)$ is the number of particles per unit volume with momentum p and internal state j, then according to (8.59),

[25]Kumar, N., 2005, *Resonance* **89** [12], pp. 2093–2100.

$$N_j(p) = \frac{4\pi}{h^3} e^\gamma (g_j e^{-E_j/kT}) e^{-p^2/2mkT} p^2 dp. \tag{8.91}$$

It may be noted that here $g_j e^{-E_j/kT}$ represents the Boltzmann distribution and $4\pi p^2 e^{-p^2/2mkT} dp/h^3$ is the Maxwell distribution.

Integrating over p, the number density of particles in the state j becomes

$$N_j = g_j e^{\gamma - E_j/kT} (2\pi mkT)^{3/2}/h^3. \tag{8.92}$$

Note that $(2\pi mkT)^{3/2}$ is the effective volume in phase space per unit volume of real space and $(2\pi mkT)^{3/2}/h^3$ is the number of phase cells in that volume.

If the particles have a spin angular momentum $\hbar/2$, then $g_j = 2$, and (8.92) gives

$$N = 2(2\pi mkT)^{3/2} e^\gamma/h^3.$$

We have seen that non-degeneerate state arises when $e^\gamma \ll 1$, and we get back (8.77). This means the number of particles is much less than the number of phase cells.

8.7.2   Partition functions

Atoms

In the case of atoms, the partition function becomes

$$Z(T) = \sum_j g_j e^{-E_j/kT}, \tag{8.93}$$

so that (8.92) gives

$$\frac{N_j}{N} = g_j e^{-E_j/kT}/Z(T). \tag{8.94}$$

Molecules

We have seen in Chapter 6 that the internal energy of a diatomic molecule consists of vibrational energy $E_v = h\nu(v + 1/2)$, with $v = 0, 1, 2, \cdots$, rotational energy $E_r = \hbar^2 J(J + 1)/2I$, with $J = 0, 1, 2, \cdots$, where ν is the vibrational frequency and I is the moment of inertia of the molecule. If we denote N_{vJ} to be the number density of molecules in the rotational state J of the vibrational state v, and g_{vJ} the corresponding weight, then (8.92) gives

$$\frac{N_{vJ}}{N} = \frac{g_{vJ} \exp[-\{h\nu(v + 1/2) - \hbar^2 J(J + 1)/2I\}/kT]}{Z(T)},$$

where the partition function $Z(T)$ is given by

$$Z(T) = \sum_{vJ} g_{vJ} \exp[-\{h\nu(v+1/2) - \hbar^2 J(J+1)/2I\}/kT]. \tag{8.95}$$

But $g_{vJ} = g_v g_J = 2J+1$. The vibrational partition function can be seen to be

$$Z_v(T) = \sum_{v=0}^{\infty} e^{-h\nu(v+1/2)/kT} = \frac{e^{-h\nu/2kT}}{1 - e^{-h\nu/kT}}. \tag{8.96a}$$

The rotational states are more closely packed, so we can replace summation over J by integration. Taking

$$J(J+1) = x, \quad (2J+1)dJ = dx,$$

the rotational partition function can be seen to be

$$Z_r(T) = \sum_{J=0}^{\infty} (2J+1) \exp[-\hbar^2 J(J+1)/2IkT]$$

$$\simeq \int_0^{\infty} \exp[-\hbar^2 x/2IkT]dx$$

$$= 2IkT/\hbar^2. \tag{8.96b}$$

Thus the partition function of (8.95) is given by

$$Z(T) = \frac{2IkTe^{-h\nu/2kT}}{\hbar^2(1 - e^{-h\nu/kT})}. \tag{8.96c}$$

8.7.3 Ionization and dissociation

We have seen in Chapter 4 that Gibb's free energy per particle, that is, the chemical potential of all the species in thermodynamic equilibrium (which can be converted into one another) must be the same. Consider a multielectron atom X and let us denote by X^{k+} the kth ionized state of the atom where k electrons have been knocked off the atom. Consider a mixture of X^{k+}, $X^{(k+1)+}$ and electrons. When $X^{(k+1)+}$ combines with an electron, it results in the ion X^{k+}. In such a case, what is the relation between their chemical potentials in equilibrium?

Consider a two-phase system $A \rightleftharpoons B$ such as water (liquid) and water vapour, or a three-phase system $A \rightleftharpoons B + C$ such as hydrogen atoms, protons and electrons. In the liquid–vapour case, a molecule from the liquid phase may go to vapour phase, and vice versa. As the particle number in one phase increases, that in the other phase decreases. Let us denote their chemical potentials by μ_A, μ_B, etc. When a molecule of phase A goes to phase B, we add energy μ_B to the system and subtract energy μ_A from it. Since there is no net change in the energy of the system, we must have

$$\mu_B - \mu_A = 0 \Rightarrow \mu_A = \mu_B.$$

When we have a three-phase system with the reaction $A \rightleftharpoons B + C$, as A is converted to B and C, we add energy $\mu_B + \mu_C$ but also subtract energy μ_A. Therefore we have

$$\mu_A = \mu_B + \mu_C.$$

Ionization

In the above case of ionized atoms and electrons, we have the reaction

$$X^{(k+1)+} + e \rightleftharpoons X^{k+}.$$

If we denote their chemical potentials by μ_e and μ_k, etc, the equilibrium condition would be

$$\mu_k = \mu_{k+1} + \mu_e. \tag{8.97}$$

Let the ground state energies of X^{k+} and $X^{(k+1)+}$ be denoted by E_0^k and E_0^{k+1} respectively. Let us denote their electronic levels by E_j^k and $E_{j'}^{k+1}$, respectively, both with respect to E_0. Finally, let $\chi_k = E_0^{k+1} - E_0^k$, as shown in Fig. 8.8, and note that it is the energy required to ionize a k-times ionized atom.

Now according to (8.92), the number densities of the three species would be

$$N_{kj} = g_{kj} \frac{(2\pi m_k kT)^{3/2}}{h^3} \exp[(\mu_k - E_{kj})/kT],$$

$$N_{(k+1)j'} = g_{(k+1)j'} \frac{(2\pi m_{k+1} kT)^{3/2}}{h^3} \exp[(\mu_{k+1} - E_{(k+1)j'} - \chi_k)/kT],$$

$$N_e = 2\frac{(2\pi m_e kT)^{3/2}}{h^3} e^{\mu_e/kT},$$

where the different m's are masses of the particles of the respective species. Taking $m_{k+1} \approx m_k$, this gives

$$\frac{N_{(k+1)j'} N_e}{N_{jk}} = \frac{2g_{(k+1)j'}}{g_j} \frac{(2\pi m kT)^{3/2}}{h^3} e^{-(E_{(k+1)j'} + \chi_k - E_j)/kT}. \tag{8.98}$$

Noting in analogy with (8.94), that

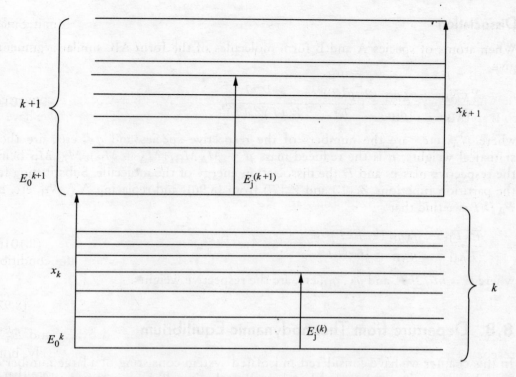

Figure 8.8 *Energy level of k times and* $(k+1)$ *times ionized atom*

$$\frac{N_{(k+1)j'}}{N_{k+1}} = \frac{g_{(k+1)j'}}{Z^{k+1}(T)} e^{-E_{(k+1)j'}/k},$$

$$\frac{N_{kj}}{N_k} = \frac{g_{kj}}{Z^k(T)} e^{-E_{kj}/kT},$$

Equation (8.98) finally reduces to

$$\frac{N_{k+1}N_e}{N_k} = \frac{2Z^{k+1}(T)}{Z^k(T)} \frac{(2\pi mkT)^{3/2}}{h^3} e^{-\chi_k/kT}. \tag{8.99}$$

This is *Saha's ionization equation*. For a neutral atom and singly ionized ion, with number densities denoted by N_I and N_{II} respectively in astrophysical notation, it takes the form

$$\frac{N_{II}N_e}{N_I} = \frac{2Z_{II}(T)}{Z_I(T)} \frac{(2\pi mkT)^{3/2}}{h^3} e^{-\chi_I/kT}, \tag{8.100}$$

with obvious meanings of Z_I and Z_{II} and χ_I being the first ionization potential.

Dissociation

When atoms of species A and B form molecules of the form AB, similar arguments give

$$\frac{N_A N_B}{N_{AB}} = \frac{g_A g_B}{g_B} \frac{(2\pi\mu kT)^{3/2}}{h^3} \frac{e^{-D/kT}}{Z_v(T) Z_r(T)}, \tag{8.101a}$$

where N_A, etc., are the numbers of the respective species and g_A, etc., are their statistical weights, μ is the reduced mass $\mu = M_A M_B/(M_A + M_B)$, M_A, M_B being the respective masses and D the dissociation energy of the molecule. Substituting for the partition functions $Z_v(T)$ and $Z_r(T)$ from (8.96), and replacing N_A, N_B, etc. by P_A/kT, we find that

$$\frac{P_A P_B}{P_{AB}} = \frac{g_A g_B}{g_{AB}} \frac{(2\pi\mu kT)^{3/2}}{16\pi^3 \hbar I} (e^x - e^{-x})e^{-D/kT}, \tag{8.101b}$$

where $x = h\nu/2kT$, and g_A, g_B, etc. are the respective weights.

8.8 Departure from Thermodynamic Equilibrium

In this chapter we have considered an isolated system consisting of a large number of particles interacting with each other through inelastic collisions so that the energy and the particle number remain constant. We discussed its final state of thermodynamic equilibrium when it attains a constant temperature throughout. From probability considerations, it was possible to find the average numbers of particles of each type in various single-particle states. This makes it possible to determine the temperature of the system of particles. For example, one can use Maxwell–Boltzmann distribution, Boltzmann equation, Saha's equation or dissociation equation, and all of them will give a unique temperature.

We will conclude this chapter with the following remark. On relaxing the conditions of isolation and largeness of the number of particles, there will occur departures from thermodynamic equilibrium. For example, inflow and/or outflow of energy and particles, introduction of electric and magnetic fields, extreme rarification, etc., can disturb the equilibrium. Such non-equilibrium states require special treatment in which the principle of detailed balancing of each reaction plays an important part.

Problems

8.1 Consider a system of four indistinguishable classical particles, each of which has energy levels 0, $h\nu$, $2h\nu$, and $3h\nu$. If the total energy of the system is (a) $3h\nu$, (b) $4h\nu$, list the accessible microstates.

8.2 Consider a system of three particles, each of which can be in any one of four energy states. Count the number of accessible states of the system if the particles satisfy (i) M–B statistics, (ii) B–E statistics, (iii) F–D statistics.

8.3 Consider translational and rotational degrees of freedom for monatomic, diatomic and polyatomic ideal gases and use the law of equipartition of energy to derive the values of C_v, C_p and their ratio $\gamma = C_p/C_v$ for them.

8.4 Consider vibrational degrees of freedom for an atom in a solid and use the law of equipartition of energy to show $C_v = 3R$ per mole when the oscillators behave classically. What will be the modification if they are Planck oscillators?

8.5 Obtain the canonical partition function for one-dimensional quantum mechanical oscillator with energy $E_n = (n + 1/2)\hbar\omega$. Hence obtain the mean energy and discuss its high and low temperature limits.

8.6 Calculate the root mean square velocity of (i) N_2 and O_2 in air at 300 K, and (ii) H and Fe in solar atmosphere at 6000 K.

8.7 A diatomic molecule can be considered as a vibrating linear harmonic oscillator with circular frequency ω. Obtain the expression for the vibrational partition function in canonical ensemble. Also obtain the average energy and the vibrational heat capacity. What are their high and low temperature limits? Estimate them for nitrogen molecules in the atmosphere at normal temperature and pressure.

8.8 Show that the mean relative velocity between two molecules of a classical ideal gas is $\sqrt{2}$ times their root mean squared velocity. Use this result to show that the mean free path, that is, the mean distance travelled by the molecule between two collisions is $1/(\sqrt{2}\pi n d^2)$, where n is the number density of molecules and d is the diameter of the molecule.

8.9 Show that the number density of Planck oscillators having frequency between ν and $\nu + d\nu$ is $u_\nu d\nu = 8\pi\nu^2 d\nu/c^3$ and their mean energy is $h\nu/(e^{h\nu/kT} - 1)$. From these obtain Planck's law for B_ν (introduced in Chapter 4). What will be the result if one uses classical law of equipartition of energy.

8.10 Show that there is no Bose–Einstein condensation for a two-dimensional ideal Bose gas. [Hint: Show that the number of particles in the ground state n_0 cannot tend to infinity in this case.]

8.11 Obtain the Planck's radiation formula for a two-dimensional photon gas in a black body enclosure at temperature T and area A. Hence obtain the Stefan–Boltzmann law in two dimensions.

8.12 How would you rate the state of degeneracy in the following cases (in the order material, density, temperature, constituent particles to be considered)? Also, estimate the average inter-particle distance in each case:

(i) Air, 1.3×10^{-3} gm/cm^3, 300 K, $N_2 + O_2$;

(ii) Copper, 9 gm/cm^3, 300 K, free electrons;

(iii) Sun's centre, 1300 gm/cm^3, 1.6×10^7 K, e, p;

(iv) He white dwarf, 6×10^5 gm/cm^3, 10^7 K, e, α;

(v) Neutron star, 5×10^{14} gm/cm^3, 5×10^8 K, n, p, e.

8.13 Classical statistics fails when the average inter-particle distance in a free gas becomes comparable to (or less than) the average de Broglie wavelength of the particles. Write down this condition for N free particles of mass m in a volume V at a temperature T. Estimate numerically the value of N/V when, say $T = 1$ K, when the cross-over occurs for a hydrogen molecular gas.

8.14 Referring to Section 8.6.2, prove that the fraction of Bose particles condensed in the ground state when $T < T_c$ is given by

$$\frac{\overline{n}_0}{V} = 1 - \left(\frac{T}{T_c}\right)^{3/2}.$$

Also, relate it to v/v_c and λ/λ_c.

Elements of Nuclear and Particle Physics

WE begin this chapter with the discovery of the electron in 1897 and of the nucleus in the beginning of the twentieth century. We describe Rutherford's alpha particle scattering experiment and its results. This leads us to describe how it gave rise to the structure of the atom, and then to the structure of the nucleus. Then we discuss strong and weak forces and beta-stability. This is followed by nuclear reactions, reaction rates and thermonuclear reactions. We then discuss the proliferation of 'elementary' or 'fundamental' particles, whose number has gone up to about 200 in just over a hundred years. We give a glimpse of the present state in this area. Finally we conclude the chapter with some applications of nuclear reactions and nucleosynthesis in astrophysics.

9.1 Discovery of the Nucleus

We review in this section the discovery of the electron in 1897, followed by Rutherford's alpha particle scattering experiments leading to the discovery of the nucleus.

9.1.1 Discovery of electron

Atomic nature of matter postulated by Dalton in 1803 was firmly established by the end of the nineteenth century. Chemists had determined the atomic weights A, now called *mass numbers*, of all the known 92 elements with respect to hydrogen and oxygen, in which O (oxygen) was assigned a mass of 16. The elements were neatly arranged in the periodic table by Mendelief with increasing mass number and grouped into eight columns according to the valency and chemical properties of the individual elements.

That the atoms are not the ultimate indivisible constituents of matter came to light in 1896–97. Firstly, A. H. Becquerel discovered the phenomenon of radioactivity which showed that uranium atoms could break into smaller components. Secondly, J. J. Thompson discovered that all elements contain negatively charged particles which are now known as electrons. Thompson was studying the bending of cathode rays under the influence of electric and magnetic fields. He found that irrespective of the material of the cathode, the emitted particles had an identical ratio of charge to mass, e/m. The charge of the electron was measured by R. A. Millikan in 1909 in his cloud chamber experiment. The result indicated that the electron has a charge of $e = 4.802 \times 10^{-10}$ esu (electrostatic unit) and a mass $m = 9.107 \times 10^{-28}$ gm.

Thompson also studied the 'canal' rays passing through a hole in the cathode. They consisted of positively charged particles, or ions. But, for them the ratio of e/M, where M is the mass of these particles, depended upon the gas in the discharge tube. As the ions were produced by the removal of an electron from the atom, they had a positive charge equal to the charge of the electron. It was found that e/M was inversely proportional to the mass number of the element, and it had a maximum value for hydrogen. Considering the hydrogen ion to be the basic constituent of all atoms, it was given the name of *proton*. Now the mass of the proton is found to be 1836.11 times that of an electron. Thus all matter was considered to be made of electrons and protons. As the atom itself is electrically neutral, it would have equal number of electrons and protons, their number being equal to the mass number of the element.

9.1.2 Rutherford's experiments

Radioactivity was studied by Rutherford, Soddy and Villard between 1900–1902. They classified the emitted radiation into three categories:

(i) alpha (α) radiation consisting of positively charged particles that have low penetrating power of less than 1 mm of aluminium;

(ii) beta (β) radiation made of negatively charged particles that have intermediate penetrating power of a few mm of aluminium file (they were just the electrons discovered by Thompson) and;

(iii) gamma (γ) radiation which is electrically neutral and had the highest penetrating power of tens of mm of aluminium foil. These were found to be highly energetic photons with energy in the MeV range.

Rutherford used α particles from radon to study the structure of gold atoms. Alpha particles from source R (see Fig. 9.1a) were collimated by passing them through a slit S and then allowed to strike a gold foil G. It was found that they were scattered in all directions. The large-angle scattering of particles not only indicated that they are being repelled by the positive charge in the gold atom but also that the positive charge must be concentrated in a very small region in the atom.

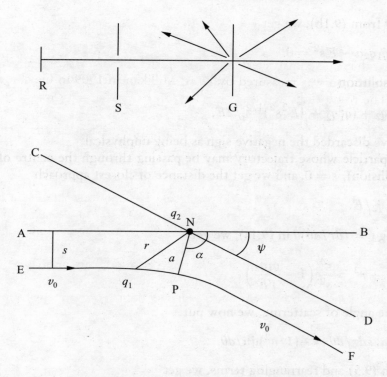

Fig. 9.1 *Rutherford's alpha particle scattering: (a) alpha particles from a source R striking a gold foil G; (b) the alpha particle following the trajectory EPF, with the atom centred at N.*

Figure 9.1(b) shows the hyperbolic trajectory EPF of a particle directed towards the atom N with velocity v_0 and impact parameter s. ANB and CND are the asymptotes to the hyperbola. Let us calculate a, the distance of the closest approach at vertex P, and the scattering angle, $\psi = \angle$ BND.

Energy E of the alpha particle and its angular momentum l are two constants of integration. Let (r, θ) be the coordinates of a point in the trajectory with respect to N. Also let q_1 be the charge of the α particle and q_2 that of the target atom, and m the mass of the α particle. Then,

$$E = \frac{1}{2}mv_0^2 = \frac{1}{2}m(\dot{r}^2 + r^2\dot{\theta}^2) + \frac{q_1q_2}{r}, \tag{9.1a}$$

$$l = mr^2\dot{\theta} = mv_0s = (2mE)^{1/2}s. \tag{9.1b}$$

At P, $r = a, \dot{r} = 0$; so (9.1a) gives

$$\frac{1}{2}ma^2\dot{\theta} = E - \frac{q_1q_2}{r}.$$

Then using $\dot{\theta}$ from (9.1b), we get

$$Ea^2 - q_1 q_2 a - E s^2 = 0. \tag{9.2}$$

This gives a solution

$$a = [q_1 q_2 + (q_1^2 q_2^2 + 4E^2 s^2)^{1/2}]/2E, \tag{9.3}$$

where we have discarded the negative sign as being unphysical.

For an α particle whose trajectory may be passing through the centre of the atom (head-on collision), $s = 0$, and we get the distance of closest approach

$$a_0 = q_1 q_2 / E. \tag{9.4}$$

Now, putting $\dot{r} = (dr/d\theta)\dot{\theta}$ in (9.1a), we get

$$\left(\frac{dr}{d\theta}\right)^2 + r^2 = \frac{r^4}{s^2}\left(1 - \frac{q_1 q_2}{rE}\right). \tag{9.5}$$

To obtain the angle of scattering, we now put

$$r = 1/u, \quad dr/d\theta = -(1/u^2)du/d\theta.$$

Using this in (9.5) and rearranging terms, we get

$$\frac{du}{d\theta} = \left[\frac{1}{s^2}\left(1 - \frac{q_1 q_2 u}{E}\right) - u^2\right]^{1/2}.$$

Integrating this results in

$$\theta - \theta_0 = \int_{u(a)}^{u} \left[\frac{1}{s^2}\left(1 - \frac{q_1 q_2 u}{E}\right) - u^2\right]^{-1/2} du,$$

where $\theta_0 = \theta(a)$ is the polar coordinate at the closest approach, so that θ is the angle measured from NP. This gives $u(a) = 1/a$, and

$$\theta = \cos^{-1}\{(u + q_1 q_2/2Es^2)/[(1 + q_1^2 q_2^2/4E^2 s^2)/s^2]^{1/2}\}. \tag{9.6}$$

Using (9.3), it can be checked that $\theta = 0$ for $u = 1/a$.

It can be seen from Fig. 9.1 (b) that the angle of scattering is related to α by

$$\psi = \pi - 2\alpha. \tag{9.7}$$

The angle α shown in Fig. 9.1(b) is the value of θ when $r \to \infty$, or $u = 0$. This gives

$$\alpha = \cos^{-1}\{q_1 q_2 / (2s^2 E^2 + q_1^2 q_2^2)^{1/2}\}. \tag{9.8}$$

When $s \to 0, a \to a_0 = q_1 q_2 / E$, and $\alpha \to 0, \psi \to \pi$. It is found that, generally, $\psi > \pi/2$ when $s < q_1 q_2 / 2E$.

Initially, Rutherford did not know the values of q_1 and q_2; on taking them equal to e times the atomic weight, he found $a_0 \approx 10^{-11}$ cm for the radius of the nucleus. This is a thousand times smaller than the estimated sizes of atoms and molecules. Later experiments involving scattering of α particles, protons, and electrons of high energies ≈ 20 MeV indicated that the charge on the nucleus is equal to e times the atomic number Z of the element. Their radii were seen to follow the relation

$$R \approx r_0 A^{1/3}, \quad r_0 = 1.5 \text{ fm}, \tag{9.9}$$

where A is the mass number of the element.

9.1.3 Structure of the atom

As the atom is electrically neutral and there is a positive charge Z on the nucleus, there must be Z electrons outside the nucleus. They are the ones responsible for the chemical properties of the element. Thus the periodic table is based on Z and not on A. This discovery removed some anomalies in the periodic table; for example, argon with $Z = 18, A = 40$ was moved to a place before potassium with $Z = 19, A = 39$.

In 1913 Bohr used this model for hydrogen atom in which one electron moved around one proton in some specified orbit. It was so successful that it was soon extended to other atoms. It led to the discovery of electron spin and other developments such as Pauli exclusion principle, matrix and wave mechanics, with the crowning glory of Dirac's theory of relativistic electron in 1927. We have discussed these developments in detail in Chapter 5. It is worth noting that the existence of Dirac's predicted *positron*, which is a particle of charge e and mass m_e (the electronic mass), was discovered by C. D. Anderson in 1932 in cosmic ray showers. A positron is now represented by e^+ to distinguish it from an electron which is denoted by e^-.

9.2 Structure of the Nucleus

9.2.1 General properties

We have already seen that the nuclei have a radius of the order of femtometers and they have a positive charge of Ze. The other properties of nuclei include their mass, composition, angular momentum and magnetic moment. We discuss them here.

Mass

This is determined by the mass spectrograph of Aston. It is found that the mass of a nucleus is much greater than the mass of the electron, and is in fact A times the mass of a proton. Also $A > Z$ in general and $A \approx 2Z$ for light nuclei except for hydrogen. The density of the nucleus including that of a proton comes out to be of the order of 10^{14} gm/cm^3.

Composition

Originally, it was thought that the nucleus is made up of A protons and $A - Z$ electrons. But with the development of quantum mechanics, it was realised that it is not possible to confine an electron within the nuclear size. The hypothesis is also untenable for $_7N^{14}$ which was found to obey Bose–Einstein statistics. If this nucleus had $A = 14$ protons and $Z = 7$ electrons, all with spin 1/2, it would have an half-odd-integral spin and follow Fermi–Dirac statistics. Heisenberg suggested the existence of a neutral particle of the same mass as a proton. Thus the nucleus has Z protons and $A - Z$ neutral particles. This was later called *neutron* and denoted by n, while a proton is denoted by p. It was discovered by Chadwick in 1932. It has a mass of $m_n = 1838.55 m_e \approx m_p$, where m_p is the mass of the proton. A *nucleon* indicated a proton or a neutron. Thus we would say that a nucleus contains Z protons and $N = A - Z$ neutrons, or A nucleons.

   Different nuclei would have different number of protons and neutrons, which are mainly responsible for their physical properties. Nuclei having same Z or N or A are given special names, as described here.

(a) Elements with the same Z and different A are called *isotopes*, for example $_6C^{12}$, $_6C^{13}$, $_6C^{14}$. Their neutral atoms have similar chemical properties because each of them has 6 electrons.

(b) Elements with the same N and different A are called *isotones*, for example $_6C^{13}$, $_7N^{14}$.

(c) Elements with the same A and different Z and N are called *isobars*, they have similar nuclear properties, for example $_6C^{14}$, $_7C^{14}$.

(d) Different excited states of the same nucleus in an element are known as *isomers*.

Angular momentum

p and n have spin 1/2 and obey Pauli exclusion principle and Fermi–Dirac statistics. Therefore each elementary cell in phase space can contain at most 2 protons and 2 neutrons. Spin and orbital angular momenta make up the total angular momentum I of the nucleus. The nucleus can have $2I + 1$ orientations in space, with the projection of the total angular momentum along the chosen axis equal to $I, I-1, \cdots, -I$. Hence I can be determined by observing the number of levels in Stern–Gerlach experiment. It is seen generally that $I = 0$ for even-A nucleus, except when Z and N are both odd. Such nuclei are $_1\text{H}^2$, $_3\text{Li}^6$, $_5\text{B}^{10}$, $_7\text{N}^{14}$ and are found to have $I = 1$. Usually, I is small, less than 9/2, with the nucleus Lu^{176} with $I = 7$ being an exception.

The total nuclear angular momentum I interacts with the total electronic angular momentum J to give rise to hyperfine structure of atomic lines. We have $\mathbf{I} + \mathbf{J} = \mathbf{F}$, with F going from $I + J$ to $|I - J|$. For example, in the ground state of neutral hydrogen, we have $J = I = 1/2$, so that $F = 0$ or 1; transition between these two states of 1s^1 level give rise to the 21-cm line of hydrogen in the radio spectrum.

Magnetic moment

Although neutron has no charge, it was found that it has a magnetic moment. It was assigned a spin angular momentum $s = \hbar/2$. The magnetic moment of the proton and neutron are found to be

$$\mu_{\text{p}} = 2.7934\mu_{\text{B}}, \quad \mu_{\text{n}} = -1.9135\mu_{\text{B}}.$$

This gives rise to a conjecture that although the neutron is electrically neutral as a whole, it must have some structure. Writing the magnetic moment μ of the proton and neutron (as a multiple of Bohr magneton) as $\mu = gs$ where s is the spin, we have

$$g_{\text{p}} = 5.59, \quad g_{\text{n}} = -3.83.$$

We may write the magnetic moment for a nucleus in general as $\mu = gI$; then the spectroscopic splitting factor g for the nucleus is to be determined experimentally.

9.2.2 Nuclear forces

Among the four basic interactions, the gravitational interaction among nucleons is too weak to cause any observable effects. We discuss here the other three interactions.

Coulomb repulsion

Any two protons in the nucleus have an electrostatic energy equal to e^2/r, where r is of the order of 10^{-13} cm (1 fm). Since there are Z protons each interacting with $Z-1$ other protons, the total electrostatic energy of the nucleus can be approximately taken to the $Z^2 e^2/R$, where R can be taken to be the radius of the nucleus. Then the electrostatic energy per nucleon becomes $Z^2 e^2/RA$. As we have seen earlier, $R \simeq r_0 A^{1/3}$, where $r_0 = 1.5 \times 10^{-13}$ cm. Substituting the numerical values and converting the energy to MeV units and taking approximately $Z \simeq A/2$, the electrostatic energy per nucleon is seen to be about $0.24A^{2/3}$ MeV.

Strong force

The existence of several protons in a nucleus in spite of Coulomb repulsion indicates that there is another strong attractive interaction which keeps them together. The existence of neutrons together with protons, including a deuteron, indicates that this strong attraction acts not only between p − p but also p − n, and perhaps n − n. This force must have a short range of about 1 fm and must be independent of charge. The potential of the strong force decreases sharply at $r \approx 10^{-13}$ cm, and gives rise to the potential well shown in Fig. 9.2. In order to consider the shape of the potential curve near $r \approx 0$, Yukawa argued in 1935 as follows. Here curve C is Coulomb repulsion which behaves as $1/r$. If we were considering the Coulomb type of attractive force, the potential curve will be like the curve C′ in Fig. 9.2, where $V(r) \propto -1/r$. In order to obtain a large drop in potential, Yukawa put

$$V(r) = V_0 \frac{e^{-r/\beta}}{r/\beta}, \tag{9.10}$$

which would result in a curve like Y in Fig. 9.2. Here β represents the effective range of the attractive potential.

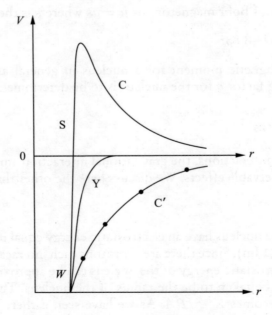

Fig. 9.2   *Nuclear potential. Curve C: Coulomb repulsion, S: strong force, C′: Coulomb attraction, Y: Yukawa potential (schematic).*

Now, we know that electric charges produce electromagnetic field which gives rise to radiation, or photons. It is therefore conceived that electrically charged particles interact with each other by exchange of photons. Similarly, we can conceive of the strong interaction as mediated by some particle. An exchange of this particle leads to the binding of the nucleons. This particle would have to be neutral for causing attractive interaction between two protons or two neutrons, charged for causing it between a proton and a neutron.

Since the range of electromagnetic force is infinite, the exchange particles, that is, photons, have zero rest mass. But the strong range has a very short range, suggesting that the exchanged particle has a finite mass. Suppose the range β is equal to the de Broglie wavelength of the particle. Taking the momentum p of a particle of mass m as mc, we may put

$$\beta = \frac{h}{mc} \Rightarrow \frac{m}{m_{\mathrm{e}}} = \frac{h}{\beta c m_{\mathrm{e}}}.$$

Taking β to be of the order of a femtometer, the mass of the particle comes out to be about 300 times the electron mass, that is,

$$m/m_{\mathrm{e}} \simeq 290.$$

Thus the strong interaction would operate through exchange of such particles which were called π mesons, or later, pions. They were discovered in primary cosmic rays by C. F. Powell in 1947. Three charge varieties of this particle were soon discovered, and they have charges $\pm e$ and 0; they are denoted by π^+, π^- and π^0 and were found to have rest mass of 273, 273 and 264 times the electronic mass, respectively. The exchange between nucleons occurs as

$$\mathrm{p} \xrightarrow{\pi^+} \mathrm{n}, \ \mathrm{n} \xrightarrow{\pi^-} \mathrm{p}, \ \mathrm{p} \xrightarrow{\pi^0} \mathrm{p}, \ \mathrm{n} \xrightarrow{\pi^0} \mathrm{n}.$$

Pions were found to have short lives of 2.55×10^{-8} sec for $\pi^{\pm}$ and about 2×10^{-6} sec for π^0. They decay into corresponding muons (μ) and μ-neutrinos (ν_μ). The muons have a mass of about $207m_{\mathrm{e}}$ while the μ-neutrinos have a very low mass of keV or eV.

To get an idea of the energy involved in strong interaction, we make use of the mass defect, Δ, that is the total mass of the Z protons plus N neutrons, minus the mass of the nucleus. If the mass of the nucleus is denoted by M_{N}, then

$$\Delta = Zm_{\mathrm{p}} + Nm_{\mathrm{n}} - M_{\mathrm{N}}. \tag{9.11}$$

By knowing the masses, we could obtain the total binding energy $c^2\Delta$ of the nucleus as well as the average binding energy per nucleon, $B = c^2\Delta/A$.

Masses and the corresponding rest energies are conventionally denoted in terms of the atomic mass unit, amu, which is defined as 1/12th of the mass of a carbon atom. It is taken as

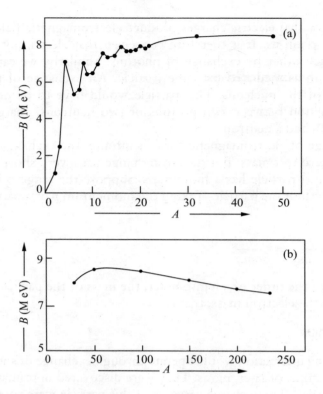

Fig. 9.3   *Binding energy per nucleon as a function of A*

$$1 \text{ amu} = 1.66 \times 10^{-24} \text{ gm} = 931 \text{ MeV}.$$

Physicists use mass units as well as energy units and amu interchangeably. In terms of these, the masses of p and n are found to be

$$m_{\text{p}} = 1.00759 \text{ amu} = 938.1 \text{ MeV},$$

$$m_{\text{n}} = 1.00893 \text{ amu} = 939.3 \text{ MeV}.$$

Using these masses and other nuclear masses in (9.11), it is seen that the binding energy (BE) per nucleon follows a pattern as shown in Fig. 9.3. We can see that the binding energy:

- increases sharply at $A = 4$ and is 7.1 MeV for He^4;

- has subsequent sharp maxima at $A = 8, 12, 16, 20$, etc., and for these, it is 7.1 MeV for Be^8, 7.7 MeV for C^{12}, 8.0 MeV for O^{16} and 8.1 MeV for Ne^{20};

- is about 6 MeV for lighter nuclei (apart from the above peaks) and slowly rises;

- is about 8.5 MeV over a wide flat range from $A = 40$ to 150 in the middle region and is slowly decreasing;

- has a broad maximum for Fe^{56};

- drops to about 7.5 MeV for larger A.

Thus, in an overall sense the binding energy is about 8 MeV per nucleon. This indicates that the strong nuclear force is saturated. The total exchange energy as well as the volume of the nucleus are proportional to A. It indicates that a nucleus has a certain energy per unit volume, much like a liquid drop. This suggest a *liquid drop model* for atomic nucleus. The reduction in B for lighter nuclei could then be ascribed to smaller surface-to-volume ratio. The smaller value of B for heavier nuclei is caused by Coulomb repulsion. For $A = 211$, for example, the electrostatic energy per nucleon comes out to be $0.24A^{2/3}$ MeV $= 8.5$ MeV. Hence nuclei with $A > 211$ should be unstable due to excess of Coulomb repulsion; this is generally true.

Weak interaction

This is manifested by the process of β-decay in which electrons or positions are emitted by the nucleus. Since the nuclei do not contain these particles, they must be produced in the process of emission itself by transitions between nuclear energy levels. It is found that the energy of β-rays for a given reaction is not constant; it shows a spectrum with a sharp maximum cutoff at some energy $E_{\max}$. Also angular momentum is not conserved as inferred from the I-values of the parent and daughter nuclei. Hence, in 1927, Pauli suggested the existence of a third particle with charge zero and spin 1/2 and extremely low rest mass (in fact, 'massless'). This is the electron neutrino ν_e, which is difficult to detect. Neutrinos from reactors were discovered in 1956 and neutrinos coming from the interior of sun and the supernova of 1987 have now been detected.

It is now known that weak interaction and electromagnetic interaction become components of a larger unified interaction at high energies, which acts over a short range of 10^{-14} cm. At that distance it acts through the exchange of W^+, W^- and Z^0, all heavy bosons with spin 1 and masses in the range of 100 GeV. Their interaction with a nucleus can be represented by reactions like those shown in Fig. 9.4. In these reactions the following processes may take place: (a) A neutron decays into a proton and a W^- particle, which in turn decays into an electron and an electron-antineutrino; (b) a proton decays into a neutron and a W^+ particle, which then decays into a positron and an electron-neutrino; (c) an electron gives rise to an electron and a Z^0 particle which decays into a muon-neutrino-antineutrino pair. These reactions can take place only inside or near a nucleus. The strength of the weak force is about 10^{-13} times that of the strong force.

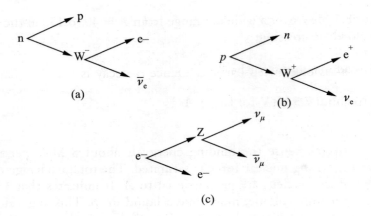

Fig. 9.4   *Weak decays that can take place inside a nucleus*

9.2.3   Stability of nuclei

Here we will discuss the stability of a nucleus against the three decays, α, β, γ.

Alpha stability

In view of the opposing Coulomb and exchange forces, the potential energy of two nuclei X and Y can have three forms, as shown in Fig. 9.5.

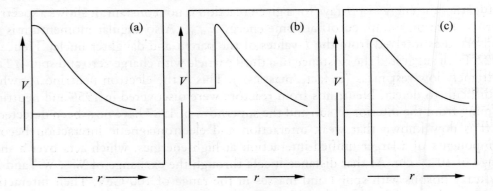

Fig. 9.5   *Three kinds of potential curves between two nuclei as a function of distance r between them*

In case the potential between X and Y is of the form A, the nuclei X and Y cannot combine.

In case the potential is of the form B, X + Y will form Z which will be unstable.

In case the potential is of the form C, X+Y will form A which will be stable.

The instability of case B arises on account of the possibility of Coulomb barrier penetration which will be discussed later. The penetration probability increases (a) for small height of potential barrier which is roughly given by $0.96 Z_X Z_Y / A^{1/3}$ MeV, and (b) for smaller mass of emitted particle. Hence α-decay is the only natural process. Fission into two nearly equal parts has very small probability even when possible; it can occur when induced by bombardment of particles, mostly neutrons.

Since α-decay is a random process governed by the laws of probability, it follows the same law as that of radioactivity. If $N(0)$ is the number of nuclei of species A at time $t = 0$ capable of emitting an α-particle at some time or the other, then the number of nuclei $N(t)$ at time t which have still not decayed is given by

$$N(t) = N(0)e^{-t/\tau},$$

where $\tau \ln 2$ is the half life of the species A.

Beta stability

Here protons and neutrons exchange roles by one of the following processes:

(a) $n \rightarrow p + e^- + \bar{\nu}_e$,

(b) $p \rightarrow n + e^+ + \nu_e$,

(c) $p + e^- \rightarrow n + \nu_e$,

(d) $n + e^+ \rightarrow p + \bar{\nu}_e$.

Reaction (a) indicates a neutron decaying into a proton, an electron and an antineutrino. In free state this reaction is possible spontaneously because of the higher mass of the neutron as compared to the particles after reaction. Reactions (b) and (c) are not possible energetically in free state. Reaction (d), although possible, is very rare due to non-availability of positrons.

Inside nuclei, we observe the following processes:

(a) $_Z X \rightarrow {}_{Z+1} Y + e^- + \bar{\nu}_e$,

(b) $_Z X \rightarrow {}_{Z-1} Y + e^+ + \nu_e$,

(c) $_Z X + e^- \rightarrow {}_{Z-1} Y + \nu_e$.

Reaction (c) here is known as K-capture. For these reactions to occur, the nuclear masses must satisfy:

(a) $M(Z, N) > M(Z + 1, N - 1) + m_e \qquad - \beta^-$ decay;

(b) $M(Z, N) > M(Z - 1, N + 1) + m_e$ $- \beta^+$ decay;

(c) $M(Z, N) + m_e > M(Z - 1, N + 1)$ $-$ K capture.

It is found that for every A, only one (Z, N) combination is stable. For example, for $A = 14$, $_7\text{N}^{14}$ is stable but $_6\text{C}^{14}$ is not.

Gamma radiation

It occurs when a nucleus is found in an excited state, denoted by X*, of X. Then X* $\rightarrow$ X + γ.

Among the three types of neutral decays, α-decay is the fastest process (because it involves strong interaction) with half lives of the order of 10^{-20} seconds, γ-decay is medium (because it involves electromagnetic interaction) with half lives of the order of 10^{-14} secconds, while β-decay is the slowest process (involving weak interaction) with half lives of the order of 10^{-1} to 1 seconds.

9.3   Nuclear Reactions

We shall begin this section by briefly introducing artificial nuclear reactions. Then we shall discuss the two famous models of a nucleus, the liquid drop model and nuclear shell model. We shall briefly discuss reaction rate, followed by thermonuclear reactions, and applications of nuclear reactions.

9.3.1   Artificial nuclear reactions

A nuclear reaction in which the mass of reactant particles is smaller than that of product particles cannot take place spontaneously. Such a reaction can be triggered to happen by supplying energy, for example, by shooting a projectile particle towards the target with sufficient energy to overcome the mass difference. Such a reaction is called an *artificial nuclear reaction.*

Such reactions were first produced by Rutherford by bombarding nuclei with protons and alpha particles. Representing heavy particles by capital letters and lighter particles by small letters, we can write a general reaction where a nucleus A is bombarded by a target a as

$$A + a \rightarrow B + b.$$

This reaction is called an (a, b) reaction. For this reaction to happen, we must have

(i)   Conservation of charge: $Z_A + Z_a = Z_B + Z_b$, where Z_A is the charge of A, etc..

(ii)   Conservation of baryons: $A_A + A_a = A_B + A_b$, where A_A is the number of nucleons in A, etc..

(iii) Conservation of angular momentum: $I_A + I_a = I_B + I_b$, where I_A is the angular momentum of A, etc.

(iv) $M_A + M_a > M_B + M_b$, where M_A is the mass of A, etc.

For example, in the reaction

$$_6C^{12} + {}_1H^1 \rightarrow {}_7N^{13} + \gamma,$$

the mass of the reacting particles is $12.00382 + 1.00812 = 13.01194$ amu, which is just slightly more than that of $_7N^{13}$ which is 13.00988 amu. Also, the total charge on either side is 7, the total baryon number of either side is 13, and the spin is 1/2.

9.3.2   Liquid drop model

In order to understand these reactions, Bohr put forward his *liquid drop model*. We have already seen that all nuclei have the same density and the nuclear forces are saturated, that is, they are proportional to A instead of A^2. In other words, they act on pairs of particles and not on all particles together. These properties are common to those of liquids, so we assume that the nucleus is like a liquid drop. When a projectile a strikes the target A, it becomes a part of the drop, forming a compound nucleus C which is in a state of excitation C^*. The excited nucleus develops different modes of oscillations and breaks in several ways, or decays along several channels as indicated below:

$$A + a \rightarrow C^* \rightarrow \left.\begin{array}{l} A + a \\[4pt] B + b \end{array}\right\} \text{small particle emission,}$$

$$\rightarrow C + \gamma \qquad - \text{gamma decay,}$$

$$\rightarrow D + e^- \text{ or } e^+ \quad - \beta \text{ decay,}$$

$$\rightarrow E + F \qquad - \text{fission.}$$

For example,

$$_7N^{15} + {}_1p^1 \rightarrow [{}_8O^{16}]^* \rightarrow {}_6C^{12} + \alpha, \qquad \rightarrow {}_8O^{16} + \gamma,$$

the two channels being known as (p, α) and (p, γ) reactions, respectively.

The liquid drop model can explain the fission of uranium as follows:

$$_{92}U^{235} + {}_0n^1 \rightarrow C^* \rightarrow {}_{56}Ba^{138-144} + {}_{36}Kr^{96-90} + {}_0n^1 + {}_0n^1,$$

where the products barium and krypton nuclei may occur in different isotopic states, as indicated. The two (or occasionally three) neutrons emitted in the reaction can split other uranium atoms in the vicinity and cause *chain reaction*. The energy released in the process can be used for producing an atom bomb or to generate electricity in a controlled reactor.

9.3.3 Nuclear shell model

In 1948, Maria Mayer pointed out that nuclei containing certain *magic numbers* of protons or neutrons are more stable than the neighbouring nuclei. The magic numbers are 2, 8, 20, 50, 82 and 126. For example,

(i) He^4 (α particle) with 2 protons and 2 neutrons is one of the most stable nuclei. Elements made with integral numbers of alphas like: B^8, C^{12}, O^{16}, Ne^{20}, Mg^{24} and Si^{28} are also stable.

(ii) O has 8 protons and it has three stable isotopes, O^{16}, O^{17} and O^{18}.

(iii) Ca with 20 protons has six stable isotopes: Ca^{40}, Ca^{42}, Ca^{43}, Ca^{44}, Ca^{46} and Ca^{48}.

(iv) Sn with 50 protons has ten stable isotopes.

(v) Xe^{136}, Sm^{144}, Ba^{138}, Le^{139}, Ce^{140} and Pi^{141} having 82 neutrons are all abundant stable nuclei.

(vi) Pb^{208} with 82 protons and 126 neutrons is also the most abundant lead isotope.

These numbers remind us of the number of electrons in the K, L, M, N etc shells of atoms, and they indicate that there ought to be similar shell structure in nuclei. It would arise from the quantum mechanically allowed orbitals of nucleons in the deep potential well of the strong force. The shape of the potential well is not exactly known. Some typical potential functions used for the purpose are:

(i) Square-well potential,

$$V(r) = -V_0, \quad \text{for } r < b,$$

$$V(r) = 0, \quad \text{for } r > b. \tag{9.12}$$

(ii) Yukawa potential of (9.10).

(iii) Harmonic oscillator potential,

$$V(r) = -V_0 + kr^2. \tag{9.13}$$

Out of these, the harmonic oscillator potential is found to reproduce the energy levels corresponding to the magic number nuclei.

Let us now consider the case of an excited nucleus C*. In Fig. 9.6, curve abcd represents the potential of the nucleus C, E is the energy of the projectile a and E_r is a bound state of C. Then Γ_a, the energy range of capture of a by A, is the same as the energy range of its penetrating the potential barrier in C. If v_i is the internal velocity and r_1 is the radius of the inner wall of the potential, then it is found that the probability per unit time of its penetration is $\Gamma_a/\hbar$, where

$$\Gamma_a = \frac{6v_i\hbar}{r_1} \exp\left\{ \frac{4}{\hbar}(2m_a Z_A Z_a e^2 r_1)^{1/2} - \frac{\pi Z_A Z_a e^2}{\hbar}\left(\frac{2m_a}{E}\right)^{1/2} \right\}. \tag{9.14}$$

We see that penetration probability increases with (i) larger E which would correspond to smaller thickness of the barrier (unless it is exactly rectangular at the top), (ii) smaller product $Z_A Z_a$, that is, smaller height of the barrier, and (iii) smaller m_a, that is, for a lighter projectile. It is obvious that for neutrons there is no barrier, and the probability of penetration is unity.

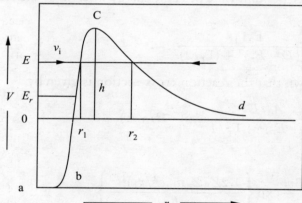

Fig. 9.6   *Barrier penetration*

9.3.4   Reaction rate

The reaction rate $R(a, b)$ for (a, b) reaction will depend on

(i) the target area of A,
(ii) probability of capture of a,
(iii) relative probability for the reaction $C \rightarrow B + b$,
(iv) the incident flux of projectile particles.

If v is the relative velocity of a with respect to A, m_a the mass of a and λ_a the de Broglie wavelength of a and E its energy, then the target area of A as seen by the projectile a will be $d\sigma = \pi\lambda_a^2$, where

$$\lambda_a = \hbar/\sqrt{2m_aE}. \tag{9.15}$$

If E_r and Γ_r are the energy and width of a bound state (we assume only one bound state), then the probability (per unit time) of depopulation (of the bound state of C) is $\Gamma_r/\hbar$. This should be equal to the total probability (per unit time) of breaking up of C into various channels. Thus, if Γ_i is the width of the ith channel, we should have $\Gamma_r = \sum_i \Gamma_i$.

The density of states (the weight) in the interval Γ_r is given within a good approximation by

$$\frac{\Gamma_r/2}{(E - E_r)^2 + (\Gamma_r/2)^2}.$$

Then the probability per unit time per unit energy interval of the reaction $A + a \rightarrow C \rightarrow B + b$ will be

$$G(a, b) = \frac{\Gamma_a\Gamma_b}{2\hbar[(E - E_r)^2 + (\Gamma_r/2)^2]}. \tag{9.16}$$

Then it can be shown that the reaction cross-section is given by

$$\sigma(a, b) = \frac{A_0/E}{(E - E_r)^2 + \Gamma_r^2/4} \exp(-B/\sqrt{E}), \tag{9.17}$$

where

$$A_0 = \frac{\pi\hbar^3\Gamma_b}{2m_a}\frac{6v_i}{r_1}\exp\left\{\frac{4(2Z_AZ_am_ae^2/r_1)^{1/2}}{\hbar}\right\},$$

$$B = \pi\sqrt{2m_a}Z_AZ_ae^2/\hbar. \tag{9.18}$$

If $N_a(v)dv$ is the number of projectile particles of species a having velocities between v and $v + dv$ per unit volume, the reaction rate $R(a, b)$ will be given by

$$R(a, b) = \int_0^\infty \sigma(v; a, b)N_a(v)dv$$

per target particle A. In laboratory experiments involving particle accelerators, v has a very small range; then

$$R(a, b) = \sigma(v; a, b)N_av,$$

where N_a is the total number of projectile particles per unit volume in a particular velocity range.

9.3.5 Thermonuclear reactions

In this case, $N_a(v)dv$ is given by Maxwell's law. Using $E = m_a v^2/2$ in Maxwell's speed distribution, we can write

$$N_a(E)dE = \frac{2\pi N_a}{(\pi kT)^{3/2}} E^{1/2} e^{-E/kT} dE.$$

Then the reaction rate comes out to be

$$R(a, b) = \frac{3\sqrt{2\pi}\hbar^3 N_a v_i}{(m_a kT)^{3/2} r_1} \exp\left\{\frac{4(2Z_A Z_a m_a e^2/r_1)^{1/2}}{\hbar}\right\} \int_0^\infty \frac{\exp(-B\sqrt{E} - E/kT)}{(E - E_r)^2 + \Gamma_r^2/4}. \quad (9.19)$$

The two limiting cases when E is far from E_r and when it is close to E_r are of particular interest. We shall discuss these here.

Non-resonance case

When E is far away from E_r, the variation of $E - E_r$ is not important; hence the term containing $E - E_r$ can be taken outside the integral sign. Then

$$R(a, b) = \frac{A'}{(kT)^{3/2}} \int_0^\infty \exp(-E/kT - B/E^{1/2})dE, \quad (9.20)$$

where

$$A' = \frac{3\sqrt{2\pi}\hbar^3 N_a v_i \exp\{4\sqrt{2m_a}(Z_A Z_a e^2 r_1)^{1/2}/\hbar\}}{m_a^{3/2} r_1\{(E - E_r)^2 + \Gamma_r^2/4\}}. \quad (9.21)$$

Now the integrand has two factors with opposite variations as shown in Fig. 9.7. The integrand has a maximum at $E_m = (BkT/2)^{2/3}$, giving $E_m/kT = (B^2/4kT)^{1/3}$. If the maximum is sharp, we can put

$$E = E_m + F, \quad |F| \ll E_m.$$

Then we see that

$$\frac{E}{kT} + \frac{B}{E^{1/2}} \simeq 3\left(\frac{B^2}{4kT}\right)^{1/3} + \frac{3F^2}{(16B^2 k^5 T^5)^{1/3}}.$$

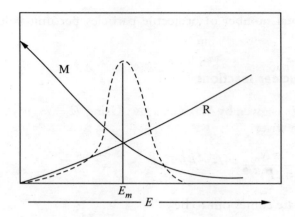

Fig. 9.7 *Competition between number of particles and reaction cross-section; M − Maxwell's law $(n \propto e^{-E/kT})$, R − reaction cross-section $(R \propto e^{-B/\sqrt{E}})$.*

Therefore the reaction rate becomes

$$R(\text{a, b}) = \frac{A' \exp(-3(B^2/4kT)^{1/3})}{(kT)^{3/2}} \, I,$$

where

$$I = \int_{-E_\text{m}}^{\infty} \exp[-3F^2/(16B^2k^5T^5)^{1/3}]dF. \tag{9.22}$$

We may let E_m tend to infinity without much error. Then it is seen that

$$I = [\pi(16B^2k^5T^5)^{1/3}/3]^{1/2}.$$

Defining a dimensionless variable

$$\tau = 3(B^2/4kT)^{1/3}, \tag{9.23}$$

we see that

$$R(\text{a, b}) \propto \tau^2 e^{-\tau}. \tag{9.24}$$

Substituting the values of e, k and $\hbar$ in CGS-Gaussian units, it can be seen that

$$\tau = 4270(Z_A^2 Z_a^2 \mu_a/T)^{1/3}, \tag{9.25}$$

where μ_a is the reduced mass of the reacting particles in amu and T is in K.

To see the temperature dependence of the reaction rate, we write $R(\text{a, b}) \propto T^n$. Then we have

$$n = \frac{d(\ln R)}{d(\ln T)} = (2 - \tau)\frac{d(\ln \tau)}{d(\ln T)}.$$

But (9.23) shows that $\tau \propto T^{-1/3}$. This gives

$$n = (\tau - 2)/3. \tag{9.26}$$

An estimate of τ and n is provided in Section 9.5.2 in the context of stellar energy.

Resonance case

When E is close to E_r, we can replace E by E_r in the exponential and write (9.19) as

$$R(\mathrm{a, b}) \propto \frac{e^{-E_r/kT}}{(kT)^{3/2}} \int_0^\infty \frac{dE}{(E - E_r)^2 + \Gamma_r^2/4}.$$

The temperature-dependence of the reaction rate can be obtained as follows. Let us write the temperature-dependent part of $R(\mathrm{a, b})$ as

$$R(\mathrm{a, b}) \propto e^{-E_r/kT}/(kT)^{3/2} \propto T^n,$$

where n is to be obtained. This gives us

$$n \ln T = \text{ constant} - (3/2) \ln T - E_r/kT,$$

so that

$$n = -3/2 + E_r/kT. \tag{9.27}$$

Generally, R will have two maxima, one near E_r and the other near E_m. If $E_r < 0$ or $E_r \gg kT$, then the maximum near E_r will be unimportant; this is true for most cases of astrophysical interest. In other cases we have to compare the two maxima to determine their relative importance. If $E_r \approx kT$, then $n \simeq 0$, and the reaction rate does not much depend on the temperature.

9.3.6 Applications of nuclear reactions

Artificial nuclear reactions have been used for several applications as under. All of them involve advanced technologies.

(i) Production of atom bombs capable of enormous destruction. In fact, this was one of the first applications before other peaceful uses were discovered.

(ii) Generation of electric power in controlled nuclear reactors.

(iii) Production of several radioactive species which themselves, and radiations from them, are used for treatment of diseases such as cancer, or for preservation of food articles, medical diagnosis, etc.

(iv) Production of transuranium elements up to about $Z = 110$. They include 93 neptunium, 94 plutonium, 95 americum, 96 curium, 97 berkelium, 98 califormium, 99 einsteinium, 100 fermium, and 101 mendelvium.

9.4   Elements of Particle Physics

As high energy particle accelerators became available, a variety of collision experiments were carried out using different targets and projectiles. This brought to the fore several new particles and reactions. Generally, processes and reactions involving an energy of a few MeV are called nuclear reactions (and belong to nuclear physics), while reactions involving higher energy are called particle reactions (and belong to particle physics or high energy physics).

We shall trace the development of particle physics in four stages. They are separated by time zones or eras of fundamental understanding of physics.

9.4.1   Pre-1947 particles

The first 'elementary' subatomic particle to be discovered was the electron, e^-, in 1897, by J. J. Thompson. Another 14 particles were discovered or postulated in the next 50 years. They are listed in Table 9.1. They are divided into two categories, *baryons* (heavier particles) and *leptons* (lighter particles) and given the quantum number B or L as shown in the table. Antiparticles have the same mass, rest energy, spin and isotopic spin. They have opposite charge and baryon or lepton number. They are denoted by the same symbol as the particle but with a bar on it, except for electron (e^-) and positron (e^+).

Of these, the antiproton and antineutron were discovered at the Lawrence Radiation Laboratory, Berkeley, in 1957, though they were postulated in the 1930s. The neutrinos ν_e and ν_μ were discovered in 1956 and 1961, respectively.

Most of these particles are sub-nuclear particles. Electrons may be sub-nuclear in β decay) or subatomic (as in photo or thermionic emission).

Almost all nuclear reactions observed until about 1947 were found to follow some laws:

(i) **Conservation of charge,**

(ii) **Conservation of mass/energy,**

(iii) **Conservation of parity.**

It was therefore taken for granted that conservation of parity is a basic law for all processes. Parity is a quality of the wavefunction $\psi(\mathbf{r})$ and of a process or reaction.

Table 9.1 *First list of elementary particles. The mass is in units of* m_e, *rest energy in MeV, charge in e, spin and isotopic spin in units of* $\hbar$; π^+ *and* π^- *are antiparticles of each other and so are* μ^+ *and* μ^-; π^0 *is its own antiparticle.*

Particle	Symbol	Mass	Rest energy	Charge	Spin	B or L	T
Electron	e^-	1	0.511	−1	1/2	$L=1$	0
Positron	e^+	1	0.511	+1	1/2	$L=-1$	0
Proton	p	1836	938.27	+1	1/2	$B=1$	1/2
Antiproton	$\bar{p}$	1836	938.27	−1	1/2	$B=-1$	1/2
Neutron	n	1839	939.57	0	1/2	$B=1$	1/2
Antineutron	$\bar{n}$	1839	939.57	0	1/2	$B=-1$	1/2
Pions	π^+	273.2	139.6	+1	0	$B=0$	1
	π^-	273.2	139.6	−1	0	$B=0$	1
	π^0	264.3	135.0	0	0	$B=0$	1
Muons	μ^+	206.8	105.66	+1	1/2	$L=1$	0
	μ^-	206.8	105.66	−1	1/2	$L=-1$	0
e-neutrino	ν_e	~ 0	$< 17 eV$	0	1/2	$L=1$	0
μ-neutrino	ν_μ	~ 0	$< 250 eV$	0	1/2	$L=1$	0
Antineutrinos	$\bar{\nu}_e$	~ 0	$< 17 eV$	0	1/2	$L=-1$	0
	$\bar{\nu}_\mu$	~ 0	$< 250 eV$	0	1/2	$L=-1$	0

Consider a phenomenon or reaction which is taking place. We look at it in a mirror which takes its wavefunction $\psi(x, y, z)$ to $\psi(x, y, -z)$. If the mirror image of the phenomenon is allowed in nature (that is, it does not violate laws of physics and can take place), then the phenomenon is said to conserve parity (parity = +1), otherwise it is said to violate parity (parity = −1).

As the strong force between protons and neutrons is charge-independent, they were considered to be two states of one particle, called *nucleon*. This was achieved by Heisenberg by introducing a new quantum number, the *isospin* or isotopic spin. Heisenberg assigned an isotopic spin $T = 1/2$ for the nucleon, with two possible projections or orientations in the isospin space; thus proton corresponds to $T_z = 1/2$ and neutron to $T_z = -1/2$.

Around this time, particles with rest mass higher than a pion or neutron were discovered in high energy reactions. One such particle was the K particle, in different charge states. Particles having rest energy equal to or higher than the proton were called hadrons, while the lighter ones were called mesons. It was found that if an additional quantum number, called the baryon number B, is introduced, then the charge Q, the isospin projection T_z and baryon number B are related by

$$Q = T_z + B/2. \tag{9.28}$$

Similarly, the pions were given the isospin $T = 1$ and baryon number $B = 0$. Then π^+, π^0 and π^- can be identified with $T_z = +1, 0, -1$, respectively, giving them the charges indicated by the superscripts. The antiproton and antineutron were assigned baryon number $B = -1$; and $T_z = -1/2$ for antiproton and $T_z = 1/2$ for antineutron. The lighter particles, electron, muon and neutrinos, were assigned a *lepton number* $L = 1$ and their antiparticles were given $L = -1$. Thus we now have two more conservation laws:

(iv) **Conservation of baryon number,**

(v) **Conservation of lepton number.**

As an example, annihilation of particle–antiparticle pair produces photons for which $Q = B = L = 0$. Also β^--decay is associated with the production of e^- and $\bar{\nu}_e$, and β^+-decay is associated with the production of e^+ and ν_e, thus conserving Q, B and L in each process.

9.4.2 From 1947 to 1961

Observations of primary cosmic rays at high altitudes and experiments with more and more powerful accelerators produced evidence for a variety of short lived particles. All of them were heavier than the pions and many of them heavier than the neutron. All of them were found not to obey some of the conservation laws, particularly the conservation of parity. They were called *strange* particles. The strange particles are listed in Table 9.2. The table also lists the strangeness quantum number S about which we will soon talk.

K mesons decay into pions, muons and electron or positron. Hyperons Λ^0, Σ^+ and Σ^- decay into nucleons and pions, while other hyperons decay into Λ^0 and p, n and γ. All the strange particles in Table 9.2 have their antiparticles. The kaon K^- is the antiparticle of K^+ and $\overline{K^0}$ is the antiparticle of K^0. The heavier particles Ξ in different charge states are called cascade particles. Together with the particles listed in Table 9.1, the number of fundamental building blocks of all matter in 1961 grew to 31.

In order to understand the strange behaviour of particles, Gellmann and Nishijima proposed a strangeness scheme in which all the particles were assigned new quantum number S called *strangeness*. The charge of the particle is now given by

$$Q = T_z + (B + S)/2 = T_z + Y/2, \tag{9.29}$$

which defines $Y = B + S$. It is called the *hypercharge*. All the earlier known particles listed in Table 9.1 were assigned $S = 0$. Antiparticles, in addition to having opposite charge and opposite baryon/lepton number, have opposite strangeness. The complete classification scheme is shown in Table 9.3.

Table 9.2 *The second list of elementary particles. The units of mass, energy and charge are as in Table 9.1*

Particle	Symbol	Rest energy	Charge	Spin	B	S	T
Heavy measons	K^+	493.9	1	0	0	1	1/2
	K^0	497.8	0	0	0	1	1/2
Hyperons	Λ^0	1115.4	0	1/2	1	−1	0
	Σ^+	1189.4	1	1/2	1	−1	1
	Σ^0	1191.5	0	1/2	1	−1	1
	Σ^-	1196.0	−1	1/2	1	−1	1
	Ξ^0	1311	0	1/2	1	−2	1/2
	Ξ^-	1318.4	−1	1/2	1	−2	1/2

Table 9.3 *Strangeness scheme of fundamental particles*

Type	S	B	Y	T	T_z	Q	Particle	Spin
Mesons	1	0	1	1/2	−1/2	0	K^0	0
				1/2	1		K^+	0
	0	0	0	1	−1	−1	π^-	0
				1	0	0	π^0	0
				1	1	1	π^+	0
Nucleons	0	1	1	1/2	1/2	1	p	1/2
				1/2	−1/2	0	n	1/2
Hyperons	−1	1	0	0	0	0	Λ^0	1/2
	−1	1	0	1	−1	−1	Σ^-	1/2
				1	0	0	Σ^0	1/2
				1	1	1	Σ^+	1/2
	−2	1	−1	1/2	−1/2	−1	Ξ^-	1/2
				1/2	1/2	0	Ξ^0	1/2

While the conservation laws listed earlier hold good for strange particles also, their reactions obey the following selection rules:

(vi) $\Delta S = 0$ **for strong reactions, which are fast;** $\Delta S = \pm 1$ **for weak reactions, which are slow.**

For example, decay reactions

$$K^+ \to \pi^+ + \pi^0, \ \Lambda^0 \to p + \pi^-, \Sigma^+ \to p + \pi^0, \Xi^0 \to \Lambda^0 + \pi^0$$

obey $\Delta S = \pm 1$, but reactions

$\pi^- + p \rightarrow K^0 + \Lambda_0, \ K^- + p \rightarrow \Xi^0 + K^0, \ K^- + p \rightarrow \Lambda^0 + \pi^0, \ K^- + n \rightarrow \Lambda^0 + \pi^-$

obey $\Delta S = 0$.

9.4.3   Quark revolution of 1961

The large number of elementary particles indicated that they are composed of still more fundamental particles. In 1961, deep inelastic scattering of energetic electrons by protons indicated that the proton has a structure, with three centres of fractional charge. So Gellmann suggested that hadrons (mesons, nucleons and hyperons) are composed of three fundamental particles called *quarks*. They are denoted by the symbols and names u (up), d (down) and s (strange) quarks, which have the properties given in Table 9.4.

Table 9.4   *Properties of u, d, s quarks*

Quark	Symbol	T	T_z	B	S	Y	Q	Spin
up	u	1/2	1/2	1/3	0	1/3	2/3	1/2
down	d	1/2	−1/2	1/3	0	1/3	−1/3	1/2
strange	s	0	0	1/3	−1	−2/3	−1/3	1/2

The antiquarks are represented by $\bar{u}, \bar{d}, \bar{s}$, respectively. For them, T_z, B, S, Y and Q have values opposite to the respective quark. The mesons are formed with a combination of one quark and one antiquark. On the other hand, baryons are formed by a combination of three quarks as indicated below.

Mesons

They form a singlet and an octet with the various quantum numbers $[T_z, B, S, Y, Q]$ shown in brackets.

Singlet: $\frac{1}{\sqrt{3}}(u\bar{u} + d\bar{d} + s\bar{s}) : [0, 0, 0, 0, 0], \quad \eta_0$ - singlet

Octet:

$$d\bar{s} : \quad [-1/2, 0, 1, 1, 0], \quad K^0 \ \Bigg\} \quad \text{doublet}$$
$$u\bar{s} : \quad [1/2, 0, 1, 1, 1], \quad K^+$$

$$d\bar{u} : \quad [-1, 0, 0, 0, -1], \quad \pi^-$$
$$\frac{1}{\sqrt{2}}(u\bar{u} - d\bar{d}) : \quad [0, 0, 0, 0, 0], \quad \pi^0 \ \Bigg\} \quad \text{triplet}$$
$$u\bar{d} : \quad [1, 0, 0, 0, 1], \quad \pi^+$$

$$\frac{1}{\sqrt{6}}(u\bar{u} + d\bar{d} - 2s\bar{s}) : \quad [0, 0, 0, 0, 0], \quad \eta \quad \text{– singlet}$$

$$s\bar{u} : \quad [-1/2, 0, -1, -1, -1], \quad K \ \Bigg\} \quad \text{doublet}$$
$$s\bar{d} : \quad [1/2, 0, -1, -1, 0], \quad K^0$$

We do not observe η_0 and η_8 separately, but their mixtures as η and η'. All these are called pseudoscalar mesons because their wavefunctions do not change sign on rotation, but do so on mirror reflection. They correspond to the lowest orbit of the two quarks with $l = 0$. We have another set for the next higher orbit with $l = 1$. They are denoted by $\varphi^0, K^{*^0}, K^{*+}, \rho^-, \rho^0, \rho^+, \phi_8, K^{*-}$ and $\overline{K^{*0}}$. They are vector mesons because their wavefunctions transform like ordinary vector on rotation and reflection.

Baryons

As they are formed by three quarks, their spin is either 1/2 or 3/2. The spin-1/2 baryons form a singlet and an octet, while spin-3/2 baryons form a decuplet, as shown below.

Singlet: $\dfrac{1}{\sqrt{6}}(uds + dsu + sud - dus - usd - sdu) : [0, 1, -1, 0, 0], \Lambda' - $ singlet

Octet:

$$
\begin{aligned}
&\text{udd}: &&[-1/2, 1, 0, -1, 0], &&\text{n} \\
&\text{uud}: &&[1/2, 1, 0, 1, 1], &&\text{p}
\end{aligned} \quad \Big\} \text{ doublet (939 MeV)}
$$

$$
\begin{aligned}
&\text{dds}: &&[-1, 1, -1, 0, -1], &&\Sigma^- \\
&\text{uds}: &&[0, 1, -1, 0, 0], &&\Sigma^0 \\
&\text{uus}: &&[1, 1, -1, 0, 1], &&\Sigma^+
\end{aligned} \quad \Big\} \text{ triplet (1189 MeV)}
$$

$$
\begin{aligned}
&\text{uds}: &&[0, 1, -1, 0, 0], &&\Lambda_8 \quad \text{singlet}
\end{aligned}
$$

$$
\begin{aligned}
&\text{dss}: &&[-1/2, 1, -2, -1, -1], &&\Xi^- \\
&\text{uss}: &&[1/2, 1, -2, -1, 0], &&\Xi^0
\end{aligned} \quad \Big\} \text{ doublet (1321 MeV)}
$$

Decuplet

$$
\begin{aligned}
&\text{ddd}: &&[-3/2, 1, 0, 1, -1], &&\Delta^- \\
&\text{udd}: &&[-1/2, 1, 0, 1, 0], &&\Delta^0 \\
&\text{uud}: &&[1/2, 1, 0, 1, 1], &&\Delta^+ \\
&\text{uuu}: &&[3/2, 1, 0, 1, 2], &&\Delta^{++}
\end{aligned} \quad \Big\} \text{ quartet (1232 MeV)}
$$

$$
\begin{aligned}
&\text{dds}: &&[-1, 1, -1, 0, -1], &&\Sigma^{*-} \\
&\text{uds}: &&[0, 1, -1, 0, 0], &&\Sigma^{*0} \\
&\text{uus}: &&[1, 1, -1, 0, 1], &&\Sigma^{*+}
\end{aligned} \quad \Big\} \text{ triplet (1385 MeV)}
$$

$$
\begin{aligned}
&\text{dss}: &&[-1/2, 1, -2, -1, -1], &&\Xi^{*-} \\
&\text{uss}: &&[1/2, 1, -2, -1, 0], &&\Xi^{*0}
\end{aligned} \quad \Big\} \text{ doublet (1533 MeV)}
$$

$$
\begin{aligned}
&\text{sss}: &&[0, 1, -3, -2, -1], &&\Omega^{-1} \quad - \text{ singlet (1672 MeV)}
\end{aligned}
$$

Again, we observe combination of Λ' and Λ_8 as Λ^0 with a mass of 1116 MeV.

In these newly discovered particles, the combinations ddd, uuu, sss would be forbidden by the exclusion principle. In order to avoid this dilemma, the quarks are assigned another quantum number, called *colour*, with designations red (R), blue (B), and green (G).

It is now possible to represent β-decay as a result of change in the type of quark in a nucleon. Thus β-decay can be shown as in Fig. 9.8 (a) and (b). Later, Glashow and Abdus Salam suggested the possibility of interchange in the positions of u and d quarks in the proton, giving rise to neutral current decay, as shown in Fig. 9.8 (c).

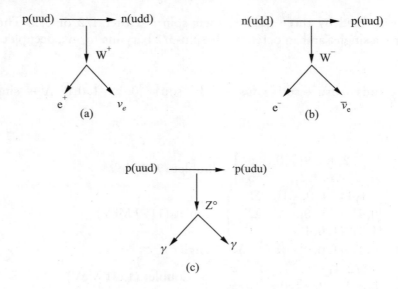

Fig. 9.8 *Beta decay and neutral current decay produced as change of quarks: (a) β^+ decay, (b) β^- decay, (c) neutral current decay.*

Neutrino oscillations and recent results

When neutrinos from the sun were detected in the laboratory, their observed number turned out to be only one-third of the estimated number calculated on the basis of sun's core temperature, hydrogen content, density, rate of fusion reaction, etc. This lead to the suggestion of neutrino oscillations, in which an electron neutrino is converted into a muon neutrino and back. What we were observing were only the electron neutrinos, not the muon neutrinos, thus explaining the discrepancy. This theory of neutrino oscillations required the existence of a fourth neutrino called the sterile neutrino which interacted only through the gravitational interaction.

A recent Fermilab experiment called MiniBooNE (for Mini Booster Neutrino Experiment) in early 2007, however, provides staunch new evidence for the idea that only three low-mass neutrino species exist. Protons from Fermilab's booster accelerator are smashed into a fixed target, creating a swarm of mesons which very quickly decay into secondary particles, among them a lot of muon neutrinos. The experiment uses 500 MeV neutrinos detected after a distance of 500 m, and cleverly discriminates between the rare electron neutrinos and the much more common muon neutrinos. After taking account of the expected background events, the MiniBooNE did not see any sign of neutrino oscillations. Further experiments have been planned with greater precision and accuracy.

9.4.4   The need for more quarks

Around 1974 physicists discovered the following new particles — φ meson of 1019.41 MeV, J/ψ meson of 3096.93 MeV, Υ meson of 9460.32 MeV, and τ^- lepton of 1.8 GeV. Also, tau-neutrino[26] ν_τ (< 70 MeV) was suggested in 1982–83. In order to account for these particles, it was necessary to postulate the existence of three more quarks with *charm* quantum number c, and *beauty* quantum number b, as listed in Table 9.5. They are respectively called the charm quark, beauty quark and the top quark.

Table 9.5   *Charm, beauty and top quarks*

Quark	Symbol	T	T_z	B	S	Y	c	b	Q	Spin
Charm	c	0	0	1/3	0	1/3	1	0	2/3	1/2
Top	t	0	0	1/3	0	1/3	0	1	2/3	1/2
Beauty	b	0	0	1/3	0	1/3	0	-1	$-1/3$	1/2

Each of them occurs in three colours, *red, blue* and *green,* and they have their antiquarks. The charge is now given by

$$Q = T_z + (Y + c + b)/2. \tag{9.30}$$

The structure of the above particles in terms of these quarks is $\varphi = (s\bar{s})$, $J/\psi = (c\bar{c})$ and $\gamma = (b\bar{b})$, but $(t\bar{t})$ has still not been found.

Now we have the following three families between quarks and leptons:

$$\begin{pmatrix} e^- & u \\ \nu_e & d \end{pmatrix}, \begin{pmatrix} \mu^- & c \\ \nu_\mu & s \end{pmatrix}, \begin{pmatrix} \tau^- & t \\ \nu_\tau & b \end{pmatrix}.$$

The masses of these six quarks are, u (300 MeV), d (300 MeV), c (1500 MeV), s (500 MeV), t (30-50 GeV) and b (4.45 GeV). The lepton masses are e$^-$ (0.511 MeV), ν_e ($\simeq 20$ eV), μ^- (105 MeV), ν_μ (< 45 eV), τ^- (1.8 GeV), and ν_τ (< 70 MeV).

---

[26]A search for these neutrinos is currently on.

472 An Overview of Basic Theoretical Physics

Quarks have not been seen outside nuclei. They interact with and bind each other through particles called *gluons* which confine them within the nucleus. Gluons act through the colour property of quarks. Just as electrically charged particles interact with each other through photons according to quantum electrodynamics (QED), quarks interact with each other through gluons according to quantum chromodynamics (QCD). As there are three colours, red (R), blue (B) and green (G), we have nine kinds of reactions, such as

$$\overline{R}R \quad \overline{R}B \quad \overline{R}G$$
$$\overline{B}R \quad \overline{B}B \quad \overline{B}G$$
$$\overline{G}R \quad \overline{G}B \quad \overline{G}G.$$

Out of these nine combinations; of the nine reactions, $\overline{R}R + \overline{B}B + \overline{G}G$ is colourless, and hence inoperative. So we are left with eight independent types of gluons; they have not been observed outside the nuclei.

9.4.5 The emerging picture

We started our search for fundamental particles with the discovery of electrons in 1897. After meeting a couple of hundred of such particles during a little over a hundred years, it is now concluded that all matter is made up of 6 leptons and 18 types of quarks which occur in three families. They interact with each other at five different levels which we will now discuss.

(i) Quarks interact with each other through massless gluons with a very strong force effective within a short distance less than 10^{-13} cm, forming hydrons.

(ii) Hydrons interact with each other through pions of mass ~ 140 MeV with another strong force upto a distance of 10^{-13} cm, forming nuclei of atoms.

(iii) Electromagnetic interaction between nucleons with a weak force carried by $W^{\pm}$ and Z^0 bosons of 100 GeV mass, which cause β-radioactivity. This force prevails over a thousandth of a femtometre (10^{-16} cm) and has a strength equal to 10^{-13} times the strong force of level (ii).

(iv) Electromagnetic interaction between charged particles acting through massless photons, which determines the structure of atoms and molecules. It has a strength equal to 1/137 of the strong force and it operates over all distances up to infinity.

(v) Finally we have the gravitational interaction between all particles, which acts through the postulated massless graviton. It is the weakest force with a strength of about 10^{-38} or 10^{-40} of the strong force and acts over all distances up to infinity. It is responsible for the formation of stars, glaxies and planetary systems.

In spite of the fact that both gravitation and electromagnetism are long range interactions, the latter does not seem to play much role in dealing with large scale structures. This is because any two masses attract each other under gravitational interaction, and so we assign all mass the same, positive sign. On the other hand, charge appears in two kinds (apart from neutral), with attraction or repulsion between them. Any bulk volume has a positive mass but a net zero charge. Therefore their regions of influence are quite distinct. At the microscopic level, gravitation could be entirely neglected.

The forces (v) to (i) become manifest at higher and higher energies and are supposed to combine into a single force at the highest temperature and densities. We also see that we have to go on adding more and more degrees of freedom (quantum numbers) as we probe deeper into the structure of the atom. It is believed that ultimately one would have to introduce 10 degrees of freedom and consider *strings* instead of points as the nature of fundamental particles in this ten-dimensional space. Such a string theory could perhaps lead to answers concerning the fundamental structure of matter.

9.5 Applications in Astrophysics

Although the domains of nuclear and particle physics and astrophysics are vastly different, both nuclear and particle physics find a large number of applications in astrophysics. In fact, the discovery of many elementary particles came from observations of high energy cosmic rays before they were produced in the laboratory. Nuclear physics plays an important part in explaining the vast source of stellar energy and also in the observed abundances of cosmic elements. All kinds of high-energy particle interactions must be occurring at places in the universe which have high temperatures and high densities of matter. The entire sequence of birth, life and death of a star depends on such elementary interactions and processes. We shall study a few such aspects in this section.

9.5.1 Classification of stars

General properties

Our sun is a typical star of mass $M_\odot = 2 \times 10^{33}$ gm, radius $R_\odot = 6.96 \times 10^{10}$ cm, luminosity $L_\odot = 4 \times 10^{33}$ erg/s and a surface temperature of about 6000 K. Other stars have a range of masses from 0.1 $M_\odot$ to $100 M_\odot$, radii from 0.02 $R_\odot$ to $500 R_\odot$, luminosities from $10^{-3} L_\odot$ to $10^6 L_\odot$, and surface temperatures from 3000 K to 50000 K. The subscript $\odot$ denotes the corresponding quantities for the sun. The surface temperature manifests itself through the colour of the star and nature of its spectrum. Accordingly, the stars are classified into the following spectral types:

O – voilet colour, He$^+$ lines in the spectrum;

B – blue colour, with He lines;

A – white colour, with H lines;

F – green colour, with lines of ionized atoms;

G – yellow colour, with lines of ionized and neutral atoms;

K – orange colour, lines of neutral atoms and molecules;

M – red colour, molecular lines and bands.

The properties of stars are depicted in the Hertzsprung–Russell (HR) diagram of Fig. 9.9, where the x-axis runs towards decreasing temperature and y-axis runs towards increasing luminosity. Nearly 70% of the stars lie on a diagonal running from top left to bottom right in the HR diagram. They form the *main sequence*. Properties of the main sequence stars are listed in Table 9.6. We see that O stars have the highest mass, luminosity and surface temperature, while the M stars have the lowest mass, luminosity and surface temperature. The main sequence stars show a mass–luminosity relation of the form

$$L/L_\odot = (M/M_\odot)^{3.5}.$$

Table 9.6   *Properties of the main sequence stars*

Spectral type	Colour	Surface temperature (K)	Principal spectral lines	Mass $(M/M_\odot)$	Radius $(R/R_\odot)$	Luminosity $(L/L_\odot)$	Example
O	Violet	35000	Ionised helium	40	20	3×10^5	Zeta Piscium
B	Blue	20000	Neutral helium	17	10	1×10^4	Spica (*Chitra*)
A	White	9600	Balmer lines	3	2.5	50	Sirius (*Vyadh*)
F	Greenish	7600	Ionised metals	1.3	1.2	3	Procyon A
G	Yellow	5800	Ionised and neutral metals	1.0	1.0	1	Sun (*Surya*)
K	Orange	4500	Neutral metals Molecular bands	0.6	0.7	0.25	Epsilon Eridani
M	Red	3000	Molecular bands	0.2	0.3	0.01	Barnard's Star

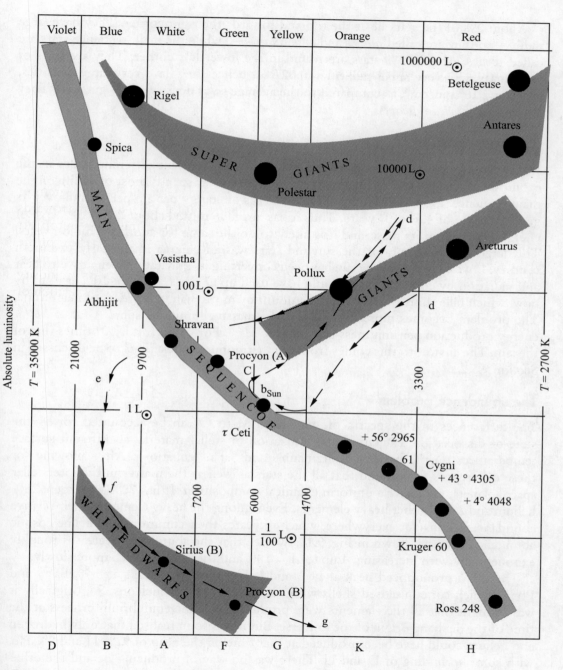

Fig. 9.9 *Properties of the main sequence stars*

About 2% of the stars lie at the top and towards the right corner. They have large radii — so they are called *giants* and *supergiants* — while the main sequence stars are called *dwars*. About 23% stars are found in the lower left corner. They are stars of solar mass with very small radius ($\sim 0.02 R_\odot$), hence they have very large densities, exceeding 10^5 gm/cm^3, as compared to the average solar density of 1.4 gm/cm^3. They are known as *white dwarfs*.

The energy problem

If the main sequence stars, giants and white dwarfs represent different stages in the evolution of a single star, we have to conclude that a star spends most of its life on the main sequence. Now the sun is producing 2 ergs of energy per gm per second over its geological age of 4.5×10^9 years. Thus it has already emitted about 2.7×10^{17} ergs of energy per gram of its mass, and it is likely to do the same for another 5 billion years. This large energy supply of the sun and stars was an enigma in the early twentieth century. If we assume that the sun has been converting its gravitational energy into heat and radiation by contraction, it would have used up $GM_\odot/R_\odot \simeq 2 \times 10^{15}$ erg/gm by now, which falls short of the required amount by more than two orders of magnitude. The problem becomes more difficult for the massive luminous stars. As $L \propto M^{3/5}$, energy production per unit mass goes as $L \propto M^{2.5}$. For O stars, it is 10^5 times that of the sun. The answer to this comes from the thermonuclear reactions to be discussed in Section 9.5.2.

The abundance problem

The differences in the spectra of stars from M to O can be accounted for by the state of dissociation, excitation and ionization of stellar material at different surface temperatures on the basis of Boltzmann's and Saha's equations. After allowing for these differences, it is found that all the stars as well as the matter in the interstellar space, planets, etc., have a uniform chemical composition. It has 70% hydrogen, 28% helium and 2% of other heavy elements. Even among the heavy elements, their relative abundance is constant everywhere, which indicates their common origin. The cosmic abundance curve is shown in Fig. 9.10. We see that the abundance decreases roughly exponentially with increasing A up to $A = 100$, and then drops much more slowly.

There is a pronounced peak at Fe^{56} and less pronounced peaks at Sr^{88}, Ba^{138} and Pb^{206}, which contain closed shells with magic numbers of nucleons. So, originally it was thought that all the elements were produced by some equilibrium process at the time of the big bang origin of the universe. But it was soon realised that only hydrogen and helium could have been produced at that time, in the ratio of 92% H and 8% He, with some sprinkling of D and Li. There was no way of producing Be and the other heavier elements because

$$_2\text{He}^4 + {}_2\text{He}^4 \rightarrow {}_4\text{Be}^8$$

is an endothermic process. So one has to synthesise the heavier elements inside stars, as suggested by some scientists in 1957.

9.5.2 Stellar energy sources

Main sequence stars

From theoretical considerations about the structure of stars, Eddington had concluded that central temperatures of stars could be as high as 10 to 30 million degree. He was also convinced that the radiation emitted by the stars ought to be derived by conversion of mass into energy according to Einstein's equation, $E = mc^2$. It was in 1939 that Bethe suggested the following *carbon–nitrogen cycle* as a source of such energy:

1. $_6\text{C}^{12} + {}_1\text{H}^1 \rightarrow {}_7\text{N}^{13} + \gamma;$

2. $_7\text{N}^{13} \rightarrow {}_6\text{C}^{13} + \text{e}^+ + \nu_\text{e};$

3. $_6\text{C}^{13} + {}_1\text{H}^1 \rightarrow {}_7\text{N}^{14} + \gamma;$

4. $_7\text{N}^{14} + {}_1\text{H}^1 \rightarrow {}_8\text{O}^{15} + \gamma;$

5. $_8\text{O}^{15} \rightarrow {}_7\text{N}^{15} + \text{e}^+ + \nu_\text{e};$

6. $_7\text{N}^{15} + {}_1\text{H}^1 \rightarrow [{}_8\text{O}^{16} + \gamma] \rightarrow {}_6\text{C}^{12} + {}_2\text{He}^4.$ (9.31)

In this process, carbon acts as a catalyst for converting four protons into an alpha particle. The mass difference between four protons and an alpha is emitted as γ radiation and other particles.

A little later, Gamow suggested another *proton–proton chain reaction* for the same purpose:

1. $_1\text{H}^1 + {}_1\text{H}^1 \rightarrow {}_1\text{H}^2 + \text{e}^+ + \nu;$

2. $_1\text{H}^2 + {}_1\text{H}^1 \rightarrow {}_2\text{He}^3;$

3. $_2\text{He}^3 + {}_2\text{He}^3 \rightarrow {}_2\text{He}^4 + {}_1\text{H}^1 + {}_1\text{H}^1.$ (9.32)

In both cases, four protons get converted into one helium nucleus, with a loss of mass equal to 0.0294 amu, which is converted into energy.

In the C–N cycle of (9.31), reaction 4 is the slowest. To obtain the temperature dependence of its reaction rate, we can apply (9.25) and (9.26). For the reaction of nitrogen nucleus and a proton in the central region of a star, we have

$$Z_\text{A} = 7, \quad Z_\text{a} = 1, \quad \mu \sim 1, \quad T \sim 2 \times 10^7,$$

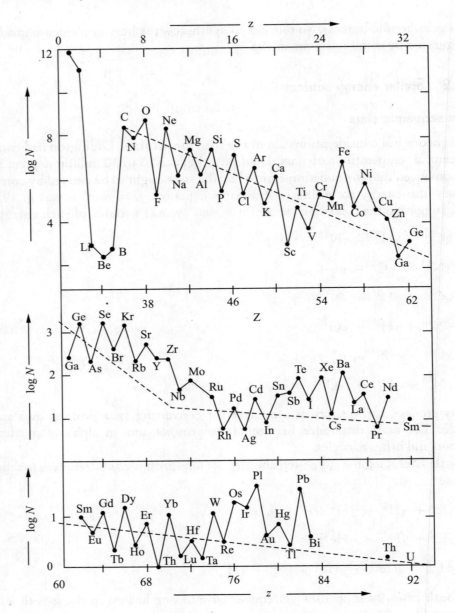

Fig. 9.10 *Cosmic abundances of elements*

in appropriate units as indicated earlier. This gives us

$$\tau \simeq 57, \quad n \simeq 18.5.$$

Thus the reaction rate goes as $T^{18.5}$. If X_H and X_{CN} are the fractions of H and CN masses in the stellar material, ρ its mean density and m_H the mass of a hydrogen atom, then the number densities of the target and projectile would be

$$N_a = \rho X_H/m_H, \quad N_A = \rho X_{CN}/m_H.$$

Therefore the energy production per unit mass would be

$$\mathcal{E} \propto \rho X_H X_{CN} T^{18.5}. \tag{9.33}$$

In the p–p chain reaction of (9.32), the first reaction is the slowest. For the reaction of two protons, we have

$$Z_A = Z_a = 1, \quad \mu = 1/2 \text{ amu}, \quad T \sim 10^7 \text{ K},$$

which gives

$$\tau = 14.8, n = 4.3.$$

Thus the reaction rate goes as $T^{4.3}$.

Since both the target and the projectile are hydrogen atoms, the energy production is given by

$$\mathcal{E} \propto \rho X_H^2 T^{4.3}. \tag{9.34}$$

A comparison of (9.33) and (9.34) shows that CN cycle is more efficient at higher temperatures and pp chain is more efficient at lower temperatures. It is found that for stars less massive than the sun for which the central temperature is less than 14 million kelvin, the pp chain reaction provides the energy. On the other hand, in the case of stars more massive than the sun for which the central temperature is more than 14 million kelvin, the energy is produced by CN cycle. In the sun itself, most contribution to energy comes from the pp chain reaction, with a small contribution from CN cycle.

Due to high power dependence on T, both pp chain reaction and CN cycle take place in the central cores of the stars while the envelope only transmits the energy for release at the surface.

Giants and supergiants

A star is formed by the contraction of an interstellar cloud of gas and dust under its own gravitation. The gravitational energy gets converted into kinetic energy, and heats up the star. When the central temperature reaches values of 10 to 30 million kelvin, thermonuclear reactions are initiated, as discussed above. The star stops contracting and a steady state is reached, with a steady output of energy. This is the main sequence stage of the star; it lasts for 10^6 years for an O star, 10^{10} years for a sun-like star and

10^{12} years for an M star. During this stage, energy is produced in the central region of the star whose composition slowly changes from hydrogen to helium.

Chandrashekar and Shoenberg showed that the helium core can grow until it contains 10% of the mass of a star. Thereafter the core contracts and the envelope expands taking the star into the giant and supergiant region of the HR diagram. The temperature of the contracting core increases and when it reaches about 20 million kelvin, the following triple alpha reaction starts producing energy, with the reactions

1. $_2\mathrm{He}^4 + {}_2\mathrm{He}^4 \rightleftharpoons {}_4\mathrm{Be}^8$;

2. $_4\mathrm{Be}^8 + {}_2\mathrm{He}^4 \rightarrow {}_6\mathrm{C}^{12} + \gamma$.

The second reaction has to take place before Be has a chance to break up into two alphas, hence it is called a triple alpha reaction.

For the second reaction, we find that

$$Z_A = 4, Z_a = 2, \quad \mu = 8/3 \,\text{amu}, \quad T \simeq 10^7 \text{ K},$$

giving

$$\tau = 89, \quad n = 29,$$

showing a very strong dependence on temperature.

As the star evolves, the temperature rises further and helium gets exhausted by successive reactions like

$$_6\mathrm{C}^{12} + {}_2\mathrm{He}^4 \rightarrow {}_8\mathrm{O}^{16}, \quad {}_8\mathrm{O}^{16} + {}_2\mathrm{He}^4 \rightarrow {}_{10}\mathrm{Ne}^{20},$$

$$_{10}\mathrm{Ne}^{20} + {}_2\mathrm{He}^4 \rightarrow {}_{12}\mathrm{Mg}^{24}.$$

Once the helium is exhausted, the core contracts, and at appropriate temperatures we have the following reactions:

$$_6\mathrm{C}^{12} + {}_6\mathrm{C}^{12} \rightarrow {}_{10}\mathrm{Ne}^{20} + {}_2\mathrm{He}^4 \rightarrow {}_{12}\mathrm{Mg}^{24},$$

$$_8\mathrm{O}^{16} + {}_8\mathrm{O}^{16} \rightarrow {}_{14}\mathrm{Si}^{28} + {}_2\mathrm{He}^4 \rightarrow {}_{16}\mathrm{S}^{32},$$

$$_6\mathrm{C}^{12} + {}_8\mathrm{O}^{16} \rightarrow {}_{12}\mathrm{Mg}^{24} + {}_2\mathrm{He}^4 \rightarrow {}_{14}\mathrm{Si}^{28}.$$

Then the addition of protons would produce all the elements upto sulphur and silicon. Stars of mass less than $8M_\odot$ explode at the end of this process as supernovae. The envelope is thrown away and the core becomes a white dwarf of mass $< 1.4 M_\odot$, which is the Chandrashekhar limit for white dwarfs. The fate of more massive stars will be discussed in the next sub-section.

9.5.3 Nucleosynthesis of elements in stars

We have noted earlier that only hydrogen and helium were produced at the big bang epoch, with some fraction of deuterium and lithium. So the first generation stars had to depend on pp chain reaction for producing energy during the main sequence stage. But when the second generation stars were formed from the material ejected during the explosions of first generation stars, they would contain carbon and nitrogen so that energy could be derived from CN cycle also. We have already seen what happens to stars of mass less than $8M_\odot$. In particular, we have noted that all elements upto S and Si have been generated. Let us now consider the fate of more massive stars.

After producing elements upto sulpur, the massive stars contract further without exploding. The central core reaches temperatures of 3 to 4 billion kelvin and densities of 10^5 gm/cm^3. At this stage, as suggested by Chandrashekhar and Henrich in 1942, protons, neutrons, α-particles, electrons and photons will be in equilibrium which will increase the abundance of more stable elements depending upon their degree of stability. In this way, it is possible to explain the observed abundance of elements up to iron, with a peak at $A = 44$ to 66.

Formation of other elements proceeds as follows. While the core is contracting, in the surrounding shell, which still contains hydrogen, the following Ne–Na cycle can operate:

$$_{10}\text{Ne}^{20} + {}_1\text{H}^1 \quad \rightarrow \quad {}_{11}\text{Na}^{21} + \gamma;$$

$$_{11}\text{Na}^{21} \quad \rightarrow \quad {}_{10}\text{Ne}^{21} + e^+ + \nu_e;$$

$$_{10}\text{Ne}^{21} + {}_1\text{H}^1 \quad \rightarrow \quad {}_{11}\text{Na}^{22} + \gamma;$$

$$_{11}\text{Na}^{22} + {}_1\text{H}^1 \quad \rightarrow \quad {}_{12}\text{Mg}^{23} + \gamma;$$

$$_{12}\text{Mg}^{23} \quad \rightarrow \quad {}_{11}\text{Na}^{23} + e^+ + \nu_e;$$

$$_{11}\text{Na}^{23} + {}_1\text{H}^1 \quad \rightarrow \quad [{}_{12}\text{Mg}^{24}]^* \rightarrow {}_{12}\text{Mg}^{24} + \gamma \rightarrow {}_{10}\text{Ne}^{20} + {}_2\text{He}^4.$$

Here, again, four protons get converted into one alpha to produce energy. But another important reaction also takes place, which is

$$_{10}\text{Ne}^{21} + {}_{12}\text{He}^4 \rightarrow {}_{12}\text{Mg}^{24} + n.$$

The neutrons so produced travel into the core and cause chain reactions which produce the elements beyond the iron peak in two ways as discussed now.

(i) If the evolution is slow (s-process), the time between the capture of two neutrons is of the order of 1 to 1000 years. This gives an opportunity for the production of β-stable elements from Fe^{56} to Ge^{70} onwards, including the magic number elements Sr^{88}, Ba^{138} and Pb^{206}.

(ii) If the evolution is rapid (r-process), the interval between capture of neutrons is of the order of 1 second and it is possible to produce β-unstable elements right upon Cf^{254}.

Ultimately when the star explodes in the form of a supernova producing a neutron star or black hole, all these elements are thrown out into the interstellar medium. So the successive generations of stars become richer and richer in heavy elements, reaching a value of 2% at the time of formation of the sun. Since then the enrichment has proceeded to about 4%.

Problems

9.1 Check the stability of $_{11}\mathrm{Na}^{22}$ nucleus against $\beta^{\pm}$ decay and K-capture in terms of its mass and the masses of the neighbouring nuclei. Take the nuclear masses in amu to be $_{11}\mathrm{Na}^{22}$: 21.99444, $_{12}\mathrm{Mg}^{22}$: 21.99985, $_{10}\mathrm{Ne}^{22}$: 21.99139.

9.2 Calculate the closest distance of approach of a 5.3 MeV alpha particle which is directed head-on towards a gold nucleus ($Z = 79$).

9.3 How far does a beam of muons with a kinetic energy of 100 GeV and half life of 2.2×10^{-6} s travel in empty space before its intensity is reduced to one half of its initial value?

9.4 Find the minimum kinetic energy in the lab system needed by an alpha particle to cause the reaction $^{14}\mathrm{N}(\alpha, \mathrm{p})\ ^{17}\mathrm{O}$. You may take the masses of the nuclides in amu to be $^{14}\mathrm{N}$: 14.00307 u, $^{4}\mathrm{He}$: 4.00260 u, $^{1}\mathrm{H}$: 1.00783 u, $^{17}\mathrm{O}$: 16.99913 u.

9.5 The Q-value of a nuclear reaction is the total rest mass energy of the reacting particles minus that of the outgoing particles. If $Q > 0$, the reaction can take place spontaneously, otherwise an energy equal to $-Q$ has to be supplied. Calculate the Q-values for the reactions $_{19}\mathrm{K}^{40}(\mathrm{p}, \alpha)\ _{18}\mathrm{Ar}^{37}$ and $_{26}\mathrm{Fe}^{57}(\mathrm{p}, \mathrm{d})\ _{26}\mathrm{Fe}^{56}$. In the case of reactions with negative Q, calculate the threshold energy in the laboratory system. Take the masses to be $m_{\mathrm{n}} = 1.00865$ amu, $m_{\mathrm{p}} = 1.00783$ amu, $m_{\mathrm{d}} = 2.0141$ amu, 1 amu $= 931.48$ MeV. Take the target to be at rest.

9.6 In the light of the conservation laws for nuclear interactions, based on charge, isospin, strangeness, and baryon number, state, giving reasons, whether or not the following reactions are possible:

$\pi^- + p \rightarrow \pi^0 + \Lambda^0$;

$\pi^+ + n \rightarrow K^0 + K^+$;

$p + K^- \rightarrow \Lambda^0 + \pi^+ + \pi^-$;

$\Sigma^+ \rightarrow p + \pi_0$;

$\pi^+ \rightarrow \mu^+ + \nu_\mu$;

$p + \bar{p} \rightarrow \gamma$;

$p + K^- \rightarrow \Sigma^+ + \pi^+ + \pi^- + \pi^0 + \pi^-$;

$\nu_\mu + p \rightarrow e^+ + n$;

$\nu_e + p = e^- + \Sigma^+ + K^+$.

$p + \bar{p} \rightarrow \pi^+ + \pi^- + \pi^0 + \pi^+ + \pi^-$;

$p + \pi^- \rightarrow p + K^-$;

$\nu_\mu + p \rightarrow \mu^+ + n$;

$\nu_e + p \rightarrow e^+ + \Lambda^0 + K^0$;

$K^+ \rightarrow \pi^+ + \pi^- + \pi^+ + \pi^- + \pi^+ + \pi^0$;

$n \rightarrow \pi^+ + e^+$;

$\pi^- + p \rightarrow \overline{\Sigma}^0 + \Lambda^0$;

$\nu_e + p \rightarrow e^+ + \Lambda^0 + K^0$;

Bibliography and References

Abhyankar, K D, *Astrophysics of the Solar System*, Universities Press, Hyderabad (1999).

Abhyankar, K D, *Astrophysics: Stars and Galaxies*, Universities Press, Hyderabad (2001).

Bates, D R, *Quantum Theory VIII - Radiation and High Energy Physics*, Academic Press, New York (1962).

Beer, R, Hutchins, R B, Norton, R H and Lambert D L, *Astrophysical Journal*, **172**, 89 (1972)

Bergman, P G, *Introduction to the Theory of Relativity*, Asia Publishing House, Delhi (1960).

Blatt, J M and Weisskopf, V F, *Theoretical Nuclear Physics*, John Wiley and Sons, New York (1952).

Close, F, Marten, M and Sutton, C, *The Particle Explosion*, Oxford University Press, Oxford (1987).

Dick, R H, and White, J P, *Introduction to Quantum Mechanics*, Prentice Hall, India (1972).

Dingle, H *Special Theory of Relativity*, John Wiley and Sons, New York (1959).

Eisberg, Robert, and Resnick, Robert, *Quantum Physics of Atoms, Molecules, Solids, Nuclei and Particles*, John Wiley and Sons, New York (1974).

Eton, L R B, *Introduction to Nuclear Theory*, Pitman and Sons, London (1955).

Feynman, R P, Leighton, R B and Sands, M, *Lectures on Physics*, Addison-Wesley, Reading (1969).

Feynman, R P, et al, *Quantum Mechanics*, Addison Wesley, Reading (1969).

Fraunfelder, H and Henley, E M, *Subatomic Physics*, Prentice-Hall, New York (1974).

Griffiths, D J, *Introduction to Electrodynamics*, Prentice-Hall of India, New Delhi (2002).

Griffiths, D J, *Introduction to Quantum Mechanics*, Pearson Education Inc., Delhi (2005).

Harrison, G R and Lord R C, *Practical Spectroscopy*, Prentice Hall, New York (1948).

Heckman, H H, and Starling, P W, *Nuclear Physics and Fundamental Practices*, Holt-Reinhardt, New York (1963).

Heitler, W, *Quantum Theory of Radiation*, Clarendon Press, Oxford (1957).

Herzberg, G, *Spectra of Diatomic Molecules*, van Nostrand and Reinherd, New York (1950).

Hulst, van de, *Uttrecht Researches*, XI, Parts I and II (1949).

Jackson, J D, *Classical Electrodynamics*, 3d ed., John Wiley and Sons, New York (1998).

Joos, G, *Theoretical Physics*, Blackie and Sons, London (1942).

Joshi, A W, *Elements of Group Theory for Physicists*, 4th ed., New Age International, New Delhi (1997).

Joshi, A W, *Matrices and Tensors in Physics*, 3d ed., New Age International, New Delhi (2005).

Katz, R, *Introduction to Special Theory of Relativity*, van Nostrand, New Jersey (1964).

Kittel, C, *Elementary Statistical Physics*, John Wiley and Sons, New York (1958).

Kompaneyets, A S, *Theoretical Physics*, Mir Publications, Moscow (1961).

Lahiri, A, and Pal, P B, *A First Book of Quantum Field Theory*, Narosa, New Delhi (2001).

Landau, L D, and Lifshitz, E M, *Classical Theory of Fields*, Pergamon Press, Oxford (1962).

Lee, J F, and Seers, F W, *Statistical Thermodynamics*, Addison Wesley, London (1963).

McCrea, W H, *Relativity Physics*, Methuen Co, London (1960).

Merill, P W, *Lines of Chemical Elements in Astronomical Spectra*, Cambridge Institute, Washington (1956).

Meyerhof, W E, *Elements of Nuclear Physics*, McGraw-Hill, New York (1989).

Mie, G, *Ann de Physik*, **24**, 377(1908).

Mukunda, N, *Pramana*, **9**, 1–14 (1974).

Narlikar, J V, *General Relativity and Cosmology*, Asia Publishing House, Delhi (1978)

Narlikar, J V, *Elements of Cosmology*, Universities Press, Hyderabad (1998).

Panofsky, W R H and Phillips, H, *Classical Electricity and Magnetism*, Addison Wesley, Massachussets (1962).

Pauli, W, *Theory of Relativity*, Pergamon Press, London (1958)

Pearse, R W B and Geydon, A C, *Identification of Molecular Spectra*, Chapman and Hall, London (1950).

Rana, N C and Joag, P S, *Classical Mechanics*, Tata McGraw-Hill, New Delhi (1992).

Reif, F, *Fundamentals of Statistical and Thermal Physics*, McGraw-Hill, Auckland (1985).

Rojansky, V, *Introduction to Quantum Mechanics*, Prentice Hall, New York (1950).

Rosser, W G V, *Introduction to the Theory of Relativity*, Butterworths, New York (1964).

Saha, M N and Srivastava, B N, *Treatise on Heat*, India Press, Kolkata (1936).

Sakurai, J J, *Invariance Principles and Elementary Particles*, Princeton University Press, Princeton (1964).

Sakurai, J J, *Modern Quantum Mechanics*, Addison Wesley, New York (1999).

Sakurai, J J, *Advanced Quantum Mechanics*, Addison Wesley, New York (1999).

Schiff, L I, *Quantum Mechanics*, McGraw-Hill, Singapore (1968).

Schwabl, F, *Quantum Mechanics*, Narosa Publishing House, New Delhi (1995).

Sears, S W, *Introduction to Thermodynamics, Kinetic Theory of Gases and Statistical Mechanics*, Addison Wesley, London (1963).

Sommerfeld, A, *Atomic Structure and Spectral Lines*, Methuen Co, London (1923).

Sommerfeld, A, *Thermodynamics and Statistical Mechanics*, Academic Press, New York (1963).

Synge, J L and Griffith, B A, *Principles of Mechanics*, 3d ed., McGraw-Hill, New York (1959).

Wangness, K R, *Introductory Topics in theoretical Physics*, John Wiley and Sons, New York (1963).

White, H E, *Introduction to Atomic Spectra*, McGraw-Hill Kogakusha (1934).

Wong, Samual, S M, *Introduction to Nuclear Physics*, Prentice Hall, London (1999).

Zeidel, A N, *Tables of Spectral Lines*, Pergamon Press, Oxford (1959).

Index